Springer Series in Advanced Manufacturing

The **Springer Series in Advanced Manufacturing** includes advanced textbooks, research monographs, edited works and conference proceedings covering all major subjects in the field of advanced manufacturing.

The following is a non-exclusive list of subjects relevant to the series:

1. Manufacturing processes and operations (material processing; assembly; test and inspection; packaging and shipping).
2. Manufacturing product and process design (product design; product data management; product development; manufacturing system planning).
3. Enterprise management (product life cycle management; production planning and control; quality management).

Emphasis will be placed on novel material of topical interest (for example, books on nanomanufacturing) as well as new treatments of more traditional areas.

As advanced manufacturing usually involves extensive use of information and communication technology (ICT), books dealing with advanced ICT tools for advanced manufacturing are also of interest to the Series.

Springer and Professor Pham welcome book ideas from authors. Potential authors who wish to submit a book proposal should contact Anthony Doyle, Executive Editor, Springer, e-mail: anthony.doyle@springer.com.

INDEXED IN SCOPUS

Ajay Kumar · Parveen Kumar · Yang Liu
Editors

Industry 4.0 Driven Manufacturing Technologies

Springer

Editors
Ajay Kumar
Department of Mechanical Engineering
School of Engineering and Technology
JECRC University
Jaipur, Rajasthan, India

Parveen Kumar
Department of Mechanical Engineering
Rawal Institute of Engineering
and Technology
Faridabad, Haryana, India

Yang Liu
Department of Management
and Engineering
Linköping University
Linköping, Sweden

ISSN 1860-5168 ISSN 2196-1735 (electronic)
Springer Series in Advanced Manufacturing
ISBN 978-3-031-68270-4 ISBN 978-3-031-68271-1 (eBook)
https://doi.org/10.1007/978-3-031-68271-1

This Springer imprint is published by the registered company Springer Nature Switzerland AG
The registered company address is: Gewerbestrasse 11, 6330 Cham, Switzerland

Preface

Modern manufacturing technologies are supported by various powerful innovative strategies like Internet of things (IoT), cyberphysical production system, artificial intelligence, big data informatics, blockchain, cybersecurity, machine learning, augmented/virtual reality, etc. of Industry 4.0. These approaches of Industry 4.0 in manufacturing have redefined the value of manufacturing and unlocked a new age for sustainability in intelligent manufacturing. Utilization of Industry 4.0 tools in manufacturing facilitates autonomous sensing, collaboration, interconnection, cognition, learning analysis, decision making and control and brings the development and implementation of human-machine collaboration to ensure resilience, compliance and accountability in manufacturing to ensure cost-effective, high-quality and efficiency and environment-friendly services. Based on the research of applications of Industry 4.0 tools in manufacturing techniques, it is analyzed that it is triggering a major change in methodology, ecosystems, means and models of manufacturing sector to reframe the architecture of intelligent manufacturing. So that manufacturing technologies can carve a better pathway to achieve both economic sustainability and energy transition across the globe for industries involved in mass customization by utilizing Industry 4.0 tools. Industry 4.0 driven manufacturing technologies exhibit work using advanced methods and enable industrial systems to work efficiently, diverse options and optimize various industrial engineering and management perspectives with an emphasis on improving green production all over the globe, reducing carbon content from environment and improving quality life of human personnel. Thus, the integration of Industry 4.0 tools with manufacturing and engineering is applied to various sectors like automobile industries, aerospace, agriculture, food industries, medical, pharmaceuticals, healthcare, etc. In recent years, the development of different strategies and techniques of manufacturing has been dramatic. Now, it is essential to combine and implement these advanced manufacturing techniques in industries with Industry 4.0 approach to divert the world toward green production.

The book will benefit various stakeholders like industries, professionals, academics, research scholars, senior graduate students and human healthcare. It will be a reference book for libraries of all technical institutions and an ideal compendium for senior graduate-level courses, etc. This book compiles overall aspects of advancements in manufacturing by bringing the latest research and development by a comprehensive range of mathematical, numerical and simulation, modeling techniques of Industry 4.0 to strengthen the engineering science and technological developments for the future. The book consists of eighteen chapters that describe perspectives of Industry 4.0 driven manufacturing technologies.

- Chapter "Evolution of Digital Twin in Manufacturing Application: Definition, Architecture, Applications, and Tools" explores the evolution of digital twin technology in manufacturing sector, examining different definitions and their respective scopes, as well as the architectures, tools and methodologies employed in the development of digital twins, and its applications in manufacturing.
- Chapter "Intelligent Feature Engineering for Monitoring Tool Health in Machining" highlights the importance of feature engineering in robust low-cost machine learning algorithms for tracking the tool condition in precision machining, showing that tool wear can be accurately inferred with low-cost algorithms of the variance of machining vibrations.
- Chapter "Experimental Investigation of Process Parameters Effects on Extrusion Blow Molding Process Using Response Surface Methodology for Industry 4.0" investigates the effect of various parameters such as extruder pressure, die temperature along with extruder speed on production time for 20 l bottle made up of low-density polyethylene (LDPE) material using extrusion molding process. Design of experiments using response surface methodology is used to obtain experimental results.
- Chapter "Finite Element Analysis and Experimental Investigation of 3D Printed Biomimetic Structures" explores the design, finite element analysis and experimental evaluation of 3D printed biomimetic structures, identifying marsh horsetail as the optimal design for biomedical scaffolds due to superior mechanical properties.
- Chapter "Industry 4.0 in Aircraft Manufacturing: Innovative Use Cases and Patent Landscape" evaluates the integration of Industry 4.0 technologies into aircraft manufacturing, exploring practical use cases, patent analysis trends, competitive dynamics and opportunities for stakeholders to leverage these innovations.
- In Chapter "Comparative Multi-criteria-Decision Making Approach for the Optimization of Abrasive Water Jet Machining Process Parameters Using MABAC," weight calculation techniques like equal weight (EW), standard deviation (STD) and entropy (ENT) were considered while calculating the rank using multi-attributive border approximation area comparison.
- Chapter "An Empirical Analysis of Factors Influencing Industry 4.0 Implementation in Manufacturing SMEs" investigates the critical factors of Industry 4.0 transformation and their relationships in the context of manufacturing SMEs in developing countries.

- Chapter "A Cost-Minimization Approach to Production and Maintenance Planning Considering Imperfect Repairs and Human Resource Constraints" presents a cost-minimization approach to production and maintenance planning that considers imperfect repairs, human resource constraints and various factors influencing decision making to optimize system availability and operational efficiency.
- Chapter "Environment-Friendly Practices for Integrating Green Business with Green Supply Chain Management: Industry 4.0 Perspectives and Beyond" covers the current research work carried out in GB and GSCM for Industry 4.0 perspectives and beyond.
- Chapter "Barrier Analysis for the Sustainable Business Practice of a Textile and Apparel Industry in Fiji Using an ISM Approach" identifies and analyzes barriers to adopting sustainable business practices using the ISM approach.
- Chapter "Strategic Design Optimization of Cutting Tools for Enhanced Manufacturing Efficiency" investigates the optimization of cutting tool design parameters, focusing on high carbon high chromium (HCHC) steel, to enhance structural characteristics and machining performance for improved durability and efficiency in manufacturing processes.
- Chapter "Exploring the Challenges of Integrating Lean Green Practices in Industry 4.0 Manufacturing Frameworks: An Empirical Study" underscore the critical importance of addressing motivational factors to propel sustainable manufacturing practices amidst Industry 4.0 advancements.
- Chapter "Robotic Arm 3D Printing: Technological Advancements and Applications" covers the built-in benefits of RA3DP, such as more design flexibility, less need for supports and the ability to print at almost any angle.
- Chapter "Elephant Swarm Water Search Algorithm-Based Optimization of a Laser Beam Machining Process" discuses a newly developed metaheuristic algorithm, i.e., elephant swarm water search algorithm (ESWSA), inspired by the behavior of social elephants, to determine the optimal combination of gas pressure (P_a), pulse width (W_p), pulse frequency (f_p) and cutting speed (S_c) during Nd:YAG laser-based straight profile cutting of thin aluminum alloy sheet.
- Chapter "Progressive Automation: Mapping the Horizon of Smart Manufacturing with RoboDK Workstations and Industry 4.0" explores the convergence of progressive automation, RoboDK workstations and Industry 4.0, clarifying the mutually beneficial connection between these elements in creating the future of production.
- Chapter "Improvising the Quality of Manifold Production Using Six-Sigma Technique for Implementation in Automobile Manufacturing Industries: A Case Study" deals with process improvement methods to enhance productivity, reduce defects, ease the methodology and save money.
- Chapter "Digital Twin Integration for Enhanced Control in FDM 3D Printing" offers a thorough grasp of the thermal dynamics associated with FDM 3D printing, highlighting the critical role that Digital Twins play in reducing difficulties and improving the capabilities of the manufacturing process.

- Chapter "Intelligent Manufacturing in Aerospace: Integrating Industry 4.0 Technologies for Operational Excellence and Digital Transformation" discusses how Industry 4.0 technologies are integrated across various facets of aircraft manufacturing, maintenance and MRO, with real-life applications spanning the industry.

The editors acknowledge the professional support received from Springer and express their gratitude for this opportunity.

Reader's observations, suggestions and queries are welcome.

Jaipur, India — Ajay Kumar
Faridabad, India — Parveen Kumar
Linköping, Sweden — Yang Liu

Acknowledgements

The editors are grateful to Springer for showing their interest to publish this book in the buzz area of Industry 4.0 driven manufacturing technologies. The editors express their personal adulation and gratitude to Anthony Doyle (Executive Editor, Engineering) Springer, for giving consent to publish our work. He undoubtedly imparted the great and adept experience in terms of systematic and methodical staff who have helped the editors to compile and finalize the manuscript. The editors also extend their gratitude to Mrs. Amudha Vijayarangan, (Project Coordinator—Total Service, Books Production), Springer, for support during her tenure.

The editors wish to thank all the chapters' authors for contributing their valuable research and experience to compile this volume. The chapters' authors, the corresponding author in particular, deserve special acknowledgments for bearing with the editors, who persistently kept bothering them for deadlines and with their remarks.

Dr. Ajay also wishes to express his gratitude to his parents, Sh. Jagdish and Smt. Kamla, and his loving brother Sh. Parveen for their true and endless support. They have made him able to walk tall before the world regardless of sacrificing their happiness and living in a small village. He cannot close these prefatory remarks without expressing his deep sense of gratitude and reverence to his life partner Mrs. Sarita Rathee for her understanding, care, support and encouragement to keep his morale high all the time. No magnitude of words can ever quantify the love and gratitude he feels in thanking his daughters, Sejal Rathee and Mahi Rathee, and son Kushal Rathee who are the world's best children.

Finally, the editors obligate this work to the divine creator and express their indebtedness to the "ALMIGHTY" for gifting them the power to yield their ideas and concepts into substantial manifestation. The editors believe that this book would enlighten the readers about each feature and characteristics of Industry 4.0 driven manufacturing technologies.

Ajay Kumar
Parveen Kumar
Yang Liu

Contents

Editors and Contributors

About the Editors

Ajay Kumar is currently serving as Professor in School of Engineering and Technology, JECRC University, Jaipur, Rajasthan, India. He received his Ph.D. in the field of Advanced Manufacturing from Guru Jambheshwar University of Science and Technology, Hisar, India, after B.Tech. (Hons.) in mechanical engineering and M.Tech. (Distinction) in manufacturing and automation. His areas of research include incremental sheet forming, artificial intelligence, sustainable materials, additive manufacturing, mechatronics, smart manufacturing, Industry 4.0, waste management and optimization techniques. He has over 120 publications in international journals of repute including SCOPUS, Web of Science and SCI indexed database and refereed international conferences. He has also co-authored and co-edited more than 20 books with ELSEVIER, CRC Press, WILEY, De Gruyter and IGI Global and conference proceedings with IOP. He has organized various national and international events including an International Conference on Mechatronics and Artificial Intelligence (ICMAI-2021) and International conference on Artificial Intelligence, Advanced Materials and Mechatronics Systems (AIAMMS-2023) as a conference chair. He has more than 20 national and international patents to his credit. He has supervised more than 8 M.Tech., Ph.D. scholars and numerous undergraduate projects/ thesis. He has a total of 15 years of experience in teaching and research. He is Guest Editor and Review Editor of the reputed journals including *Frontiers in Sustainability*. He has contributed to many international conferences/symposiums as a session chair, expert speaker and member of the editorial board. He has won several proficiency awards during the course of his career, including merit awards, best teacher awards and so on. He is an adviser of QCFI, Delhi Chapter student cell at JECRC University, and has also authored many in-house course notes, lab manuals, monographs and invited chapters in books. He has organized a series of Faculty Development Programs, international conferences, workshops and seminars for researchers, Ph.D., UG and PG-level students. He is associated with many research, academic and professional societies in various capacities.

https://scholar.google.co.in/citations?user=TmZS4JIAAAAJ&hl=en
https://www.webofscience.com/wos/author/record/D-5813-2019
Publons Profile: https://publons.com/researcher/1596469/ajay-kumar/
Research Gate: https://www.researchgate.net/profile/Ajay_Kumar349
ResearcherID: D-5813-2019
http://www.researcherid.com/rid/D-5813-2019

Parveen Kumar is currently serving as Assistant Professor and Head in the Department of Mechanical Engineering at Rawal Institute of Engineering and Technology, Faridabad, Haryana, India. He has submitted Ph.D. from National Institute of Technology, Kurukshetra, Haryana, India. He completed his B.Tech. (Hons.) from Kurukshetra University, Kurukshetra, India, and M.Tech. (Distinction) in Manufacturing and Automation from Maharshi Dayanand University, Rohtak, India. His areas of research include intelligent manufacturing systems, materials; die-less forming, design of automotive systems, additive manufacturing, CAD/CAM and artificial intelligence, machine learning and Internet of Things in manufacturing and multi-objective optimization techniques. He has over 50 publications in international journals of repute including SCOPUS, Web of Science and SCI indexed database and refereed international conferences. He has 8 national and international patents in his credit. He has supervised 2 M.Tech. scholars and numerous undergraduate projects/ theses. He has a total of 13 years of experience in teaching and research. He has co-authored/co-edited several books with ELSEVIER, CRC Press, WILEY, De Gruyter and IGI Global. He is also the book series editor of *Industrial and Manufacturing Systems and Technologies: Sustainable and Intelligent Perspectives* in CRC Press, Taylor and Francis. He has organized a series of Faculty Development Programs, workshops and seminars for researchers and UG-level students.

https://scholar.google.com/citations?hl=en&authuser=1&user=ESQ-RnY AAAAJ

Dr. Yang Liu received the M.Sc. (Tech.) degree in telecommunication engineering and D.Sc. (Tech.) degree in industrial management from the University of Vaasa, Finland, in 2005 and 2010, respectively. He is currently a tenured associate professor and doctoral supervisor with the Department of Management and Engineering, Linköping University, Sweden; a visiting faculty with the Department of Production, University of Vaasa, Finland; and a chair professor with Jinan University, China. Meanwhile, he is an adjunct/visiting professor at multiple other universities. His research interests include sustainable smart manufacturing, product service innovation, decision support system, competitive advantage, control systems, autonomous robots, signal processing and pattern recognition.

Prof. Liu has authored or co-authored more than 130 peer-reviewed scientific articles and is ranked No.1 among the top authors on "big data analytics in manufacturing." His publications have appeared in multiple distinguished journals, and some ranked as top 0.1% ESI Hot Papers and top 1% ESI Highly Cited Papers. He serves as an associate editor of the prestigious *Journal of Cleaner Production* and *Journal of Intelligent Manufacturing*, referees in over 60 SCI leading journals and external reviewer for NSERC of Canada and CONICYT of Chile.

https://scholar.google.com/citations?user=jHaIVjMAAAAJ
https://www.researchgate.net/profile/Yang_Liu57
https://www.researcherid.com/rid/C-8320-2013

Contributors

Abhishek Agarwal Department of Mechanical Engineering, College of Science and Technology, Royal University of Bhutan, Phuentsholing, Bhutan

Malika Aitzhanova School of Engineering and Digital Sciences, Nazarbayev University, Astana, Kazakhstan

Gaydaa AlZohbi Department of Mathematics and Natural Sciences, Prince Mohammad Bin Fahd University, Al Khobar, Kingdom of Saudi Arabia

Sourabh Anand University School of Information, Communication and Technology, Guru Gobind Singh Indraprastha University, Dwarka, New Delhi, India

R. Ananthanarayanan Mechanical Engineering, School of Information Technology, Engineering, Mathematics and Physics, The University of the South Pacific, Suva, Fiji

Pushpendra S. Bharti University School of Information, Communication and Technology, Guru Gobind Singh Indraprastha University, Dwarka, New Delhi, India

Vineet Bhatia Department of Technology Management, Defence Institute of Advanced Technology, Pune, India

George M. Bollas Department of Chemical and Biomolecular Engineering, Pratt & Whitney Institute of Advanced Systems Engineering, University of Connecticut, Storrs, Connecticut, USA

Shankar Chakraborty Department of Production Engineering, Jadavpur University, Kolkata, West Bengal, India

Rakesh Chaudhari Mukesh Patel School of Technology Management and Engineering, Shirpur Campus, Shri Vile Parle Kelavani Mandal Narsee Monjee Institute of Management Studies (NMIMS), Dhule, Maharashtra, India

Aditya Choksi Mechanical Engineering Department, Institute of Technology, Nirma University, Ahmedabad, India

Soham Das Department of Mechanical Engineering, Sikkim Manipal Institute of Technology, Sikkim Manipal University, Sikkim, India

Dhiraj Intelligent Systems Group CSIR-Central Electronics Engineering Research Institute, Pilani, Rajasthan, India

Dinara Dikhanbayeva School of Engineering and Digital Sciences, Nazarbayev University, Astana, Kazakhstan

Sunny Diyaley Department of Mechanical Engineering, Sikkim Manipal Institute of Technology, Sikkim Manipal University, Majitar, Sikkim, India

Gayatri University School of Information, Communication and Technology, Dwarka, New Delhi, India

Ranjan Kumar Ghadai Department of Mechanical and Industrial Engineering, Manipal Institute of Technology, Manipal Academy of Higher Education, Manipal, India

Surendra Chandrakant Ghorpade Department of Technology Management, Defence Institute of Advanced Technology, Pune, India

Ali Golestanian Department of Industrial Engineering, Sadjad University, Mashhad, Iran

P. R. C. Gopal Department of School of Management, National Institute of Technology, Warangal, Telangana, India

Spandan Guha School of Mechanical Engineering, KIIT University, Bhubaneswar, India

Taha-Hossein Hejazi Department of Industrial Engineering, College of Garmsar, Amirkabir University of Technology (Tehran Polytechnic), Tehran, Iran

Suveg V. Iyer Department of Mechanical Engineering, Birla Institute of Technology and Science Pilani, Pilani, Rajasthan, India

A. K. Jeevanantham Mechanical Engineering Department, School of Mechanical Engineering, Vellore Institute of Technology, Vellore, India

Sarvoday Joharapurkar Mukesh Patel School of Technology Management and Engineering, Shirpur Campus, Shri Vile Parle Kelavani Mandal Narsee Monjee Institute of Management Studies (NMIMS), Dhule, Maharashtra, India

Kanak Kalita Department of Mechanical Engineering, Vel Tech Rangarajan Dr. Sagunthala R&D Institute of Science and Technology, Avadi, India;
University Centre for Research and Development, Chandigarh University, Mohali, India

Sanjeev Kumar Khare Department of Technology Management, Defence Institute of Advanced Technology, Pune, India

Ajay Kumar Department of Mechanical Engineering, School of Engineering and Technology, JECRC University, Jaipur, Rajasthan, India

Aman Kumar Department of Mechanical Engineering, University Institute of Engineering and Technology, Maharshi Dayanand University, Rohtak, Haryana, India

Ashwini Kumar Department of Mechanical Engineering, School of Engineering and Technology, Manav Rachna International Institute of Research and Studies, Faridabad, Haryana, India

Mukul Kumar University School of Information, Communication and Technology, Guru Gobind Singh Indraprastha University, Dwarka, New Delhi, India

P. Kumar Mechanical Engineering, School of Information Technology, Engineering, Mathematics and Physics, The University of the South Pacific, Suva, Fiji

Parveen Kumar Department of Mechanical Engineering, Rawal Institute of Engineering and Technology, Faridabad, Haryana, India

Rajender Kumar Department of Mechanical Engineering, School of Engineering and Technology, Manav Rachna International Institute of Research and Studies, Faridabad, Haryana, India

Rajesh Kumar Department of Mechanical Engineering, School of Engineering and Technology, Manav Rachna International Institute of Research and Studies, Faridabad, Haryana, India

Sachin Kumar Department of Mechanical Engineering, University Institute of Engineering and Technology, Maharshi Dayanand University, Rohtak, Haryana, India

V. Loganathan Department of Mechanical Engineering, Rohini College of Engineering and Technology, Kanyakumari, Tamil Nadu, India

Anjaney Lokapur Mechanical Engineering Department, Institute of Technology, Nirma University, Ahmedabad, India

M. A. Makhesana Mechanical Engineering Department, Institute of Technology, Nirma University, Ahmedabad, India

B. K. Mawandiya Mechanical Engineering Department, Institute of Technology, Nirma University, Ahmedabad, India

Debasish Mishra Data Science and Information Systems, Krea University, Sri City, Andhra Pradesh, India

K. M. Patel Mechanical Engineering Department, Institute of Technology, Nirma University, Ahmedabad, India

Krishna R. Pattipati Department of Electrical and Computer Engineering, Pratt & Whitney Institute of Advanced Systems Engineering, University of Connecticut, Storrs, Connecticut, USA

D. L. Belgin Paul Department of Mechanical Engineering, Vel Tech Rangarajan Dr. Sagunthala R&D Institute of Science and Technology, Chennai, Tamil Nadu, India

Ayyappan Susila Praveen Department of Mechanical Engineering, Vel Tech Rangarajan Dr. Sagunthala R&D Institute of Science and Technology, Chennai, Tamil Nadu, India

Donya Ramezani Department of Industrial Engineering, College of Garmsar, Amirkabir University of Technology (Tehran Polytechnic), Tehran, Iran

Sarita Rathee Department of Electronics and Communication Engineering, School of Engineering and Technology, JECRC University, Jaipur, India

Kuldip Singh Sangwan Department of Mechanical Engineering, Birla Institute of Technology and Science Pilani, Pilani, Rajasthan, India

Manoj Kumar Satyarthi University School of Information, Communication and Technology, Guru Gobind Singh Indraprastha University, Dwarka, New Delhi, India

D. Seenivasagam Management Department, College of Business, Hospitality and Tourism Studies, Fiji National University, Nasinu, Fiji

Satyam Sharma Department of Mechanical Engineering, Sikkim Manipal Institute of Technology, Sikkim Manipal University, Sikkim, India

Essam Shehab School of Engineering and Digital Sciences, Nazarbayev University, Astana, Kazakhstan

Rajendra Shimpi NAMTECH Transitory Campus: Research Park, IIT Gandhinagar Campus, Palaj, Gandhinagar, Gujarat, India

Sumati Sidharth Department of Technology Management, Defence Institute of Advanced Technology, Pune, India

Hargovind Soni Department of Mechanical Engineering, National Institute of Technology Delhi, Delhi, India;
Department of Industrial Engineering, College of Science, Engineering and Technology, University of South Africa, Pretoria, South Africa

Bimal Tamang Department of Mechanical Engineering, Sikkim Manipal Institute of Technology, Sikkim Manipal University, Sikkim, India

Ali Turkyilmaz UIS School of Business and Law, University of Stavanger, Stavanger, Norway

Chandrakant Wani School of Technology Management and Engineering, Shri Vile Parle Kelavani Mandal Narsee Monjee Institute of Management Studies, Hyderabad, Telangana, India

Evolution of Digital Twin in Manufacturing Application: Definition, Architecture, Applications, and Tools

Suveg V. Iyer, Kuldip Singh Sangwan, and Dhiraj

Abstract The fourth industrial revolution (I4.0) is based on the digitalization of the manufacturing process and its allied activities. The term Industry 4.0 was introduced in 2011 as part of a technology-based manufacturing project by Germany. The I4.0 came into existence because of the advancements in Artificial Intelligence, Robotics, the Internet of Things, and computing technologies. Digital Twin (DT) is a major concept based on the I4.0 technologies and has shown a continuous increase in interest by the industry and research community. This chapter explores the evolution of DT in manufacturing and report various definitions and their scope, architectures, tools/methodologies used in developing DT, and its applications in manufacturing. The DT concept is still evolving and there is a lack of a standard definition and unified architecture for DT application in manufacturing. The interoperability of data and lifecycle maintenance of DT are two big challenges for the adoption of DT in manufacturing industries. Integration of human expertise, improvement of communication protocols, and use of cognition in DT needs to be further explored by the research community. Decision support for decentralized manufacturing, validation of DT model, creation of data lake for digitalization, and DT for inspection are areas where a lot of further research is desired.

Keywords Digital twin · Fourth industrial revolution (Industry 4.0) · Digitalization in manufacturing · IoT · Simulation · Data driven models

---

S. V. Iyer · K. S. Sangwan (✉)
Department of Mechanical Engineering, Birla Institute of Technology and Science Pilani, Pilani, Rajasthan 333031, India
e-mail: kss@pilani.bits-pilani.ac.in

Dhiraj
Intelligent Systems Group CSIR-Central Electronics Engineering Research Institute, Pilani, Rajasthan 333031, India

A. Kumar et al. (eds.), *Industry 4.0 Driven Manufacturing Technologies*, Springer Series in Advanced Manufacturing, https://doi.org/10.1007/978-3-031-68271-1_1

1 Introduction

The start of the century saw advancements in technologies like data analytics, robotics, artificial intelligence, industrial internet of things, etc. making manufacturing more digitalized and computer oriented. This development in the manufacturing sector was complementing the demand for more customized products which led to small batch production. The industrialized countries initiated strategic plans to explore the opportunity, referred to as the fourth Industrial Revolution (I4.0). The common term Industry 4.0 or Industrie 4.0 was phrased in the high-tech strategy 2011 of Germany. Some of the other similar strategies leading to the goal of digitalization are: Japan—Society 5.0, Austria—Platform Industrie 4.0, Switzerland—Industry 2025, China—Made-in-China 2025 (Li et al. 2020), United States of America—Advanced Manufacturing or Smart Manufacturing, France—Industry of the Future, Italy—Intelligent Factory, and "The Factory of Future" in Europe (Semeraro et al. 2021).

The combined benefits of Industry 4.0 enabling technologies like artificial intelligence, robotics, internet of things (Industrial Internet of Things), digital simulation, augmented reality, cloud computing, and additive manufacturing helped in developing real-time bi-directional control systems. A control system supports monitoring, analysis, decision making, and execution, leading to the development of the cyber-physical system (CPS). The CPS gained research interest in the first decade of the twenty-first century, based on which the concept of the DT was conceived. The DT's conceptual model was introduced during a Product Lifecycle Management (PLM) executive course at the University of Michigan by Prof. Grieves in 2003.

1.1 Digital Twin Terminology and History

The philosophy based on which DT was developed or has evolved can be traced back to 1969, used by NASA in the year 1970 in Apollo 13 mission. The advancement in communication technologies and the information transfer techniques has resulted in the emergence of the philosophy as a concept and its implementation in the manufacturing industry. Figure 1 shows the evolution of DT from its philosophy leading to future scope.

In 2011 the conceptual model was developed as a technological concept by Prof. Grieves and defined DT. In 2012 NASA with the US Air force used the concept in aircraft predictive maintenance and performance evaluation and designed the road map for DT making it a publicly accepted concept. Further improvement of the technology made the industries accept the concept and increased the research interest. In 2017 Gartner listed the DT in the top ten strategic Science and Technology trends list. The report by Marketsandmarkets predicts a 58% CAGR estimating a $48 billion market value by 2026 (Marketsandmarkets 2019).

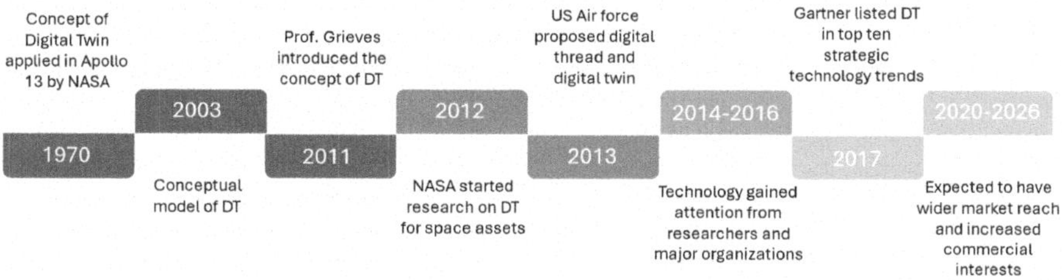

Fig. 1 Chronological evolution of digital twin

This chapter intends to address the following:

- How the digital twin has evolved in manufacturing
- Definition from industrial and academic perspectives
- Reference architecture and their modules
- Applications in the manufacturing and horizontal growth
- Tools/models available for the DT enabling technologies
- Challenges and future research scopes of DT in manufacturing.

2 Digital Twin Definition and Its Scope

The DT is based on several I4.0 enabling technologies leading to multiple definitions and varied scope. This section tries to consolidate the definitions/understanding by the major organizations using the technology and some of the reported definitions in the published articles.

The interest in the DT concept started increasing from 2016 and around the same time major organizations also showed interest and explained the understanding of the concept. Table 1 presents some of the definitions and understandings as reported by organizations and institutions and Table 2 gives a detailed view of the definitions as reported in the literature.

2.1 Comparing DT with Simulation and CPPS

The concept of DT is often considered to be the same as digital simulation. But on critical analysis of the concept, simulation is a part of DT, but both have their characteristics and targeted applications. The comparison between the Simulation and DT is presented in Table 3.

DT is also often compared with the cyber-physical system due to the similarity in architecture. DT and CPS have been categorized as engineering and scientific, respectively. In terms of mapping, the cyber part with real space shows a clear difference as CPS is one-to-many integration but DT is one-to-one integration. The

Table 1 Definition of digital twin by organizations

S. No.	Organization	Digital twin definition/understanding
1	GE	Testing and optimizing by debugging the machines in a virtual environment by integrating the physical machinery and the analytical techniques (Qi et al. 2018)
2	PTC	By extending the PLM process into the design cycle, a closed-loop process product design is created achieving product predictive maintenance (Qi et al. 2018)
3	Siemens	Accurate simulation of actual operations supported by its product life cycle data model (Qi et al. 2018)
4	Oracle	Simulating the actual complexities of physical entities through the virtual models and projection of insights into applications (Qi et al. 2018)
5	ANSYS	Gain strategic insights by combining simulation with data analysis capabilities (Qi et al. 2018)
6	Dassault	Designers and customers through the 3D experience platform interact with the product to explore its working (Qi et al. 2018)
7	SAP	Synchronized data acquisition and analysis to promote product development and innovation by building digitized models (Qi et al. 2018)
8	Altair	Virtual models of products with various physical properties are developed through virtual simulation technologies to have better characteristics (Qi et al. 2018)
9	NASA	Multi-physics, multi-scale, probabilistic simulation is integrated to mirror the life of the physical twin by using the appropriate sensor data, fleet history (Glaessgen and Stargel 2012)
10	CIRP	Digital representation of a dynamic product or service system comprising its behavioral characteristics and properties utilizing data, models, and information (Rasor et al. 2021)
11	ISO	Optimizing overall performance by converging and synchronizing the realized instance and the digital instance to provide a perspective of the physical process for the lifecycle (Wu et al. 2021)

comparison of core elements shows that DT has data and models at its core and CPS has sensors and actuators at its core. The comparison and correlation of DT and CPS are given in Tao et al. (2019b). Comparison and description of DT as a purely virtual space and CPS as the real world and virtual space are presented in Brovkova et al. (2021).

Table 2 Definition of digital twin by academia/research

S. No.	Author	Digital twin definition
1	Shafto et al. and Glaessgen et al.	Multi-physics, multi-scale, probabilistic simulation is integrated to mirror the life of the physical twin by using the appropriate sensor data, fleet history (Glaessgen and Stargel 2012; Shafto et al. 2012)
2	Tuegel et al.	The aircraft structure's cradle-to-grave model meets the requirements, with its electronic subsystems, flight controls, and propulsion system (Tuegel 2012)
3	Gockel et al.	A computer model of an aircraft structure with an ultra-realistic design to evaluate its capability in meeting the specified requirements (Gockel et al. 2012)
4	Lee et al.	Data driven cloud-based machine health condition simulation model based on its physical knowledge and analytical (Lee et al. 2013)
5	Reifsnider et al.	Models of the resources and structures controlling the life of a vehicle with ultra-high fidelity (Reifsnider and Majumdar 2013)
6	Majumdar	A structural model with material level quantitative data with high sensitivity (Majumdar et al. 2013)
7	Rosen et al.	Realistic models of the procedural behavior and its current state interacting with the real-world environment (Rosen et al. 2015)
8	Rios et al.	Physical product digital counterpart (Ríos et al. 2016)
9	Bielefeldt et al.	Computational models based on ultra-realistic multi-physics combining flight history with each aircraft (Bielefeldt et al. 2015)
10	Bazilevs et al.	Digital counterpart of the physical system incorporating fatigue damage control with a high-fidelity structural model (Bazilevs et al. 2015)
11	Schluse et al.	Virtual smart object representation with communication capabilities of real-world entities on the IoT and internet of services, acting as intelligent nodes (Schluse and Rossmann 2016)
12	Canedo et al.	Representing a real-world asset digitally focussing on the object itself (Canedo 2016)

(continued)

Table 2 (continued)

S. No.	Author	Digital twin definition
13	Gabor et al.	System simulation to predict the future states of the physical object (Gabor et al. 2016)
14	Schroeder et al.	Representing a real product virtually in the context of cyber-physical systems (Schroeder et al. 2016)
15	Kraft et al.	Digital thread enabled probabilistic multi-physics, multi-scale simulation model using the available sensor information, and input data over the lifecycle of its physical twin to replicate and predict activities/performance (Kraft 2016)
16	Bajaj et al.	Configuration-controlled repository models in multiple vendor tools with the ability to coordinate architecture along with its mechanical and electrical aspects with software verification across the system lifecycle (Bajaj et al. 2016)
17	Grieves and Vickers	Virtual information set describing the atomic level to the geometrical level of a manufactured product (Grieves and Vickers 2016)
18	Alam and El Saddik	CPS cyber layer evolving independently and integrating with the physical layer (Alam and El Saddik 2017)
19	Brenner and Hummel	A real time globally available digital copy of a physical manufacturing unit that is independently expanded and automatically updated (Brenner and Hummel 2017)
20	Ciavotta et al.	A digital avatar meshing the virtual CPS data and intelligence with its physical worlds incorporating its structure, semantics, and behaviour (Ciavotta et al. 2017)
21	Graessler and Poehler	A cyber-physical device representing its properties, preferences, work schedule, and skillset and attempts to emulate the human workforce dynamically by adopting values from a database (Graessler and Poehler 2018)
22	Zhang et al.	A realistic process model using a large volume of data to fast simulate for quick and effective assessment of the consequences, production line performance, and quality of the design of products (Zhang et al. 2017)
23	Negri et al.	An Industry 4.0 linked virtual model of a physical system exploiting a real-time synchronization of data from the field (Negri et al. 2017)

(continued)

Table 2 (continued)

S. No.	Author	Digital twin definition
24	Schleich et al.	Execution of manufacturing processes and other activities of a physical artifact through the product life cycle by a set of its virtual models having bi-directional relation with its physical layer (Schleich et al. 2017)
25	Schluse et al.	A virtual replica of an asset containing its data, function, and communication interface integrating knowledge resulting from modeling and functional data recorded during real-world operation (Schluse et al. 2017)
26	Söderberg et al.	The virtual model of a product or a production system for real-time optimization from the design to production phase (Söderberg et al. 2017)
27	Stark et al.	Digital master model of a physical unit with its discrete digital shadow and a logical linkage of the two elements (Stark et al. 2017)
28	Weber et al.	Digital representation of a physical asset with its states and functions to achieve holistic intelligence with the potential to interact with other digital twins for decentralized self-control (Weber et al. 2017)
29	Yun et al.	An accurate cyber-model of a physical machine reflecting the status and can tightly control the system (Yun et al. 2017)
30	Autiosalo	The cyber part of a cyber-physical system (Autiosalo 2018)
31	Asimov et al.	A replica of a physical installation in a virtual space capable of monitoring data to detect and forecast problems and support business decisions through its AI knowledge engine (Asimov 2018)
32	Bao et al.	Simulating a physical object with its characteristics in real time through a virtual model in the virtual space (Bao et al. 2019)
33	Lee and Kim	A near real-time digital image development of a physical object or process by combining IoT (Internet of Things) and IoS (Internet of Service) concepts to materialize helping to enhance business performance (Lee and Kim 2018)

(continued)

Table 2 (continued)

S. No.	Author	Digital twin definition
34	Haag and Anderl	A set of realistic models that can simulate a real-life individual product through its life cycle with its properties, condition, and behavior (Haag and Anderl 2018)
35	Luo et al.	A virtual prototype of a physical system with one-to-one multi-domain consistent between the intended and the actual environment with accurate data mapping for the control system (Luo et al. 2018)
36	Nikolakis et al.	A digital replica of the physical environment constraining and reproducing the physical system's actuators (Nikolakis et al. 2019)
37	Tao et al.	A set of virtual models having geometric characteristics and rules and behaviors that could imitate the life cycle process, with the ability to simulate, monitor, and control the physical system through prediction (Tao et al. 2019a)
38	Liu et al.	A living model using real-time sensory data to continuously adapt to the environment or operation and can predict the future of the respective physical assets for predictive maintenance (Liu et al. 2018)
39	Zhuang et al.	A virtual dynamic model coherent with its corresponding physical entity and capable of simulating its performance, characteristics, behavior, and life in a timely fashion (Zhuang et al. 2018)
40	Leng et al.	Digital representation of a real device synchronizing with its physical model by collecting sensory values, that can monitor and control the asset (Leng et al. 2019a)
41	Lu et al. adapted from Schleich et al.	Physical space is represented through its high-fidelity operational dynamics enabled by near real-time synchronization (Schleich et al. 2017; Lu et al. 2020)

3 Reference Architecture

The application of the I4.0 concept and subsequent development of its enabling technologies has led to defining and developing multiple frameworks for DT. The architecture as given by Grieves (Fig. 2) having the physical layer, digital layer, and the communication layer between the two (Lu et al. 2020) is the base for reference architecture. This architecture concentrated more on the data and communication between the physical and the digital space.

Table 3 Comparison between simulation and DT

Simulation	Digital twin
Based on computer graphics	Data driven graphical representation
Dynamic in nature based on unit data and mathematical reference	Dynamic with a two-way flow of current data
Implementable proof of a concept	Dynamic system behavior of a concept
Turns out to be static if no change in user input data	Continuous as real time data is fed as input
Concentrated on a single process/part for development	Applied for the entire lifecycle of the part/process

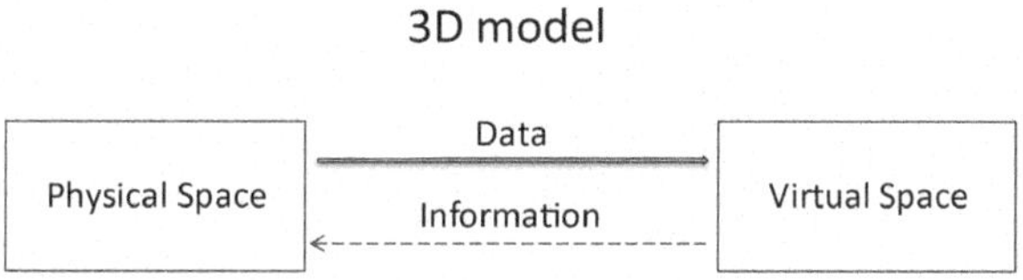

Fig. 2 Grieves DT architecture (Lu et al. 2020)

Various reference architectures have been developed by industrial initiatives, research project results, or by academicians. The common method in the development of the architectures has been to extend the existing standards for digital manufacturing, simulation, IoT, etc. ISO 23247 provides a detailed architecture keeping DT in focus whereas other reference architectures are Industry 4.0 focussed which can be adapted for the DT.

3.1 Reference Architecture Model Industrie 4.0 (RAMI 4.0) (Weber et al. 2017; Beisheim et al. 2020; Park et al. 2021a; Schweichhart 2019)

German Electrical and Electronic Manufacturers' Association (ZVEI) developed RAMI 4.0 as part of the Platform Industrie 4.0 initiative. The model integrates the value stream and the life cycle information with a multi-layered pyramid of manufacturing. The architecture is based on the standards of automation like IEC 62890, IEC 61512/ISA 95. The three dimensions or axes of the model are hierarchy levels, value stream, and layers.

- Hierarchy Levels deal with the functionality of components from product to connected world on an increasing scale.
- Value Stream (Life Cycle) classifies the life cycle status and differentiates between the type and instance.
- Layers function as the interface between the physical and cyber worlds separating the interoperability and mutual understanding of syntax and semantics.

3.2 *Stuttgart IT-Architecture for Manufacturing (SITAM) (Weber et al. 2017; Gröger et al. 2016)*

SITAM, extending through the product life cycle, was developed based on the research projects conducted in advanced manufacturing. It is a data driven architecture supporting parallel middleware components providing value-added services.

The middleware integration is considered the crux of the architecture which provides the analytics module with data mining, key performance indicator management, and predictive-prescriptive analysis. The visualization module consists of synchronization and mobile/portable display units.

3.3 *Industrial Internet Reference Architecture (Weber et al. 2017; Lin et al. 2017; Industrial Internet Consortium 2015)*

It is based on the Industrial Internet Architecture Framework by Industrial Internet Consortium (IIC). The architecture is developed as an open-source standard model (ISO/IEC/IEEE 42010:2011 Systems and Software Engineering–Architecture Description). The architecture specification adopts concern, stakeholders, and viewpoint as its frame; and views and models as its representation in analyzing Industrial Internet of Things (IIoT) systems.

The major functional domains of the architecture target business, operations, information, application, and control. The four peripheral modules are the user interface, system characteristics, crosscutting functions, and the physical systems (considered as separate entities as it is the existing machines/tools.). The IIRA architecture is more holistic as it takes into consideration the manufacturing system with its business aspects and integrates tools like ERP, MES, PLM, etc. The operational function domain consists of the monitoring and diagnostics, optimization, and deployment modules. The control module manages the CPS part of the architecture with sensing, actuating, and modeling asset management.

3.4 *ISO 23247 Digital Twin Framework for Manufacturing (Shao and Helu 2020; Digital Manufacturing Working Group (WG15) 2021)*

It is based on ISO 30141 functional entities customized for manufacturing. ISO 30141 was published by ISO with the International Electrotechnical Commission (IEC) in 2018. The framework was conceived to target the connection of heterogeneous machines in IoT. Due to the advancement in the DT concept, the standard was modified keeping manufacturing as its focus. The architecture is partitioned into four layers each defined by standards.

- The first layer (lowest) defines the physical manufacturing elements on the floor.
- The second layer is the device communication layer; monitoring and controlling the physical elements when a change of state is observed.
- The third layer is the digital layer being updated based on the signals from the communication layer.
- The fourth layer is the user interface layer to increase process efficiency.

3.5 *Activity-Resource-Type-Instance (ARTI) Architecture (Juarez et al. 2021; Anton et al. 2020; Borangiu et al. 2019)*

It is based on Product-Resource-Order-Staff Architecture (PROSA). PROSA employs the multi-agent system pattern with the holonic manufacturing system developed by PMA-KU Leuven. It is based on an intelligent being (IB) responsible for cloning the behavior of a system and an intelligent agent (IA) responsible for decision making.

3.6 *Five-Dimensional Model (Tao et al. 2019c)*

Five-Dimensional model (Tao et al. 2019c) by Fei Tao is among the highly cited and applied architectures by the peer research groups. It is based on Grieves three-dimension architecture extended with the data and services layer.

The comparison of these architectures (Table 4) shows that the characteristics vary primarily from the development point of view. The results show that the used terminologies have varied but the structure has shown similar characteristics across the reference architectures.

4 Digital Twin Applications in Manufacturing

The quantitative results show that about 20% of the total articles are based on optimization and framework for optimization. The higher number of articles on framework confirms that no unified architecture or framework is available for DT in manufacturing therefore more researchers are developing their architectures or are modifying the existing architectures for different applications. The reference architecture and the custom-developed architectures are primarily based on modules like the physical layer, virtual layer, and communication layer as given by Grieves. This denotes that the maturity level has not yet been reached from an architectural point of view.

Table 4 Comparative results of reference architectures

S. No.	Reference architecture	Developer	Characteristics	Novelty	Remarks
1	RAMI 4.0	German Electrical and Electronic Manufacturers Association (ZVEI)	Hierarchy levels life cycle and value stream layers	Interoperability because of the general framework	IEC 62890, IEC 61512/ ISA 95
2	Stuttgart IT architecture	Research projects outcome	Role based application value adding middleware product life cycle	Mobile middleware—information provisioning	
3	IIRA	Industrial Internet Consortium (IIC)	Functional domains system characteristics crosscutting functions	Business operations functional domain	ISO/IEC/ IEEE 42010:2011
4	ISO 23247	International Standards Organization	Data collection and device control core entity user entity cross entity	Detailed digital twin framework	ISO 30141
5	ARTI	Adapted from PROSA by PMA-KU Leuven	Intelligent being intelligent agent	Holon based	
6	Five-dimensional model	Fei Tao	Physical entity virtual entity connection services digital twin data		Extended from Grieves 3-D model

DT finds application in all facets of manufacturing. Eighteen fields were identified as sub-categories under manufacturing on which articles have been published. Major applications are in Optimization, Monitoring, and Data analytics (Table 5) and have scope in Manufacturing Education, Virtual commissioning, Verification, and Testing domains. Optimization finds a high number of papers covering the process, production, scheduling, performance, and capacity optimization sub-topics. The research interest on data analytics also suggests that the data acquisition methodologies are well established, and more work is being done on data analytics for optimization.

Table 5 Major applications of published papers

Application	Major sub-applications	References
Optimization	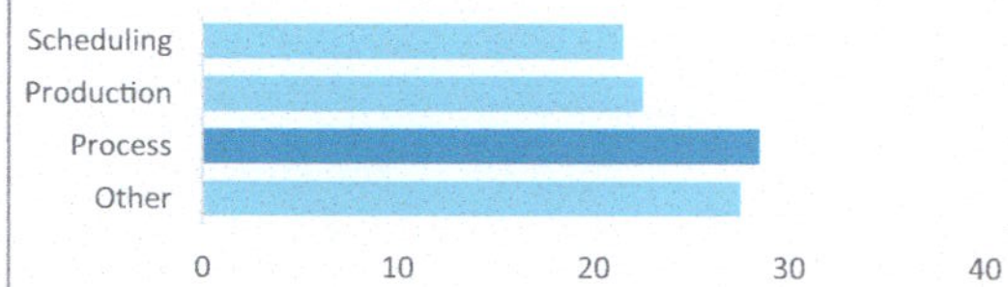	Reifsnider and Majumdar (2013), Zhang et al. (2017, 2018, 2020a, b, c), Bao et al. (2019), Vachalek et al. (2017), Yao et al. (2018), Bauer et al. (2018), Gericke et al. (2019), Gallego-García et al. (2019), Centomo et al. (2019), Gurjanov et al. (2019), Karanjkar et al. (2018), Tabar et al. (2019, 2020), Nafors et al. (2020), Feng et al. (2020), Barni et al. (2020), Dobrescu et al. (2020), Zhao et al. (2020, 2021), Ehrhardt and Hoffmann (2020), Latif and Starly (2020), Ippolito et al. (2020), Khanesar et al. (2020), Ma et al. (2020a, b), Eschemann et al. (2020), Zhu et al. (2021), Balderas et al. (2021), Mourtzis et al. (2021a, b), Ruhland et al. (2021), Magnanini et al. (2021), Vrabič et al. (2021), Zhou et al. (2021a), Wang et al. (2021a), Nguyen et al. (2021), Ur Rehman et al. (2021), Polini and Corrado (2021), Vachálek et al. (2021), Seok et al. (2021), Bai et al. (2021), Guo et al. (2021a, b), Xu et al. (2021), Eisenbarth et al. (2019), Papanagnou (2020), Flores-Garcia et al. (2020), Wang and Wu (2020), Escriche Lng et al. (2021), Zhifeng et al. (1884), Ruiz et al. (2021), Martínez-Gutiérrez et al. (2021), Yu et al. (2021), Chuang et al. (2021), Schuh et al. (2021), Koulouris et al. (2021) and Yan et al. (2021)

(continued)

Table 5 (continued)

Application	Major sub-applications	References
Monitoring	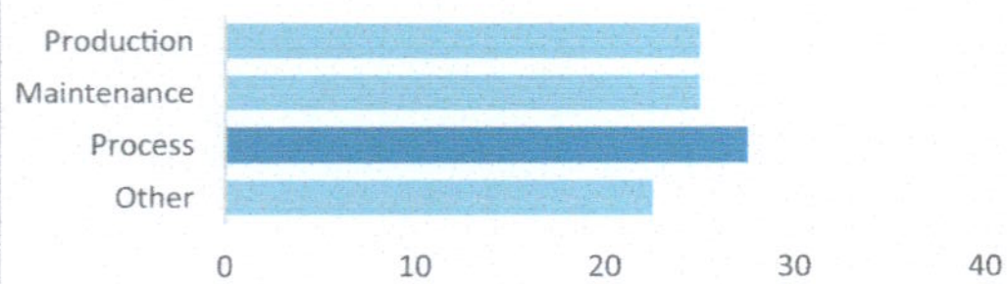	Zhuang et al. (2018), Beregi et al. (2018), Kholopov et al. (2019), Danilczyk et al. (2019), Halenar et al. (2019), Vijayakumar et al. (2019), Schützer et al. (2019), Wu et al. (2019), Liu et al. (2020a), Viola and Chen (2020), Greco et al. (2020), Borangiu et al. (2020), Stieber et al. (2020), Qi and Park (2020), He et al. (2020), Wang et al. (2020a, 2021b, c), Novák et al. (2020), Rebmann et al. (2020), Lacueva-Perez et al. (2019), Ralph et al. (2020), Feldt et al. (2020), Kuts et al. (2020), Stavropoulos et al. (2020), Yi et al. (2021), Ward et al. (2021a), Morabito et al. (2021), Wu and Li (2021), Moreno et al. (2021), Gao et al. (2021), Stavropoulos et al. (2021), Martinez et al. (2021), Duan et al. (2021), Rolo et al. (2021), Zhuang et al. (2019), Chetan et al. (2021), Ghosh et al. (2021) and Mylrea et al. (2021)
Data analytics	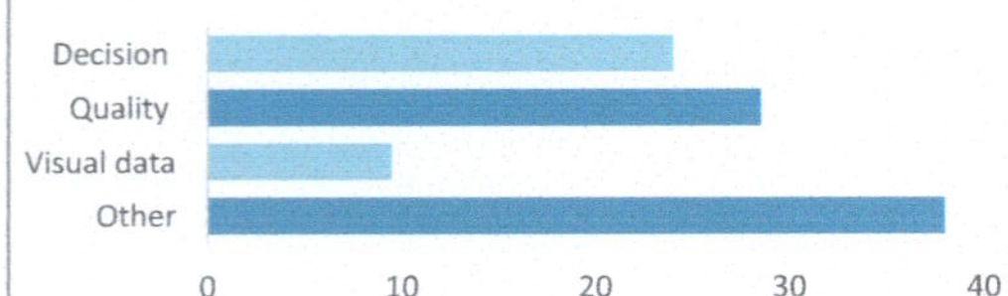	Gockel et al. (2012), Canedo (2016), Deac et al. (2017), Li et al. (2017, 2021a, b, c), Gyulai et al. (2018), Cronrath et al. (2019), Min et al. (2019), Jaensch et al. (2018), Qiao et al. (2019), Zhang and Ji (2019), Abburu et al. (2020), Ratnayake et al. (2020), Bazaz et al. (2020), Olalere and Olanrewaju (2020), Wang et al. (2020b, 2021d), Hinchy et al. (2020), Ertveldt et al. (2020), Zhao et al. (2020), Yang et al. (2020) Szabo et al. (2020), dos Santos et al. (2021), Hürkamp et al. (2021), May et al. (2021), Intizar Ali et al. (2021), Zhou et al. (2021b), Yiping et al. (2021), Son et al. (2021), Gunasegaram et al. (2021), Wang and Luo (2021), Rojek et al. (2021), Park et al. (2021b), Zotov and Kadirkamanathan (2021), Xia et al. (2021b), Rožanec et al. (2022), Kalaboukas et al. (2021), Mourtzis et al. (2020), Gramegna et al. (2020), Shahpar (2020) and Alexopoulos et al. (2020)

[a] Darker shade indicating the higher number of articles are published in the sub-application
[b] The values are in percentage terms

Monitoring and control are considered one of the basic applications of DT. The data acquisition and visualization directly support monitoring for production, maintenance, process fault, etc. The control part involves the analysis of data though not explicitly mentioned. Cloud-based DT also shows an increased interest in covering topics like the internet of machines, network of DT, etc.

5 Tools and Models Used in DT Development

Since DT is based on multiple technologies, there are many pre-established tools and models focused on each technology. Modeling, data management/analytics, simulation, 3D visualization, and database are identified as the facets where the tools have been used. Table 6 shows some of the tools used for developing DT. In some cases, modeling and simulation are performed with the same tool. But the custom-built agents in DT require standalone modeling tools capable of developing 3D digital models of machines and tools under consideration. Data, being the core element of the DT concept, is continuously researched for its communication protocols, monitoring and visualization tools, and trained data model for analysis. The data acquisition from heterogeneous machines and achieving machine-to-machine (M2M) communication is a topic of interest that is based on a central server either in a local system or in the cloud. Hence, data structuring and management become important tools with a clear definition of database features. 3D visualization is getting more traction because of the ease of use for the end-users, especially for maintenance and shop-floor quality control.

The analysis of tools and models used in DT research shows limited availability of open-source tools for the development which hinders the lifecycle maintenance of DT. The expected life term of a DT will be more than the license terms provided by the commercial software packages, which may lead to the breakdown of the model.

The variety of tools used is also less primarily due to the unavailability of free-to-use or academic versions and the research community may not be able to access the licensed version of the tools, particularly in the emerging economies and underdeveloped nations.

- The modeling of physical components into their 3D digital counterparts is performed with Siemens NX and Solidworks consistently over the years. Siemens NX is preferred more in recent years because of its in-built capability to manage simulation and virtual kernel along with modeling features. Also, the Siemens family of software tools helps for better interoperability among the software models. Other similar tools available as a package are the PTC Thingworx along with the KepserverEx data management tool, and ANSYS Twin builder with existing analysis modules.
- Communication—OPC Unified Architecture (OPC-UA) is gaining wide acceptance and is used by more research groups. It is an open-source standard for data exchange supporting the research community. Industries and machine tool manufacturers are also integrating OPCUA as it is a platform independent as well as

Table 6 Tools/models used for DT development

Functionality	Tool	References
Modeling	Solidworks	He et al. (2020), Duan et al. (2021), Tao and Zhang (2017), Sommer et al. (2020), Vladareanu et al. (2020), Cai et al. (2020), Qin et al. (2021)
	Siemens NX	Ippolito et al. (2020), Qi and Park (2020), Janda et al. (2019), Redelinghuys et al. (2020), Protic et al. (2020)
	CATIA	Mourtzis et al. (2021b), Zhou et al. (2021b), Dai et al. (2021)
Data management/ communication/ analytics	Automation ML	Schroeder et al. (2016), Bao et al. (2019), Beisheim et al. (2020), Liu et al. (2020b), Lou et al. (2019), Orive et al. (2019), Cai et al. (2019), Dittmann et al. (2020), Peng and Zhong (2020), Zhang et al. (2020d)
	DBN—Dynamic Bayesian Network	Gao et al. (2021), Li et al. (2017)
	MTConnect	Liu et al. (2020b), Hu et al. (2018), Tong et al. (2020), Review et al. (2021), Ward et al. (2021b), Hu et al. (2021)
	OPCUA	Chuang et al. (2021), Negri et al. (2019), Souza et al. (2019), Lou et al. (2019), Protic et al. (2020), Yu-Ming et al. (2020), de Andrade et al. (2021), Assad et al. (2021), Wang et al. (2021e)
	Kepware	Vijayakumar et al. (2019), Redelinghuys et al. (2020), Huynh et al. (2019), Konstantinov et al. (2021)
Simulation	AnyLogic	Beregi et al. (2018), Rolo et al. (2021), Kassen et al. (2021), Makarov et al. (2021), Meierhofer et al. (2021)
	Tecnomatix	Vachálek et al. (2021), Ruiz et al. (2021), Greco et al. (2020), Ward et al. (2021b), Ngo et al. (2018), Bambura et al. (2020), Fera et al. (2020), Xia et al. (2021a), Židek et al.(2020)
	MATLAB Simulink	Gericke et al. (2019), Feng et al. (2020), Dobrescu et al. (2020), Khanesar et al. (2020), Ma et al. (2020b), Mourtzis et al. (2021b), Polini and Corrado (2021), Ward et al. (2021a), Wang et al. (2021b), Rolo et al. (2021), Negri et al. (2019), Bolotov et al. (2019), Cattaneo and MacChi (2019), Vladareanu et al. (2020), Villalonga et al. (2020), Wagner et al. (2020), Negri et al. (2020), Zhang et al. (2020f), Zheng and Sivabalan (2020), Henson et al. (2021), Wang et al. (2021f), Suthar et al. (2021)

(continued)

Table 6 (continued)

Functionality	Tool	References
3D visualize	Unity 3D	Wu et al. (2021), Zhuang et al. (2018), He et al. (2020), Kuts et al. (2020), Konstantinov et al. (2021), Samir et al. (2019), Kuts et al. (2019), Yu-Ming et al. (2020), Jeong et al. (2019), Wang et al. (2020c), Yan and Zhang (2020), Zhang et al. (2020e), Yildiz et al. (2020), Xia et al. (2020), Leng et al. (2019b), Suthar et al. (2021), Matulis and Harvey (2021)
Database	MongoDB	Angrish et al. (2017)
	SQL	Wu et al. (2021), Vachálek et al. (2021), Zheng et al. (2018), Yan and Zhang (2020), Lin and Low (2020), Assad et al. (2021), Latsou et al. (2021)

supports interoperability of data. Other advantages of the OPCUA protocol are easy scalability due to its architecture, quick migration, and better data security. Other M2M communication protocols are the MTConnect and AutomationML which were preferred in the initial phase but on further development and acceptance of IoT, MTConnect and OPCUA were developed and are widely used.

- Simulation Software—MATLAB Simulink is the most preferred simulation software along with Siemens Tecnomatix Plant simulation and AnyLogic. The MATLAB Simulink along with its other modules provides more flexibility as data analysis and modeling can be performed within the package. AnyLogic has an academic version and is more suitable for the process simulation.
- Data storage—SQL-based databases are seen to be the most preferred. This may be due to the structured data communication and ease to access. SQL databases provide more reliability of transactions as it is structured, and query based. The NoSQL databases like MongoDB are preferred for their higher scalability and fast response, hence mostly preferred in industries managing a large volume of data compared to the research group.
- The analysis shows that 3D visualization and augmented reality-based applications are dominated by Unity 3D over the years, consistently. Unity 3D is preferred for scene development and PTC Vuforia for augmented reality application development.

6 Discussion, Conclusions and Future Research Directions

The chapter focuses on the evolution of the DT in the manufacturing domain. The focus of the chapter is on the scope, reference architectures, field of application, tools, and models used to understand the challenges in DT for manufacturing. Table 7 presents some of the similar terminologies used with their explanations and the comparative analyses. The terms are used interchangeably, but detailed analysis gives clarity on the concept and its nuances. The closest definition to DT is digital surrogate but DT must be used purely in its sense.

The following conclusions can be inferred:

- The lack of clarity in the concept and the disparity in understanding the level of integration is evident. The concept has similarities to CPS, CPPS, and simulation, which are used together but have distinctive features and applications. This confusion hinders the further research in the topic.
- The definition of DT is not yet matured as the concept has been defined in multiple perspectives and interchangeable terminologies. While multiple definitions signify progress in the field, the varying definitions hinder wider acceptance and application of the concept. The clear distinction among the DT, digital shadow, digital thread, digital model, and digital surrogate on one side and simulation, CPPS, and CPS on the other side is expected to lead to different verticals in the digitalization of manufacturing activities thereby leading to wider and deeper research and acceptance.

Table 7 Comparative results between similar terminologies used for digital twin in literature

	Digital twin	Digital shadow	Digital thread	Digital model	Digital surrogate
Explanation	The bi-directional data flow between a physical object and its digital object	Automated unidirectional data flow between the physical object and its digital object (Ehrhardt and Hoffmann 2020; Rolo et al. 2021; Kassen et al. 2021)	Data communication and data visualization throughout the product lifecycle from design to usage	Digital representation of a physical object without data communication	A model representing, a physical manufacturing system through its historical and real-time data (Ward et al. 2021b)
Digitalization	✓	✓	✓	✓	✓
Bi-directional communication	✓	✗	×[2]	✗	✗
Data analytics	✓	×[1]	×[3]	✗	×[4]
Product lifecycle	✓	✓	✓	✗	✓

✓ The feature is part of the concept
× 1. May/May not be in the concept
× 2. Considered for the whole manufacturing system
× 3. May be applied to a part of concept
× 4. It May be applied independently with the concept
✗ The feature does not belong in the concept

- The architecture of DT is developed primarily based on the application of available reference architectures. The agreement among the research community is in the usage of modules in architecture which indicates the transition towards a matured architecture.
- The concept is getting more acceptance, and the breadth of the topic is also increasing as more applications of DT are evaluated and experimented within the manufacturing domain. The digitalization of manufacturing has been regarded as essential and DT supporting digitalization helps in its further development.
- The availability of tools is still limited, and its practice is not fully explored as software development is catching up to the requirements of the manufacturing sector.

7 The Challenges Faced by DT

7.1 Lack of Unified Framework/Architecture

The deficiency of a unified framework or architecture leads to the generation of custom application-oriented architecture. This lack of standardization makes the lifecycle maintenance of the DT an issue and the disruption of the DT in companies makes the other companies in wait-and-watch mode. This makes the commercialization of the concept difficult and unsustainable. The discrepancy in architecture hinders the development and implementation when the concept matures.

7.2 Lack of Interoperability of Data

The manufacturing domain gets data from multiple sources which will be in varying formats. Since DT needs to amalgamate all the data, the contradicting data format becomes a challenge to develop a communicable DT.

7.3 Lack of Connectivity, Data Security, and Communication Protocol

Dependence on the internet and data transfer leads to higher importance for connectivity. Attaining uninterrupted connectivity in manufacturing units is still a challenge. Along with connectivity comes data security. Transfer of data through a secured channel is still not achieved efficiently. The communication between heterogeneous machines is a challenge as there is no uniform protocol implemented across the domain.

7.4 *Lifecycle Maintenance of DT*

DT is intended for the life cycle of the physical machines/assets. But due to the lack of open-source tools and changing license terms of software, maintaining the DT through the asset lifecycle is a challenging task.

8 Future Research Scope

8.1 *Integration of Human Resource Expertise with Digital Twin*

Human resource is an important aspect of Industry 4.0 systems. Integrating human resource skills and expertise effectively with the digital medium is certainly one of the upcoming goals. Contextual decision making is still not fully developed and practiced with the existing data analytics models and the role of human skill is particularly important in the automation and digitalization of the manufacturing sector. The ability to interpret and digitalize human action and decisions for machine application is still an open-ended research challenge.

8.2 *Optimization of Communication Methods*

Data acquisition is one of the prime requirements for developing DT. The communication protocol for multi-format data transfer and analysis must be developed further. Synchronization of the physical model with its digital model with minimum latency is a crucial factor in DT which require an optimized communication channel.

8.3 *Cognition in Digital Twin for Better Decision Making*

Decision-making through DT will help in enhancing the manufacturing output in terms of quantity and quality. For a better decision-making model, cognitive assistance through DT must be developed with good accuracy and scope for updating. This avoids duplication of models and can improvise the existing models.

8.4 Decentralized Manufacturing Application

DT for decentralized decision support and production systems need to be explored further. The number of published works in this direction is also minimal indicating less work has been conducted on the topic. Synchronizing the production factors to get real-time updates will be of key importance to providing decision support remotely and decentralized monitoring of production facilities.

8.5 Application of DT in Inspection

DT in quality control has been primarily targeted at additive manufacturing fault detection but has not been used extensively for geometric dimensioning and tolerance related machines. DT for contact and non-contact inspection tools are yet to be explored and has a future research scope as geometrical accuracy with lean principles is desired in manufacturing.

8.6 Development of Data Lake for Digitalization

The possibility of using a common data lake or pool of data for similar machines is required to be analyzed. For prediction models and effective data analysis, it is better to have a large source of data as it covers more width as well as depth. The foreseeable challenges, to be conquered, in the data lake, are the data format and the structuring of the data source.

8.7 Validation of DT Model

Method to validate a DT with its physical counterpart is not studied in detail and the number of articles is also few. Effective validation will be of importance as the support and control features from the DT can be applied directly only if obtained from a validated model. Also, the decision support provided by DT needs to be validated before being implemented. For this purpose, virtual commissioning along with DT might be a useful area to be explored.

References

Abburu S, Berre AJ, Jacoby M, Roman D, Stojanovic L, Stojanovic N (2020) COGNITWIN—hybrid and cognitive digital twins for the process industry. In: Proceedings of 2020 IEEE international conference on engineering, technology and innovation ICE/ITMC 2020. https://doi.org/10.1109/ICE/ITMC49519.2020.9198403

Alam KM, El Saddik A (2017) C2PS: a digital twin architecture reference model for the cloud-based cyber-physical systems. IEEE Access 5:2050–2062. https://doi.org/10.1109/ACCESS.2017.2657006

Alexopoulos K, Nikolakis N, Chryssolouris G (2020) Digital twin-driven supervised machine learning for the development of artificial intelligence applications in manufacturing. Int J Comput Integr Manuf 33(5):429–439. https://doi.org/10.1080/0951192X.2020.1747642

Angrish A, Starly B, Lee YS, Cohen PH (2017) A flexible data schema and system architecture for the virtualization of manufacturing machines (VMM). J Manuf Syst 45:236–247. https://doi.org/10.1016/j.jmsy.2017.10.003

Anton F, Borangiu T, Raileanu S, Anton S (2020) Cloud-based digital twin for robot integration in intelligent manufacturing systems. In: International conference on robotics in Alpe-Adria Danube Region RAAD 2020: advances in service and industrial robotics, vol 004, pp 2018–2020

Asimov RM (2018) Digital twin in the analysis of a big data. In: Fourth international conference and expo BIG DATA ADVANCED ANALYTICS, May 2018. [Online]. Available: https://www.researchgate.net/profile/R_Asimov/publication/325038225_DIGITAL_TWIN_IN_THE_ANALYSIS_OF_A_BIG_DATA/links/5af2cc3e458515c283797f7a/DIGITAL-TWIN-IN-THE-ANALYSIS-OF-A-BIG-DATA.pdf

Assad F, Konstantinov S, Ahmad MH, Rushforth EJ, Harrison R (2021) Utilising web-based digital twin to promote assembly line sustainability. In: Proceedings of 2021 4th IEEE international conference on industrial cyber-physical systems ICPS 2021, pp 381–386. https://doi.org/10.1109/ICPS49255.2021.9468209

Autiosalo J (2018) Platform for industrial internet and digital twin focused education, research, and innovation: Ilmatar the overhead crane. In: IEEE world forum internet things, WF-IoT 2018—proceedings, vol 2018, Jan 2018, pp 241–244. https://doi.org/10.1109/WF-IoT.2018.8355217

Bai Y, You JB, Lee IK (2021) Design and optimization of smart factory control system based on digital twin system model. Math Probl Eng 2021. https://doi.org/10.1155/2021/2596946

Bajaj M, Zwemer D, Cole B (2016) Architecture to geometry—integrating system models with mechanical design. In: AIAA SPACE and astronautics forum and exposition, Sept 2016. https://doi.org/10.2514/6.2016-5470

Balderas D, Ortiz A, Méndez E, Ponce P, Molina A (2021) Empowering digital twin for industry 4.0 using metaheuristic optimization algorithms: case study PCB drilling optimization. Int J Adv Manuf Technol 113(5–6):1295–1306. https://doi.org/10.1007/s00170-021-06649-8

Bambura R, Šolc M, Dado M, Kotek L (2020) Implementation of digital twin for engine block manufacturing processes. Appl Sci 10(18). https://doi.org/10.3390/APP10186578

Bao J, Guo D, Li J, Zhang J (2019) The modeling and operations for the digital twin in the context of manufacturing. Enterp Inf Syst 13(4):534–556. https://doi.org/10.1080/17517575.2018.1526324

Barni A, Pietraroia D, Züst S, West S, Stoll O (2020) Digital twin based optimization of a manufacturing execution system to handle high degrees of customer specifications. J Manuf Mater Process 4(4). https://doi.org/10.3390/jmmp4040109

Bauer H, Brandl F, Lock C, Reinhart G (2018) Integration of Industrie 4.0 in lean manufacturing learning factories. Procedia Manuf 23(2017):147–152. https://doi.org/10.1016/j.promfg.2018.04.008

Bazaz SM, Lohtander M, Varis J (2020) The prediction method of tool life on small lot turning process—development of digital twin for production. Procedia Manuf 51(2019):288–295. https://doi.org/10.1016/j.promfg.2020.10.041

Bazilevs Y, Deng X, Korobenko A, Di Scalea FL, Todd MD, Taylor SG (2015) Isogeometric fatigue damage prediction in large-scale composite structures driven by dynamic sensor data. J Appl Mech Trans ASME 82(9). https://doi.org/10.1115/1.4030795

Beisheim N, Kiesel M, Linde M, Ott T (2020) Using AutomationML and graph-based design languages for automatic generation of digital twins of cyber-physical systems. Adv Transdiscipl Eng 12:135–142. https://doi.org/10.3233/ATDE200070

Beregi R, Szaller Á, Kádár B (2018) Synergy of multi-modeling for process control. IFAC-PapersOnLine 51(11):1023–1028. https://doi.org/10.1016/j.ifacol.2018.08.473

Bielefeldt B, Hochhalter J, Hartl D (2015) Computationally efficient analysis of SMA sensory particles embedded in complex aerostructures using a substructure approach. In: ASME 2015 conference on smart materials, adaptive structures and intelligent systems SMASIS 2015, vol 1, pp 1–10. https://doi.org/10.1115/SMASIS2015-8975

Bolotov MA, Pechenin VA, Ruzanov NV, Grachev IA (2019) Information model and software architecture for the implementation of the digital twin of the turbine rotor. J Phys Conf Ser 1368(5). https://doi.org/10.1088/1742-6596/1368/5/052013

Borangiu T, Oltean VE, Raileanu S, Anton F (2019) Embedded digital twin for ARTI-type control of semi-continuous production processes. In: International workshop on service orientation in holonic and multi-agent manufacturing, pp 20–23

Borangiu T, Raileanu S, Silisteanu A, Anton S, Anton F (2020) Smart manufacturing control with cloud-embedded digital twins. In: 2020 24th international conference on system theory, control and computing. ICSTCC 2020—proceedings, pp 915–920. https://doi.org/10.1109/ICSTCC50638.2020.9259684

Brenner B, Hummel V (2017) Digital twin as enabler for an innovative digital shopfloor management system in the ESB logistics learning factory at Reutlingen—university. Procedia Manuf 9:198–205. https://doi.org/10.1016/j.promfg.2017.04.039

Brovkova M, Molodtsov V, Bushuev V (2021) Implementation specifics and application potential of digital twins of technological systems. Int J Adv Manuf Technol 117(7–8):2279–2286. https://doi.org/10.1007/s00170-021-07141-z

Cai H, Zhang W, Zhu Z (2019) Quality management and analysis of aircraft final assembly based on digital twin. In: Proceedings of 2019 11th international conference on intelligent human-machine systems and cybernetics IHMSC 2019, vol 1, pp 202–205. https://doi.org/10.1109/IHMSC.2019.00054

Cai Y, Wang Y, Burnett M (2020) Using augmented reality to build digital twin for reconfigurable additive manufacturing system. J Manuf Syst 56(May):598–604. https://doi.org/10.1016/j.jmsy.2020.04.005

Canedo A (2016) Industrial IoT lifecycle via digital twins. In: 2016 international conference on hardware/software codesign and system synthesis CODES+ISSS 2016, p 2974008. https://doi.org/10.1145/2968456.2974007

Cattaneo L, MacChi M (2019) A digital twin proof of concept to support machine prognostics with low availability of run-to-failure data. IFAC-PapersOnLine 52(10):37–42. https://doi.org/10.1016/j.ifacol.2019.10.016

Centomo S, Panato M, Fummi F (2019) Cyber-physical systems integration in a production line simulator. In: IFIP/IEEE international conference on very large scale integration VLSI-SoC, vol 2018, Oct 2018, pp 237–242. https://doi.org/10.1109/VLSI-SoC.2018.8644836

Chetan M, Yao S, Griffith DT (2021) Multi-fidelity digital twin structural model for a sub-scale downwind wind turbine rotor blade. Wind Energy 24(12):1368–1387. https://doi.org/10.1002/we.2636

Chuang W, Guanghui Z, Junsheng W (2021) Smart cyber-physical production system enabled workpiece production in digital twin job shop. Adv Mech Eng 13(9):1–15. https://doi.org/10.1177/16878140211040888

Ciavotta M, Alge M, Menato S, Rovere D, Pedrazzoli P (2017) A microservice-based middleware for the digital factory. Procedia Manuf 11(June):931–938. https://doi.org/10.1016/j.promfg.2017.07.197

Cronrath C, Aderiani AR, Lennartson B (2019) Enhancing digital twins through reinforcement learning. In: IEEE international conference on automation science and engineering, vol 2019, Aug 2019, pp 293–298. https://doi.org/10.1109/COASE.2019.8842888

Dai S, Zhao G, Yu Y, Zheng P, Bao Q, Wang W (2021) Ontology-based information modeling method for digital twin creation of as-fabricated machining parts. Robot Comput Integr Manuf 72:102173. https://doi.org/10.1016/j.rcim.2021.102173

Danilczyk W, Sun Y, He H (2019) ANGEL: an intelligent digital twin framework for microgrid security. In: 51st North American power symposium NAPS 2019, pp 37–42. https://doi.org/10.1109/NAPS46351.2019.9000371

de Andrade MAN, Lepikson HA, Tosta Machado CA (2021) A new framework and methodology for digital twin development. In: 2021 14th IEEE international conference on industry applications INDUSCON 2021—proceedings, pp 134–138. https://doi.org/10.1109/INDUSCON51756.2021.9529701

Deac GC, Deac CN, Popa CL, Ghinea M, Cotet CE (2017) Machine vision in manufacturing processes and the digital twin of manufacturing architectures. In: Annals of DAAAM for 2011 & proceedings of the 22nd international DAAAM symposium, pp 733–736. https://doi.org/10.2507/28th.daaam.proceedings.103

Digital Manufacturing Working Group (WG15) (2021) Digital twin framework for manufacturing

Dittmann S, Zhang P, Glodde A, Dietrich F (2020) Towards a scalable implementation of digital twins—a generic method to acquire shopfloor data. Procedia CIRP 96:157–162. https://doi.org/10.1016/j.procir.2021.01.069

Dobrescu R, Chenaru O, Florea G, Geampalia G, Mocanu S (2020) Hardware-in-loop assessment of control architectures. In: 2020 24th international conference on system theory, control and computing ICSTCC 2020—proceedings, pp 880–885. https://doi.org/10.1109/ICSTCC50638.2020.9259636

dos Santos CH, Gabriel GT, do Amaral JVS, Montevechi JAB, de Queiroz JA (2021) Decision-making in a fast fashion company in the industry 4.0 era: a digital twin proposal to support operational planning. Int J Adv Manuf Technol 116(5–6):1653–1666. https://doi.org/10.1007/s00170-021-07543-z

Duan JG, Ma TY, Zhang QL, Liu Z, Qin JY (2021) Design and application of digital twin system for the blade-rotor test rig. J Intell Manuf. https://doi.org/10.1007/s10845-021-01824-w

Ehrhardt JM, Hoffmann CT (2020) The digital shadow: developing a universal model for the automated optimization of cyber-physical production systems based on real-time data. Procedia CIRP 93:304–310. https://doi.org/10.1016/j.procir.2020.03.069

Eisenbarth D, Soffel F, Wegener K (2019) Geometry-based process adaption to fabricate parts with varying wall thickness by direct metal deposition. In: International conference of progress in digital and physical manufacturing, no 25498, pp 16–18

Ertveldt J, Guillaume P, Helsen J (2020) MiCLAD as a platform for real-time monitoring and machine learning in laser metal deposition. Procedia CIRP 94:456–461. https://doi.org/10.1016/j.procir.2020.09.164

Eschemann P, Borchers P, Feeken L, Stierand I, Zernickel JS, Neumann M (2020) Towards digital twins for optimizing the factory of the future. In: Modelling and simulation 2020—European simulation and modelling conference ESM 2020, Oct 2021, pp 208–215

Escriche Lng S et al (2021) A digital twin based approach for simulation and emulation of an automotive paint workshop. SAE technical paper, no 2021, pp 22–24. https://doi.org/10.4271/2021-01-0240

Feldt J, Kourouklis T, Kontny H, Wagenitz A (2020) Digital twin: revealing potentials of real-time autonomous decisions at a manufacturing company. Procedia CIRP 88:185–190. https://doi.org/10.1016/j.procir.2020.05.033

Feng X, Zhao Z, Zhang C (2020) Simulation optimization framework for online deployment and adjustment of reconfigurable machines in job shops. In: IEEE international conference on industrial engineering and engineering management, Dec 2020, pp 731–735. https://doi.org/10.1109/IEEM45057.2020.9309782

Fera M et al (2020) Towards digital twin implementation for assessing production line performance and balancing. Sensors 20(97):1–18

Flores-Garcia E, Kim G-Y, Ynag J, Wiktorsson M, Do Noh S (2020) Analyzing the characteristics of digital twin and discrete event simulation in cyber physical systems

Gabor T, Belzner L, Kiermeier M, Beck MT, Neitz A (2016) A simulation-based architecture for smart cyber-physical systems. In: Proceedings—2016 IEEE international conference on autonomic computing ICAC 2016, pp 374–379. https://doi.org/10.1109/ICAC.2016.29

Gallego-García S, Reschke J, García-García M (2019) Design and simulation of a capacity management model using a digital twin approach based on the viable system model: case study of an automotive plant. Appl Sci 9(24). https://doi.org/10.3390/app9245567

Gao X, Liu P, Zhang Q, Gao D, Huang X (2021) Analysis and application of manufacturing data driven by digital twins. J Phys Conf Ser 1983(1). https://doi.org/10.1088/1742-6596/1983/1/012104

Gericke GA, Kuriakose RB, Vermaak HJ, Mardsen O (2019) Design of digital twins for optimization of a water bottling plant. In: IECON proceedings (industrial electronics conference), Oct 2019, pp 5204–5210. https://doi.org/10.1109/IECON.2019.8926880

Ghosh AK, Ullah AS, Teti R, Kubo A (2021) Developing sensor signal-based digital twins for intelligent machine tools. J Ind Inf Integr 24:100242. https://doi.org/10.1016/j.jii.2021.100242

Glaessgen EH, Stargel DS (2012) The digital twin paradigm for future NASA and U.S. air force vehicles. In: Collection of technical papers—AIAA/ASME/ASCE/AHS/ASC structures, structural dynamics & materials conference, pp 1–14. https://doi.org/10.2514/6.2012-1818

Gockel BT, Tudor AW, Brandyberry MD, Penmetsa RC, Tuegel EJ (2012) Challenges with structural life forecasting using realistic mission profiles. In: Collection of technical papers—AIAA/ASME/ASCE/AHS/ASC structures, structural dynamics & materials conference, Apr 2012, pp 1–11. https://doi.org/10.2514/6.2012-1813

Graessler I, Poehler A (2018) Integration of a digital twin as human representation in a scheduling procedure of a cyber-physical production system. In: IEEE international conference on industrial engineering and engineering management, Dec 2018, pp 289–293. https://doi.org/10.1109/IEEM.2017.8289898

Gramegna N, Greggio F, Bonollo F (2020) Smart factory competitiveness based on real time monitoring and quality predictive model applied to multi-stages production lines. In: IFIP advances in information and communication technology IFIP, vol 592, pp 185–196. https://doi.org/10.1007/978-3-030-57997-5_22

Greco A, Caterino M, Fera M, Gerbino S (2020) Digital twin for monitoring ergonomics during manufacturing production. Appl Sci 10(21):1–20. https://doi.org/10.3390/app10217758

Grieves M, Vickers J (2016) Digital twin: mitigating unpredictable, undesirable emergent behavior in complex systems. Transdiscipl Perspect Complex Syst New Find Approaches 89(9):85–113. https://doi.org/10.1007/978-3-319-38756-7_4

Gröger C et al (2016)"The data-driven factory leveraging big industrial data for agile, learning and human-centric manufacturing. In: ICEIS 2016—proceedings of the 18th international conference on enterprise information systems, vol 1, pp 40–52. https://doi.org/10.5220/0005831500400052

Gunasegaram DR et al (2021) Towards developing multiscale-multiphysics models and their surrogates for digital twins of metal additive manufacturing. Addit Manuf 46. https://doi.org/10.1016/j.addma.2021.102089

Guo H, Zhu Y, Zhang Y, Ren Y, Chen M, Zhang R (2021a) A digital twin-based layout optimization method for discrete manufacturing workshop. Int J Adv Manuf Technol 1307–1318. https://doi.org/10.1007/s00170-020-06568-0

Guo H, Chen M, Mohamed K, Qu T, Wang S, Li J (2021b) A digital twin-based flexible cellular manufacturing for optimization of air conditioner line. J Manuf Syst 58(PB):65–78. https://doi.org/10.1016/j.jmsy.2020.07.012

Gurjanov AV, Zakoldaev DA, Shukalov AV, Zharinov IO (2019) Formation principles of digital twins of cyber-physical systems in the smart factories of industry 4.0. IOP Conf Ser Mater Sci Eng 483(1):1–5. https://doi.org/10.1088/1757-899X/483/1/012070

Gyulai D, Pfeiffer A, Nick G, Gallina V, Sihn W, Monostori L (2018) Lead time prediction in a flow-shop environment with analytical and machine learning approaches. IFAC-PapersOnLine 51(11):1029–1034. https://doi.org/10.1016/j.ifacol.2018.08.472

Haag S, Anderl R (2018) Digital twin—proof of concept. Manuf Lett 15:64–66. https://doi.org/10.1016/j.mfglet.2018.02.006

Halenar I, Juhas M, Juhasova B, Borkin D (2019) Virtualization of production using digital twin technology. In: Proceedings of 2019 20th international Carpathian control conference ICCC 2019, pp 7–11. https://doi.org/10.1109/CarpathianCC.2019.8765940

He Y, Zhang N, Wang A (2020) Digital twin process and simulation operation control technology for intelligent manufacturing unit. IOP Conf Ser Mater Sci Eng 836(1):1–7. https://doi.org/10.1088/1757-899X/836/1/012010

Henson CM, Decker NI, Huang Q (2021) A digital twin strategy for major failure detection in fused deposition modeling processes. Procedia Manuf 53(2020):359–367. https://doi.org/10.1016/j.promfg.2021.06.039

Hinchy EP, Carcagno C, O'Dowd NP, McCarthy CT (2020) Using finite element analysis to develop a digital twin of a manufacturing bending operation. Procedia CIRP 93:568–574. https://doi.org/10.1016/j.procir.2020.03.031

Hu L et al (2018) Modeling of cloud-based digital twins for smart manufacturing with MT connect. Procedia Manuf 26:1193–1203. https://doi.org/10.1016/j.promfg.2018.07.155

Hu Z, Fang X, Zhang J (2021) A digital twin-based framework of manufacturing workshop for marine diesel engine. Int J Adv Manuf Technol 117(11–12):3323–3342. https://doi.org/10.1007/s00170-021-07891-w

Hürkamp A, Lorenz R, Ossowski T, Behrens BA, Dröder K (2021) Simulation-based digital twin for the manufacturing of thermoplastic composites. Procedia CIRP 100:1–6. https://doi.org/10.1016/j.procir.2021.05.001

Huynh BH, Akhtar H, Sett MK (2019) A universal methodology to create digital twins for serial and parallel manipulators. In: Proceedings of IEEE international conference on systems, man and cybernetics, vol 2019, Oct 2019, pp 3104–3109. https://doi.org/10.1109/SMC.2019.8914195

Industrial Internet Consortium (2015) Industrial internet reference architecture. Technical report, pp 1–101. [Online]. Available: http://www.iiconsortium.org/IIRA.htm

Intizar Ali M, Patel P, Breslin JG, Harik R, Sheth A (2021) Cognitive digital twins for smart manufacturing. IEEE Intell Syst 36(2):96–100. https://doi.org/10.1109/MIS.2021.3062437

Ippolito D, Constantinescu C, Rusu CA (2020) Enhancement of human-centered workplace design and optimization with exoskeleton technology. Procedia CIRP 91:243–248. https://doi.org/10.1016/j.procir.2020.02.173

Jaensch F, Csiszar A, Scheifele C, Verl A (2019) Digital twins of manufacturing systems as a base for machine learning. In: Proceedings of the 2018 25th international conference on mechatronics and machine vision in practice M2VIP 2018, pp 1–6. https://doi.org/10.1109/M2VIP.2018.8600844

Janda P, Hajicek Z, Bernardin P (2019) Implementation of the digital twin methodology. In: Annals of DAAAM for 2011 & proceedings of the 22nd international DAAAM symposium, vol 30, no 1, pp 533–538. https://doi.org/10.2507/30th.daaam.proceedings.072

Jeong Y, Flores-Garcia E, Wiktorsson M (2020) A design of digital twins for supporting decision-making in production logistics. In: Proceedings of winter simulation conference, vol 2020, no 2019, Dec 2020, pp 2683–2694. https://doi.org/10.1109/WSC48552.2020.9383863

Juarez MG, Botti VJ, Giret AS (2021) Digital twins: review and challenges. J Comput Inf Sci Eng 21(3). https://doi.org/10.1115/1.4050244

Kalaboukas K, Rožanec J, Košmerlj A, Kiritsis D, Arampatzis G (2021) Implementation of cognitive digital twins in connected and agile supply networks—an operational model. Appl Sci 11(9). https://doi.org/10.3390/app11094103

Karanjkar N, Joglekar A, Mohanty S, Prabhu V, Raghunath D, Sundaresan R (2019) Digital twin for energy optimization in an SMT-PCB assembly line. In: Proceedings of 2018 IEEE international conference on internet of things and intelligence systems IOTAIS 2018, pp 85–89. https://doi.org/10.1109/IOTAIS.2018.8600830

Kassen S, Tammen H, Zarte M, Pechmann A (2021) Concept and case study for a generic simulation as a digital shadow to be used for production optimisation. Processes 9(8). https://doi.org/10.3390/pr9081362

Khanesar MA, Bansal R, Martínez-Arellano G, Branson DT (2020) XOR binary gravitational search algorithm with repository: industry 4.0 applications. Appl Sci 10(18):1–32. https://doi.org/10.3390/APP10186451

Kholopov VA, Antonov SV, Kurnasov EV, Kashirskaya EN (2019) Digital twins in manufacturing. Russ Eng Res 39(12):1014–1020. https://doi.org/10.3103/S1068798X19120104

Konstantinov S, Assad F, Azam W, Vera D, Ahmad B, Harrison R (2021) Developing web-based digital twin of assembly lines for industrial cyber-physical systems. In: Proceedings of 2021 4th IEEE international conference on industrial cyber-physical systems ICPS 2021, pp 219–224. https://doi.org/10.1109/ICPS49255.2021.9468227

Koulouris A, Misailidis N, Petrides D (2021) Applications of process and digital twin models for production simulation and scheduling in the manufacturing of food ingredients and products. Food Bioprod Process 126:317–333. https://doi.org/10.1016/j.fbp.2021.01.016

Kraft EM (2016) The US air force digital thread/digital twin—life cycle integration and use of computational and experimental knowledge. In: 54th AIAA aerospace sciences meeting, Jan 2016, pp 1–22. https://doi.org/10.2514/6.2016-0897

Kuts V, Otto T, Tahemaa T, Bondarenko Y (2019) Digital twin based synchronised control and simulation of the industrial robotic cell using virtual reality. J Mach Eng 19(1):128–145

Kuts V, Cherezova N, Sarkans M, Otto T (2020) Digital twin: industrial robot kinematic model integration to the virtual reality environment. J Mach Eng 20(2):53–64

Lacueva-Perez FJ, Hermawati S, Amoraga P, Salillas-Martinez R, Del Hoyo Alonso R, Lawson G (2020) SHION: towards an interactive digital twin supporting shopfloor operations on real time. IEEE Internet Comput 7801:1–10. https://doi.org/10.1109/MIC.2020.3047349

Latif H, Starly B (2020) A simulation algorithm of a digital twin for manual assembly process. Procedia Manuf 48(2019):932–939. https://doi.org/10.1016/j.promfg.2020.05.132

Latsou C, Farsi M, Erkoyuncu JA, Morris G (2021) Digital twin integration in multi-agent cyber physical manufacturing systems. IFAC-PapersOnLine 54(1):811–816. https://doi.org/10.1016/j.ifacol.2021.08.096

Lee H, Kim T (2018) Smart factory use case model based on digital twin. ICIC Express Lett Part B Appl 9(9):931–936. https://doi.org/10.24507/icicelb.09.09.931

Lee J, Lapira E, Bagheri B, Kao H (2013) Recent advances and trends in predictive manufacturing systems in big data environment. Manuf Lett 1(1):38–41. https://doi.org/10.1016/j.mfglet.2013.09.005

Leng J, Zhang H, Yan D, Liu Q, Chen X, Zhang D (2019a) Digital twin-driven manufacturing cyber-physical system for parallel controlling of smart workshop. J Ambient Intell Humaniz Comput 10(3):1155–1166. https://doi.org/10.1007/s12652-018-0881-5

Leng J et al (2019b) Digital twin-driven rapid reconfiguration of the automated manufacturing system via an open architecture model. Robot Comput Integr Manuf 63(December):2020. https://doi.org/10.1016/j.rcim.2019.101895

Li C, Mahadeven S, Ling Y, Wang L, Choze S (2017) A dynamic Bayesian network approach for digital twin. In: 19th AIAA non-deterministic approaches conference 2017. https://doi.org/10.2514/6.2017-1566

Li P, Zhu H, Luo L (2020) Digital twin technology in intelligent manufacturing. In: Proceedings of 2020 2nd international conference on artificial intelligence and advanced manufacture AIAM 2020, pp 195–200. https://doi.org/10.1109/AIAM50918.2020.00046

Li J, Pang D, Zheng Y, Le X (2021a) Digital twin enhanced assembly based on deep reinforcement learning. In: 2021 11th international conference on information science and technology ICIST 2021, pp 432–437. https://doi.org/10.1109/ICIST52614.2021.9440555

Li X, Wang L, Zhu C, Liu Z (2021b) Framework for manufacturing-tasks semantic modeling and manufacturing-resource recommendation for digital twin shop-floor. J Manuf Syst 58(PB):281–292. https://doi.org/10.1016/j.jmsy.2020.08.003

Li Y, Chen J, Hu Z, Zhang H, Lu J, Kiritsis D (2021c) Co-simulation of complex engineered systems enabled by a cognitive twin architecture. Int J Prod Res. https://doi.org/10.1080/00207543.2021.1971318

Lin WD, Low MYH (2020) Concept design of a system architecture for a manufacturing cyber-physical digital twin system. In: IEEE international conference on industrial engineering and engineering management, vol 2020, Dec 2020, pp 1320–1324. https://doi.org/10.1109/IEEM45057.2020.9309795

Lin S-W et al (2017) The industrial internet of things volume G1: reference architecture. In: Industrial internet consortium white paper, version 1. Seiten, p 58

Liu Z, Meyendorf N, Mrad N (2018) The role of data fusion in predictive maintenance using digital twin. AIP Conf Proc 1949(April):2018. https://doi.org/10.1063/1.5031520

Liu J, Yu D, Bi X, Hu Y, Yu H, Li B (2020a) The research of ontology-based digital twin machine tool modeling. In: 2020 IEEE 6th international conference on computer and communications ICCC 2020, pp 2130–2134. https://doi.org/10.1109/ICCC51575.2020.9344997

Liu C, Jiang P, Jiang W (2020b) Web-based digital twin modeling and remote control of cyber-physical production systems. Robot Comput Integr Manuf 64:101956. https://doi.org/10.1016/j.rcim.2020.101956

Lou X, Guo Y, Gao Y, Waedt K, Parekh M (2019) An idea of using digital twin to perform the functional safety and cybersecurity analysis. In: Lecture notes in informatics (LNI), proceedings—series of the gesellschaft für informatik, vol 295, pp 283–294. https://doi.org/10.18420/inf2019_ws32

Lu Y, Liu C, Wang KIK, Huang H, Xu X (2020) Digital twin-driven smart manufacturing: connotation, reference model, applications and research issues. Robot Comput Integr Manuf 61:101837. https://doi.org/10.1016/j.rcim.2019.101837

Luo W, Hu T, Zhu W, Tao F (2018) Digital twin modeling method for CNC machine tool. In: ICNSC 2018—15th IEEE international conference on networking, sensing and control, no 51405270, pp 1–4. https://doi.org/10.1109/ICNSC.2018.8361285

Ma J et al (2020a) A digital twin-driven production management system for production workshop. Int J Adv Manuf Technol 110(5–6):1385–1397. https://doi.org/10.1007/s00170-020-05977-5

Ma Y et al (2020b) Digital twin enhanced optimization of manufacturing service scheduling for industrial cloud robotics. In: IEEE international conference on industrial informatics, vol 2020, July 2020, pp 469–476. https://doi.org/10.1109/INDIN45582.2020.9442235

Magnanini MC et al (2021) A digital twin-based approach for multi-objective optimization of short-term production planning. IFAC-PapersOnLine 54(1):140–145. https://doi.org/10.1016/j.ifacol.2021.08.077

Majumdar PK, Haider MF, Reifsnider K (2013) Multi-physics response of structural composites and framework for modeling using material geometry. In: 54th AIAA/ASME/ASCE/AHS/ASC structures, structural dynamics and materials conference, pp 1–8. https://doi.org/10.2514/6.2013-1577

Makarov VL, Bakhtizin AR, Beklaryan GL, Akopov AS (2021) Digital plant: methods of discrete-event modeling and optimization of production characteristics. Bus Inform 15(2):7–20. https://doi.org/10.17323/2587-814X.2021.2.7.20

Marketsandmarkets (2019) Digital twin market by technology, type (product, process, and system), application (predictive maintenance), industry (aerospace & defense, automotive & transportation, healthcare), and geography—global forecast to 2026

Martinez S et al (2021) A digital twin demonstrator to enable flexible manufacturing with robotics: a process supervision case study. Prod Manuf Res 9(1):140–156. https://doi.org/10.1080/21693277.2021.1964405

Martínez-Gutiérrez A, Díez-González J, Ferrero-Guillén R, Verde P, Álvarez R, Perez H (2021) Digital twin for automatic transportation in industry 4.0. Sensors 21(10). https://doi.org/10.3390/s21103344

Matulis M, Harvey C (2021) A robot arm digital twin utilising reinforcement learning. Comput Graph 95:106–114. https://doi.org/10.1016/j.cag.2021.01.011

May MC, Overbeck L, Wurster M, Kuhnle A, Lanza G (2021) Foresighted digital twin for situational agent selection in production control. Procedia CIRP 99:27–32. https://doi.org/10.1016/j.procir.2021.03.005

Meierhofer J et al (2021) Digital twin-enabled decision support services in industrial ecosystems. Appl Sci 11(23). https://doi.org/10.3390/app112311418

Min Q, Lu Y, Liu Z, Su C, Wang B (2019) Machine learning based digital twin framework for production optimization in petrochemical industry. Int J Inf Manage 49:502–519. https://doi.org/10.1016/j.ijinfomgt.2019.05.020

Morabito L, Ippolito M, Pastore E, Alfieri A, Montagna F (2021) A discrete event simulation based approach for digital twin implementation. IFAC-PapersOnLine 54(1):414–419. https://doi.org/10.1016/j.ifacol.2021.08.164

Moreno T, Almeida A, Ferreira F, Caldas N, Toscano C, Azevedo A (2021) Digital twin for manufacturing equipment in industry 4.0. Adv Transdiscipl Eng 15:362–367. https://doi.org/10.3233/ATDE210062

Mourtzis D, Angelopoulos J, Siatras V (2020) Cycle time estimation model for hybrid assembly stations based on digital twin, no 723711

Mourtzis D, Togias T, Angelopoulos J, Stavropoulos P (2021a) A digital twin architecture for monitoring and optimization of fused deposition modeling processes. Procedia CIRP 103:97–102. https://doi.org/10.1016/j.procir.2021.10.015

Mourtzis D, Angelopoulos J, Panopoulos N (2021b) Equipment design optimization based on digital twin under the framework of zero-defect manufacturing. Procedia Comput Sci 180(2019):525–533. https://doi.org/10.1016/j.procs.2021.01.271

Mylrea M et al (2021) BioSecure digital twin: manufacturing innovation and cybersecurity resilience

Nafors D, Johansson B, Gullander P, Erixon S (2020) Simulation in hybrid digital twins for factory layout planning. In: Proceedings of winter simulation conference, vol 2020, Dec 2020, pp 1619–1630. https://doi.org/10.1109/WSC48552.2020.9384075

Negri E, Fumagalli L, Macchi M (2017) A review of the roles of digital twin in CPS-based production systems. Procedia Manuf 11:939–948. https://doi.org/10.1016/j.promfg.2017.07.198

Negri E, Assiro G, Caioli L, Fumagalli L (2019) A machine state-based digital twin development methodology. In: Summer school F. Turco-industrial systems engineering, vol 1, pp 34–40

Negri E, Berardi S, Fumagalli L, Macchi M (2020) MES-integrated digital twin frameworks. J Manuf Syst 56:58–71. https://doi.org/10.1016/j.jmsy.2020.05.007

Ngo D, Guerra-Zubiaga D, González-Badillo G, Vatankhah RB (2018) Towards a digital twin for cloud manufacturing—case study

Nguyen HX, Trestian R, To D, Tatipamula M (2021) Digital twin for 5G and beyond. IEEE Commun Mag 59(2):10–15. https://doi.org/10.1109/MCOM.001.2000343

Nikolakis N, Alexopoulos K, Xanthakis E, Chryssolouris G (2019) The digital twin implementation for linking the virtual representation of human-based production tasks to their physical counterpart in the factory-floor. Int J Comput Integr Manuf 32(1):1–12. https://doi.org/10.1080/0951192X.2018.1529430

Novák P, Vyskocil J, Wally B (2020) The digital twin as a core component for industry 4.0 smart production planning. IFAC-PapersOnLine 53:10803–10809. https://doi.org/10.1016/j.ifacol.2020.12.2865

Olalere IO, Olanrewaju OA (2020) Optimising production through intelligent manufacturing. E3S Web Conf 152:3–6. https://doi.org/10.1051/e3sconf/202015203012

Orive D, Iriondo N, Burgos A, Saráchaga I, Álvarez ML, Marcos M (2019) Fault injection in digital twin as a means to test the response to process faults at virtual commissioning. In: IEEE international conference on emerging technologies and factory automation ETFA, vol 2019, Sept 2019, pp 1230–1234. https://doi.org/10.1109/ETFA.2019.8869334
Papanagnou C (2020) A digital twin model for enhancing performance measurement in assembly lines
Park KT, Yang J, Do S (2021a) VREDI: virtual representation for a digital twin application in a work-center-level asset administration shell. J Intell Manuf 32(2)
Park KT, Son YH, Ko SW, Noh SD (2021b) Digital twin and reinforcement learning-based resilient production control for micro smart factory. Appl Sci 11(7):12–14. https://doi.org/10.3390/app11072977
Peng G, Zhong H (2020) Data exchange of digital twins based on AML in space science experiment equipment. IOP Conf Ser Mater Sci Eng 816(1). https://doi.org/10.1088/1757-899X/816/1/012021
Polini W, Corrado A (2021) Digital twin of stone sawing processes. Int J Adv Manuf Technol 112(1–2):121–131. https://doi.org/10.1007/s00170-020-06384-6
Protic A, Jin Z, Marian R, Abd K, Campbell D, Chahl J (2020) Implementation of a bi-directional digital twin for industry 4 labs in academia: a solution based on OPC UA. In: IEEE international conference on industrial engineering and engineering management, vol 2020, Dec 2020, pp 979–983. https://doi.org/10.1109/IEEM45057.2020.9309953
Qi B, Park HS (2020) Data-driven digital twin model for predicting grinding force. IOP Conf Ser Mater Sci Eng 916(1). https://doi.org/10.1088/1757-899X/916/1/012092
Qi Q, Tao F, Zuo Y, Zhao D (2018) Digital twin service towards smart manufacturing. Procedia CIRP 72:237–242. https://doi.org/10.1016/j.procir.2018.03.103
Qiao Q, Wang J, Ye L, Gao RX (2019) Digital twin for machining tool condition prediction. Procedia CIRP 81:1388–1393. https://doi.org/10.1016/j.procir.2019.04.049
Qin H, Wang H, Zhang Y, Lin L (2021) Constructing digital twin for smart manufacturing. In: Proceedings of the 2021 IEEE 24th international conference on computer supported cooperative work in design CSCWD 2021, pp 638–642. https://doi.org/10.1109/CSCWD49262.2021.9437791
Ralph BJ, Schwarz A, Stockinger M (2020) An implementation approach for an academic learning factory for the metal forming industry with special focus on digital twins and finite element analysis. Procedia Manuf 45:253–258. https://doi.org/10.1016/j.promfg.2020.04.103
Rasor R, Göllner D, Bernijazov R, Kaiser L, Dumitrescu R (2021) Towards collaborative life cycle specification of digital twins in manufacturing value chains. Procedia CIRP 98:229–234. https://doi.org/10.1016/j.procir.2021.01.035
Ratnayake D, Lohit P, Singh B, Mishra VP (2020) Analysis of machine learning algorithms in smart manufacturing. In: ICRITO 2020—IEEE 8th international conference on reliability, infocom technologies and optimization (trends and future directions), pp 707–712. https://doi.org/10.1109/ICRITO48877.2020.9198017
Rebmann A, Knoch S, Emrich A, Fettke P, Loos P (2020) A multi-sensor approach for digital twins of manual assembly and commissioning. Procedia Manuf 51:549–556. https://doi.org/10.1016/j.promfg.2020.10.077
Redelinghuys AJH, Basson AH, Kruger K (2020) A six-layer architecture for the digital twin: a manufacturing case study implementation. J Intell Manuf 31:1383–1402. https://doi.org/10.1007/s10845-019-01516-6
Reifsnider K, Majumdar P (2013) Multiphysics stimulated simulation digital twin methods for fleet management. In: Collection of technical papers—AIAA/ASME/ASCE/AHS/ASC structures, structural dynamics & materials conference, pp 1–11
Review UMA, Wu L, Leng J (2021) SS symmetry digital twins-based smart design and control
Ríos J, Hernández JC, Oliva M, Mas F (2015) Product avatar as digital counterpart of a physical individual product: literature review and implications in an aircraft. Adv Transdiscipl Eng 2:657–666. https://doi.org/10.3233/978-1-61499-544-9-657

Rojek I, Mikołajewski D, Dostatni E (2021) Digital twins in product lifecycle for sustainability in manufacturing and maintenance. Appl Sci 11(1):1–19. https://doi.org/10.3390/app11010031

Rolo GR, Rocha AD, Tripa J, Barata J (2021) Application of a simulation-based digital twin for predicting distributed manufacturing control system performance. Appl Sci 11(5):1–19. https://doi.org/10.3390/app11052202

Rosen R, Von Wichert G, Lo G, Bettenhausen KD (2015) About the importance of autonomy and digital twins for the future of manufacturing. IFAC-PapersOnLine 28(3):567–572. https://doi.org/10.1016/j.ifacol.2015.06.141

Rožanec JM et al (2022) Actionable cognitive twins for decision making in manufacturing. Int J Prod Res 60(2):452–478. https://doi.org/10.1080/00207543.2021.2002967

Ruhland P, Li Y, Coutandin S, Fleischer J (2021) Production of hybrid tubular metal-fiber preforms: development of a digital twin for the draping process. Procedia CIRP 99:437–442. https://doi.org/10.1016/j.procir.2021.03.062

Ruiz JCS, Bru JM, Escoto RP (2021) Smart digital twin for ZDM-based job-shop scheduling. In: 2021 IEEE international workshop on metrology for industry 4.0 & IoT (MetroInd4.0 & IoT)—proceedings, pp 510–515. https://doi.org/10.1109/MetroInd4.0IoT51437.2021.9488473

Samir K, Maffei A, Onori MA (2019) Real-time asset tracking; a starting point for digital twin implementation in manufacturing. Procedia CIRP 81:719–723. https://doi.org/10.1016/j.procir.2019.03.182

Schleich B, Anwer N, Mathieu L, Wartzack S (2017) Shaping the digital twin for design and production engineering. CIRP Ann Manuf Technol 66(1):141–144. https://doi.org/10.1016/j.cirp.2017.04.040

Schluse M, Rossmann J (2016) From simulation to experimentable digital twins. In: IEEE international symposium on systems engineering, pp 1–6

Schluse M, Atorf L, Rossmann J (2017) Experimentable digital twins for model-based systems engineering and simulation-based development. In: 11th annual IEEE international systems conference SysCon 2017—proceedings. https://doi.org/10.1109/SYSCON.2017.7934796

Schroeder GN, Steinmetz C, Pereira CE, Espindola DB (2016) Digital twin data modeling with AutomationML and a communication methodology for data exchange. IFAC-PapersOnLine 49(30):12–17. https://doi.org/10.1016/j.ifacol.2016.11.115

Schuh G, Kelzenberg C, Helbig J, Frey C (2021) Operational implementation of digital production twins in single and small batch production. In: ACM international conference proceeding series, pp 72–79. https://doi.org/10.1145/3463858.3463859

Schützer K, de Andrade Bertazzi J, Sallati C, Anderl R, Zancul E (2019) Contribution to the development of a digital twin based on product lifecycle to support the manufacturing process. Procedia CIRP 84:82–87. https://doi.org/10.1016/j.procir.2019.03.212

Schweichhart K (2019) RAMI 4.0 reference architectural model for Industrie 4.0, vol 66, no 2. InTech, p 15. [Online]. Available: https://ec.europa.eu/futurium/en/system/files/ged/a2-schweichhart-reference_architectural_model_industrie_4.0_rami_4.0.pdf

Semeraro C, Lezoche M, Panetto H, Dassisti M (2021) Digital twin paradigm: a systematic literature review. Comput Ind 130. https://doi.org/10.1016/j.compind.2021.103469

Seok MG, Cai W, Park D (2021) Hierarchical aggregation/disaggregation for adaptive abstraction-level conversion in digital twin-based smart semiconductor manufacturing. IEEE Access 9:71145–71158. https://doi.org/10.1109/ACCESS.2021.3073618

Shafto M et al (2012) Modeling, simulation, information technology & processing roadmap. Technol Area 11:1–38

Shahpar S (2020) Building digital twins to simulate manufacturing variation. In: Proceedings of ASME turbo expo, vol 2A-2020, pp 4–5. https://doi.org/10.1115/GT2020-15263

Shao G, Helu M (2020) Framework for a digital twin in manufacturing: scope and requirements. Manuf Lett 24:105–107. https://doi.org/10.1016/j.mfglet.2020.04.004

Söderberg R, Wärmefjord K, Carlson JS, Lindkvist L (2017) Toward a digital twin for real-time geometry assurance in individualized production. CIRP Ann Manuf Technol 66(1):137–140. https://doi.org/10.1016/j.cirp.2017.04.038

Sommer M, Stjepandic J, Stobrawa S, Von Soden M (2020) Automated generation of a digital twin of a manufacturing system by using scan and convolutional neural networks. Adv Transdiscipl Eng 12:363–372. https://doi.org/10.3233/ATDE200095

Son YH, Park KT, Lee D, Jeon SW, Noh SD (2021) Digital twin–based cyber-physical system for automotive body production lines. Int J Adv Manuf Technol 115(1–2):291–310. https://doi.org/10.1007/s00170-021-07183-3

Souza V, Cruz R, Silva W, Lins S, Lucena V (2019) A digital twin architecture based on the industrial internet of things technologies. In: 2019 IEEE international conference on consumer electronics ICCE 2019, pp 1–2. https://doi.org/10.1109/ICCE.2019.8662081

Stark R, Kind S, Neumeyer S (2017) Innovations in digital modeling for next generation manufacturing system design. CIRP Ann Manuf Technol 66(1):169–172. https://doi.org/10.1016/j.cirp.2017.04.045

Stavropoulos P, Papacharalampopoulos A, Athanasopoulou L (2020) A molecular dynamics based digital twin for ultrafast laser material removal processes. Int J Adv Manuf Technol 108(1–2):413–426. https://doi.org/10.1007/s00170-020-05387-7

Stavropoulos P, Papacharalampopoulos A, Michail CK, Chryssolouris G (2021) Robust additive manufacturing performance through a control oriented digital twin. Metals (Basel) 11(708)

Stieber S et al (2020) Towards real-time process monitoring and machine learning for manufacturing composite structures. In: IEEE symposium on emerging technologies and factory automation ETFA, vol 2020, Sept 2020, pp 1455–1458. https://doi.org/10.1109/ETFA46521.2020.9212097

Suthar B, Bongale A, Kumar S (2021) Three degrees of freedom robotic arm and its digital twin using Simulink—a bibliometric analysis. Libr Philos Pract 2021:1–36

Szabo G, Peto J, Nemeth L, Vidacs A (2020) Information gain regulation in reinforcement learning with the digital twins' level of realism. In: IEEE international symposium on personal, indoor and mobile radio communications, PIMRC, vol 2020, Aug 2020. https://doi.org/10.1109/PIMRC48278.2020.9217201

Tabar RS, Wärmefjord K, Söderberg R (2019) A method for identification and sequence optimisation of geometry spot welds in a digital twin context. Proc Inst Mech Eng Part C J Mech Eng Sci 233(16):5610–5621. https://doi.org/10.1177/0954406219854466

Tabar RS, Warmefjord K, Soderberg R, Lindkvist L (2020) Efficient spot welding sequence optimization in a geometry assurance digital twin. J Mech Des Trans ASME 142(10). https://doi.org/10.1115/1.4046436

Tao F, Zhang M (2017) Digital twin shop-floor: a new shop-floor paradigm towards smart manufacturing. IEEE Access 5:20418–20427. https://doi.org/10.1109/ACCESS.2017.2756069

Tao F et al (2019a) Digital twin-driven product design framework. Int J Prod Res 57(12):3935–3953. https://doi.org/10.1080/00207543.2018.1443229

Tao F, Qi Q, Wang L, Nee AYC (2019b) Digital twins and cyber-physical systems toward smart manufacturing and industry 4.0: correlation and comparison. Engineering 5(4):653–661. https://doi.org/10.1016/j.eng.2019.01.014

Tao F, Zhang H, Liu A, Nee AYC (2019c) Digital twin in industry: state-of-the-art. IEEE Trans Ind Inform 15(4):2405–2415. https://doi.org/10.1109/TII.2018.2873186

Tong X, Liu Q, Pi S, Xiao Y (2020) Real-time machining data application and service based on IMT digital twin. J Intell Manuf 31(5):1113–1132. https://doi.org/10.1007/s10845-019-01500-0

Tuegel EJ (2012) The airframe digital twin: some challenges to realization. In: Collection of technical papers—AIAA/ASME/ASCE/AHS/ASC structures, structural dynamics & materials conference, Apr 2012, pp 1–8. https://doi.org/10.2514/6.2012-1812

Ur Rehman A, Naranje V, Salunkhe S, Sankaram MVM (2021) Digital twin for additive manufacturing: a critical tool for the future. In: Proceedings of 2nd IEEE international conference computational intelligence and knowledge economy, ICCIKE 2021, pp 494–499. https://doi.org/10.1109/ICCIKE51210.2021.9410721

Vachalek J, Bartalsky L, Rovny O, Sismisova D, Morhac M, Loksik M (2017) The digital twin of an industrial production line within the industry 4.0 concept. In: Proceedings of 2017 21st international conference on process control, PC 2017, pp 258–262. https://doi.org/10.1109/PC.2017.7976223

Vachálek J, Šišmišová D, Vašek P, Fit'ka I, Slovák J, Šimovec M (2021) Design and implementation of universal cyber-physical model for testing logistic control algorithms of production line's digital twin by using color sensor. Sensors 21(5):1–12. https://doi.org/10.3390/s21051842

Vijayakumar K, Dhanasekaran C, Pugazhenthi R, Sivaganesan S (2019) Digital twin for factory system simulation. Int J Recent Technol Eng 8(1):63–68

Villalonga A, Negri E, Fumagalli L, MacChi M, Castaño F, Haber R (2020) Local decision making based on distributed digital twin framework. IFAC-PapersOnLine 53(2):10568–10573. https://doi.org/10.1016/j.ifacol.2020.12.2806

Viola J, Chen YQ (2020) Digital twin enabled smart control engineering as an industrial AI: a new framework and case study. In: 2nd international conference on industrial artificial intelligence IAI 2020. https://doi.org/10.1109/IAI50351.2020.9262203

Vladareanu L et al (2020) Digital twin in 5G digital era developed through cyber physical systems. IFAC-PapersOnLine 53(2):10885–10890. https://doi.org/10.1016/j.ifacol.2020.12.2822

Vrabič R, Erkoyuncu JA, Farsi M, Ariansyah D (2021) An intelligent agent-based architecture for resilient digital twins in manufacturing. CIRP Ann 70(1):349–352. https://doi.org/10.1016/j.cirp.2021.04.049

Wagner R, Haefner B, Biehler M, Lanza G (2020) Digital DNA in quality control cycles of high-precision products. CIRP Ann 69(1):373–376. https://doi.org/10.1016/j.cirp.2020.03.020

Wang P, Luo M (2021) A digital twin-based big data virtual and real fusion learning reference framework supported by industrial internet towards smart manufacturing. J Manuf Syst 58:16–32. https://doi.org/10.1016/j.jmsy.2020.11.012

Wang Y, Wu Z (2020) Digital twin-based production scheduling system for heavy truck frame shop. Proc Inst Mech Eng Part C J Mech Eng Sci (58):1–12. https://doi.org/10.1177/0954406220913306

Wang Q, Jiao W, Zhang YM (2020a) Deep learning-empowered digital twin for visualized weld joint growth monitoring and penetration control. J Manuf Syst 57:429–439. https://doi.org/10.1016/j.jmsy.2020.10.002

Wang T, Cheng J, Yang Y, Esposito C, Snoussi H, Tao F (2020b) Adaptive optimization method in digital twin conveyor systems via range-inspection control. IEEE Trans Autom Sci Eng 1–9. https://doi.org/10.1109/TASE.2020.3043393

Wang PY, Liu WC, Liu N, You YP (2020c) Digital twin-driven system for roller conveyor line: design and control. J Ambient Intell Humaniz Comput 11(11):5419–5431. https://doi.org/10.1007/s12652-020-01898-z

Wang T, Li J, Kong Z, Liu X, Snoussi H, Lv H (2021a) Digital twin improved via visual question answering for vision-language interactive mode in human-machine collaboration. J Manuf Syst 58:261–269. https://doi.org/10.1016/j.jmsy.2020.07.011

Wang JF, Huang YQ, Tang DL (2021b) A digital twin simulator for real time energy saving control of serial manufacturing system. In: 2021 IEEE international conference on real-time computing and robotics, RCAR 2021, pp 720–725. https://doi.org/10.1109/RCAR52367.2021.9517579

Wang KJ, Lee YH, Angelica S (2021c) Digital twin design for real-time monitoring—a case study of die cutting machine. Int J Prod Res 59(21):6471–6485. https://doi.org/10.1080/00207543.2020.1817999

Wang T, Li J, Deng Y, Wang C, Snoussi H, Tao F (2021d) Digital twin for human-machine interaction with convolutional neural network. Int J Comput Integr Manuf 34(7–8):888–897. https://doi.org/10.1080/0951192X.2021.1925966

Wang Z, Feng W, Ye J, Yang J, Liu C (2021e) A study on intelligent manufacturing industrial internet for injection molding industry based on digital twin. Complexity 2021. https://doi.org/10.1155/2021/8838914

Wang Y, Cao Y, Wang FY (2021f) Anomaly detection in digital twin model. In: Proceedings of 2021 IEEE 1st international conference on digital twins and parallel intelligence DTPI 2021, pp 208–211. https://doi.org/10.1109/DTPI52967.2021.9540116

Ward R et al (2021a) Machining digital twin using real-time model-based simulations and lookahead function for closed loop machining control. Int J Adv Manuf Technol 117(11–12):3615–3629. https://doi.org/10.1007/s00170-021-07867-w
Ward R, Soulatiantork P, Finneran S, Hughes R, Tiwari A (2021b) Real-time vision-based multiple object tracking of a production process: industrial digital twin case study. Proc Inst Mech Eng Part B J Eng Manuf 235(11):1861–1872. https://doi.org/10.1177/09544054211002464
Weber C, Königsberger J, Kassner L, Mitschang B (2017) M2DDM—a maturity model for data-driven manufacturing. Procedia CIRP 63:173–178. https://doi.org/10.1016/j.procir.2017.03.309
Wu Z, Li J (2021) A framework of dynamic data driven digital twin for complex engineering products: the example of aircraft engine health management. Procedia Manuf 55:139–146. https://doi.org/10.1016/j.promfg.2021.10.020
Wu P, Qi M, Gao L, Zou W, Miao Q, Liu LL (2019) Research on the virtual reality synchronization of workshop digital twin. In: Proceedings of 2019 IEEE 8th joint international information technology and artificial intelligence conference ITAIC 2019, pp 875–879. https://doi.org/10.1109/ITAIC.2019.8785552
Wu Q, Mao Y, Chen J, Wang C (2021) Application research of digital twin-driven ship intelligent manufacturing system: pipe machining production line. J Mar Sci Eng 9(3). https://doi.org/10.3390/jmse9030338
Xia L, Lu J, Zhang H (2020) Research on construction method of digital twin workshop based on digital twin engine. In: Proceedings of 2020 IEEE international conference on advances in electrical engineering and computer applications, AEECA 2020, pp 417–421. https://doi.org/10.1109/AEECA49918.2020.9213649
Xia K et al (2021a) A digital twin to train deep reinforcement learning agent for smart manufacturing plants: environment, interfaces and intelligence. J Manuf Syst 58:210–230. https://doi.org/10.1016/j.jmsy.2020.06.012
Xia M, Shao H, Williams D, Lu S, Shu L, de Silva CW (2021b) Intelligent fault diagnosis of machinery using digital twin-assisted deep transfer learning. Reliab Eng Syst Saf 215:107938. https://doi.org/10.1016/j.ress.2021.107938
Xu Z et al (2021) Digital twin-driven optimization of gas exchange system of 2-stroke heavy fuel aircraft engine. J Manuf Syst 58:132–145. https://doi.org/10.1016/j.jmsy.2020.08.002
Yan Q, Zhang H (2020) Real-time multi-agent-based decision-making approach for dynamic machine tool selection problem. In: ACM international conference proceeding series. https://doi.org/10.1145/3424978.3425033
Yan J, Liu Z, Zhang C, Zhang T, Zhang Y, Yang C (2021) Research on flexible job shop scheduling under finite transportation conditions for digital twin workshop. Robot Comput Integr Manuf 72:102198. https://doi.org/10.1016/j.rcim.2021.102198
Yang R, Mo Q, Huang Z, Zhang Y (2020) Transfer learning for the design of a digital twins-based automatic relay production line. J Phys Conf Ser 1682(1). https://doi.org/10.1088/1742-6596/1682/1/012028
Yao F, Keller A, Ahmad M, Ahmad B, Harrison R, Colombo AW (2018) Optimizing the scheduling of autonomous guided vehicle in a manufacturing process. In: Proceedings of IEEE 16th international conference on industrial informatics, INDIN 2018, pp 264–269. https://doi.org/10.1109/INDIN.2018.8471979
Yi L, Glatt M, Ehmsen S, Duan W, Aurich JC (2021) Process monitoring of economic and environmental performance of a material extrusion printer using an augmented reality-based digital twin. Addit Manuf 48. https://doi.org/10.1016/j.addma.2021.102388
Yildiz E, Møller C, Bilberg A (2020) Virtual factory: digital twin based integrated factory simulations. Procedia CIRP 93:216–221. https://doi.org/10.1016/j.procir.2020.04.043
Yiping G, Xinyu L, Gao L (2021) A deep lifelong learning method for digital twin-driven defect recognition with novel classes. J Comput Inf Sci Eng 21(3):1–9. https://doi.org/10.1115/1.4049960
Yu H, Han S, Yang D, Wang Z, Feng W (2021) Job shop scheduling based on digital twin technology: a survey and an intelligent platform. Complexity 2021. https://doi.org/10.1155/2021/8823273

Yu-Ming Q, Bing X, San-Peng D (2020) Research on intelligent manufacturing flexible production line system based on digital twin. In: Proceedings of 2020 35th youth academic annual conference of Chinese association of automation YAC 2020, pp 854–862. https://doi.org/10.1109/YAC51587.2020.9337500

Yun S, Park JH, Kim WT (2017) Data-centric middleware based digital twin platform for dependable cyber-physical systems. In: International conference on ubiquitous and future networks ICUFN, pp 922–926. https://doi.org/10.1109/ICUFN.2017.7993933

Zhang C, Ji W (2019) Digital twin-driven carbon emission prediction and low-carbon control of intelligent manufacturing job-shop. Procedia CIRP 83:624–629. https://doi.org/10.1016/j.procir.2019.04.095

Zhang H, Liu Q, Chen X, Zhang D, Leng J (2017) A digital twin-based approach for designing and multi-objective optimization of hollow glass production line. IEEE Access 5:26901–26911. https://doi.org/10.1109/ACCESS.2017.2766453

Zhang M, Zuo Y, Tao F (2018) Equipment energy consumption management in digital twin shop-floor: a framework and potential applications. In: ICNSC 2018—15th IEEE international conference on networking, sensing and control, pp 1–5. https://doi.org/10.1109/ICNSC.2018.8361272

Zhang YF, Shao YQ, Wang JF, Li SQ (2020a) Digital twin-based production simulation of discrete manufacturing shop-floor for onsite performance analysis. In: IEEE international conference on industrial engineering and engineering management, vol 2020, Dec 2020, pp 1107–1111. https://doi.org/10.1109/IEEM45057.2020.9309928

Zhang C, Zhou G, Hu J, Li J (2020b) Deep learning-enabled intelligent process planning for digital twin manufacturing cell. Knowledge-Based Syst 191:105247. https://doi.org/10.1016/j.knosys.2019.105247

Zhang Z, Guan Z, Gong Y, Luo D, Yue L (2020c) Improved multi-fidelity simulation-based optimisation: application in a digital twin shop floor. Int J Prod Res 60(3):1016–1035. https://doi.org/10.1080/00207543.2020.1849846

Zhang H, Yan Q, Wen Z (2020d) Information modeling for cyber-physical production system based on digital twin and AutomationML. Int J Adv Manuf Technol 107(3–4):1927–1945. https://doi.org/10.1007/s00170-020-05056-9

Zhang Z, Lu J, Xia L, Wang S, Zhang H, Zhao R (2020e) Digital twin system design for dual-manipulator cooperation unit. In: Proceedings of 2020 IEEE 4th information technology, networking, electronic and automation control conference ITNEC 2020, pp 1431–1434. https://doi.org/10.1109/ITNEC48623.2020.9084652

Zhang K et al (2020f) Digital twin-based opti-state control method for a synchronized production operation system. Robot Comput Integr Manuf 63:101892. https://doi.org/10.1016/j.rcim.2019.101892

Zhao Z, Wang S, Wang Z, Wang S, Ma C, Yang B (2020) Surface roughness stabilization method based on digital twin-driven machining parameters self-adaption adjustment: a case study in five-axis machining. J Intell Manuf. https://doi.org/10.1007/s10845-020-01698-4

Zhao L, Fang Y, Lou P, Yan J, Xiao A (2021) Cutting parameter optimization for reducing carbon emissions using digital twin. Int J Precis Eng Manuf 22(5):933–949. https://doi.org/10.1007/s12541-021-00486-1

Zhao P et al (2020) The modeling and using strategy for the digital twin in process planning. IEEE Access 8:41229–41245. https://doi.org/10.1109/ACCESS.2020.2974241

Zheng P, Sivabalan AS (2020) A generic tri-model-based approach for product-level digital twin development in a smart manufacturing environment. Robot Comput Integr Manuf 64:101958. https://doi.org/10.1016/j.rcim.2020.101958

Zheng P, Lin TJ, Chen CH, Xu X (2018) A systematic design approach for service innovation of smart product-service systems. J Clean Prod 201:657–667. https://doi.org/10.1016/j.jclepro.2018.08.101

Zhifeng LIU, Yueze Z, Caixia Z, Jun YAN, Shiyao GUO (1884) Real-time workshop digital twin scheduling platform for discrete manufacturing. J Phys Conf Ser 1:2021. https://doi.org/10.1088/1742-6596/1884/1/012006

Zhou H, Yang C, Sun Y (2021a) Intelligent ironmaking optimization service on a cloud computing platform by digital twin. Engineering 7(9):1274–1281. https://doi.org/10.1016/j.eng.2021.04.022

Zhou X et al (2021b) Intelligent small object detection based on digital twinning for smart manufacturing in industrial CPS. IEEE Trans Ind Inform 18(2):1377–1386. https://doi.org/10.1109/TII.2021.3061419

Zhu Z, Xi X, Xu X, Cai Y (2021) Digital twin-driven machining process for thin-walled part manufacturing. J Manuf Syst 59:453–466. https://doi.org/10.1016/j.jmsy.2021.03.015

Zhuang C, Liu J, Xiong H (2018) Digital twin-based smart production management and control framework for the complex product assembly shop-floor. Int J Adv Manuf Technol 96(1–4):1149–1163. https://doi.org/10.1007/s00170-018-1617-6

Zhuang C, Miao T, Liu J, Xiong H (2019) The connotation of digital twin, and the construction and application method of shop-floor digital twin. Robot Comput Integr Manuf 68:2021. https://doi.org/10.1016/j.rcim.2020.102075

Židek K, Pitel' J, Adámek M, Lazorík P, Hošovský A (2020) Digital twin of experimental smart manufacturing assembly system for industry 4.0 concept. Sustainability 12(9):1–16. https://doi.org/10.3390/su12093658

Zotov E, Kadirkamanathan V (2021) CycleStyleGAN-based knowledge transfer for a machining digital twin. Front Artif Intell 4:1–14. https://doi.org/10.3389/frai.2021.767451

Intelligent Feature Engineering for Monitoring Tool Health in Machining

Debasish Mishra, Krishna R. Pattipati, and George M. Bollas

Abstract This chapter explores the development of a data-driven digital twin to detect the health of cutting tools in precision machining processes. A set of cutting tool run-to-failure machining tests was conducted in a semi-production CNC milling machine by changing the cutting parameter settings. Real-time data was acquired from various sensors during the tests. The sensory data were processed to explore features that correlate with the deteriorating health of the cutting tool. Specifically, we processed the signals in the time–frequency domain using wavelets to derive informative features. Our analysis presents evidence that a pool of features exists in higher wavelet subspaces that are informative of the cutting tool condition. The variance and kurtosis of the acquired signals explain the changes to the cutting tool condition. A comparative study performed between time–frequency (wavelets) and other domains suggests benefits of feature engineering with time–frequency analyses. An intuitive explanation of the features informative of the tool condition is shown by exploring the frequency spectrum of the signals. A strong linear correlation of 0.97 was obtained between the chosen feature and tool wear. A supervised machine learning-based monitoring system is developed that exploits the variability of information in different wavelet subspaces to forecast the tool wear. Vibration and force signals are found to be the most informative sensors of tool wear progression.

D. Mishra
Data Science and Information Systems, Krea University, 5655 Central Expressway, Sri City, Andhra Pradesh 517646, India
e-mail: debasish.mishra@krea.edu.in

K. R. Pattipati
Department of Electrical and Computer Engineering, Pratt & Whitney Institute of Advanced Systems Engineering, University of Connecticut, 371 Fairfield Way, Storrs, Connecticut 06269, USA
e-mail: krishna.pattipati@uconn.edu

G. M. Bollas (✉)
Department of Chemical and Biomolecular Engineering, Pratt & Whitney Institute of Advanced Systems Engineering, University of Connecticut, 159 Discovery Drive, Storrs, Connecticut 06269, USA
e-mail: george.bollas@uconn.edu

A. Kumar et al. (eds.), *Industry 4.0 Driven Manufacturing Technologies*, Springer Series in Advanced Manufacturing, https://doi.org/10.1007/978-3-031-68271-1_2

This chapter highlights the importance of feature engineering and a robust low-cost machine learning algorithm for tracking the tool condition.

Keywords Machining · Tool health monitoring · Signal processing · Feature engineering · Machine learning · Wavelets · Regression

1 Introduction

The global emphasis on reducing the carbon footprint and energy consumption in manufacturing processes reflects a crucial shift toward sustainability. This endeavour has led manufacturers to explore alternative mechanisms, particularly data-driven automated systems, to achieve advanced control and decision-making support (Wang et al. 2018). The implementation of data-driven automated systems aims to reduce dependence on human resources and mitigate the possibility of errors in manufacturing processes. These systems utilize robust monitoring algorithms that interact with the entirety or a subset of available information. Key attributes of these systems include the collection of sensory data from the manufacturing process, the concomitant processing of this data to derive meaningful insights, followed by the application of machine learning (ML) algorithms for automated decision-making.

1.1 Focus of This Study

This study presents a monitoring system developed to predict the tool wear in machining processes, aiming to assist manufacturers in deciding the optimum tool replacement time. Machining processes are integral to manufacturing, involving the precise removal of material from a workpiece using a cutting tool (Dimla 2000). This intricate interplay between the cutting tool and the workpiece manifests as forces that are crucial for achieving the desired geometries.

During the material removal process, the shearing action of the cutting tool on the workpiece induces mechanical deformation, resulting in a yield strength reduction of the workpiece (Yıldırım et al. 2019). To facilitate efficient material removal, cutting tool materials are strategically chosen to be superior to the workpiece. In the initial stages, a healthy cutting tool experiences stable cutting forces. However, as machining progresses, the sharpness of the tool surface diminishes, leading to adherence, bluntness, and wear. This wear is critical to monitor, as it alters the dynamics of the machining process. Instead of cutting, the tool may rub against the workpiece, causing sporadic deviations in the signal amplitudes (Salur et al. 2019). The developed monitoring system focuses on capturing and analyzing these deviations in real time, providing a predictive tool wear assessment. As we delve into the details of this tool condition monitoring system, it becomes evident that its implementation holds the key to proactive maintenance strategies and improved overall machining performance.

Since the quality of workpiece is crucial, the condition of the cutting tool emerges as an important factor in machining processes. Machining with a worn tool not only compromises workpiece quality, but also contributes to increased scrap rates and energy consumption (Kothuru et al. 2019). The direct link between the tool condition and the final product highlights the critical need for predictive tool wear monitoring. Beyond its impact on workpiece quality, tool wear is also detrimental to the overall machine operation (Salur et al. 2020). The use of a worn tool substantially increases the cutting forces, leading to potential wear and tear on machinery components (Abellan-Nebot and Romero Subirón 2010). Moreover, tool wear gives rise to various workpiece and system-level faults. These faults result in various performance issues such as poor surface finish, reduced machining accuracy, high scrap and energy consumption, and decreased tool life (Kothuru et al. 2019). Thus, tracking the condition of the cutting tool is essential to avoid these performance issues. Indeed, a robust tool condition monitoring systems is vital for sustained and efficient machining.

The measurement of flank wear is a common criterion for determining the tool life in machining processes. This type of wear initiates and propagates on the cutting edges of the tool, and has a direct impact on the surface texture and on the cutting geometry, leading to a deterioration in the overall performance of the tool (Ranjan et al. 2020). Tool life is assumed to end upon reaching a certain value of flank wear (Kious et al. 2010). Flank wear is measured by analyzing the cutting edge under an optical microscope or advanced equipment such as a scanning electron microscope. However, the conventional method of intermittently removing the cutting tool from the machine to measure the flank wear poses challenges, particularly in terms of production throughput. To address this interruption, sensors are integrated into the Computer Numerical Control (CNC) machines to collect machining data, enabling an indirect inference of the tool condition without halting the production process.

The exploration of various sensors for monitoring tool wear has been an important topic in prior research, encompassing a diverse range of devices, such as current and power sensors (Khajavi et al. 2016; Yang et al. 2022), force dynamometers (Awasthi et al. 2022; Han et al. 2022), thermocouples (Wanigarathne et al. 2005), thermal imagers, accelerometers (Mishra et al. 2023a, b), microphones (Han et al. 2021), and acoustic emission sensors (Pimenov et al. 2023; Kuntoğlu and Sağlam 2021). These varied sensors play a crucial role in collecting real-time data, which is then utilized to develop ML and deep learning (DL) models for predicting the tool condition. The accuracy of these models relies on the information contained in the sensory data, making sensing an invaluable mechanism for enhancing the precision and efficiency of machining operations. An overview of the recent monitoring systems developed in prior research efforts for estimating the tool condition is presented in Table 1.

Table 1 Modeling in prior research works

S. No.	Signal	Features	Modeling	References
Tool wear estimation				
1	*T*	Model learned	Stacked sparse autoencoder	He et al. (2021)
2	*V*	Trigonometric functions	Artificial neural network (ANN)	Javed et al. (2015)
3	*V*	Distance metrics	Interactive multiple model	Yang et al. (2024)
4	*F*	Statistical and energy	Gaussian process regression (GPR)	Kong et al. (2018)
5	*V*	Statistical, frequency, and time–frequency	Neuro-fuzzy network	Zhang et al.(2016)
6	*F*, *V*	Wavelet packet and spectral subtraction	Convolutional neural networks (CNN)	Aghazadeh et al. (2018)
7	*V*, *F*, *A*	Model learned	Long short-term memory (LSTM)	Liu et al. (2021)
8	*V*, *F*, *A*	Statistical	Hidden semi-Markov model	Lin et al. (2022)
9	*V*, *F*, *A*	Statistical	Deep-LSTM	Zhao et al. (2016)
10	*V*, *F*, *A*	Holistic and local	Holistic–local LSTM	Chan et al. (2022)
11	*V*, *F*, *A*	Spatiotemporal	Time-distributed convolutional LSTM	Qiao et al. (2018)
12	*V*, *F*, *A*	Temporal	Transformer NN	Liu et al. (2020)
13	*V*, *F*, *A*	Statistical and model learned	Gated recurrent unit network (GRU)	Zhao et al. (2018)
14	*C*, *F*, *V*, *S*	Statistical	ANN and GPR	Han et al. (2022)
Tool wear classification				
15	*F*	Model learned from Gramian angular summation field images	CNN	Martínez-Arellano et al. (2019) and Gouarir et al. (2018)
16	*V*	Distance metrics	Jenks natural breaks (JNB)	Mishra et al. (2023b)
17	*A*	Frequency	Hidden Markov model	Lu and Wan (2013)
18	–	Model learned from workpiece images	CNN	Kumar et al. (2021)
19	*V*	Model learned from images of signals	CNN	Naveen Venkatesh et al. (2022)

(continued)

Table 1 (continued)

S. No.	Signal	Features	Modeling	References
20	*T*	Model learned from thermographic images	CNN	Brili et al. (2021)
21	*V*, *F*	Statistical and entropy	Restricted Boltzmann machine	Li et al. (2020)
22	*V*, *F*	Model learned from color recurrence plots	Graph neural network	Zhou et al. (2022)
23	*V*, *F*	Fractal	Support vector machine	Guo et al. (2020)

C current/power, *V* vibration, *F* cutting force, *A* acoustic emission, *T* temperature, *S* acoustic sound signals

1.2 Our Propositions

In this article, we present two key propositions that highlight our contribution to tool condition monitoring in machining operations. Our primary focus is on the accurate identification of the health of cutting tools through the analysis of sensory data generated during machining. To achieve this, we advocate for the extraction and computation of tool condition-relevant features from sensory data, emphasizing the need for explainability in the context of the machining process. Our investigation spans various sensor configurations, including mechanical, electrical, and thermal setups, to determine the most effective signal suite for tool condition monitoring.

The first aspect of our argument highlights the importance of explainable features, showcasing how the features contribute to an understanding of the tool condition in a machining process. We identify features from the sensory data that are not only informative of the tool condition but also interpretable, ensuring their practical applicability in real-world manufacturing scenarios. The second aspect revolves around the feasibility of utilizing robust and cost-effective ML algorithms to model the tool condition. Unlike many prior research works that delve into neural networks or DL solutions, we advocate robustness, practical implementation and computational efficiency of the ML algorithms. Deep learning models are powerful, but demand extensive high-quality and high-volume data, and high computational resources, rendering them less practical and prone to lack of robustness to out of sample distributions that are typical in manufacturing settings. Moreover, their black-box modeling limits the explainability of their decisions to machine operators.

Instead, we showcase that ML algorithms can effectively model tool condition using informative and explainable features because of their ability to discern patterns and relationships. We emphasize the cost-effectiveness of ML solutions, making them highly suitable for the challenges of the manufacturing sector. We aim to contribute to the development of a robust and practical tool condition monitoring strategy involving our propositions for enhanced performance in machining operations.

2 Materials and Experiments

This section presents the details of run-to-failure machining tests conducted on new and healthy cutting tools.

2.1 Run-to-Failure Tests

Machining processes involve a variety of methods, such as turning, milling, drilling, shaping, boring, and grinding. Each of these methods aims to achieve a desired geometrical shape and size by removing material from the workpiece using a cutting tool. In our experimental tests, we focused on the milling process, specifically targeting the face milling of a stainless steel workpiece.

In Fig. 1, a schematic depiction of the workpiece and tool configuration during face milling tests is presented. The workpiece is clamped on the machine bed, while the cutting tool is mounted onto the spindle, enabling rotational motion. Furthermore, the spindle can also traverse in the Z-direction to facilitate the interaction of the tool with the workpiece during the tests. The machine bed can move horizontally in the X- and Y-directions relative to the spindle to execute the cutting process and achieve the desired geometric features and dimensions. To start the cutting process, the spindle speed and feed velocity are set to rotate the tool and move the machine bed, respectively. Additionally, the machine is instructed on the depth of penetration of the tool into the workpiece, which involves two parameters: axial and radial depths of cut. The synchronized execution of these movements results in the shearing of the workpiece, thereby leading to material removal.

We employed a stainless steel workpiece composed of AISI 4340 steel with a diameter of 177.8 mm, as depicted in Fig. 2a. The milling process was executed utilizing a two-flute circular cutter with a 20 mm diameter and 39 mm length. The cutting operation was progressive, following a spiral trajectory, that was initiated from the outer edges and gradually moved towards the center of the workpiece. The cutting operation continued until the tool sheared material from the workpiece, penetrating the surface to achieve a depth of 10.16 mm, consequently forming a boss, as sketched in Fig. 2b.

The experimental procedures involved using a semi-production CNC milling machine (HAAS, Mini Mill) to conduct the machining tests. This machine is equipped with a high-speed spindle system that rotates at a maximum speed of 6000 rpm. Its axes can traverse 406 mm $\times$ 305 mm $\times$ 254 mm. The machining parameters used in the tests are listed in Table 2, which includes the tool's rotational speed and feed velocity. These parameters were systematically varied at two levels while maintaining a constant chip load.

The machining operation commenced after axially feeding the tool by 2.54 mm and radially by 10.16 mm into the workpiece, as shown in Fig. 2a. This initial

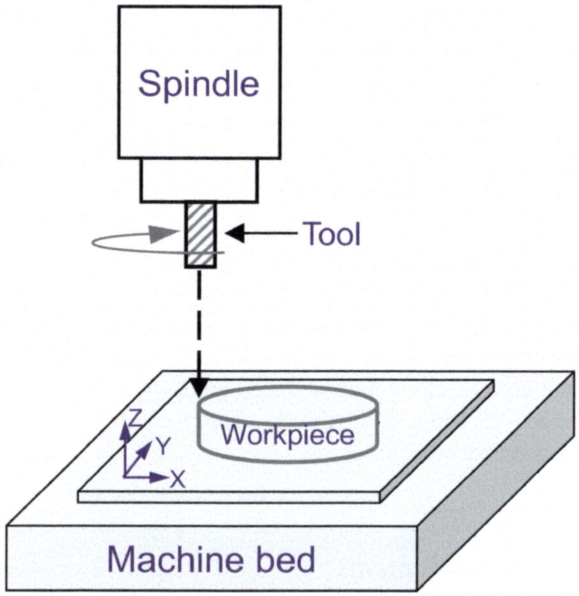

Fig. 1 Schematic diagram of workpiece and tool arrangement in milling tests

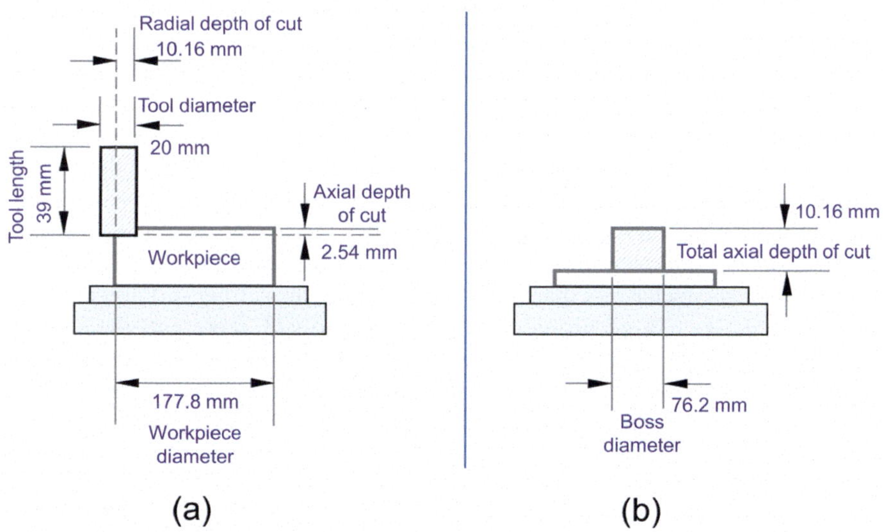

Fig. 2 Schematic of cutting specifications: **a** before machining and **b** after machining

Table 2 Machining parameters for the tests

Test run #	Machining parameters	
	Tool rotational speed (rpm)	Feed velocity (mm/min)
Run #1–16	2330	710
Run #17–28	3184	970.5

tool positioning set the stage for the subsequent material removal process. The tool continuously removed material in four consecutive passes, gradually reaching a depth of 10.16 mm from the surface of the workpiece, as depicted in Fig. 2b.

The experiments began with the utilization of a new and healthy cutting tool, and the tests continued until reaching a predefined end of tool life, set at a wear limit of 0.4 mm. The choice of 0.4 mm as the threshold for tool life termination is based on the industrially acceptable wear limit.

2.2 *Tool Wear Measurements*

To evaluate the extent of tool wear, the tool was inspected under an optical microscope following each test run. Figure 3 presents the tool wear measurements, marking the end-of-tool-life at 0.4 mm. The tool wear curve unfolds in three stages, each revealing different characteristics of tool degradation across its useful life. Stage 1 comprises Run #1 to Run #9 and represents the break-in period. During this phase, the initially sharp and undamaged tool efficiently shears material from the workpiece, resulting in a relatively low rate of change in tool wear. The sharpness of the tool diminishes with progression in machining. Stage 2 spans from Run #10 to Run #16, representing the uniform wear period. In this stage, the partially affected tool continues to remove material from the workpiece at a near-constant wear rate. Although the rate of change in tool wear is somewhat less than the break-in period, the cutting surface continues to be affected. Stage 3, occurring during Runs #17 to #28, represents the accelerated wear state. Here, an increase in spindle speed and feed rate contributes to the acceleration in tool wear. The higher frictional heating and plastic deformation in this phase lead to a rapid rise in the rate of change in tool wear, ultimately leading to tool failure.

2.3 *Sensory Data Acquisition*

We installed sensors of different configurations on the CNC machine to monitor the machining process. These configurations included mechanical, electrical, and thermal setups, each having a specific purpose for capturing the various aspects of the machining dynamics. In the mechanical configuration, a piezoelectric dynamometer

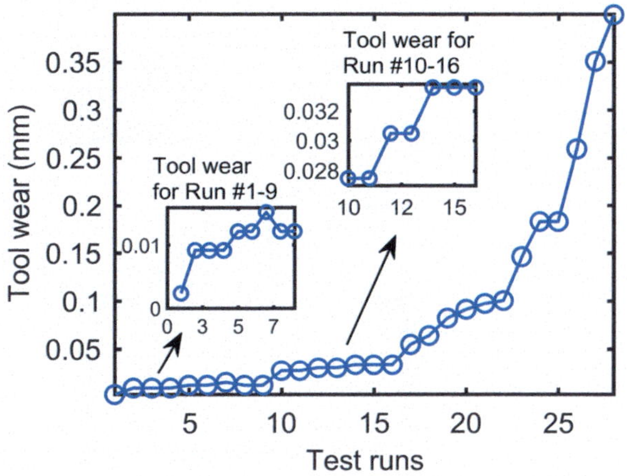

Fig. 3 Experimentally measured tool wear

(KISTLER) was mounted onto the machine spindle, and an accelerometer (TE, 4030) was attached to the spindle housing. These sensors were positioned to capture the precise cutting force and vibration data during the machining operation. In the electrical configuration, we integrated a current sensor (AIM DYNAMICS, AIMH021), a voltage sensor (MAGNELAB, DVT-100), and a power sensor into the main control panel of the CNC machine. The current sensor utilized a Hall Effect transducer operating within the range of 0–200 A, while the voltage sensor operated within the range of 0–100 V. The power sensor was built into the CNC machine and recorded the total power consumed during each machining test. These sensors monitored the energy aspects of the machining process. In the thermal configuration, we employed an infrared temperature sensor (OMEGA, OS101E) with an operating range of − 18 to 538 °C. This sensor was utilized to capture the thermal history of the machining process. The force signals were recorded at a sampling frequency of 6250 Hz, while the other signals were recorded at a frequency of 16 Hz.

3 Methods

This section introduces the methods employed for processing the sensory data and the machine learning algorithm utilized in the tool wear modeling. Figure 4 presents a schematic diagram outlining the key steps in the methodology.

3.1 Signal Processing

The acquired signals were processed using the wavelet transformation (WT) method. Wavelets are small waves, having null amplitude at the beginning and the end, along

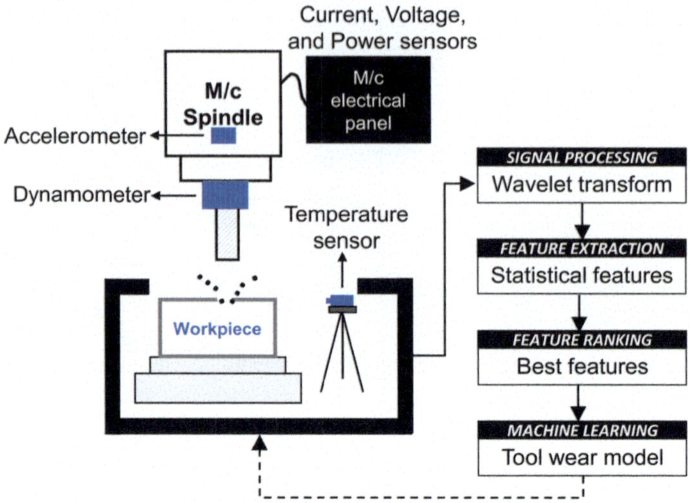

Fig. 4 Schematic diagram for workflow

with finite length, multiple zero crossings, and a zero mean. Their shorter duration, instantaneous change in amplitudes, and fast decaying nature make them particularly well-suited as basis functions for analyzing signals with localized features. The application of WT to a time-series signal is presented in Eq. (1), providing a mathematical representation of the transformation process. This method enables the decomposition of a signal into its constituent wavelet components, revealing both high- and low-frequency details (called scale s in wavelet and multi-resolution analysis).

$$W(s, \tau, f(t), \varphi(t)) = \frac{1}{\sqrt{s}} \int_{-\infty}^{\infty} f(t)\psi^*\left(\frac{t-\tau}{s}\right)dt \tag{1}$$

$$\varphi_{(s,\tau)}(t) = \frac{1}{\sqrt{s}}\psi\left(\frac{t-\tau}{s}\right) \tag{2}$$

In Eq. (1), $f(t)$ represents the time-series signal, ψ is the analyzing wavelet function, and s and τ denote scale and time parameters, respectively. The wavelet function acts as a window on the signal and facilitates the analysis of fast- and slow-changing components by computing the wavelet coefficients. The unique property of wavelets to stretch and compress signals enables the detection of discontinuities in real-world signals, such as abrupt changes or spikes. The stretched version is adept at investigating slow-changing details, while the compressed version is best suited for exploring fast-changing components. These variations stem from a mother wavelet, as presented in Eq. (2). This method of signal transformation is known as continuous WT (CWT), which provides a detailed representation of time and scale information in the time-series signal.

Although CWT is effective, it can yield redundant outputs due to computations across all scales and position (time). Alternatively, discrete wavelet transform (DWT) mitigates redundancy by computing wavelet coefficients at fixed values only. DWT discretizes parameters s and τ as j and k, extracting wavelet functions from a mother wavelet, as expressed in Eq. (3). This transformation acts as a band-pass filter, with the wavelet function focusing on finer or coarser details depending on the scaling function. We implemented the Mallat pyramidal algorithm for DWT (Stéphane 2009), utilizing smoothing (low-pass) and non-smoothing (high-pass) filters. The coefficients are computed by taking $s_0 = 2$ and $\tau_0 = 1$.

$$D(j,k) = \frac{1}{\sqrt{s_0^j}} \int_{-\infty}^{\infty} f(t)\psi^*\left(\frac{t - ks_0^j}{s_0^j}\right)dt \tag{3}$$

$s = s_0^j$ and $\tau = ks_0^j\tau_0$; $(j > 1)$ and $(k > 0)$.

Figure 5 presents the block diagram of DWT implementation. The DWT decomposes the original time-series signal, denoted as $f(t)$, into two components: approximation (A_1) and detail (D_1). This process iteratively extends across subsequent levels, revealing different aspects of the signal. The approximation component comprises coefficients representing the slow-changing aspects in $f(t)$, while the detail component unveils coefficients highlighting fast-changing portions.

In our study, we utilized DWT to process the acquired signals, aiming to extract informative features from the resulting wavelet coefficients. We investigated the ability of DWT to split a signal into distinct components, effectively reducing the overall noise and facilitating the analysis of temporal features. A Daubechies mother wavelet function (Db4) was used to decompose signals up to six levels, and coefficients were extracted from each signal. These coefficients were then examined for their informativeness regarding tool wear. This multi-level decomposition enabled a detailed exploration of the signal components across various scales.

3.2 *Feature Extraction*

We performed a rigorous analysis of the acquired signals by computing various statistical features from the wavelet coefficients extracted at each level. These features, including mean, median, maximum, variance, kurtosis, and skewness, were calculated for both approximation and detail coefficients. The objective was to capture a comprehensive set of descriptors that reflect the diverse characteristics of the signals. The relationship between these computed features and actual tool wear values was assessed using the Pearson correlation coefficient, which is defined in Eq. (4) (Hammond et al. 2014). This coefficient (r) measures the association between the features and tool wear values, providing insights into the relevance of each feature in the context of tool wear progression.

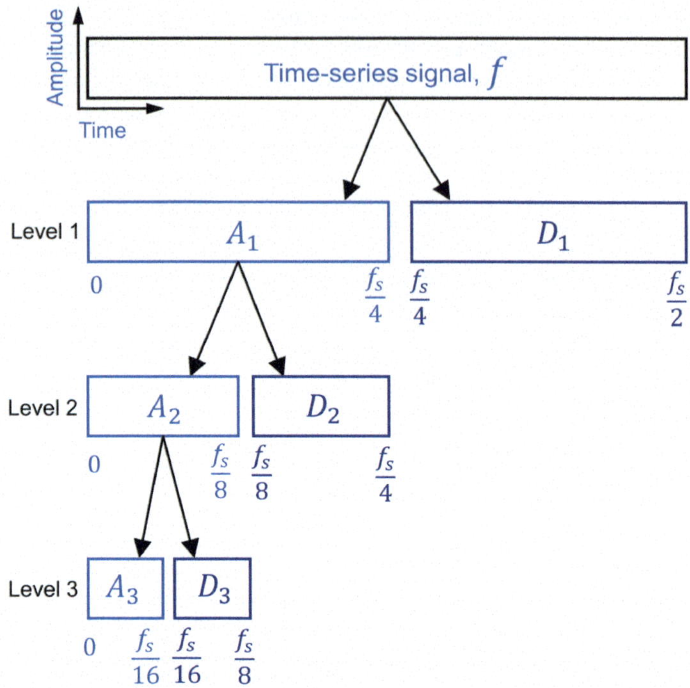

Fig. 5 Block diagram of DWT

$$r = \frac{\sum(F_i - \overline{F})(TW_i - \overline{TW})}{\sqrt{\sum(F_i - \overline{F})^2 \sum(TW_i - \overline{TW})^2}} \tag{4}$$

where F_i and TW_i represent the values of the feature and tool wear, respectively, and $\overline{F}$ and $\overline{TW}$ are their means.

Further, a comparative study was conducted between the time, frequency, and time–frequency domains to ascertain the most effective signal processing approach for tool wear monitoring. Statistical features were extracted from the time domain and frequency domain signals, and the r value between these features and tool wear was determined. This comparative analysis facilitated the selection of the most informative signal processing approach based on the calculated r values for each domain. This comprehensive methodology aims to identify the key features and domains contributing to accurate tool wear modeling.

3.3 Feature Explanation to Tool Condition

We delved into the dominant features extracted from the wavelets to discover their correlation with the evolving state of the cutting tool. The basis for associating these features with the condition of the tool lies in exploring the frequency spectrum of the signal.

In manufacturing operations, rotating components inherently possess resonating frequencies or naturally occurring frequencies. Anomalies in the system disturb the ideal condition, giving rise to high-amplitude harmonics. We anticipated that deviations in machining signals be indicative of the deteriorating tool condition.

To explore this phenomenon, we initially transformed the signal to the Fourier domain using the fast Fourier transform (FFT) (Bachman et al. 2000). The Fourier transformation utilizes sine and cosine functions to analyze the similarity between the signal and the analyzing function, revealing the different frequencies present in the signal and their respective amplitude.

In our approach, we focused on the frequency band in the spectrum corresponding to a dominant feature identified from the wavelets. This targeted analysis allowed us to examine specific frequency components associated with the identified features. To map feature information to the tool condition, we randomly selected a few machining test runs and analyzed their spectra. This comprehensive analysis aimed to unveil frequency-based insights into the tool wear, providing a deeper understanding of the dynamic changes in the machining signals as the tool condition evolves.

3.4 Machine Learning for Tool Wear Prediction

We utilized the most correlated features, identified through the Pearson correlation coefficient, to develop a ML model for predicting the tool wear. Given our emphasis on identifying the most informative features for tool wear, we opted for a cost-effective ML model to automate the prediction process. To achieve this goal, we developed polynomial regression models to forecast the tool wear using the derived signal features. The optimization of these ML models involved experimentation with various polynomial features to capture the relationship between features and tool wear.

In the modeling process, the ability to predict future instances is crucial for monitoring the progressive health condition of cutting tools in machining. To address this, we employed a training-validation-testing approach. Features corresponding to initial machining tests were utilized for training the model, features from the current sample were used for validation, and future samples were reserved for testing purposes. This approach is schematically presented in Fig. 6 and it ensures that the ML model is adept at predicting the progression of tool wear.

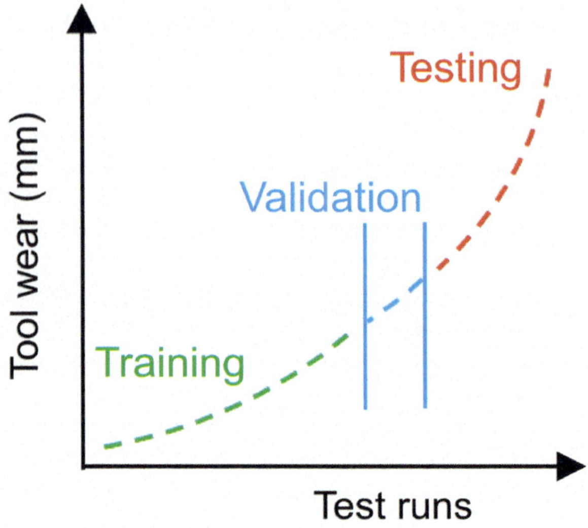

Fig. 6 Schematic of model development

4 Results

This section presents the outcomes after the application of signal processing, feature extraction, and ML model development methods.

4.1 Signal Processing

Figure 7c, d present the CWT maps derived from vibration signals corresponding to the cutting tool in healthy and worn-out conditions, respectively. These maps present wavelet coefficients, revealing distinct patterns that signify the health status of the cutting tool.

The CWT map in Fig. 7c for the tool in healthy and undamaged condition shows nominal variations in machining vibrations. The consistent and uniform appearance of the wavelet coefficients indicates stable machining conditions, which is attributed to the healthy condition of the tool. This healthy state allows for efficient removal of material from the workpiece. However, a significant deviation in the amplitude of the CWT map becomes apparent in the case of a worn-out tool condition, as presented in Fig. 7d. The worn-out tool experiences higher resistance from the workpiece due to diminished sharpness, leading to increased stresses and subsequently higher machining vibrations. Similar observations were also identified in other machining signals, which reinforces the consistency of the relationship between wavelet coefficients and tool wear in the machining process.

Furthermore, the amplitude of the wavelet coefficients is found to vary across frequencies, as is evident in Fig. 7. Instead of examining the entire map of wavelet

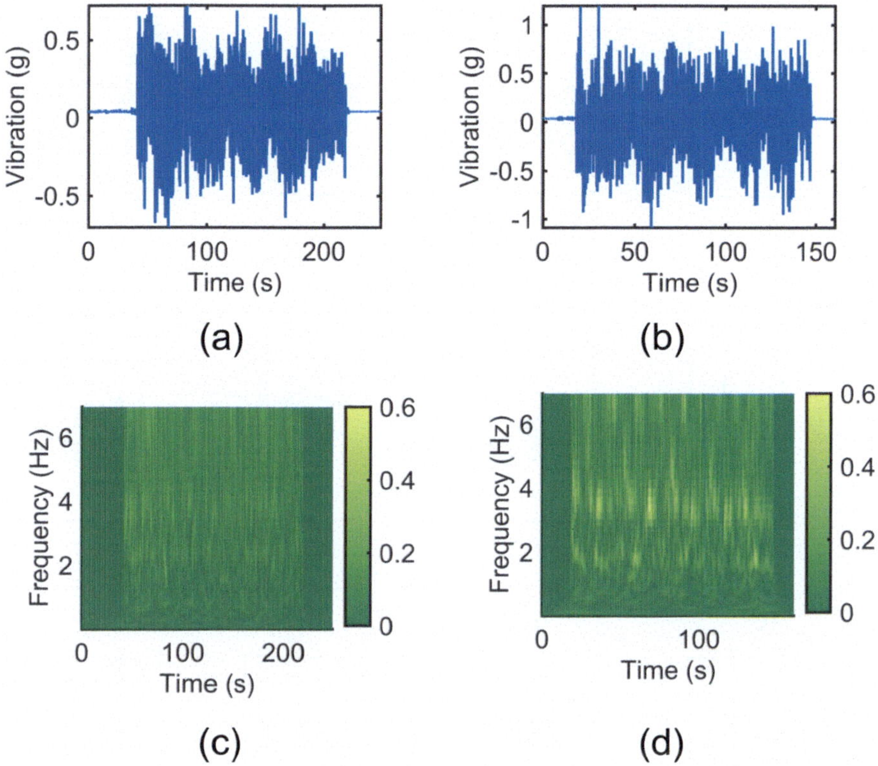

Fig. 7 Machining vibrations during: **a** Run #1 and **b** Run #28, and CWT maps for machining vibrations of: **c** Run #1 and **d** Run #28

coefficients, we conducted a detailed exploration of separate frequency bands resulting from the decomposition of the signal using DWT. Figure 8a–f and g–l show the detail coefficients, D_1 to D_6, in various wavelet subspaces with respect to the tool health states.

To distinguish between these two health state conditions, the scale of wavelet coefficients at each decomposition level is kept consistent in Fig. 8a–f and g–l. Clearly, abrupt and unsteady changes can be seen in the coefficients corresponding to the worn-out condition in Fig. 8g–l. These changes are indicative of the inability of the cutting tool to withstand the machining stresses due to the disruption of cutting edges. As a result, the tool experiences rubbing, leading to higher cutting forces.

Moreover, the non-uniform deformation across the tool cutting edge results in variations in cutting forces at different frequencies. Therefore, the wavelet subspaces serve as effective tools to investigate these differences, as presented in Fig. 8a–f and g–l. For instance, a significant difference can be observed in D_2 subspaces representing the healthy and worn-out conditions in Fig. 8b and h, respectively. The wavelet coefficients amplify as a result of machining with a worn-out tool.

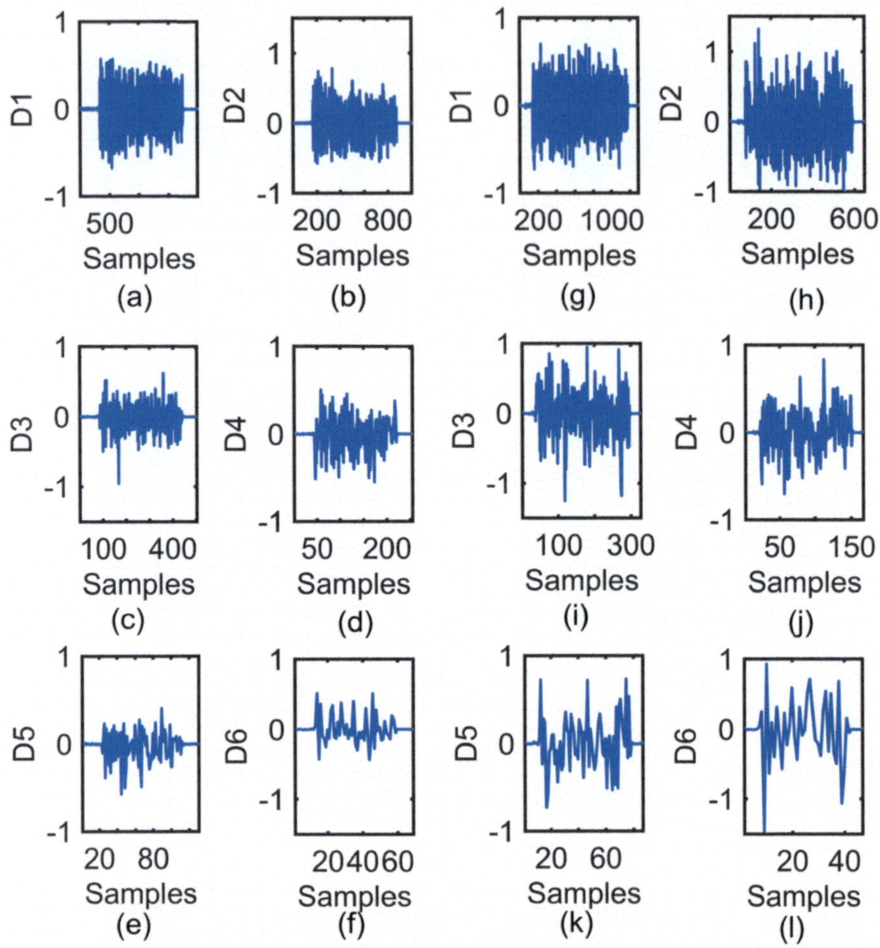

Fig. 8 Detail wavelet coefficients of machining vibrations: **a–f** Run #1 and **g–l** Run #28

4.2 Feature Extraction

In Fig. 9a, we highlight features computed from the detail wavelet coefficients, like variance and kurtosis, specifically those having a correlation coefficient higher than 0.9 with the tool wear values. The cutting forces can be observed to be more informative indicators of tool condition, with several features derived from force signals having higher correlation coefficients with tool wear.

Among these, the variance and kurtosis of wavelet coefficients stand out as strong indicators of tool wear. In Fig. 9b, the kurtosis of the D_5 subspace of cutting force signals is found to be the most informative feature, having a correlation coefficient of 0.973 with tool wear values. This finding highlights the sensitivity of kurtosis to

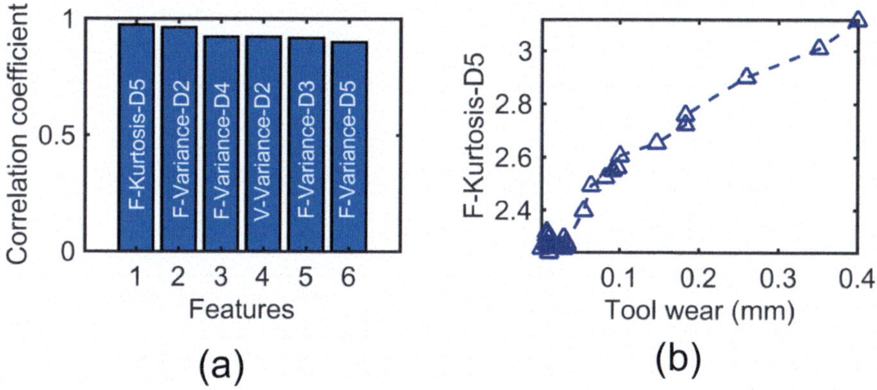

Fig. 9 **a** Features extracted from detail coefficients having correlation coefficient 0.9 and higher and **b** feature having the highest correlation coefficient. F-Kurtosis-D5 is the kurtosis of force signal in the D_5 subspace, F-Variance-Dn represents the variance of force signal in the D_n subspace

changes in the cutting tool condition, offering a precise feature for monitoring the tool wear progression. Overall, the prominence of cutting forces in capturing tool wear dynamics suggests their crucial role in assessing the health of the cutting tool. The high correlation coefficients signify a strong association between these features and the actual tool wear values.

Figure 10a presents features computed from the approximation coefficients that have correlation coefficients higher than 0.9 with the actual tool wear values. Once again, variance of the coefficients is observed to be the most correlated measure of tool wear, with the variance of A_6 in the cutting force signals having the highest correlation coefficient of 0.911.

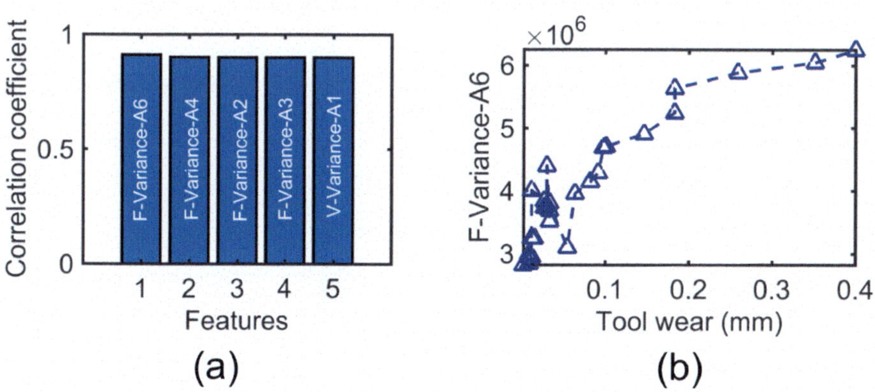

Fig. 10 **a** Features extracted from approximation coefficients having correlation coefficient 0.9 and higher and **b** feature having the highest correlation coefficient

The analysis of both approximation and detail coefficients reveals that variance and kurtosis are the most informative statistical features, exhibiting a strong relationship with the progressive condition of the cutting tool. In contrast, other statistical features, such as mean, median, and maximum of wavelet coefficients, are comparatively less informative in this regard. Furthermore, cutting forces and machining vibrations prove to be more sensitive indicators of tool condition compared to other signals.

Figure 11a, b present features extracted from the temperature and current signals having the highest correlation coefficients with tool wear values. These features, specifically the variance of D_3 and variance of D_4 subspaces extracted from temperature and current signals, exhibit weaker relationships with tool wear, with correlation coefficients of 0.83 and 0.78, respectively. The relatively poor correlation of current signals may be attributed to susceptibility to noise from the surrounding electronic devices and the age of the machines. In the case of temperature signals, despite continuous degradation of the cutting tool, temperature change during the cutting process is a slower phenomenon. The variance of recorded temperature data exhibits differences across test runs, potentially allowing for clustering and classification into healthy and worn states. However, tracking progressing tool wear solely based on temperature features poses challenges due to the non-monotonic nature of the feature values.

In Fig. 12, we compare the informativeness of statistical features derived from different domains: time, frequency, and wavelets. In the time domain, variance and kurtosis of cutting force signals were identified as the most effective features. Similarly, in the frequency domain, the amplitude of the 20th harmonic of the force signals was identified as a notable feature. However, features computed from wavelet coefficients, specifically variance and kurtosis, have higher correlation coefficients compared to features obtained from the time and frequency domain analysis.

The signal transformation using wavelets benefits the feature extraction process, as presented in Fig. 12a. Wavelets show a 9.32% and 8.12% improvement over the

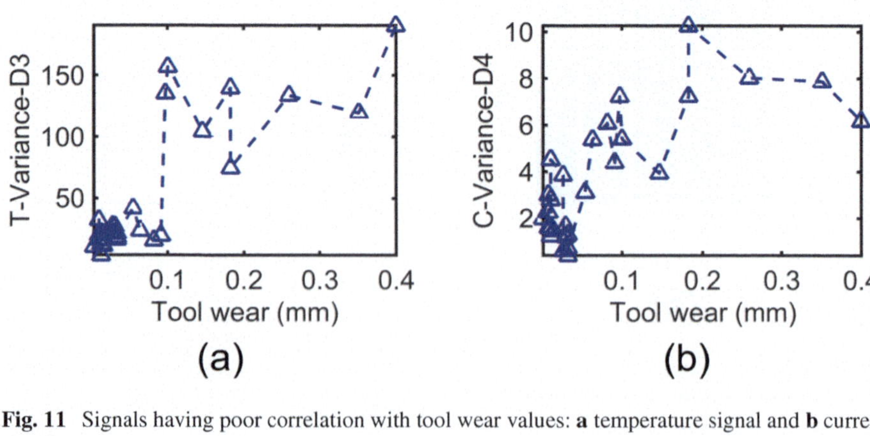

Fig. 11 Signals having poor correlation with tool wear values: **a** temperature signal and **b** current signal

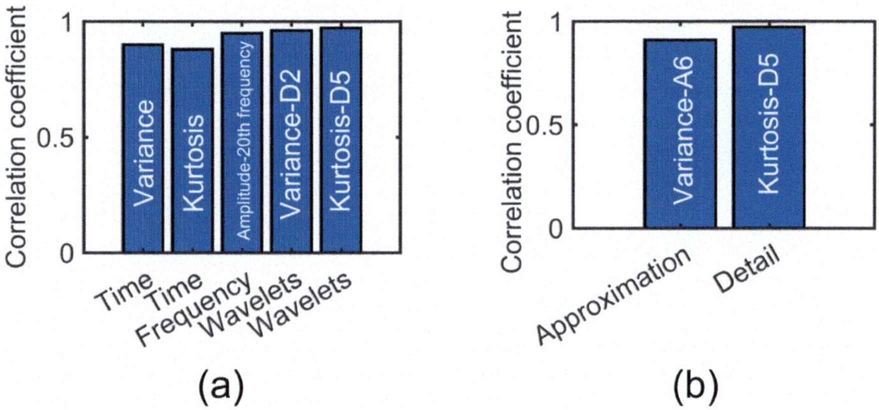

Fig. 12 **a** Correlation coefficient of variance and kurtosis features extracted using time domain, frequency domain, and wavelets and **b** correlation coefficient of features extracted from approximation and detail coefficients

time domain approach for variance and kurtosis features, respectively. Additionally, wavelets also offer a 1.26 and 2.42% advantage over the frequency domain approach with these two features. This is due to the multi-resolution analysis facilitated by wavelet filters that allows for compressing and scaling over time series signals, capturing more reliable and tool wear-sensitive information during the feature engineering process.

Furthermore, Fig. 12b presents the benefit of features derived from detail coefficients over approximation coefficients in terms of their association with tool condition. Features computed from detail coefficients exhibit 6.81% higher correlation than those derived from approximation coefficients. Detail coefficients capture de-noised smaller details, reflecting the effects of progressive tool wear in the signals. Consequently, features of detail coefficients were prioritized for further investigation and modeling.

4.3 Feature Explanation in Relationship to Tool Condition

At this point, cutting forces and machining vibrations are identified as highly informative indicators of tool condition. Specifically, the variance of these signals across wavelet subspaces is identified as a strong indicator of tool health. To further interpret the feature indication for tool health, we explored the frequency spectrum of these signals. The most correlated features identified are the variance and kurtosis of cutting forces and the variance of vibration signals derived from wavelet coefficients.

In this exploration, machining tests were considered in two groups: healthy and worn-out. We selected a subset of machining tests randomly in each group, considering increasing tool wear values. Runs #1, 5, and 12 were included in the healthy group, as the tool wear was lower. Runs #18, 21, 25, and 28 were part of the worn-out group, where wear values accelerated towards tool failure.

4.3.1 Kurtosis of Cutting Forces

The frequency spectrum of force signals is presented in Fig. 13, with the spectra concentrated in the D_5 subspace. The force signals were sampled at 6250 Hz, so the highest frequency component present in the signal is 3125 Hz, which is half of the sampling frequency. Therefore, D_5 subspace covers the range from 97.65 to 195.32 Hz, as depicted in Fig. 13.

Figure 13a–c present the spectra for Runs #1, #5, and #12, and (d–g) correspond to spectra of Runs #18, #21, #25, and #28. Within this spectrum, the third harmonic of the fundamental frequency was observed as a dominating factor with respect to the tool condition. The spindle rotation during Runs #1, #5, and #12 was 2330 rpm, resulting in a fundamental frequency of 38.83 Hz. The peaks in Fig. 13a–c are observed at 116 Hz, representing the third harmonic of the fundamental frequency. Interestingly, this component shifts to 159 Hz in Fig. 13d–g for Runs #18, #21, #25, and #28. This shift is attributed to the spindle rotation being 3184 rpm during these runs, making the fundamental frequency 53.06 Hz and the peak observed at 159 Hz, which is also the third harmonic.

This observation confirms that the force signature is sensitive to cutting settings. Another crucial observation is the amplitude of the peaks in this frequency band, which is indicative of the progressive condition of the cutting tool. The test runs designated in the healthy group have a lower amplitude at the third harmonic than the worn-out group. Despite the increase in the amplitude from healthy to worn-out condition, the shift is not uniform. For instance, the amplitude is ≈ 61 for both Runs #12 and #28, representing different tool conditions. Therefore, the feature, derived from the maximum value of cutting forces, exhibits a comparatively lower correlation coefficient of 0.88 with the tool wear values.

A superior feature computed from the cutting forces is the kurtosis in the D_5 subspace, having a correlation coefficient of 0.973 with the tool wear values. To investigate the dominance of the kurtosis feature, two extreme cases of cutting tool health are presented: Runs #1 and #25 representing the healthy and worn-out conditions, respectively. Figure 14a, b present a zoomed view of the frequency spectrum of Runs #1 and #25 presented in Fig. 13.

Kurtosis is a statistical measure that describes the shape of the signal. Specifically, it provides insights into the peakedness of a distribution. The distribution tails in Fig. 14a, b result in positive kurtosis. However, the kurtosis value is lower when the tool is healthy and undamaged, indicating a more normal or Gaussian distribution for a healthy tool. This also indicates that the cutting forces during machining follow a more regular and predictable pattern, leading to lower kurtosis. Additionally, the

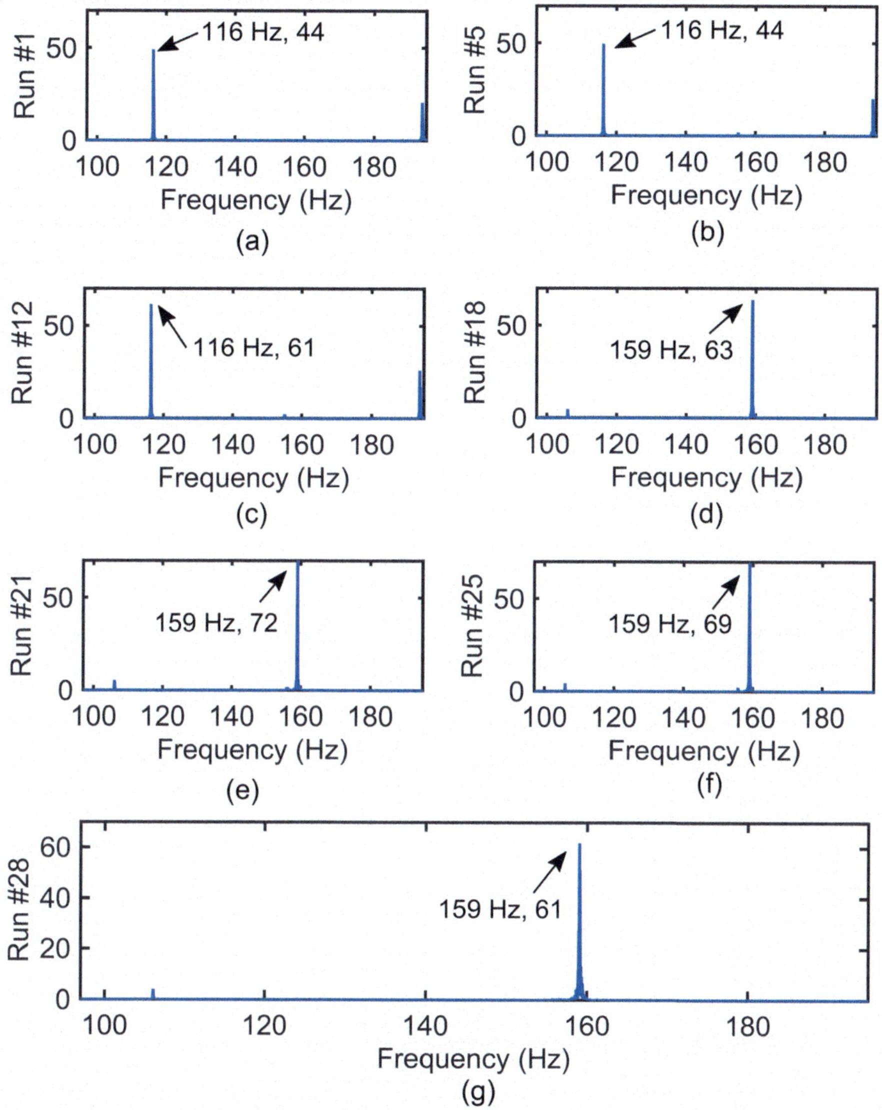

Fig. 13 Frequency spectrum of force signals, with D_5 subspace association with progressing test runs: **a** Run #1, **b** Run #5, **c** Run #12, **d** Run #18, **e** Run #21, **f** Run #25, and **g** Run #28

small peak in the force distribution for the healthy condition indicates that the cutting forces are relatively stable and consistent. However, a higher kurtosis is observed as the tool wears out due to irregularities, machining vibrations, or unexpected variations in the cutting process, resulting in force signals with more outliers. The shape of the cutting forces is affected by the change in the tool condition, especially in the low-frequency range.

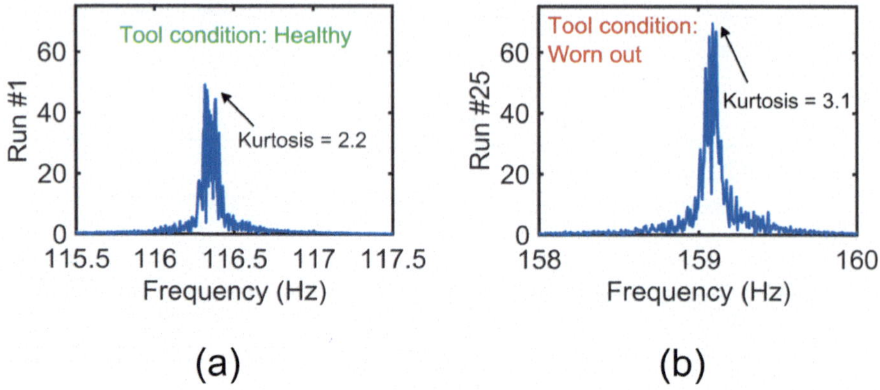

Fig. 14 Kurtosis feature informativeness with respect to tool conditions: **a** healthy and **b** worn out

4.3.2 Variance of Cutting Forces

The variance of cutting force signals in the D_2 subspace exhibits a strong association with the tool wear values, having a correlation coefficient of 0.96. Figure 15 shows the frequency spectrum of cutting force signals, with the spectra concentrated in the D_2 subspace, ranging from 781.25 to 1565.5 Hz. In this range, an increasing jitter is evident in the spectrum as the tool condition deteriorates. A healthy tool tends to maintain more stable cutting forces during the machining process, resulting in a lower variance in the cutting forces. For instance, when the tool is healthy, the amplitude at 850 Hz is 1.01. Runs #1, #5, and #12 in the healthy group exhibit peaks with lower amplitudes, which progressively increases to 2.16 as the tool wear accelerates in Run #28. The increased variability in cutting forces is attributed to the transient changes occurring at the tool-workpiece interface as the tool wears out, leading to higher fluctuations. This can be observed in Fig. 15 as the increasing jitter in the spectrum caused by the progression of tool wear.

4.3.3 Variance of Machining Vibrations

The variance of vibration signals in the D_2 subspace is highly correlated with the tool wear values, having a correlation coefficient of 0.922. Figure 16 depicts the frequency spectrum of vibration signals, with the spectra concentrated in the D_2 subspace, ranging from 2 to 4 Hz. The vibration signal is sampled at a frequency of 16 Hz, making 8 Hz the maximum frequency content in the signal. A significant change in the amplitude of peaks in this frequency range can be observed as the tool wears. Although a similar variation is also observable in other frequency components, differences within the range of 2–4 Hz are most prominent. This observation indicates that the low frequency range of the vibration signal is a more informative indicator of progressing tool wear. As there is a consequent increase in amplitude and jitter due to the progressive tool condition, the variance in this frequency range increases.

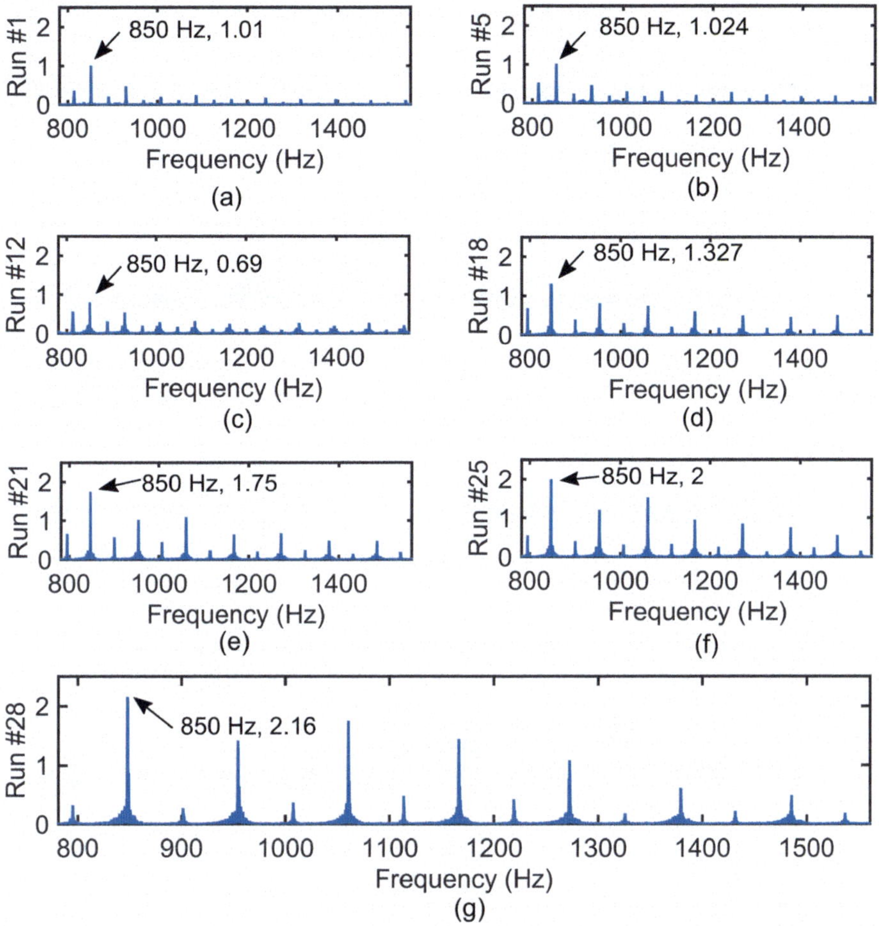

Fig. 15 Frequency spectrum of force signals with D_2 subspace association with progressing test runs: **a** Run #1, **b** Run #5, **c** Run #12, **d** Run #18, **e** Run #21, **f** Run #25, and **g** Run #28

Additionally, it is noteworthy that there is a significant difference in the sampling rates of cutting forces and machining vibrations. The vibration signals were recorded at a much lower sampling rate of 16 Hz. However, critical information associated with tool wear is found in lower frequencies of vibration signals. This suggests that the progression of tool wear is relatively slow and occurs over longer time intervals, and thus, lower sampling rates can still capture the evolving trends. Similarly, the kurtosis feature of cutting forces, found to have the highest correlation coefficient with tool wear values, is also located in the D_5 subspace, representing a lower frequency range.

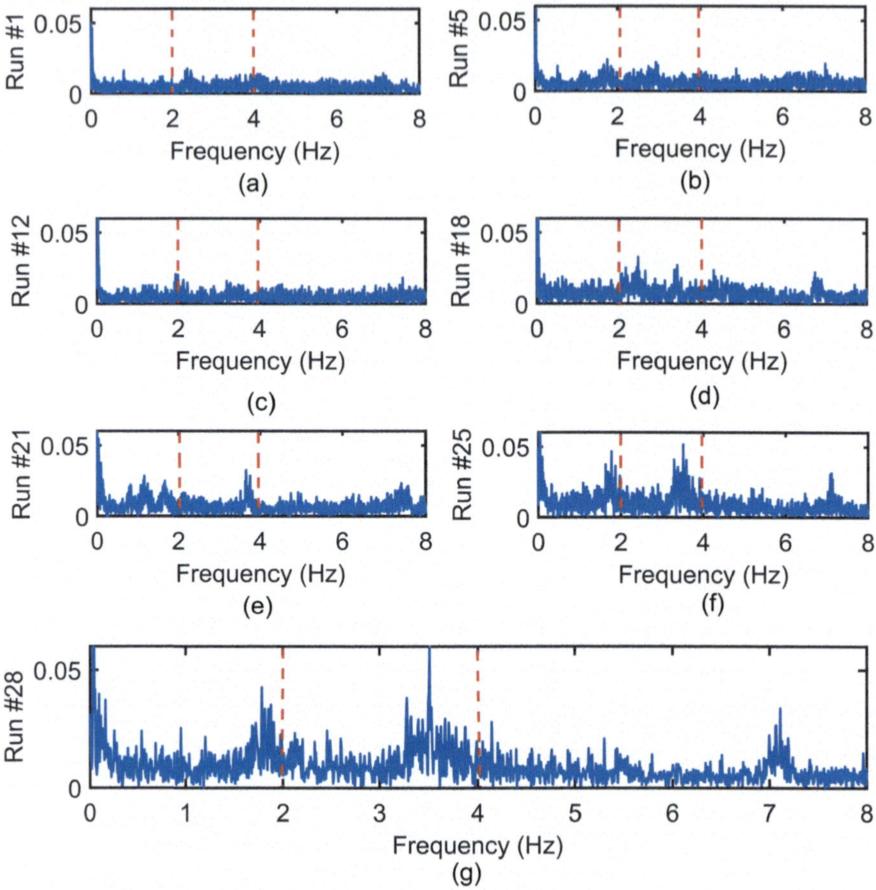

Fig. 16 Frequency spectrum of vibration signals, with D_2 subspace in the highlighted region, presenting association with progressing test runs: **a** Run #1, **b** Run #5, **c** Run #12, **d** Run #18, **e** Run #21, **f** Run #25, and **g** Run #28

4.4 Tool Wear Prediction

Three separate regression models were developed utilizing the following features: (a) kurtosis of cutting forces in the D_5 subspace, (b) variance of cutting forces in the D_2 subspace, and (c) variance of machining vibrations in the D_2 subspace. In addition, four more regression models were developed using the following feature combinations: (d) kurtosis of cutting forces in the D_5 subspace and variance of cutting forces in the D_2 subspace, (e) kurtosis of cutting forces in the D_5 subspace and variance of machining vibrations in the D_2 subspace, (f) variance of cutting forces in the D_2 subspace and variance of machining vibrations in the D_2 subspace, and (g) kurtosis of cutting forces in the D_5 subspace, variance of cutting forces in the D_2 subspace, and variance of machining vibrations in the D_2 subspace. The model training involved the use of features from Runs #1–24, with features from Run #25 reserved for validation, and features from the remaining test runs utilized for testing the model. This ensured the mutual exclusivity of training and testing data, allowing for assessment of the usefulness of the features in predicting the tool wear. Figure 17 presents the prediction of the regression models during the training phase. To address the dynamics and non-linearity in tool wear forecasting, polynomial features of degree 2, 3, and 4 were added. The optimized model was selected based on the validation sample, as presented in Fig. 18.

Table 3 presents the predictions generated by the regression models, and their performance is assessed through the calculation of the mean absolute error (MAE) relative to the actual tool wear values. Among the developed models, the one utilizing the variance feature of machining vibrations has the smallest MAE, recorded at 0.007 mm. Following closely are the models developed with variance and kurtosis features of cutting forces.

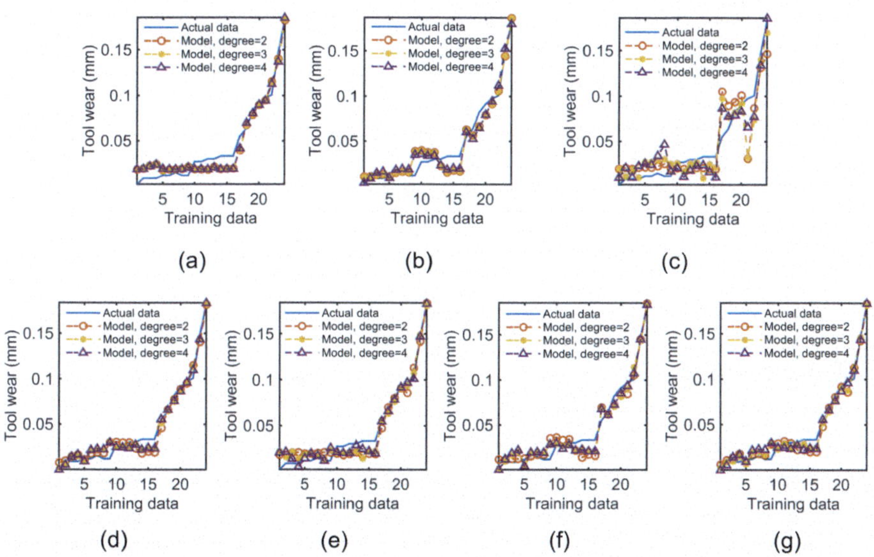

Fig. 17 Model prediction during training using: **a** kurtosis of cutting forces in D_5 subspace, **b** variance of cutting forces in D_2 subspace, **c** variance of machining vibrations in D_2 subspace, **d** kurtosis of cutting forces in D_5 subspace and variance of cutting forces in D_2 subspace, **e** kurtosis of cutting forces in D_5 subspace and variance of machining vibrations in D_2 subspace, **f** variance of machining vibrations in D_2 subspace and variance of cutting forces in D_2 subspace, and **g** kurtosis of cutting forces in D_5 subspace, variance of cutting forces in D_2 subspace, and variance of machining vibrations in D_2 subspace

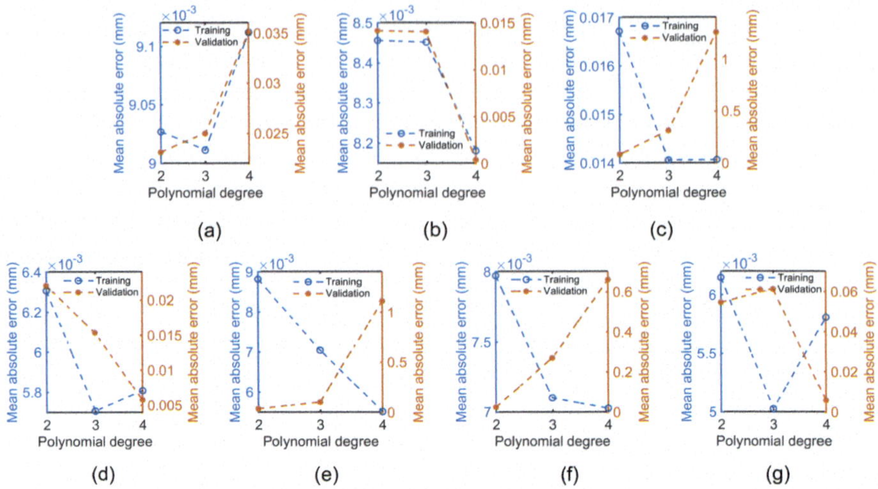

Fig. 18 Model validation: **a** kurtosis of cutting forces in D_5 subspace, **b** variance of cutting forces in D_2 subspace, **c** variance of machining vibrations in D_2 subspace, **d** kurtosis of cutting forces in D_5 subspace and variance of cutting forces in D_2 subspace, **e** kurtosis of cutting forces in D_5 subspace and variance of machining vibrations in D_2 subspace, **f** variance of machining vibrations in D_2 subspace and variance of cutting forces in D_2 subspace, and **g** kurtosis of cutting forces in D_5 subspace, variance of cutting forces in D_2 subspace, and variance of machining vibrations in D_2 subspace

This outcome highlights the efficacy of a straightforward and intuitive ML algorithm such as polynomial regression in accurate modeling of tool wear prediction. The utilization of essential features derived from cutting forces and machining vibrations contributes to the success of the model. These statistical features reveal the underlying dynamics associated with tool deterioration during the machining process.

5 Conclusions

This study focused on developing an effective tool condition monitoring system using sensory data recorded during machining operations. The primary objectives were to identify informative features related to tool wear and employ a cost-effective ML algorithm for tool wear prediction. The wavelet transformation method was employed to process the acquired signals, and various statistical features were computed. The following key findings and conclusions can be highlighted:

- The features extracted from cutting forces and machining vibrations, especially the variance and kurtosis in specific frequency subspaces, demonstrated strong correlation with the progressive condition of the cutting tool.

Table 3 Forecasting by regression model

Feature	Test run #	Tool wear (mm)		
		Actual	Predicted	MAE
Kurtosis-F-D_5	Run #26	0.2595	0.3101	0.05
	Run #27	0.3511	0.4072	0.056
	Run #28	0.4	0.516	0.115
Variance-F-D_2	Run #26	0.2595	0.3085	0.049
	Run #27	0.3511	0.3633	0.012
	Run #28	0.4	0.4453	0.045
Variance-V-D_2	Run #26	**0.2595**	**0.2945**	**0.034**
	Run #27	**0.3511**	**0.3753**	**0.024**
	Run #28	**0.4**	**0.407**	**0.007**
Kurtosis-F-D_5 and variance-F-D_2	Run #26	0.2595	0.3114	0.051
	Run #27	0.3511	0.4105	0.059
	Run #28	0.4	0.5206	0.12
Kurtosis-F-D_5 and variance-V-D_2	Run #26	0.2595	0.3281	0.068
	Run #27	0.3511	0.4348	0.083
	Run #28	0.4	0.5493	0.149
Variance-F-D_2 and variance-V-D_2	Run #26	0.2595	0.3386	0.079
	Run #27	0.3511	0.389	0.037
	Run #28	0.4	0.4814	0.081
Kurtosis-F-D_5, variance-F-D_2, and variance-V-D_2	Run #26	0.2595	0.3324	0.072
	Run #27	0.3511	0.4618	0.11
	Run #28	0.4	0.5768	0.176

F force, *V* vibration
Bold signifies a least MAE

- The wavelet domain proved to be more effective in capturing the informative features related to tool wear, showcasing its superiority over time and frequency domains. Features extracted through wavelet transformation were found to have about 9% more correlation over the time domain features and about 2% more over the frequency domain features.
- The frequency spectra of cutting forces and machining vibrations reveal meaningful information associated with the dynamics of the machining process. Although the force spectra are sensitive to variations in the cutting settings, the amplitude of spectra is affected by the cutting tool conditions.
- Machining vibrations show significant variance associated with the tool condition at lower frequencies.
- Tool wear can be accurately modeled using low-cost algorithms with the variance of machining vibrations exhibiting the lowest MAE among the developed models.

6 Limitations and Future Scope

The following highlights the limitations of this study and the scope for future works:

- The strong correlation observed between the kurtosis of cutting forces in the D_5 subspace and tool wear values highlights its significance for tool condition monitoring. However, this correlation is specific to this particular wavelet subspace and does not extend to coefficients in other wavelet subspaces. As a result, the utility of kurtosis of cutting forces as a feature for tool condition monitoring may be limited.
- Although the correlation of variance with tool wear is more prominent across various wavelet subspaces of cutting force and machining vibration signals, further validation is necessary. A more extensive set of tool run-to-failure tests would enhance the robustness of this observation and provide a more comprehensive understanding of the variance as an indicator of tool wear.
- In this study, the machining parameters during the tool's run-to-failure tests were varied in two levels. Changes in the machining settings could affect the sensor data differently, impacting the generalizability of the extracted features of tool health.
- The findings are based on a specific set of machining conditions, including machining parameters, CNC machine, workpiece, and tool material. Generalizing the results to different machining scenarios will require additional validation across a broader range of cutting settings and machines.

Statements and Declarations **Funding** We gratefully acknowledge the Air Force Research Laboratory, Materials and Manufacturing Directorate (AFRL/RXMS) for support via Contract No. FA8650-20-C-5206.

References

Abellan-Nebot JV, Romero Subirón F (2010) A review of machining monitoring systems based on artificial intelligence process models. Int J Adv Manuf Technol 47(1–4):237–257. https://doi.org/10.1007/s00170-009-2191-8

Aghazadeh F, Tahan A, Thomas M (2018) Tool condition monitoring using spectral subtraction and convolutional neural networks in milling process. Int J Adv Manuf Technol 98(9–12):3217–3227. https://doi.org/10.1007/s00170-018-2420-0

Awasthi U, Wang Z, Mannan N, Pattipati KR, Bollas GM (2022) Physics-based modeling and information-theoretic sensor and settings selection for tool wear detection in precision machining. J Manuf Process 81:127–140. https://doi.org/10.1016/j.jmapro.2022.06.027

Bachman G, Narici L, Beckenstein E (2000) Fourier and wavelet analysis. Springer, New York, NY. https://doi.org/10.1007/978-1-4612-0505-0

Brili N, Ficko M, Klančnik S (2021) Tool condition monitoring of the cutting capability of a turning tool based on thermography. Sensors 21(19):6687. https://doi.org/10.3390/s21196687

Chan Y-W, Kang T-C, Yang C-T, Chang C-H, Huang S-M, Tsai Y-T (2022) Tool wear prediction using convolutional bidirectional LSTM networks. J Supercomput 78(1):810–832. https://doi.org/10.1007/s11227-021-03903-4

Dimla DE (2000) Sensor signals for tool-wear monitoring in metal cutting operations—a review of methods. Int J Mach Tools Manuf 40(8):1073–1098. https://doi.org/10.1016/S0890-6955(99)00122-4

Gouarir A, Martínez-Arellano G, Terrazas G, Benardos P, Ratchev S (2018) In-process tool wear prediction system based on machine learning techniques and force analysis. Proc CIRP 77(Hpc):501–504. https://doi.org/10.1016/j.procir.2018.08.253

Guo J, Li A, Zhang R (2020) Tool condition monitoring in milling process using multifractal detrended fluctuation analysis and support vector machine. Int J Adv Manuf Technol 110(5–6):1445–1456. https://doi.org/10.1007/s00170-020-05931-5

Hammond FM, Malec JF, Nick TG, Buschbacher RM (eds) (2014) Handbook for clinical research. Springer Publishing Company, New York, NY. https://doi.org/10.1891/9781617050992

Han S, Mannan N, Stein DC, Pattipati KR, Bollas GM (2021) Classification and regression models of audio and vibration signals for machine state monitoring in precision machining systems. J Manuf Syst 61:45–53. https://doi.org/10.1016/j.jmsy.2021.08.004

Han S, Yang Q, Pattipati KR, Bollas GM (2022) Sensor selection and tool wear prediction with data-driven models for precision machining. J Adv Manuf Process **4**(4). https://doi.org/10.1002/amp2.10143

He Z, Shi T, Xuan J, Li T (2021) Research on tool wear prediction based on temperature signals and deep learning. Wear 478–479:203902. https://doi.org/10.1016/j.wear.2021.203902

Javed K, Gouriveau R, Zerhouni N, Nectoux P (2015) Enabling health monitoring approach based on vibration data for accurate prognostics. IEEE Trans Ind Electron 62(1):647–656. https://doi.org/10.1109/TIE.2014.2327917

Khajavi MN, Nasernia E, Rostaghi M (2016) Milling tool wear diagnosis by feed motor current signal using an artificial neural network. J Mech Sci Technol 30(11):4869–4875. https://doi.org/10.1007/s12206-016-1005-9

Kious M, Ouahabi A, Boudraa M, Serra R, Cheknane A (2010) Detection process approach of tool wear in high speed milling. Measurement 43(10):1439–1446. https://doi.org/10.1016/j.measurement.2010.08.014

Kong D, Chen Y, Li N (2018) Gaussian process regression for tool wear prediction. Mech Syst Signal Process 104:556–574. https://doi.org/10.1016/j.ymssp.2017.11.021

Kothuru A, Nooka SP, Liu R (2019) Application of deep visualization in CNN-based tool condition monitoring for end milling. Proc Manuf 34:995–1004. https://doi.org/10.1016/j.promfg.2019.06.096

Kumar MP, Dutta S, Murmu NC (2021) Tool wear classification based on machined surface images using convolution neural networks. Sādhanā 46(3):130. https://doi.org/10.1007/s12046-021-01654-9

Kuntoğlu M, Sağlam H (2021) Investigation of signal behaviors for sensor fusion with tool condition monitoring system in turning. Measurement 173:108582. https://doi.org/10.1016/j.measurement.2020.108582

Li G, Wang Y, He J, Hao Q, Yang H, Wei J (2020) Tool wear state recognition based on gradient boosting decision tree and hybrid classification RBM. Int J Adv Manuf Technol 110(1–2):511–522. https://doi.org/10.1007/s00170-020-05890-x

Lin M, Wanqing S, Chen D, Zio E (2022) Evolving connectionist system and hidden semi-Markov model for learning-based tool wear monitoring and remaining useful life prediction. IEEE Access 10:82469–82482. https://doi.org/10.1109/ACCESS.2022.3196016

Liu H, Liu Z, Jia W, Lin X, Zhang S (2020) A novel transformer-based neural network model for tool wear estimation. Meas Sci Technol 31(6):065106. https://doi.org/10.1088/1361-6501/ab7282

Liu X, Liu S, Li X, Zhang B, Yue C, Liang SY (2021) Intelligent tool wear monitoring based on parallel residual and stacked bidirectional long short-term memory network. J Manuf Syst 60:608–619. https://doi.org/10.1016/j.jmsy.2021.06.006

Lu MC, Wan BS (2013) Study of high-frequency sound signals for tool wear monitoring in micromilling. Int J Adv Manuf Technol 66(9–12):1785–1792. https://doi.org/10.1007/s00170-012-4458-8

Martínez-Arellano G, Terrazas G, Ratchev S (2019) Tool wear classification using time series imaging and deep learning. Int J Adv Manuf Technol 104(9–12):3647–3662. https://doi.org/10.1007/s00170-019-04090-6

Mishra D, Han S, Pattipati KR, Bollas GM (2023a) Explainable symbolic regression model for tool wear diagnosis. In: 2023 9th international conference on control, decision and information technologies (CoDIT). IEEE, pp 2139–2144. https://doi.org/10.1109/CoDIT58514.2023.10284293

Mishra D, Awasthi U, Pattipati KR, Bollas GM (2023b) Tool wear classification in precision machining using distance metrics and unsupervised machine learning. J Intell Manuf. https://doi.org/10.1007/s10845-023-02239-5

Naveen Venkatesh S et al. (2022) Transfer learning-based condition monitoring of single point cutting tool. Comput Intell Neurosci 2022:1–14. https://doi.org/10.1155/2022/3205960

Pimenov DY, Bustillo A, Wojciechowski S, Sharma VS, Gupta MK, Kuntoğlu M (2023) Artificial intelligence systems for tool condition monitoring in machining: analysis and critical review. J Intell Manuf 34(5):2079–2121. https://doi.org/10.1007/s10845-022-01923-2

Qiao H, Wang T, Wang P, Qiao S, Zhang L (2018) A time-distributed spatiotemporal feature learning method for machine health monitoring with multi-sensor time series. Sensors 18(9):2932. https://doi.org/10.3390/s18092932

Ranjan J et al (2020) Artificial intelligence-based hole quality prediction in micro-drilling using multiple sensors. Sensors 20(3):885. https://doi.org/10.3390/s20030885

Salur E, Aslan A, Kuntoglu M, Gunes A, Sahin OS (2019) Experimental study and analysis of machinability characteristics of metal matrix composites during drilling. Compos Part B Eng 166:401–413. https://doi.org/10.1016/j.compositesb.2019.02.023

Salur E, Aslan A, Kuntoglu M, Gunes A, Şahin ÖS (2020) Optimization of cutting forces during turning of composite materials. Acad Platf J Eng Sci 423–431. https://doi.org/10.21541/apjes.631260

Wang J, Ma Y, Zhang L, Gao RX, Wu D (2018) Deep learning for smart manufacturing: methods and applications. J Manuf Syst 48:144–156. https://doi.org/10.1016/j.jmsy.2018.01.003

Wanigarathne PC, Kardekar AD, Dillon OW, Poulachon G, Jawahir IS (2005) Progressive tool-wear in machining with coated grooved tools and its correlation with cutting temperature. Wear 259(7–12):1215–1224. https://doi.org/10.1016/j.wear.2005.01.046

Stéphane M (2009) A wavelet tour of signal processing. Elsevier. https://doi.org/10.1016/B978-0-12-374370-1.X0001-8

Yang Q, Pattipati KR, Awasthi U, Bollas GM (2022) Hybrid data-driven and model-informed online tool wear detection in milling machines. J Manuf Syst 63:329–343. https://doi.org/10.1016/j.jmsy.2022.04.001

Yang Q, Mishra D, Awasthi U, Bollas GM, Pattipati KR (2024) Tool wear and remaining useful life estimation in precision machining using interacting multiple model. J Manuf Syst 74:367–386. https://doi.org/10.1016/j.jmsy.2024.04.001

Yıldırım ÇV, Sarıkaya M, Kıvak T, Şirin Ş (2019) The effect of addition of hBN nanoparticles to nanofluid-MQL on tool wear patterns, tool life, roughness and temperature in turning of Ni-based Inconel 625. Tribol Int 134:443–456. https://doi.org/10.1016/j.triboint.2019.02.027

Zhang C, Yao X, Zhang J, Jin H (2016) Tool condition monitoring and remaining useful life prognostic based on a wireless sensor in dry milling operations. Sensors 16(6):795. https://doi.org/10.3390/s16060795

Zhao R, Wang J, Yan R, Mao K (2016) Machine health monitoring with LSTM networks. In: 2016 10th international conference on sensing technology (ICST). IEEE, 2016, pp 1–6

Zhao R, Wang D, Yan R, Mao K, Shen F, Wang J (2018) Machine health monitoring using local feature-based gated recurrent unit networks. IEEE Trans Ind Electron 65(2):1539–1548. https://doi.org/10.1109/TIE.2017.2733438
Zhou Y et al (2022) A new tool wear condition monitoring method based on deep learning under small samples. Measurement 189:110622. https://doi.org/10.1016/j.measurement.2021.110622

Experimental Investigation of Process Parameters Effects on Extrusion Blow Molding Process Using Response Surface Methodology for Industry 4.0

Rajendra Shimpi, Chandrakant Wani, Rakesh Chaudhari, Sarvoday Joharapurkar, Hargovind Soni, and Ajay Kumar

Abstract Plastic has changed the human life by frequent usage in daily needs due to its practical features compared to other materials. In addition, the ability to get easily molded and low production cost make it better a choice than other materials. Industry 4.0 concepts and technologies are playing vital role in manufacturing sector. Among the available processes extrusion molding is popularly used for producing the plastic products. The production time and quality of the product in extrusion blow molding is greatly affected by material selection, pressure, temperature, cooling time and extrusion speed. The optimum parameters selection is very crucial to increase the productivity with low cost of the product. The present work investigates the effect of various parameters such as extruder pressure, die temperature along with extruder speed on

R. Shimpi (✉)
NAMTECH Transitory Campus: Research Park, IIT Gandhinagar Campus, Palaj, Gandhinagar, Gujarat 382055, India
e-mail: rajendrashimpi9@gmail.com

C. Wani
School of Technology Management and Engineering, Shri Vile Parle Kelavani Mandal Narsee Monjee Institute of Management Studies, Hyderabad, Telangana, India

R. Chaudhari · S. Joharapurkar
Mukesh Patel School of Technology Management and Engineering, Shirpur Campus, Shri Vile Parle Kelavani Mandal Narsee Monjee Institute of Management Studies (NMIMS), Dhule, Maharashtra, India

H. Soni (✉)
Department of Mechanical Engineering, National Institute of Technology Delhi, Delhi, India
e-mail: hargovindsoni2002@gmail.com

Department of Industrial Engineering, College of Science, Engineering and Technology, University of South Africa, Pretoria, South Africa

A. Kumar (✉)
Department of Mechanical Engineering, School of Engineering and Technology, JECRC University, Jaipur, Rajasthan 303905, India
e-mail: ajay.kumar30886@gmail.com

A. Kumar et al. (eds.), *Industry 4.0 Driven Manufacturing Technologies*, Springer Series in Advanced Manufacturing, https://doi.org/10.1007/978-3-031-68271-1_3

production time for 20-L bottle made up of low-density polyethylene (LDPE) material using extrusion molding process. Design of experiments using response surface methodology is used to obtain experimental results. The main and interaction effect of process parameters are studied successfully using central composite design (CDD). The individual main effect graphs and 3D surface plots are used to understand the effect of input parameters on the process timing.

Keywords Extrusion blow molding · Production time · RSM · Central composite design · ANOVA

1 Introduction

Plastic play's vital role in daily life and used in numerous human needs. The widespread application of plastic is a due to lightweight, durability and low cost. It can be readily molded into products of any shape and size and available in different colors. The production of plastics has drastically increased over the last 60 years. Plastics, resins and fibers found significant application in packaging industry where these are used for preservation and safety food items ensuring the quality (Raheem 2013). Plastic bottles' applications run a huge business in every corner of the world. The plastics bottles are widely used to store variety of liquid products such as water, milk, oil and cold drinks. The percentage of bottle and containers is 75% in total plastic products used in daily applications (Belcher 2011). The huge demand of plastic bottles and containers need the rapid and efficient production system. The numbers of manufacturing processes are available now a day to fulfil this demand.

Extrusion blow molding process is effectively used for manufacturing the good quality bottles and containers made with plastic and other fibers. In this process the melted plastic is forced to extrude into a hollow tube called as a parison which latter is captured by cold metal mold (Cooper 2013). The air blown into the hallow tube helps to inflate it into hollow shape of bottle and container. The process can produce shampoo and milk bottles, water cans and drums used in industrial applications (Singh and AlMangour 2023). The performance of extrusion blow molding process depends on other process parameters such as material selection, pressure, temperature, cooling time, and extrusion speed. All these parameters have significant effect of the quality of product.

Therefore, the selection of optimal parameters is very important for ensuring maximum and efficient productivity enhancing the quality of the product. The investigations of extrusion blow molding process parameters such as barrel temperature, cooling cycle time and extruder speed was carried out to achieve higher productivity and quality of the product. It has been observed the barrel temperature is significant factor affecting extrusion blow molding. These optimum operating process parameters are always achieved through modelling and simulation as well as experimental investigation followed by optimization using various techniques. The optimal parameters if achieved by experimental investigations need to alter the operational conditions, machine set-up and tools. The optimization by experimental investigation is expensive and creates unnecessary interruption in the process due to implementation of new process parameters (Kumar et al. 2019, 2020, 2023, 2024a; Kumar and Gulati 2021).

Number of researchers has carried out experimental investigations and simulations for optimization of molding process parameters. The effect of various input and output parameters on extrusion blow molding process has been explored by number of researchers. The optimization carried out on process control parameters of the High-density polyethylene (HDPE) pipe using Taguchi technique exhibited the highest value of signal to noise (S/N) ratio as optimum parameter with minimum variance. The confirmation run conducted on final process showed significant improvement in withstanding pressure. The experimental investigations for an improvement of HDPE plastic quality using six sigma and Taguchi methods have reported that processes parameters such as extruder temperature, extruder and winder rotation has significant effect quality industry appliance and household products. This method found useful in shortening the overall development time of the process with significant saving of material. The combined method is very efficient and fruitful to obtain optimum thickness distribution of parison in case of blow molded components. This has helped to improve the thickness of the product formed by extrusion blow molding process.

The optimum use of raw material in components formed by blow molding process is of increasing interest in recent innovations. In present practices containers are divided into different sections depending on their change in shape and reducing the thickness as per the functional requirement of the section. In other words, the thickness is considered as function of geometry of the component. The numerical simulations based on finite element analysis results are carried out to estimate the performance of the industrial bottle of specific thickness. The results of the investigation validated the optimization process with potential to recognize the most precarious areas for the material application. The optimization of the extrusion-blow molding process using Soft Computing and Taguchi Method is successfully used to calculate the optimal die gap in case of extrusion blow molding processes. Taguchi method is one of the effective methods to optimize the process factors (Kumar and Gulati 2018, 2019). It is also attempted to impart uniform thickness of the product later than the prison inflation by timely varying the die gap opening. The proposed approach has yielded satisfactory results on overall performance of the product after variation in die gap opening (Yu et al. 2004). The research is also conducted to propose blow molding simulation approach to calculate thickness of complex shape containers and bottles. Shape of full bottle is simulated with bottle of rectangular cross section to compare with cross section methodology. The FEM results are verified by comparison with experimental results of extrusion molding process. The deviation in the results yielded by cross section technique is less than the full bottle simulation when compared with experimental results. The cross-section technique found as a satisfactory approach for estimating thickness of parison that flows through extrusion die (Suvanjumrat et al. 2018).

The implementation of concurrent engineering approach is explored to decrease the total weight of plastic dumbbell with improvement in thickness distribution. The product is developed using simulation of concurrent engineering method as well as experimental work. This implementation has significantly reduced the development cycle time. In addition, the man-hours also effectively reduced for the same quality of the product. More than double time span is saved for the development cycle (Attar et al. 2008). It is observed that cooling phase has high impact on process cycle time and properties of extrusion blow molded products. The optimization of different process parameters such as inflating pressure, cooling time, melt and mold temperatures, and die gap was carried out for high density polyethylene (HDPE) and a metallocene polyethylene (MPE) rectangular shape bottle. The use of various cooling methods and temperature allocation in the product is also explored. It was found that the amount of cooling time can be effectively reduced by using optimal parison value. It is also revealed that the lesser cooling time helps to improve stress free cooling of the product (Bendada et al. 2005).

In extrusion blow molding the experimental investigations are mostly conducted on PVC pipes and thin films. Relatively less work has been done on bottles are made of HDPE, polypropylene (PP) resins and low-density polyethylene (LDPE) material (Singh et al. 2017). Different techniques such as Finite Element Analysis (FEA), Taguchi method, Response Surface Methodology (RSM), Grey-Taguchi Method, Artificial Neural Network (ANN), Geometrical Algorithm Method (AGM), Energy Gap Method (GAP) are successfully used for optimization of blow molding process. In blow molding process, product thickness, screw speed, cooling time and barrel temperature are considered as optimization variable parameters. The present research attempts to investigate the effect of input parameters such as extruder pressure, die temperature and extruder speed on production cycle time for each bottle formed by blow molding process (Table 1).

2 Central Composite Design Method and Response Surface Methodology

To determine the cause-and-effect relationships among the factors of influence in a model, design of experiment is used. In this approach, experiments are designed for investigating effect of one factor keeping another factor constant. However, in this classical design of experiment it is not possible to figure out the exact interactions between various factors and thus error in response against the factors being investigated can be observed. In addition, this approach doesn't yield accurate results. This drawback can be overcome by reducing space domain for the factor under consideration or by reducing amount of parameters levels. In these instances, conclusion

Table 1 Outcome of introduction

S. No.	Outcomes
1	The performance of extrusion blow molding process depends on other process parameters such as material selection, pressure, temperature, cooling time, and extrusion speed. All these parameters have significant effect of the quality of product
2	Response surface methodology (RSM) successfully used for optimization of blow molding process for Industry 4.0
3	Industry 4.0 facilitates greater customization and flexibility in manufacturing. Through RSM, manufacturers can quickly adapt the extrusion blow molding process parameters to accommodate varying product designs and specifications, meeting diverse customer needs without compromising on efficiency or quality
4	The integration of RSM in the EBM process not only enhances the understanding and control of process parameters but also aligns with the core principles of Industry 4.0
5	Optimizing the EBM process through RSM contributes to sustainability by minimizing material usage and energy consumption. This aligns with Industry 4.0's goals of creating more sustainable and resource-efficient manufacturing practices

certainty gets decreased. For resolving this issue, method of factorial experiment design was developed which allowed consideration of varying amounts of parameters in a pattern (Montgomery and Runger 2020). All feasible arrangements of factor amounts in a example can be included in a full factorial design but number of experiments also increases as factors increase. This method is suitable for first order model involving linear effect of factor. For ensuring a better quality of results, three or more factor levels are required but in this case many experiments need to be performed resulting in increased cost and longer time. Box and Wilson proposed a Central Composite Design method (CCD method) as a remedy to these issues. This approach is very helpful in providing much information for trio-level factorial strategy with fewer runs than a full factorial strategy resulting into saving of experimentation time and cost (Montgomery and Runger 2020; Diler and Ipek 2012).

After developing design of experiment and conducting experimental runs, RSM has been implemented for finding the influence of system parameters on response variable. Response surface methodology popularly known as RSM is a statistical technique used for modelling and analysis of a problem with an objective of finding the effect of system factors on the response and to optimize it. In most of the problems of RSM, nature of the relationship among independent variables and response variable isn't known. Therefore, an appropriate approximation of functional relationship between input variable and response is found initially (Montgomery and Runger 2020).

3 Experimental Design

In the present study, the experiments are designed based on central composite design method. In Fig. 1, each axis represents the extrusion blow molding process parameters such as extruder pressure, die temperature and extruder speed whereas while each point on the cube represents levels of the selected parameters. The CCD consists of 2k + 2k + m runs, in which k represents number of factors and first 2k represent number of the factorial points at the corners of the cube, another 2k in the above equation is number of the axial points on the axis of each design parameter at the calculated distance of $\pm \alpha$ ($\alpha = 2k/4 = 1.68179$ for $k = 3$) from the cube centre whereas m represents amount of the centre facts at the block centre. There is no need to replicate the whole experimental design as centre point is replicated to find the experimental error. It is mostly recommended in CDD is to take six centre points including three components. The sum of investigates recommended in the current study are $8 + 6 + 6 = 20$ (Diler and Ipek 2012).

In the present study extruder pressure, die temperature and extruder speed have been considered as process parameters referring work done by other researchers. Table 2 gives the range of input parameters and their levels with coded value.

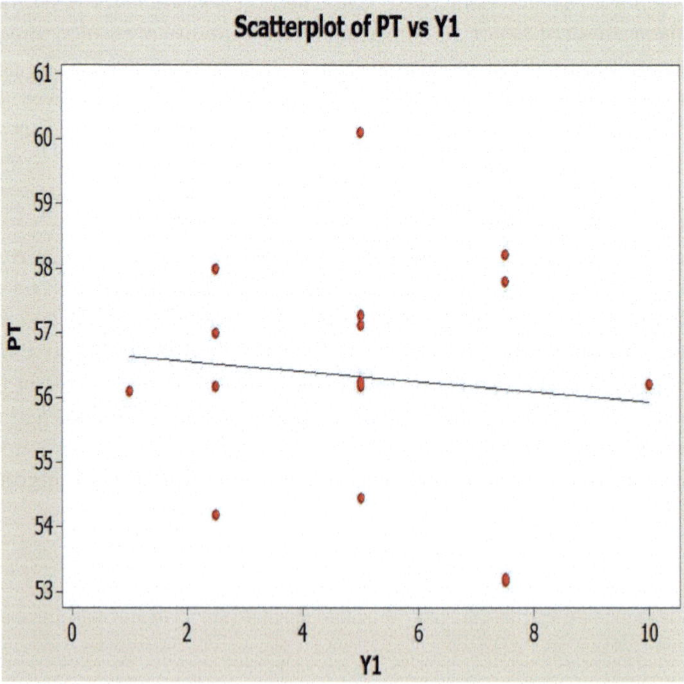

Fig. 1 Main effect graph of production time versus pressure

Table 2 Components and uncoded amounts of parameters at different coded levels

Factors	Symbols		Uncoded values of coded levels				
	Uncoded	Coded	– 1.6818	– 1	0	1	1.6818
Pressure (bar)	Y_1	X_1	1	2.5	5	7.5	10
Temperature (°C)	Y_2	X_2	160	162	165	170	178
Extruder speed (rpm)	Y_3	X_3	100	200	300	400	500

Table 3 shows details of experiments of design parameters and trial outcome. Each factor is coded as – 1, 0 and 1, for low, middle and high levels respectively. The minimal and extreme levels are coded as – 1.68179 and 1.68179. This coding helps to facilitate the computation for regression analysis in CDD.

The plan of the experiment table is prepared after selection of process parameters' range for run in order 20. In Table 3, details of plan of experiments for all process

Table 3 Plan of the experiment table

S. No.	Coded values			Uncoded values			Response variable
	X_1	X_2	X_3	Y_1	Y_2	Y_3	Production time (PT) (in s)
1	– 1	– 1	– 1	2.5	162	200	56.18
2	1	– 1	– 1	7.5	162	200	53.17
3	– 1	1	– 1	2.5	178	200	54.19
4	1	1	– 1	7.5	178	200	53.21
5	– 1	– 1	1	2.5	162	400	58.00
6	1	– 1	1	7.5	162	400	57.80
7	– 1	1	1	2.5	178	400	57.00
8	1	1	1	7.5	170	400	58.20
9	– 1.68	0	0	1	165	300	56.10
10	1.68	0	0	10	165	300	56.21
11	0	– 1.68	0	5	160	300	57.12
12	0	1.68	0	5	160	300	56.20
13	0	0	– 1.68	5	165	100	54.45
14	0	0	1.68	5	165	500	60.10
15	0	0	0	5	165	300	57.28
16	0	0	0	5	165	300	56.25
17	0	0	0	5	165	300	56.25
18	0	0	0	5	165	300	56.18
19	0	0	0	5	165	300	56.25
20	0	0	0	5	165	300	56.24

parameters and production time as a response variable with 20 sets of combinations have been given. Table 3 is for 20-L container and response variable as cycle time for one bottle.

4 Materials and Methods

In the present experimental study, manufacturing of 20-L bottles using extrusion blow molding has been considered. A semi-rigid and translucent polymer low-density polyethylene (LDPE) material is selected for the bottle. The reason for selecting the LDPE material is its high degree of short and long side-chain branching. The raw material is placed in hopper for processing and then forwarded to heater for melting. The melted metal then forced to pass through the extruder and screw and subsequently flows towards the mouth were the parison formation takes place. After the formation of parison, the mold gets closed and remaining air is blown through the pin. The bottle at the same time is cooled in the mold by the cooling channel. The bottle comes out of the mold after successful cooling. The extra part formed along with the bottle is removed to obtained finished product.

Factors and their levels are calculated by CDD method and according to the run order mentioned in Table 3. The effect of extruder pressure (Y_1), die temperature (Y_2) and extruder speed (Y_3) on production or cycle time (PT) is observed as a response variable. As per the standard industrial practice for extrusion blow molding process for 20-L bottle, parameter range for extruder pressure is selected as 1–10 bar. The range of temperature kept as 1600–1780 °C. The extruder speed is varied from 100 to 500 rpm. Experiments were conducted for observing production time for one bottle of 20-L size with various combinations of Y_1, Y_2 and Y_3.

Based on design of experiment and first order model of input parameters, all the statistical analysis for the data obtained through experimental run data has been performed using MINITAB software (Kumar et al. 2024b).

5 Results and Discussion

The experimental investigations were carried out as per the design of experiments developed using RSM method. The results obtained in statistical analysis are tabulated to determine the significance of the developed model. In addition, the analysis of variance (ANOVA) was carried out to find out the significant factors and their relations with each other. The adequacy of the model is estimated to elaborate the optimum process parameters.

Tables 4 and 5 shows the results obtained during ANOVA. The value of predicted target factor (R^2 Pred.) model is 0.932 which is almost close to 1. The obtained value is in seasonally agreeing with the adjusted determination coefficient (R^2 Adj.) which is equal to 0.871. It indicates that sample variation of 93.20% is attributed to the factors and their interactions. *P*-value is measured as 0.023 which is less than standard value (< 0.05) for interactions of Y_1 and Y_3. It implies that process parameters extruder pressure (Y_1) and extruder speed (Y_3) are significant parameters affecting the cycle time in extrusion blow molding process while manufacturing of 20-L bottle of LDPE material.

Model summary is:

$$S = 0.5984 \quad \text{R-Sq} = 93.2\% \quad \text{R-Sq (adj)} = 87.1\%$$

Based on above analysis, the regression equation excluding the insignificant process parameters involved in blow molding process while manufacturing of 20-L bottle is established as below.

The regression equation: $PT = 56.15 + 3.54\ Y_3 + 2.24\ (Y_1 * Y_3)$.

Table 4 ANOVA of CCD model

Source	DF	Seq. SS	Adj. SS	Adj. MS	F	*P*
Regression	9	49.043	49.043	5.4492	15.22	0.000
Linear	3	44.341	39.713	13.2377	36.97	0.000
Square	3	2.060	1.252	0.4175	1.17	0.371
Interaction	3	2.642	2.642	0.8805	2.46	0.123
Residual error	10	3.581	3.581	0.3581		
Lack of fit	4	2.242	2.242	0.5605	2.51	0.150
Pure error	6	1.339	1.339	0.2231		
Total	19	52.623				

Table 5 Response surface regression analysis

Term	Coef.	SE coef.	T	*P*
Constant	56.1542	0.3263	172.079	0.000
Y1	– 0.2864	0.3260	– 0.879	0.400
Y2	– 0.3674	0.2724	– 1.349	0.207
Y3	3.5429	0.3463	9.414	0.000
Y1 * Y1	– 0.1915	0.4601	– 0.416	0.686
Y2 * Y2	– 0.1748	0.4561	– 0.383	0.710
Y3 * Y3	0.7728	0.4726	1.635	0.133
Y1 * Y2	0.5893	0.5072	1.162	0.272
Y1 * Y3	2.2453	0.8373	2.682	0.023
Y2 * Y3	0.5984	0.5609	1.067	0.311

It is predicted from ANOVA analysis as well as interaction effect graphs of selected process parameters that it is important to correlate the effect of pressure, temperature and extruder speed on production cycle time. In this regards the individual main effect of every significant factor need to explain. Moreover, their interaction is also required to confer using 3D response surface graph (Cochran 1957).

From individual main effect graph, it was inferred that when extruder pressure, die temperature and extruder speed are investigated for their effects on production time, interactions of extruder pressure and extruder speed has significant effect on the response (production time) of blow molding process. From the main effect plot of production time versus different process parameters such as extruder pressure, die temperature and extruder speed are shown in Figs. 1, 2 and 3 respectively. It is evident from the main effect graphs shown the extruder speed significantly affects the response of the process as compared to extruder pressure and die temperature.

The 3D surface plots are also drawn to establish the relationship between production time as dependent variable and combine effect of two out of three independent variables. The plot in Fig. 4 depicts the relation between the production time and two process parameters viz. pressure and temperature. It indicates that the production time has significantly more at low pressure applied during the process when consider at low temperature, but it showed much letter production time at low pressure and high temperature. In contrast the production time is less at low temperature and slowly

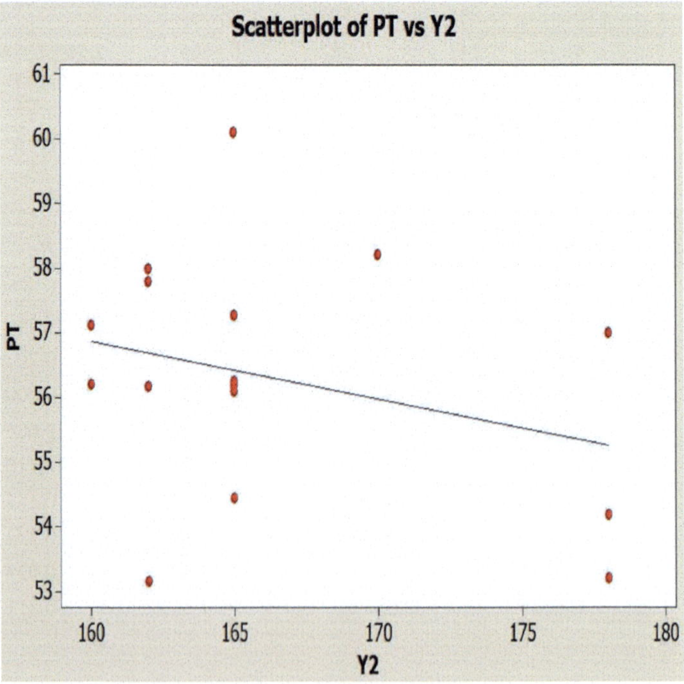

Fig. 2 Main effect graph of production time versus temperature

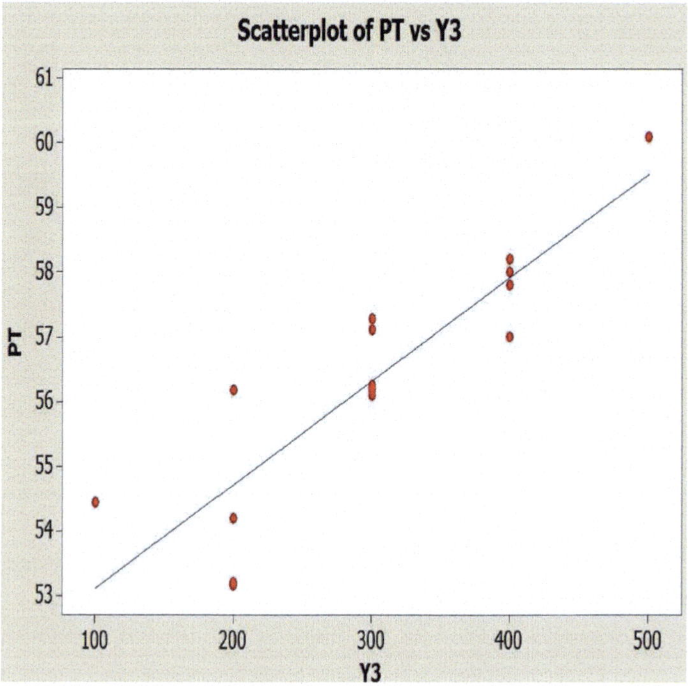

Fig. 3 Main effect graph of production time versus extruder speed

increases with the increase in temperature. It is attributed to the working condition implied by these parameters while extruding the plastic material. At low pressure the extrusion of plastic takes more time increasing overall production time whereas with increasing in pressure the extrusion will be faster requiring lesser production time. In the same way at moderate temperature the extruded product will take lesser time for solidification lowering production cycle time (Chaudhari and Ingle 2019).

A shown in Fig. 5 when investigated in combination with extruder speed the temperature does have much impact on production time. The required production time has drastically increased with increase in extruder speed at all temperature. This is also due to low rate of solidification of the product at these temperature and extruder speed.

A shown in Fig. 6 when investigated in combination with extruder speed the pressure the production time has shown decrease with increase in pressure at low extruder speed whereas drastic increase in production time is observed at higher pressure and higher extruder speed. Moreover, the production time has shown abrupt increase with increasing temperature at high pressure.

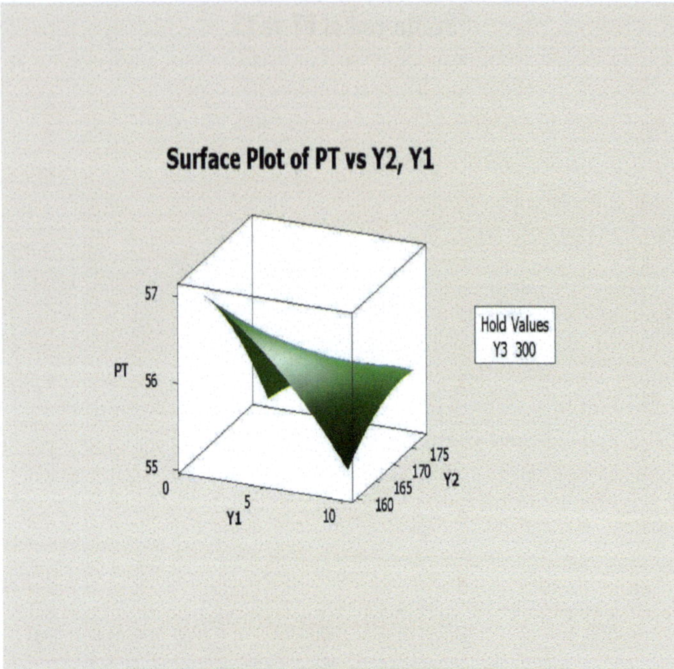

Fig. 4 Surface plot of production time versus pressure and temperature

6 Future Scope

The future scope of using response surface methodology (RSM) to experimentally investigate the effects of process parameters on the extrusion blow molding process within the Industry 4.0 framework is highly promising. This approach will enable the integration of advanced data analytics, IoT sensors, and machine learning algorithms to optimize and automate the process in real-time, enhancing product quality and operational efficiency. By leveraging big data and predictive analytics, manufacturers can achieve precise control over process parameters, leading to reduced waste and energy consumption. Additionally, incorporating sustainable materials and practices, along with advanced quality control techniques and human–machine collaboration, will further drive innovation and adaptability in the extrusion blow molding industry, ensuring it meets the evolving demands for customization and environmental responsibility.

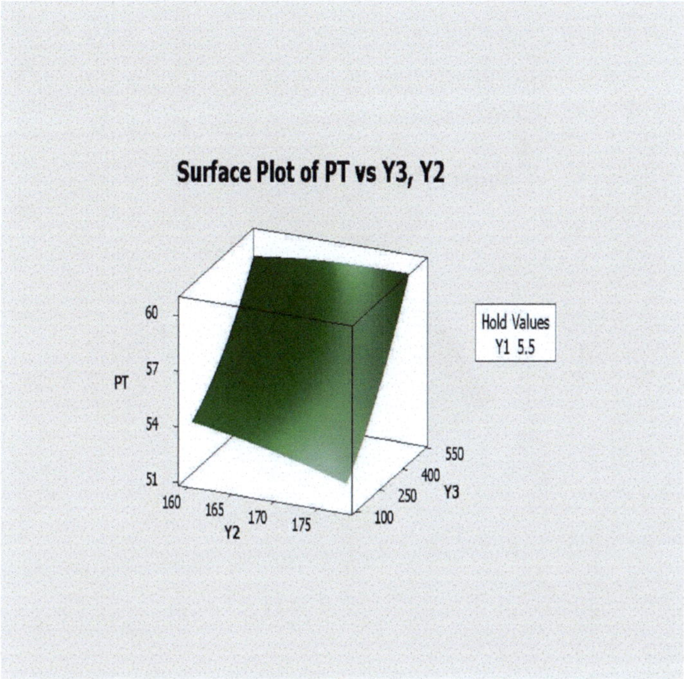

Fig. 5 Surface plot of production time versus temperature and extruder speed

7 Conclusions

The CCD is found successful in studying the main and interaction effect of process parameters pressure, die temperature and extruder peed on production time extrusion blow molding process used to form 20-L bottle made of LPDE material. As per the ANOVA analysis extruder pressure and speed are revealed as most significant parameters affecting the overall production time of the process. The individual main effect graph exhibited that extruder pressure and speed have significant influence the overall production time. As per the 3D surface plot of the production time is needed is high at low pressure and low temperature whereas it is very low at higher temperature at same pressure. At all temperature values the drastic increase in production time is seen with increase in extruder speed. The maximum production time is required at high pressure and high extruder speed when the effect of parameters was studied together.

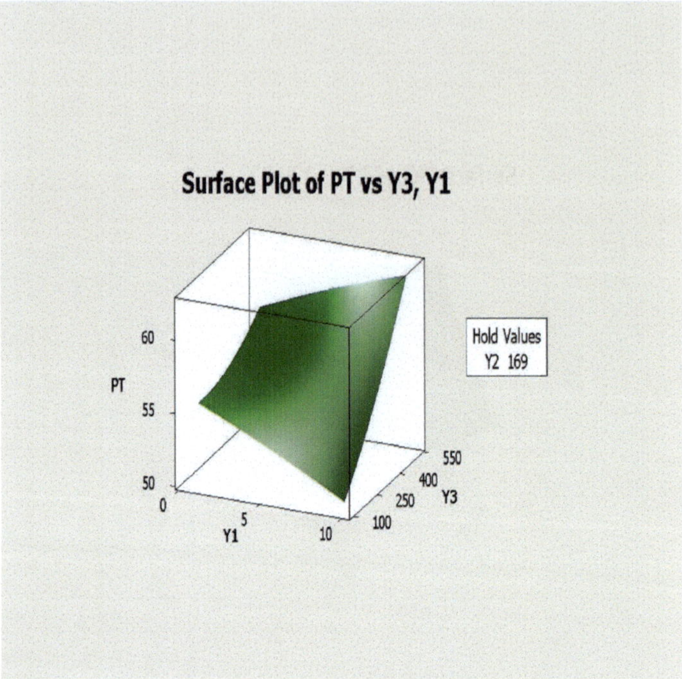

Fig. 6 Surface plot of production time versus pressure and extruder speed

References

Attar A, Bhuiyan N, Thomson V (2008) Manufacturing in blow molding: time reduction and part quality improvement. J Mater Process Technol 204(1–3):284–289. https://doi.org/10.1016/j.jmatprotec.2007.11.040

Belcher SL (2011) Blow molding. In: Applied plastics engineering handbook. William Andrew Publishing, pp 267–288. https://doi.org/10.1016/B978-1-4377-3514-7.10016-9

Bendada A, Erchiqui F, Kipping A (2005) Understanding heat transfer mechanisms during the cooling phase of blow molding using infrared thermography. NDT & E Int 38(6):433–441. https://doi.org/10.1016/j.ndteint.2004.11.007

Chaudhari R, Ingle A (2019) Experimental investigation of dissimilar metal weld of SA335 P11 and SA312 TP304 formed by gas tungsten arc welding (GTAW). Trans Indian Inst Met 72(5):1145–1152. https://doi.org/10.1007/s12666-019-01587-2

Cochran WG (1957) Analysis of covariance: its nature and uses. Biometrics 13(3):261–281. https://doi.org/10.2307/2527916

Cooper TA (2013) Developments in plastic materials and recycling systems for packaging food, beverages and other fast-moving consumer goods. In: Trends in packaging of food, beverages and other fast-moving consumer goods (FMCG), pp 58–107 https://doi.org/10.1533/9780857098979.58

Diler EA, Ipek R (2012) An experimental and statistical study of interaction effects of matrix particle size, reinforcement particle size and volume fraction on the flexural strength of Al–SiCp composites by P/M using central composite design. Mater Sci Eng A 548:43–55. https://doi.org/10.1016/j.msea.2012.03.066

Kumar A, Gulati V (2018) Experimental investigations and optimization of forming force in incremental sheet forming. Sādhanā 43:1–15

Kumar A, Gulati V (2019) Experimental investigation and optimization of surface roughness in negative incremental forming. Measurement 131:419–430

Kumar A, Gulati V (2021) Optimization and investigation of process parameters in single point incremental forming. Indian J Eng Mater Sci (IJEMS) 27(2):246–255

Kumar K, Kumar A, Singh V (2019) Optimization of process parameters for erosion wear in slurry pipeline. In: Advances in engineering design: select proceedings of FLAME 2018. Springer Singapore, pp 131–140

Kumar A, Kumar D, Kumar P, Dhawan V (2020) Optimization of incremental sheet forming process using artificial intelligence-based techniques. In: Nature-inspired optimization in advanced manufacturing processes and systems. CRC Press, pp 113–130

Kumar A, Mittal RK, Haleem A (eds) (2023) Advances in additive manufacturing artificial intelligence, nature-inspired, and biomanufacturing. Elsevier. https://doi.org/10.1016/C2020-0-03877-6

Kumar A, Shrivastava VK, Kumar P, Kumar A, Gulati V (2024a) Predictive and experimental analysis of forces in die-less forming using artificial intelligence techniques. Proc Inst Mech Eng Part E J Process Mech Eng. https://doi.org/10.1177/09544089241235473

Kumar A, Kumar P, Sharma N, Srivastava AK (eds) (2024b) 3D printing technologies: digital manufacturing, artificial intelligence, industry 4.0. Walter de Gruyter GmbH & Co KG. https://doi.org/10.1515/9783111215112

Montgomery DC, Runger GC (2020) Applied statistics and probability for engineers. Wiley. https://scholar.google.com/scholar?hl=en&as_sdt=0%2C5&q=D.+C.+Montgomery%2C+Design+and+Analysis+of+Experiments%2C+8th+Editio.+John+Wiley+%26+Sons%2C+Inc.%2C+2020.&btnG

Raheem D (2013) Application of plastics and paper as food packaging materials—an overview. Emir J Food Agric 177–188. https://doi.org/10.9755/ejfa.v25i3.11509

Singh H, AlMangour B (eds) (2023) Handbook of smart manufacturing: forecasting the future of industry 4.0. CRC Press. https://doi.org/10.1201/9781003333760

Singh N, Hui D, Singh R, Ahuja IPS, Feo L, Fraternali F (2017) Recycling of plastic solid waste: a state of art review and future applications. Compos Part B Eng 115:409–422. https://doi.org/10.1016/j.compositesb.2016.09.013

Suvanjumrat C, Ploysook N, Rugsaj R (2018) Extrusion blow molding simulation using cross-section technique for complex shape bottles. Eng J 22(2):169–183. https://doi.org/10.4186/ej.2018.22.2.169

Yu JC, Chen XX, Hung TR, Thibault F (2004) Optimization of extrusion blow molding processes using soft computing and Taguchi's method. J Intell Manuf 15:625–634. https://doi.org/10.1023/B:JIMS.0000037712.33636.41

Finite Element Analysis and Experimental Investigation of 3D Printed Biomimetic Structures

D. L. Belgin Paul, Ayyappan Susila Praveen, and V. Loganathan

Abstract Biomimetic structures are widely used in the fields of medicine, automobiles, military equipment, and aerospace due to their high mechanical and thermal properties. It also has good energy absorption and damping capabilities. Fabrication of these biomimetic structures using conventional manufacturing techniques is very challenging because of their complex lattice architecture. Therefore, additive manufacturing (AM) is a reliable and advanced method for the fabrication of biomimetic structures by layer-by-layer deposition of the materials. In this work, four biomimetic structures like cactus, marsh horsetail, equisetum arvense, and spider web were designed as multilayered scaffolds using Fusion 360 software. The finite element analysis was done to evaluate the compression stress distribution and deformation of the modelled scaffolds using a Fusion 360 simulation. The results obtained in the various samples were interpreted with different lattice designs. The best biomimetic structure among the four has been identified and ordered as marsh horsetail > equisetum arvense > spider web > cactus for the manufacturing of scaffolds for biomedical applications.

Keywords Biomimetic design · Finite element analysis · Model validation · 3D printing · Experimental validation

1 Introduction

Nature has inspired mankind in many ways throughout his journey to innovation. The bio-cellular structural designs available in nature have great potential for structural applications. Biomimetic structures are ones that have been mimicked or copied

D. L. B. Paul (✉) · A. S. Praveen
Department of Mechanical Engineering, Vel Tech Rangarajan Dr. Sagunthala R&D Institute of Science and Technology, Chennai, Tamil Nadu 600062, India
e-mail: dlbelginpaul@gmail.com

V. Loganathan
Department of Mechanical Engineering, Rohini College of Engineering and Technology, Kanyakumari, Tamil Nadu 629401, India

A. Kumar et al. (eds.), *Industry 4.0 Driven Manufacturing Technologies*, Springer Series in Advanced Manufacturing, https://doi.org/10.1007/978-3-031-68271-1_4

from cellular structures present in nature. Examples of a few cellular structures are honeycombs, leaves and stems of plants or trees, bones of animals and humans, etc. These cellular structures are chosen for many applications in different fields. Mainly, cellular structures are preferred for energy-absorbing and strength-demanding applications (Xue and Hutchinson 2006; Wadley et al. 2007; Chen et al. 2008; Ajdari et al. 2008, 2011). In recent decades, these cellular structures have been very much admired and used by many researchers in advanced manufacturing processes like 3D printing (Prasanth et al. 2021; Kucewicz et al. 2018). Biomimetic structures have high mechanical and thermal properties (Chen et al. 2020). It also provides the engineers with simple designs with high strength. Biomimetic structure-based parts are engineered and used in many fields, such as automotive sectors (Furukawa et al. 2020; Syu et al. 2020), construction (Liu et al. 2019a), military (Li et al. 2021), medical (Zhang et al. 2020a), etc. In the biomedical field, scaffolds are meant to provide the strength and shape for the cells to grow during proliferation. These are the extracellular matrix (ECM) in their native environment (Oxford et al. 2019). They offer a good strength environment for the cells, tensile strength while stretching, toughness, and elastic properties to the tissues. It also provides recreation and remodelling of its structure during pathological effects (Schönherr and Hausser 2000; Muschler et al. 2004; Yim et al. 2005; Hersel et al. 2003; Chew et al. 2008). Polymers are one of the materials used for manufacturing synthetic biomaterial-based scaffolds. Polylactic acid (PLA) is a biodegradable material that is used to develop scaffolds out of it for implants in medical applications. A detailed review of PLA is given in these articles. The readers can look into it for detailed knowledge about PLA material in the biomedical industry (Farah et al. 2016; Silva et al. 2018; Bergström and Hayman 2016).

The modelling of these biomimetic designs is done using various techniques. The cross-section of the biomimetic designs is taken, from which the desired scaffold structures can be made using numerous CAD programs. The other method is to use a 3D scanning technique to fetch the shape and dimension needed to recreate the structural design. Though the designs are very complex, it is very difficult to manufacture the biomimetic structures using conventional manufacturing techniques. Smart manufacturing has found its way to emerge in manufacturing complex structures through advanced techniques such as digital, artificial intelligence, and machine learning (Srivastava et al. 2023; Mittal et al. 2022; Kumar et al. 2024; Singh and Al Mangour 2023). Additive manufacturing (AM) has been developed in recent decades and is prominent and efficient in the manufacturing of these typical lattice-structured biomimetic designs. AM is the process of fabricating three-dimensional things by depositing material layer by layer, following a digital model (Kumar et al. 2023a). This method is capable of accommodating a diverse array of materials, such as polymers, metals, ceramics, and composites (Kumar et al. 2023b). The selection of these materials is determined by the specific qualities desired for the end result. The process of creating biomimetic structures, which imitate natural systems, begins with preprocessing. This involves utilizing computer-aided design (CAD) tools to build intricate geometries and then optimizing the model to ensure accuracy and mechanical efficiency. During the process of printing, materials are meticulously placed layer by layer in order to achieve the intended structure. Post-processing processes are essential for improving the quality of the final product. The processes involved in the process may encompass curing (for polymers), sintering or annealing (for metals and ceramics), as well as different surface treatments such as polishing or coating. These post-processing methods enhance the mechanical strength, longevity, and surface quality of the printed biomimetic structures, guaranteeing their functionality, resilience, and adherence to the necessary standards (Kumar et al. 2023c, 2023d). There are many AM techniques available for the fabrication of scaffolds and other biomedical-related structures, such as fused deposition modelling (FDM), stereolithography (SLA), direct ink writing (DIW), robocasting, selective laser sintering (SLS), and much more. Manufacturing with these processes has its own advantages and limitations (Mota et al. 2015). Finite element analysis (FEA) is a numerical method of analysing and simulating any designed model. Many loading conditions can be used for simulations, like structural, thermal, and fluidic problems (Niu et al. 2017; Beyer and Figueroa 2016; Kladovasilakis et al. 2020; Lei et al. 2019). FEA is very useful in simulating the prototypes of any model multiple times with various loading conditions before fabricating the actual product. It also helps the engineers make design changes, make appropriate design modifications, and conduct physical testing of the fabricated product (Zhu et al. 2021).

Raffaella et al. used the FEA model for analysing the orthotropic distribution of a human bone. The porous bone structures are modelled from CT scan images, and the FEA analysis helps to design new prosthetics for the patients (Raffaella et al. 2016). Alemayehu et al. (2024) developed a biomimetic voronoi porous structure mimicking trabecular bone using nTOP and creo softeares. FEA was utilised for evaluating the stress, strain, and deformation of the model during various loading conditions. They changed the porous lattice of the voronoi models and assessed the change in mechanical behaviour, which matched the actual bone (Kumar et al. 2023d). Wang et al. have used a human tibia as a mimetic design to develop a novel crash box. The response surface methodology was adopted to find the optimal design. The results show that the optimal design has enhanced energy absorption characteristics (Wang et al. 2018). Khalil et al. developed an innovative approach to biomimetic design that includes L-systems generated and distributed along their principal stress lines. L9 experimental runs were performed to analyse the numerical simulation of the designs. The optimal design was seen to have better specific strength by reducing the weight of the structure (Al Khalil et al. 2022). Table 1 gives a summary of a few pieces of literature that use biomimetic design and finite element analysis.

There are many software programmes available on the market to do this mathematical simulation mputerized. One of those is Fusion 360 software, which helps to model and simulate the design using the same software. In this work, four biomimetic structures, such as cactus, marsh horsetail, equisetum arvense, and spider web, were modelled and simulated using Fusion 360 software. Cactus is normally found in deserts and areas where there is drought. It comes under the family Cactaceae. It occurs in many different shapes and sizes. An octagonal cactus with a honeycomb infill structure is used for this study. Marsh horstail is a small grassy plant that is abundantly found in nature. Its structure is mainly adapted to absorb more water.

Table 1 Summary of literatures

Technique	Material	Application	References
Material jetting process	Vero white resin	Mechanical structures	Al Khalil et al. (2022)
Extrusion-based	GelMA/nHA hydrogels	Repair of osteochondral defects	Liu et al. (2019b)
Selective laser melting	Ti6Al4V alloy	Porous implants	Zhang et al. (2018)
Fused deposition modelling	R4600 Resin	Structural applications	Liu et al. (2023)
3D printing	Hyper-elastic resin	Energy absorption applications	Vafaeefar et al. (2024)

Equisetum arvense is another inspiring tree leaf available in nature. This is also called field horsetail, which is abundantly found in arctic regions. The final one is the spider web structure, inspired by the spider silk web extruded by the spiders to attract and catch insects. Though there are many FEA studies carried out for structural applications, the adaptability of biomimetic structures has not been fully explored for biomedical applications. In this work, all the above-mentioned 3D printable and biopolymer-based four biomimetic structures were modelled and simulated numerically with the help of Fusion 360 software and compared with each other to identify the best structure for making scaffolds out of them for biomedical applications.

2 Materials and Method

3D printable PLA in wire form was used to simulate and fabricate the structure using fused deposition modelling. Table 2 provides the detailed material properties used in the current study. Figure 1 shows the methodology used in this study for selecting the featured biomimetic design among the four. The modelling of cellular structures is done using Fusion 360 software and the. stl file was fed into the computer to generate G-codes for controlling the 3D printer coordinates.

Table 2 Properties of PLA material

Properties	Value
Chemical formula	$(C_3H_4O_2)n$
Filament diameter	1.75 mm
Melting point	157–170 °C
Young's modulus	3500 GPa
Density	1.240 g/cm^3

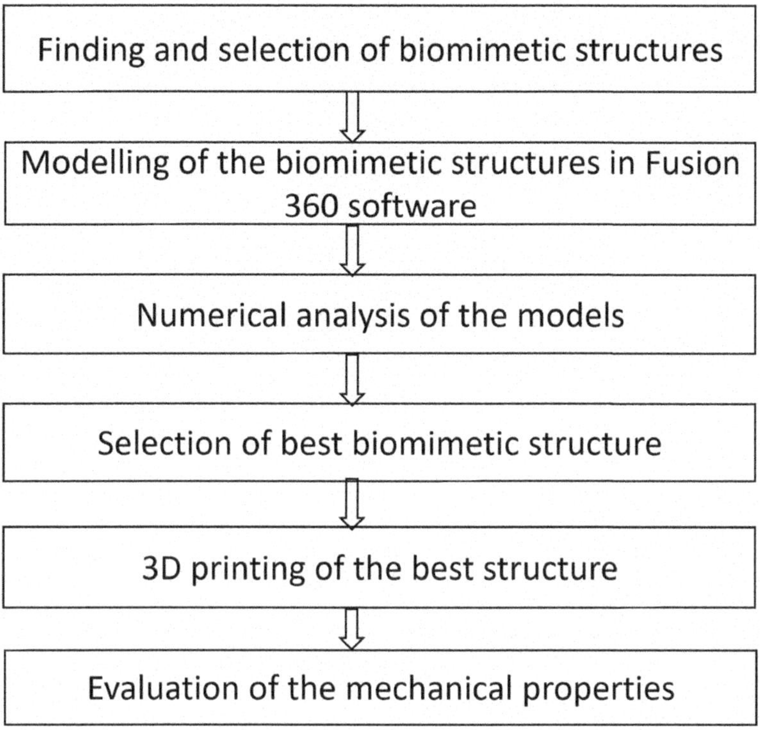

Fig. 1 Methodology adopted in the study

2.1 Modeling the Biomimetic Structures

The cross sections of the identified four biomimetic structures were taken, and the CAD model was created using Fusion 360 software. These structures are then analyzed for errors at edges during segmentation and meshing. The models were designed with a diameter of 60 mm and a height of 60 mm to maintain uniformity among the four designs. The file is then saved in STL format for further 3D printing. Figure 2 shows the four biomimetic structures designed using Fusion 360.

2.2 FEA Modeling and 3D Printing

Numerical analysis using the finite element method is one of the finest phenomena for structural strength analysis. Fusion 360 software was also used for the simulation of the structures. The boundary conditions applied are that no pre-stress is applied to the explicit dynamics. Also, gravity is disabled, and the load is applied at the top face of the model by arresting the bottom face movement against the ground (Zhang

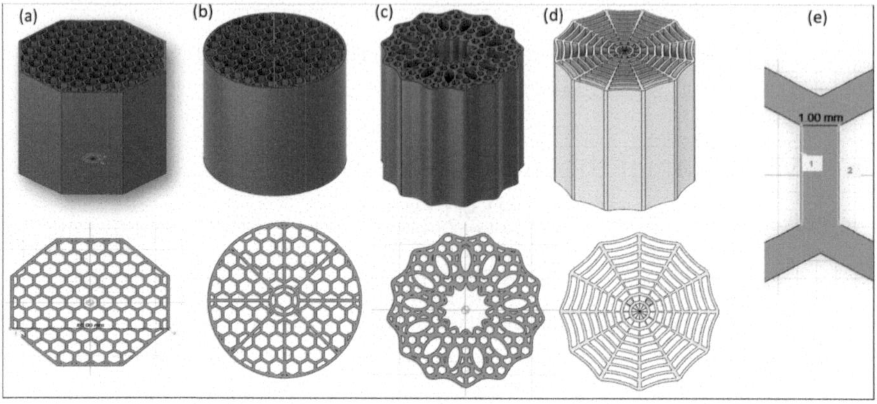

Fig. 2 Biomimetic topologies modeled using Fusion 360 **a** Cactus **b** Marsh horsetail, **c** Equisetum arvense and **d** Spider web **e** Line width dimension

Table 3 Mesh conditions

Properties	Value
Model size	10%
Element order	Parabolic
Min. element size % of average size	20
Mesh quality	Fine
Gravity	Disabled
No. of elements	92,228
No. of nodes	164,607

et al. 2020b). The discretization of the model is done by triangular elements, and a continuous model is assumed when meshing. Table 3 shows the mesh conditions used in the present FEA analysis. The average number of elements and nodes taken during the meshing and analysis of the structures is 92228 and 164,607, respectively.

3D printing of the best biomimetic structure was done using a FDM 3D printer from Fabforge. Initially, the model from the CAD design is exported as a.STL file and imported into software called Cura. From there, the model is sliced, and the G codes are generated for layer-by-layer deposition of the material to get the finished specimen. The finished part is then post-processed by cleaning and taken for further testing. An illustration of the 3D printing process is given in Fig. 3.

2.3 *Compression Test*

A Universal Testing Machine (WDW-100) was used to perform experimentation to find the compression strength of the selected 3D-printed scaffold. The dimensions

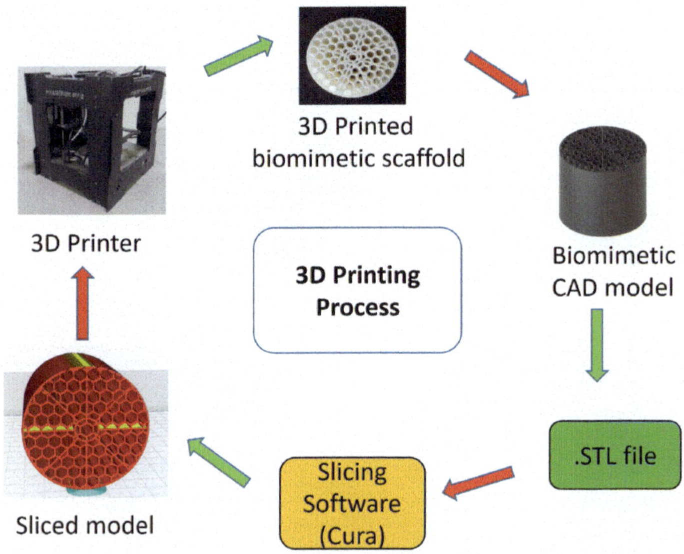

Fig. 3 Schematic of 3D printing process

of the samples for compressive strength are prepared as the same as the sample size used for numerical simulation (i.e., 60*60 mm).

3 Results and Discussion

The four biomimetic structures, such as cactus, marsh horsetail, equisetum arvense, and spider web, were imported into the Fusion 360 simulation workbench and discretized into a number of triangular nodes to form a mesh. The reported highest numbers of nodes and elements are 164607 and 92228, respectively. The meshing was done by setting the mesh quality to 'fine'. Figure 4 shows the meshed surfaces of all four biomimetic designs. The FEA analysis was mainly focused on the Von-Mies stress distribution, total deformation, and factor of safety of the analysed biomimetic structures with respect to the applied load.

3.1 FEA Analysis of Biomimetic Models

Distinct mechanical behaviour was found while investigating the results of FEA. Figure 5 illustrates the stress distribution when the biomimetic structures are loaded between 1 and 5 kN. Notably lower stress levels are seen in cactus and marsh horsetail when loaded between 1 and 5 kN, with values ranging from 1.545 to 7.724 MPa.

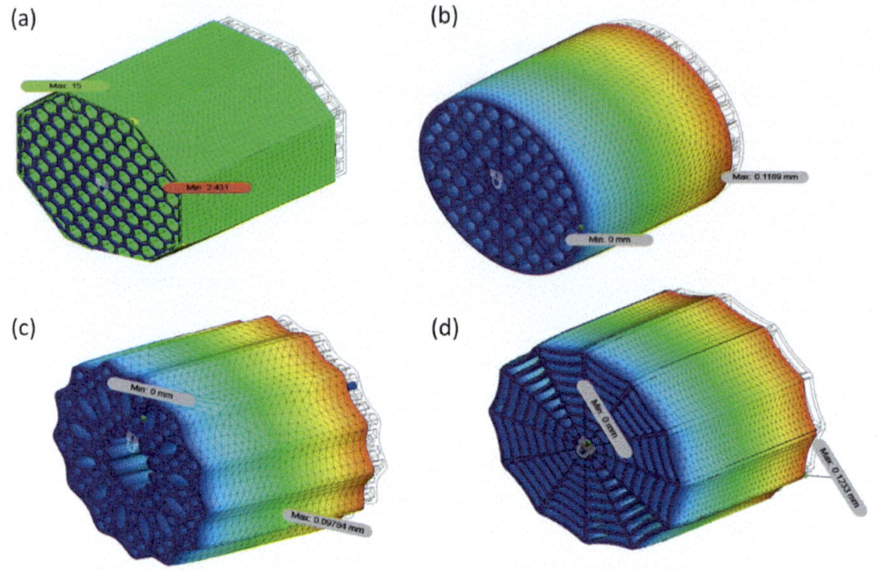

Fig. 4 Discrete models of the biomimetic structures: **a** Cactus **b** Marsh horsetail, **c** Equisetum arvense and **d** Spider web

Whereas, the equisetum arvense exhibited a stress level of 2.014 to 10.88 MPa, and the spider web biomimetic structure exhibited a stress value ranging from 2.175 to 10.07 MPa. Figure 6 shows the stress-strain curve of the modelled biomimetic structures. These variations in stress distribution attribute to the importance of naturally occurring biomimetic design for engineering applications. Moving forward Figure 7 shows the FEA results for displacement due to compressive loads, and a graph was plotted between applied load and displacement as shown in Figure 8. All the designs were loaded uniformly between 1 and 5 kN, and the displacements of the corresponding structures were noted. It was seen that the cactus structure deformed from values ranging from approximately 0.02546–0.1273 mm, and the marsh horsetail displayed 0.02339–0.1169 mm. These two biomimetic structures provided comparable deformations under the same loading conditions. The equisetum arvense displayed a deformation range of 0.01957–0.09784 mm, which indicated a consistent structural response with moderate deformation. A resilient structural deformation was observed in the spider web structure, with a displacement range of 0.02466–0.1233 mm. The relatively lower deformation occurred in the equisetum arvense design structure.

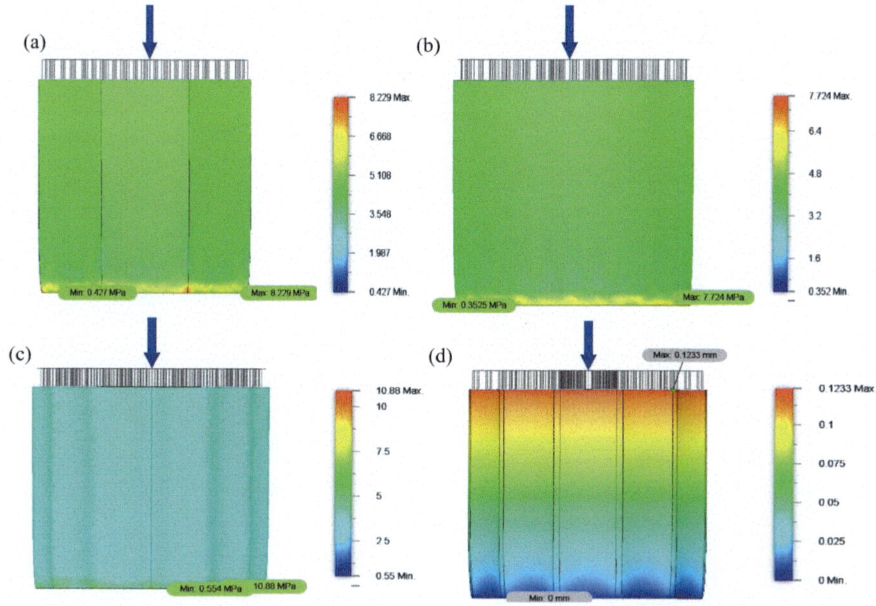

Fig. 5 Stress distribution on the biomimetic structures: **a** Cactus **b** Marsh horsetail, **c** Equisetum arvense and **d** Spider web

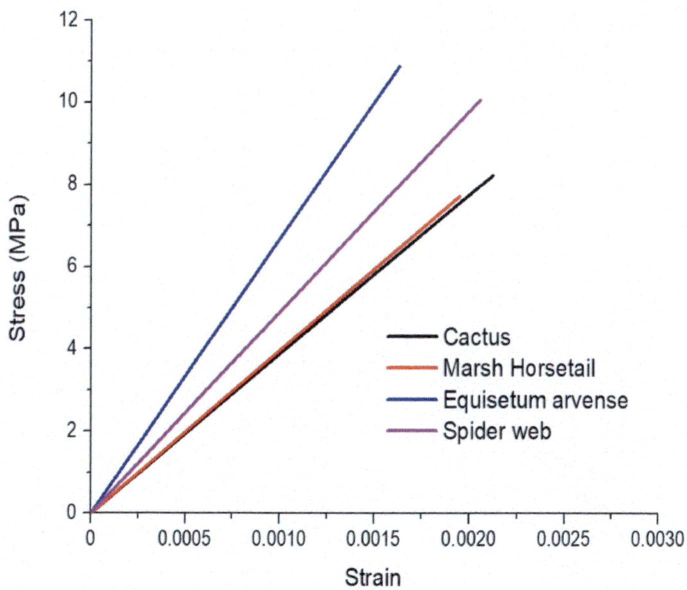

Fig. 6 Stress vs. strain from the simulations

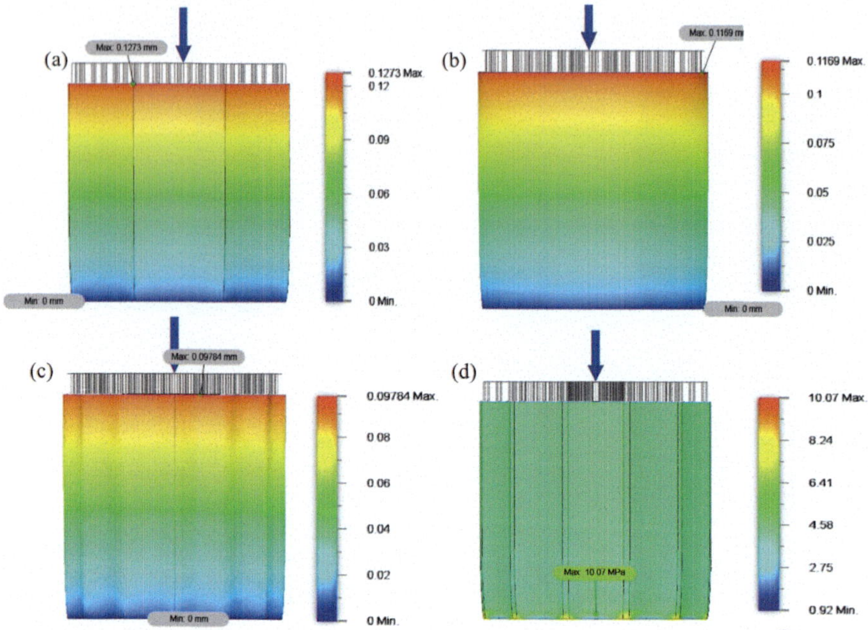

Fig. 7 Deformation on the biomimetic structures: **a** Cactus **b** Marsh horsetail, **c** Equisetum arvense and **d** Spider web

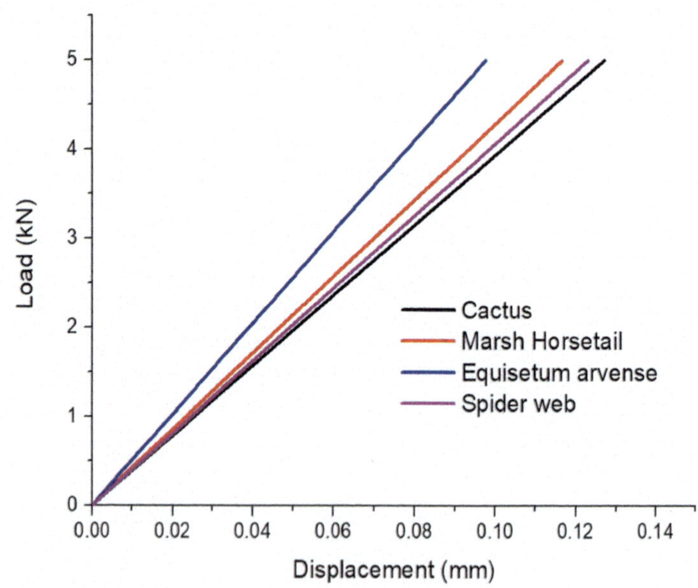

Fig. 8 Load versus displacement from the simulations results

3.2 Selection of the Best Biomimetic Design

The summary of the simulation results is given in Table 4. In the process of selecting the best biomimetic structure from the four various designs, an additional criteria was included, which is known as the Factor of Safety (FOS). Figure 9 shows the FOS values of the biomimetic structures when loaded at 5 kN.

It is a critical parameter that represents the margin of safety between the maximum stress and the normal stress distribution. Among the four biomimetic structures, the highest FOS was seen in the marsh horsetail structure at 2.589, whereas the lowest value was seen in the equisetum arvense. Cactus and spider web structures follow rank 3 and 4 with FOS values of 2.431 and 1.987, respectively. Comparing the results from Figs. 6, 7, and 8, it was observed that the stress versus load graph shows that Equisetum arvense has more stress with an increase in load compared to other biomimetic structures. The displacement versus load graph also shows equisetum arvense has less displacement compared with other biomimetic structures, but the factor of safety for equisetum arvense at 5 kN load is very marginal (1.839). Therefore, ideally, the Marsh horsetail structure is more suitable for structural application

Table 4 Summary of the simulation results

Biomimetic structure	Load (kN)	Displacement (mm)	Stress (MPa)	Strain	FOS
Cactus	1	0.02546	1.646	0.00042	12.15
	2	0.05092	3.291	0.00085	6.077
	3	0.07638	4.937	0.00127	4.051
	4	0.1018	6.583	0.00170	3.038
	5	0.1273	8.229	0.00212	2.431
Marsh Horsetail	1	0.02339	1.545	0.00039	12.95
	2	0.04677	3.089	0.00078	6.474
	3	0.07015	4.634	0.001169	4.316
	4	0.09354	6.179	0.001559	3.237
	5	0.1169	7.724	0.001948	2.589
Equisetum arvense	1	0.01957	2.175	0.000326	9.1915
	2	0.03914	4.351	0.000652	4.597
	3	0.0587	6.526	0.000978	3.065
	4	0.07827	8.702	0.001305	2.298
	5	0.09784	10.88	0.001631	1.839
Spider web	1	0.02466	2.014	0.000411	9.93
	2	0.04932	4.027	0.000822	4.967
	3	0.07398	6.04	0.001233	3.311
	4	0.09864	8.054	0.001644	2.483
	5	0.1233	10.07	0.002055	1.987

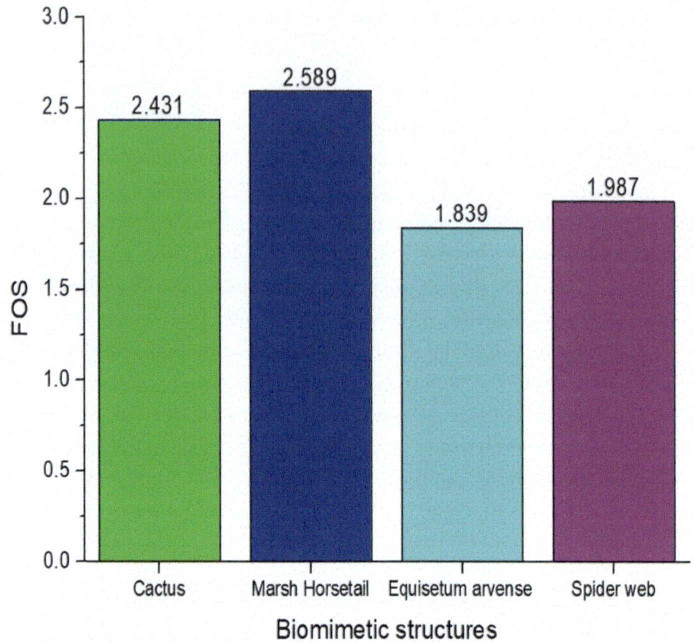

Fig. 9 Factor of safety for 5 kN load

than other structures because of its high FOS (2.589) at 5 kN load and because the displacement is very less next to the Equisetum arvense structure.

3.3 3D Printing of the Marsh Horsetail Structure

The selected Marsh horsetail biomimetic structure was 3D printed using the fused deposition modelling method. The top and front views of the printed samples are shown in Figure 10. The sample is also subjected to a compression test using a universal testing machine. The results show that the structure's simulation results have better agreement with those of the 3D-printed structure. The maximum displacement the structure could withstand before actual porous structure collapse was recorded to be ~ 0.125 mm at 500 N, as shown in Figure 11.

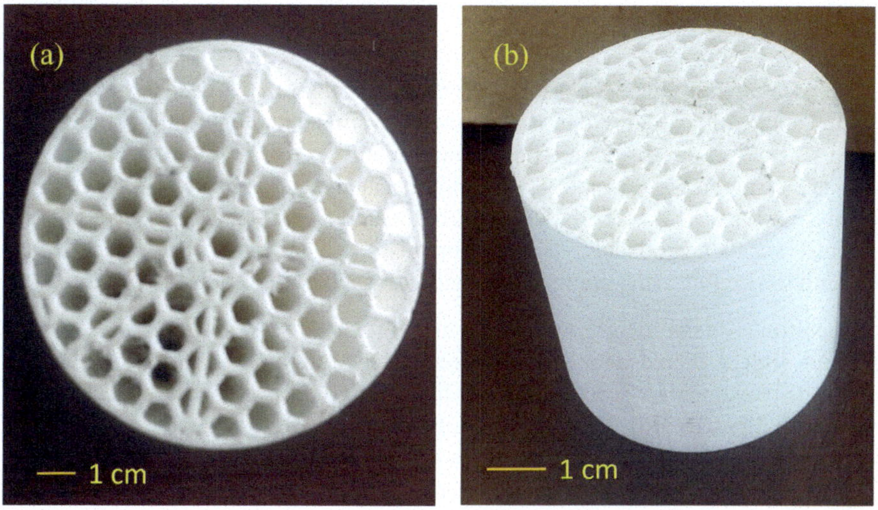

Fig. 10. 3D printed marsh horsetail biomimetic structure **a** top view **b** front view

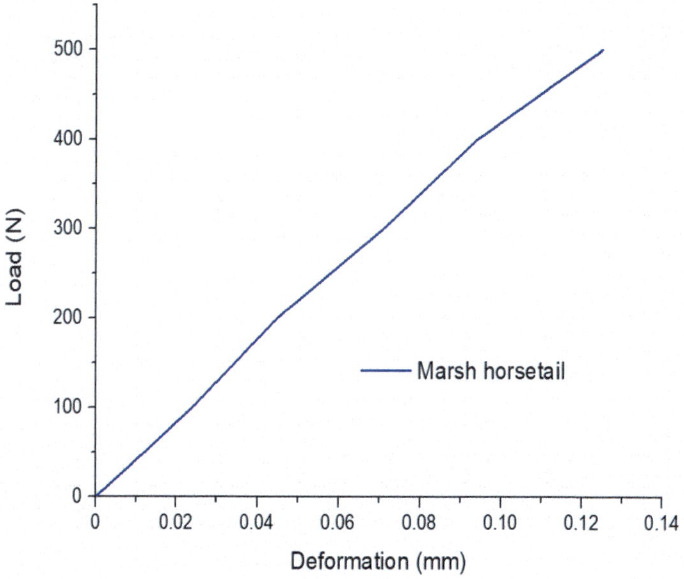

Fig. 11 Compression behavior of Marsh horsetail structure

4 Conclusion

The present study underscores the importance of biomimetic design structures in various applications, particularly in the field of biomedical applications for developing scaffolds. Four different types of biomimetic designs available in nature were recreated using Fusion 360 software. Numerical analysis was carried out to select the best design for scaffolds for tissue regeneration and bone applications. The selected structure was then validated by 3D printing the design and evaluating its mechanical properties. The following are the conclusions arrived at from the study:

- The stress versus load graph shows that Equisetum arvense experiences more stress with an increase in load compared to other biomimetic structures.
- The displacement vs. load graph also shows Equisetum arvense has less displacement compared with other biomimetic structures. But the factor of safety for Equisetum arvense at a 5 kN load was very marginal (1.839).
- Therefore, ideally, the Marsh horsetail structure is more suitable for scaffolds and structural applications than other biomimetic designs because of its high factor of safety (2.589) and minimal displacement at the same load as the Equisetum arvense structure.
- The 3D-printed Marsh horsetail structure also performed in agreement with the numerical simulation when tested under compression load. Therefore, we conclude that the Marsh horsetail biomimetic structure holds greater potential for practical implementation in scaffolds and in other structural applications.

Future scope of the work

The future scope of this work involves the integration of smart materials and adaptable structures. It can enhance performance by providing responsive behavior to physiological circumstances. Examining the biocompatibility and enduring stability of these structures is essential for their effective use in tissue engineering and regenerative medicine. Engaging in collaboration with biologists and medical researchers will yield more profound understanding of how to imitate intricate biological processes and guarantee their effectiveness within living organisms. Furthermore, by utilizing sophisticated manufacturing methods such as multi-material and nano-scale 3D printing, the accuracy and efficiency of biomimetic structures may be greatly improved. This, in turn, creates opportunities for personalized medicine and better treatment solutions.

Declaration of Competing Interest The authors of this article affirm that they do not possess any competing financial interests or personal relationships that might have influenced the findings presented.

References

Ajdari A, Nayeb-Hashemi H, Vaziri A (2011) Dynamic crushing and energy absorption of regular, irregular and functionally graded cellular structures. Int J Solids Struct 48(3–4):506–516. https://doi.org/10.1016/j.ijsolstr.2010.10.018

Ajdari A, Nayeb-Hashemi H, Canavan P, Warner G (2008) Effect of defects on elastic–plastic behavior of cellular materials. Mater Sci Eng Struct Mater Prop Microstruct Process 487(1–2):558–567. https://doi.org/10.1016/j.msea.2007.10.050

Al Khalil M, Belkebir H, Lebaal N, Demoly F, Roth S (2022) A biomimetic design method for 3D-printed lightweight structures using L-systems and parametric optimization. Appl Sci (basel, Switz) 12(11):5530. https://doi.org/10.3390/app12115530

Alemayehu DB, Todoh M, Huang S-J (2024) Advancing 3D dental implant finite element analysis: Incorporating biomimetic trabecular bone with varied pore sizes in Voronoi lattices. J Funct Biomater 15(4):94. https://doi.org/10.3390/jfb15040094

Bergström JS, Hayman D (2016) An overview of mechanical properties and material modeling of polylactide (PLA) for medical applications. Ann Biomed Eng 44(2):330–340. https://doi.org/10.1007/s10439-015-1455-8

Beyer C, Figueroa D (2016) Design and analysis of lattice structures for additive manufacturing. J Manufact Sci Eng 138(12). https://doi.org/10.1115/1.4033957

Chen P, Lin A, Lin Y, Seki Y, Stokes A, Peyras J, Olevsky E, Meyers M, Mckittrick J (2008) Structure and mechanical properties of selected biological materials. J Mech Behav Biomed Mater 1(3):208–226. https://doi.org/10.1016/j.jmbbm.2008.02.003

Chen G, Chukwunenye N, Jones GF, Li CH (2020) Biomimetic structures by leaf vein growth mechanism for pool boiling heat transfer enhancements. Int J Heat Mass Transf 155(119699):119699. https://doi.org/10.1016/j.ijheatmasstransfer.2020.119699

Chew SY, Mi R, Hoke A, Leong KW (2008) The effect of the alignment of electrospun fibrous scaffolds on Schwann cell maturation. Biomaterials 29(6):653–661. https://doi.org/10.1016/j.biomaterials.2007.10.025

da Silva D, Kaduri M, Poley M, Adir O, Krinsky N, Shainsky-Roitman J, Schroeder A (2018) Biocompatibility, biodegradation and excretion of polylactic acid (PLA) in medical implants and theranostic systems. Chem Eng J (Lausanne, Switzerland: 1996) 340:9–14. https://doi.org/10.1016/j.cej.2018.01.010

Farah S, Anderson DG, Langer R (2016) Physical and mechanical properties of PLA, and their functions in widespread applications—a comprehensive review. Adv Drug Deliv Rev 107:367–392. https://doi.org/10.1016/j.addr.2016.06.012

Furukawa K, Ochiai M, Hashimoto H, Kotani S (2020) Bearing characteristic of journal bearing applied biomimetics. Tribol Int 150(106345):106345. https://doi.org/10.1016/j.triboint.2020.106345

Hersel U, Dahmen C, Kessler H (2003) RGD modified polymers: biomaterials for stimulated cell adhesion and beyond. Biomaterials 24(24):4385–4415. https://doi.org/10.1016/s0142-9612(03)00343-0

Kladovasilakis N, Tsongas K, Tzetzis D (2020) Finite element analysis of orthopedic hip implant with functionally graded bioinspired lattice structures. Biomimetics (basel, Switzerland) 5(3):44. https://doi.org/10.3390/biomimetics5030044

Kucewicz M, Baranowski P, Małachowski J, Popławski A, Płatek P (2018) Modelling, and characterization of 3D printed cellular structures. Mater Des 142:177–189. https://doi.org/10.1016/j.matdes.2018.01.028

Kumar A, Kumar P. Mittal RK, Singh H (2023a) Printing file formats for additive manufacturing technologies. In: Advances in additive manufacturing artificial intelligence, nature-inspired, and biomanufacturing. Elsevier, pp 87–102. https://doi.org/10.1016/B978-0-323-91834-3.00006-5

Kumar A, Kumar P, Mittal RK, Gambhir V (2023b) Materials processed by additive manufacturing techniques. In: Advances in additive manufacturing. Elsevier, pp 217–233. https://doi.org/10.1016/B978-0-323-91834-3.00014-4

Kumar A, Kumar P, Mittal RK, Singh H (2023c) Preprocessing and postprocessing in additive manufacturing. In: Advances in additive manufacturing. Elsevier, pp 141–165. https://doi.org/10.1016/B978-0-323-91834-3.00005-3

Kumar A, Kumar P, Singh H, Haleem A, Mittal RK (2023d) Integration of reverse engineering with additive manufacturing. In: Advances in additive manufacturing. Elsevier, pp 43–65. https://doi.org/10.1016/B978-0-323-91834-3.00028-4

Kumar A, Kumar P, Sharma N, Srivastava AK (eds) (2024) 3D printing technologies: digital manufacturing, artificial intelligence, Industry 4.0. Walter de Gruyter GmbH & Co KG. https://doi.org/10.1515/9783111215112

Lei H, Li C, Meng J, Zhou H, Liu Y, Zhang X, Wang P, Fang D (2019) Evaluation of compressive properties of SLM-fabricated multi-layer lattice structures by experimental test and μ-CT-based finite element analysis. Mater Des 169(107685):107685. https://doi.org/10.1016/j.matdes.2019.107685

Li N, Zhuang J, Zhu Y, Su G, Su Y (2021) Fluid dynamics of a self-propelled biomimetic underwater vehicle with pectoral fins. J Ocean Eng Sci 6(2):160–169. https://doi.org/10.1016/j.joes.2020.08.002

Liu L, Xu Y, Li S, Xu M, He Y, Shi Z, Li B (2019a) A novel strategy for simultaneously improving the fire safety, water resistance and compatibility of thermoplastic polyurethane composites through the construction of biomimetic hydrophobic structure of intumescent flame retardant synergistic system. Compos Part B Eng 176(107218):107218. https://doi.org/10.1016/j.compositesb.2019.107218

Liu J, Li L, Suo H, Yan M, Yin J, Fu J (2019b) 3D printing of biomimetic multi-layered GelMA/nHA scaffold for osteochondral defect repair. Mater Des 171(107708):107708. https://doi.org/10.1016/j.matdes.2019.107708

Liu L, Li L, Guo C, Ge Y, Chen Y, Zhang L (2023).The design of a biomimetic hierarchical thin-walled structure inspired by a lotus leaf and its mechanical performance analysis. Materials 16(11). https://doi.org/10.3390/ma16114116

Mittal RK, Haleem A, Kumar A (eds) (2022) Advances in additive manufacturing: artificial intelligence, nature-inspired, and biomanufacturing. Elsevier. https://doi.org/10.1016/C2020-0-03877-6

Mota C, Puppi D, Chiellini F, Chiellini E (2015) Additive manufacturing techniques for the production of tissue engineering constructs: additive manufacturing techniques for the production of tissue engineering constructs. J Tissue Eng Regen Med 9(3):174–190. https://doi.org/10.1002/term.1635

Muschler GF, Nakamoto C, Griffith LG (2004) Engineering principles of clinical cell-based tissue engineering. J Bone Joint Surg Am 86(7):1541–1558. https://doi.org/10.2106/00004623-200407000-00029

Niu J, Choo HL, Sun W (2017) Finite element analysis and experimental study of plastic lattice structures manufactured by selective laser sintering. Proc Inst Mech Eng Part L J Mater Des Appl 231(1–2):171–178. https://doi.org/10.1177/1464420716662296

Oxford JT, Reeck JC, Hardy MJ (2019) Extracellular matrix in development and disease. Int J Mol Sci 20(1):205. https://doi.org/10.3390/ijms20010205

Prasanth AS, Ramesh, Kumar N, Radhakrishnan K, Krishna P (2021) An experimental and numerical study on the tensile and compressive behavior of 3D printed biomimetic structures. Mater Today Proc 46:550–554. https://doi.org/10.1016/j.matpr.2020.11.111

Raffaella A, Petrescu FIT, Petrescu RVV, Antonio A (2016) Biomimetic finite element analysis bone modeling for customized hybrid biological prostheses development. Am J Appl Sci 13(11):1060–1067. https://doi.org/10.3844/ajassp.2016.1060.1067

Schönherr E, Hausser H-J (2000) Extracellular matrix and cytokines: a functional unit. Dev Immunol 7(2–4):89–101. https://doi.org/10.1155/2000/31748

Singh H, Al Mangour B (eds) (2023) Handbook of smart manufacturing: forecasting the future of industry 4.0. CRC Press. https://doi.org/10.1201/9781003333760

Srivastava AK, Kumar A, Kumar P, Gautam P, Dogra N (2023) Research Progress in metal additive manufacturing: Challenges and Opportunities. Int J Interact Des Manuf (IJIDeM). https://doi.org/10.1007/s12008-023-01661-6

Syu MH, Guan YJ, Lo WC, Fuh YK (2020) Biomimetic and porous nanofiber-based hybrid sensor for multifunctional pressure sensing and human gesture identification via deep learning method. Nano Energy 76(105029):105029. https://doi.org/10.1016/j.nanoen.2020.105029

Vafaeefar M, Moerman KM, Vaughan TJ (2024) Experimental and computational analysis of energy absorption characteristics of three biomimetic lattice structures under compression. J Mech Behav Biomed Mater 151(106328):106328. https://doi.org/10.1016/j.jmbbm.2023.106328

Wadley H, Dharmasena K, Queheillalt D, Chen Y, Dudt P, Knight D, Kiddy K, Xue Z, Vaziri A (2007) Dynamic compression of square honeycomb structures during underwater impulsive loading. J Mech Mater Struct 2(10):2025–2048. https://doi.org/10.2140/jomms.2007.2.2025

Wang C, Li Y, Zhao W, Zou S, Zhou G, Wang Y (2018) Structure design and multi-objective optimization of a novel crash box based on biomimetic structure. Int J Mech Sci 138–139:489–501. https://doi.org/10.1016/j.ijmecsci.2018.01.032

Xue Z, Hutchinson JW (2006) Crush dynamics of square honeycomb sandwich cores. Int J Numer Meth Eng 65(13):2221–2245. https://doi.org/10.1002/nme.1535

Yim E, Reano R, Pang S, Yee A, Chen C, Leong K (2005) Nanopattern-induced changes in morphology and motility of smooth muscle cells. Biomaterials 26(26):5405–5413. https://doi.org/10.1016/j.biomaterials.2005.01.058

Zhang B, Pei X, Zhou C, Fan Y, Jiang Q, Ronca A, D'Amora U, Chen Y, Li H, Sun Y, Zhang X (2018) The biomimetic design and 3D printing of customized mechanical properties porous Ti6Al4V scaffold for load-bearing bone reconstruction. Mater Des 152:30–39. https://doi.org/10.1016/j.matdes.2018.04.065

Zhang C, Zhang L, Liu L, Lv L, Gao L, Liu N, Wang X, Ye J (2020a) Mechanical behavior of a titanium alloy scaffold mimicking trabecular structure. J Orthop Surgery Res 15(1). https://doi.org/10.1186/s13018-019-1489-y

Zhang B, Guo L, Chen H, Ventikos Y, Narayan RJ, Huang J (2020b) Finite element evaluations of the mechanical properties of polycaprolactone/hydroxyapatite scaffolds by direct ink writing: effects of pore geometry. J Mech Behav Biomed Mater 104(103665):103665. https://doi.org/10.1016/j.jmbbm.2020.103665

Zhu Y, Joralmon D, Shan W, Chen Y, Rong J, Zhao H, Xiao S, Li X (2021) 3D printing biomimetic materials and structures for biomedical applications. Bio-Des Manufa 4(2):405–428. https://doi.org/10.1007/s42242-020-00117-0

Industry 4.0 in Aircraft Manufacturing: Innovative Use Cases and Patent Landscape

Vineet Bhatia, Ajay Kumar, Sumati Sidharth, Sanjeev Kumar Khare, Surendra Chandrakant Ghorpade, Parveen Kumar, and Gaydaa AlZohbi

Abstract The aerospace industry stands at the precipice of a digital revolution, with Industry 4.0 (I4.0) innovations poised to usher in transformative breakthroughs. These cutting-edge technologies possess the capacity to profoundly reshape conventional aircraft manufacturing methodologies. This research endeavor seeks to evaluate the prospects of integrating I4.0 solutions into the aircraft production realm and to identify the associated benefits and challenges. Furthermore, subsequent studies aim to delve into the adoption of I4.0 technologies within the aviation sector through a comprehensive patent analytics approach. This research will employ a systematic patent analysis methodology, encompassing data collection, categorization, and trend analysis of patents pertaining to I4.0 enabling technologies in aircraft manufacturing. The study will also explore real-world use cases and examine the impact of I4.0 technologies on design, production, quality assurance, supply chain management, and maintenance practices within the industry. By leveraging a patent analytics lens, this research initiative will uncover insights into the intellectual property landscape, illuminating the competitive dynamics and technological trajectories that shape the future of aircraft production. The findings will provide a valuable roadmap for stakeholders, enabling them to navigate the complexities and seize the opportunities presented by the convergence of I4.0 and aviation manufacturing. The aviation industry has seen a surge in innovation, as evidenced by the identification of 40,258 patents in the I4.0 and aircraft manufacturing domain between

---

V. Bhatia · S. Sidharth · S. K. Khare · S. C. Ghorpade
Department of Technology Management, Defence Institute of Advanced Technology, Girinagar, Pune, India

A. Kumar (✉)
Department of Mechanical Engineering, School of Engineering and Technology, JECRC University, Jaipur Rajasthan 303905, India
e-mail: ajay.kumar30886@gmail.com

P. Kumar
Department of Mechanical Engineering, Rawal Institute of Engineering and Technology, Faridabad, Haryana 121004, India

G. AlZohbi
Department Of Mathematics and Natural Sciences, Prince Mohammad Bin Fahd University, Al Khobar 34754, Kingdom of Saudi Arabia

A. Kumar et al. (eds.), *Industry 4.0 Driven Manufacturing Technologies*, Springer Series in Advanced Manufacturing, https://doi.org/10.1007/978-3-031-68271-1_5

2012 and 2023. The majority of these patents focus on advancements in metallic (B33Y10/00) and non-metallic (B33Y30/00) materials, highlighting the industry's focus on cutting-edge technologies. The United States and the European Patents lead in filings, indicating global interest in aerospace development. Major corporations such as RTX Corporation, General Electric, Boeing, and Qualcomm are key contributors, demonstrating their significant involvement in driving technological progress within the industry. The integration of I4.0 technologies into aircraft manufacturing offers a valuable opportunity for the aviation industry's digital transformation. Patent analytics emerges as a vital tool for navigating the technological and competitive landscape, ensuring the protection and optimization of innovations in the aviation manufacturing sector.

Keywords Industry 4.0 technologies · Aircraft manufacturing · Patent insights · Digital transformation

1 Introduction

An innovative era of production and manufacturing has been brought about by Industry 4.0, which has revolutionized traditional processes by integrating cutting-edge technologies. One industry poised to reap the benefits of this digital revolution is aerospace manufacturing (Sharma et al. 2023; Lineberger et al. 2024). This research primarily deals with the possibilities of integrating I4.0 technologies into aircraft manufacturing and addressing the challenges that come with this transformation from traditional manufacturing to I4.0.

The convergence of Industry 4.0 (I4.0) technologies encompasses a spectrum of cutting-edge tools, such as IoT, Industrial internet of things (IIoT), machine learning (ML), robotics, additive manufacturing (3D printing), AI, BDA, DT, and cloud computing (CC) technologies (Batista et al. 2024; Kumar et al. 2023a; Rani et al. 2023).

Integrating these technologies into aircraft manufacturing processes presents vast opportunities, promising heightened efficiency, improved product quality, streamlined supply chains, and greater responsiveness to customer needs.

The potential for integrating these technologies into aircraft manufacturing is extensive. Leveraging AI and ML techniques can enhance design processes, assist in production planning, and fortify quality control efforts, ultimately resulting in improved aircraft performance and reduced development timelines. A case study demonstrating the implementation of Industry 4.0 techniques in an assembly cell designed for automated drilling in aeronautical structures by employing neural networks to assess hole drilling quality, achieving 95% accuracy in real product tests.

The rise of Industry 4.0 and IoT innovations has propelled Digital Twins (DT) into various sectors, including aerospace, where they optimize daily tasks. Despite existing solutions in industrial domains, the aerospace sector is yet to fully utilize

DT advantages, especially in maintenance, where they streamline operations by gathering crucial status data, as evidenced by this article's systematic mapping analysis, revealing insights into DT modeling approaches and commonly used tools in aircraft operation and maintenance.

Robotics and automation hold the potential for more efficient assembly lines, reduced human errors, and heightened safety, while additive manufacturing provides rapid prototyping, customized component manufacturing, and light weight component construction (Kumar et al. , 2024; Tadesse et al. 2024; Yadav et al. 2023; Sharma et al. 2024).

However, alongside these exciting possibilities, we must also confront certain challenges. The interconnectedness of Industry 4.0 systems gives rise to security concerns, requiring strong cybersecurity measures to protect sensitive data and maintain privacy. Integrating different systems and technologies is a complex and challenging due to interoperability issues.

Also additional efforts are required in in training and upskilling workforce to capitalize on the potential of these novel technologies (Sharma et al. 2024).

Finally, when implementing Industry 4.0 technologies, we need to think about the costs and make sure there is a clear return on investment (ROI) (Oberheitmann 2020).Through the analysis of actual case studies and the application of best practices from the aerospace industry, we can gain significant understanding regarding how to effectively incorporate I4.0 technologies.

2 Evolution of Industry 4.0 (I4.0)

Mechanization, improved transportation, and the usage of water and steam power were all brought about by the first industrial revolution. Significant changes were brought about by the 2nd industrial revolution, which helped the rise of large corporations, the widespread use of electricity, and improvements in mass production techniques. Third revolution marked the digital age with electronics, computers, and the internet. Currently, the fourth revolution integrates digital technologies such as AI, IoT, and robotics into industries, enabling smart factories and new business models. These revolutions have revolutionized productivity, economic growth, and living standards, profoundly impacting societies and paving the way for the modern world. Evolution on industry 4.0 is paced at Fig. 1

3 Enabling Technologies for Industry 4.0

Industry 4.0 remains a burgeoning field, often referred to in various ways within both scholarly and non-scholarly literature. The following paragraphs will elucidate some of the key definitions prevalent in this domain. I4.0 refers to fourth stage of

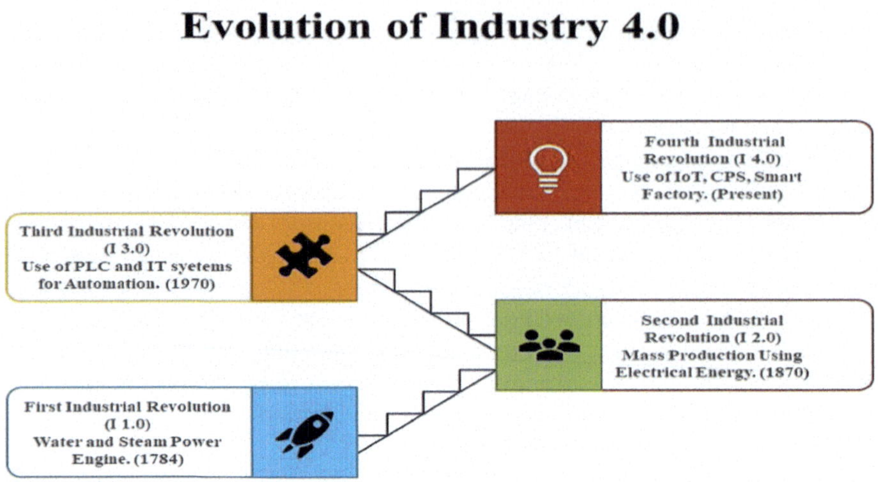

Fig. 1 Emergence of Industry 4.0

industrialization, categorized by the amalgamation of advanced enabling technologies like IoT, AI, and cloud computing into manufacturing processes to create smart factories and enhance production efficiency (Heiselberg 2013). I4.0 entails applying automation and data-driven insights to streamline workflows, raise the calibre of output, and improve client interactions (Culot et al. 2020). Advanced technologies of I4.0 (Martinelli et al. 2021; Sigov et al. 2022; Bigliardi et al. 2020) include Artificial Intelligence (AI), Robotics and Automation, Blockchain, Digital Twins, Augmented and Virtual Reality (AR/VR), Big Data and Analytics, Cyber-Physical Systems (CPS), Internet of Thing (IoT),Smart Manufacturing, Cloud Computing and Additive Manufacturing. These technologies enhance operational efficiency and open up new opportunities for innovation, thus creating smart factories (Osterrieder et al. 2020) and digital enterprises. The I4.0 enablers and technologies have been classified and categorized for improved understanding and implementation (Mabkhot et al. 2021). These key enablers and element technologies are depicted in Fig. 2 and are discussed in following paragraphs.

3.1 Industrial Internet of Things (IIoT)

The two primary enabling technologies for I4.0 include Internet of Things (IoT) and the Industrial Internet of Things (Industrial IoT), IoT is a "network of physical objects", including appliances, cars, devices, and other things, that have network connectivity, software, and sensors installed. These attributes facilitate the establishment of connections and data communication between the objects. Within IoT, a

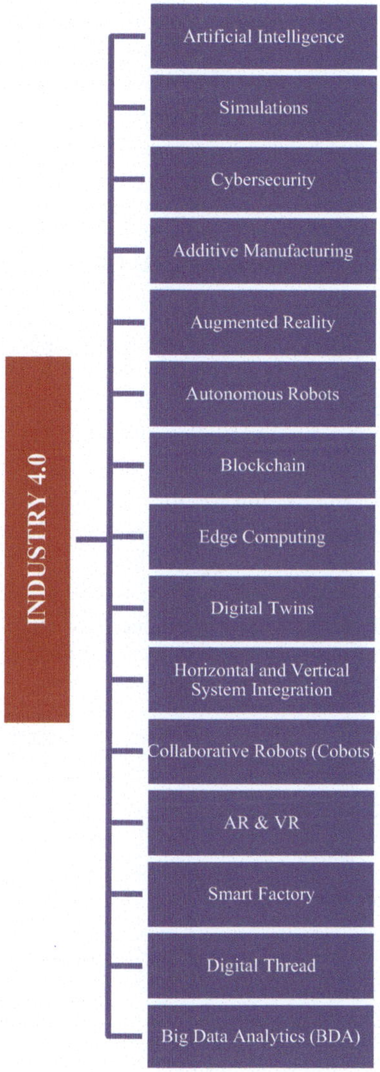

Fig. 2 Enabling technologies of Industry 4.0

"thing" can be any of two things: a digital entity existing in the virtual world or a physical object existing in the real world. These things are capable of existing, moving, and functioning in the temporal and spatial dimensions. Furthermore, they can be uniquely identified within the IoT network (Lier 2011). IoT devices can be used in various applications such as tracking assets, real-time supply chain management, and energy efficiency.

IIoT, a subset of IoT, concentrates on connecting devices and systems within industrial environments. It leverages machine learning and big data technologies, utilizing sensor data form machines and equipment's, Machine-To-Machine Communication

(M2M), and automation (Younan et al. 2020). The underlying principle of IIoT is that intelligent machines possess superior precision and consistency compared to humans when it comes to data capture and communication. This data empowers companies to detect inefficiencies and issues earlier, leading to time and cost savings while enhancing business intelligence initiatives.

3.2 IoT *Versus* IIOT

Although IoT and IIoT share same fundamental concept, there are significant differences between the two. While IIoT can be considered a subset of the broader IoT, it focuses on specific industrial applications. The primary distinction lies in the target end-users and the design parameters required for each (Deshpande and Jogdand 2020) IoT primarily caters to retail customers, offering devices such as smart bulbs, voice assistants, and robotic vacuums. On the other hand, IIoT is designed for industries, where the main objective is to collect and measure data to facilitate an intelligent ecosystem among machinery and processes (Alabadi et al. 2022).

While functionality is often assumed to be the differentiating factor, the key disparity lies in the specific design parameters that IIoT systems necessitate. The differences between IoT and IIOT are elaborated in Table 1

Table 1 Comparison between IoT and IIoT

Perspective	IIoT	IoT
Applications and use cases	Geared towards industrial production objectives	Focused on enhancing end-users' quality of life
Evaluation of effectiveness	Assessed using production metrics like overall equipment effectiveness (OEE), productivity, and downtime	Evaluated based on user experience, value, and ease of use
Uptime requirements	Stringent uptime requirements for uninterrupted industrial processes	Nominal uptime requirements for everyday consumer applications
Precision and reliability	Requires high precision and reliability in harsh environments	Reliability requirements are generally less demanding
Scale and volume of data	Deals with significantly larger volumes of data	Typically handles lower volumes of data
Security	Essential to prevent potential losses worth millions	Considered extremely important for safeguarding user privacy

3.3 Big Data and Analytics (BDA)

BDA refers to "The process of collecting, managing, processing, analyzing and visualizing continuously evolving data in terms of volume, velocity, value, variety and veracity"(Marjani et al. 2017). In Industry 4.0, the industrial ecosystem thrives on the harmonious interaction between two vital elements: the physical world comprising users and infrastructure, and the virtual world consisting of cloud-based algorithms and autonomous systems. These components are seamlessly interconnected through advanced communication technologies, facilitating efficient and autonomous operations. However, to fully unlock the potential of these industrial ecosystems, it is crucial to effectively collect, analyse, and store the huge volumes of data generated by smart devices and sensors. Big data analytics (BDA) plays important role in driving predictive manufacturing and timely detection of anomalies and system failures, resulting in improved product quality (Aujla et al. 2022). BDA algorithms and data-driven insights help manufacturers to anticipate potential issues, optimize production processes, and ensure consistent and reliable output (Sharma and Pandey 2020). BDA is transforming the landscape of Industry 4.0 by leveraging diverse input data sources, utilizing powerful software tools and techniques, and enabling valuable use cases (Ur Rehman et al. 2019).Data sources in I4.0 include sensors, manufacturing machine logs, production control systems, enterprise systems (ERP), and external data. Software Tools like Apache Hadoop, Apache Spark, and Tableau provide scalable storage, processing, and visualization capabilities (Duan and Xu 2021).

Methods such as Machine learning (ML), text analytics, and predictive modelling are employed to derive valuable insights from data. By leveraging these techniques, organizations can make informed decisions based on data, enhance operational efficiency, and foster innovation within the I4.0 era.

Figure 3 presents a taxonomy that has been developed based on various factors in the field of BDA for I4.0.

The taxonomy takes into account data sources, analytics tools, analytics techniques, and use cases of I4.0.

3.4 Cloud Computing

Cloud Computing is widely recognized as a vast, flexible, and decentralized pool of storage, services, and computational capabilities accessible to users upon request via the Internet (Alouffi et al. 2021; Alwada et al. 2015). Traditionally, cloud computing resources reside within a large-scale data storage facility, overseen and administered by a third-party provider. This provider furnishes the cloud users with computing infrastructure, enabling universal access from any location through the Internet (Stergiou et al. 2018). Contrarily, edge computing refers to moving computational power closer to the network's edge, in close proximity to data sources or endpoints. This

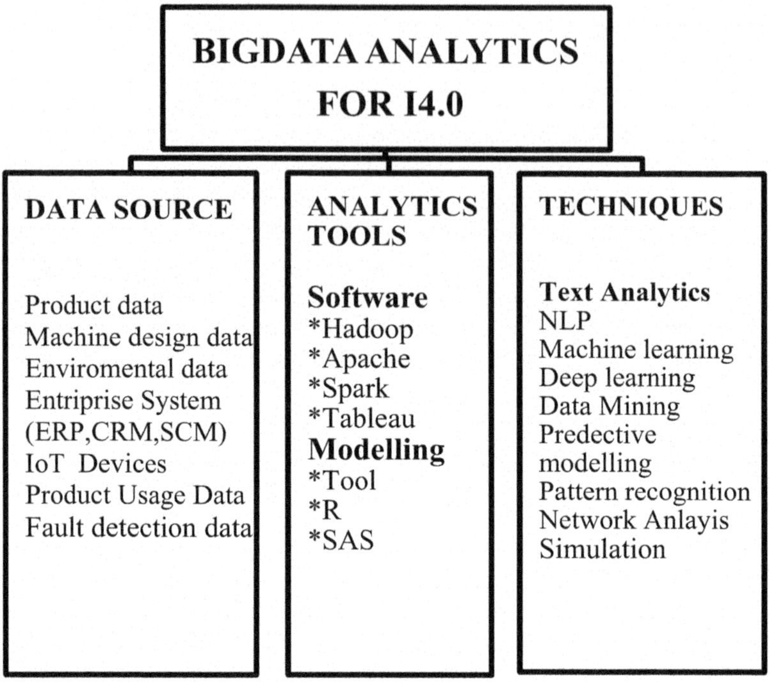

Fig. 3 BDA taxonomy for industry 4.0

approach minimizes latency and bandwidth requirements by processing data locally, without the need to transmit it to the cloud. Edge computing is especially beneficial for real-time applications in Industry 4.0, where immediate decision-making and low latency are critical (Nain et al. 2022). Cloud computing offers practical solutions for delivering computing resources through various models, namely Software as a Service (SaaS), Platform as Service (PaaS), and Infrastructure as Service (IaaS) (Mohammed and Zeebaree 2021).

3.5 Additive Manufacturing (AM)

A variety of technologies are used in additive manufacturing, commonly mentioned to as "3D" printing, to create objects by adding material rather than removing or milling it from a solid block. It includes building products layer by layer using fine powder or liquid materials, including metals (Srivastava et al. 2023), plastics, and composites (Aggoune et al. 2024; Ajay et al. 2023) Table 2 illustrates the major additive manufacturing (AM) processes (Kumar et al. 2023b; Sehrawat et al. 2022) and technologies, categorized based on the materials (Kumar et al. 2024a) they utilize (Goyal et al. 2024a).

Table 2 AM processes

AM technology	Brief description	AM process	Base materials
Stereolithography (SLA)	Employs a laser or UV light to gradually solidify liquid resin	Photopolymerization (Vat Photopolymerization)	SLA primarily uses various types of liquid photopolymer resins, which can be tailored for specific properties like strength, flexibility, or transparency
Fused deposition modelling (FDM)	Employs a heated nozzle to release and shape thermoplastic material	Material extrusion	Common filament materials used in FDM include various thermoplastics, such as acrylonitrile butadiene styrene (ABS), polylactic acid (PLA), polyamide (nylon), and others
Selective laser sintering (SLS)	Utilizes a laser to combine materials in powder form in a targeted manner	Powder bed fusion	SLS can use a variety of powdered materials, including thermoplastics (e.g., nylon, polystyrene), metals (e.g., stainless steel, titanium alloys), and ceramics (e.g., alumina, zirconia)
Selective laser melting (SLM)	Utilizes a laser of high power to melt and bond metal powders together	Powder bed fusion	SLM typically uses various metal alloys in powder form, including stainless steel, aluminum alloys, titanium alloys, and nickel-based superalloys
Electron beam melting (EBM)	Metal powders are melted and fused using an electron beam	Powder bed fusion	EBM primarily uses various metal alloys in powder form, including titanium alloys, cobalt-chrome alloys, and nickel-based superalloy
PolyJet	Jets liquid photopolymer materials that are cured with UV light	Material jetting	PolyJet uses various types of liquid photopolymer resins as base materials, including rigid, flexible, and support materials. These materials can be combined to create parts with different properties or colors
Digital light processing (DLP)	Cures liquid resin layer by layer using a digital light projector	Vat Photopolymerization	DLP primarily uses various types of liquid photopolymer resins as base materials, which can be tailored for specific properties like strength, flexibility, or transparency

(continued)

Table 2 (continued)

AM technology	Brief description	AM process	Base materials
Laminated object manufacturing (LOM)	Layers and bonds sheets of material using heat or adhesive	Sheet lamination	Common materials used in LOM include paper, plastic sheets (e.g., PVC, polyester), and composite materials (e.g., fiber-reinforced plastics)
Binder jetting	Deposition of binder onto layers of powdered material to bind them together	Binder jetting	Commonly used base materials include various metals, ceramics, and polymers in powder form, such as stainless steel, aluminum, silicon carbide, and nylon

Additive manufacturing (AM) enables the storage of parts as design files in virtual inventories, allowing them to be produced on-demand. This model, known as distributed additive manufacturing (Kumar et al. 2023c; Durão et al. 2017) offering several benefits. Firstly, it reduces transportation distances, resulting in cost savings. By eliminating the need for physical parts in inventory, it also simplifies inventory management. Instead of storing physical components, digital files can be stored, managed, and retrieved as needed (Kumar et al. 2023d, 2023e). This approach streamlines the manufacturing process and promotes efficient resource utilization.

Additive manufacturing (AM) has revolutionized multiple industries which includes aerospace and defense (A&D), automobile sector, Healthcare Architecture, consumer goods, energy, education, art, electronics, robotics, and even the food industry by enabling rapid prototyping, customization and complex component production (Industrial Applications of 3D Printing: The Ultimate Guide 2023; Kumar et al. 2024b; Bhatia and Sehgal 2023).

3.6 Horizontal and Vertical Integration

Three forms of integration are included in Industry 4.0: End-to-End engineering integration, Vertical integration, and Horizontal integration (Sony 2018). To provide better goods and services, several organisations in the value chain work together and integrate their value networks through a process known as horizontal integration (Foidl and Felderer 2016). This inter-cooperation through digitization creates an efficient digitized ecosystem. Vertical integration focuses on integrating hierarchical subsystems within an organization, connecting various informational subsystems to the ERP system. By integrating various components, organizations can establish a manufacturing system that is adaptable and reconfigurable. This integration allows for the seamless operation of smart machines, which can autonomously adjust to different product requirements with the aid of efficient big data management. In complete value chain, End-to-End engineering integration (Sehrawat et al. 2022) is essential to the development of customized goods and services.

3.7 Digital Twins (DT)

A digital twin, as described in reference (Sharma et al. 2022), "pertains to a virtual counterpart or duplication of a tangible object, system, operation, or service". It encapsulates all pertinent information, properties, and attributes of the initial entity, enabling continuous monitoring, examination, and replication. Digital twins serve as invaluable tools for extracting knowledge, enhancing efficiency, and supporting choices across different sectors and fields.

Digital twins encompass three primary components, distinguished by their respective roles (Li et al. 2021). The tangible facet incorporates sensors, data transmission technologies, and robust computing infrastructure, facilitating continuous data gathering and synchronization. The intangible facet involves analytical models, encompassing both physics-based and data-informed models, coupled with AI applications for data manipulation and interpretation. The linkage component centres on data transmission technologies and interfaces for human–machine interaction, fostering communication between the tangible and intangible facets and enabling engagement with the digital twin.

Digital twins have varied applications (Top 10 Applications Use Cases for Digital Twins | Unity 2024; Hasan 2024), including pre-production product simulation, physical entity performance analysis, and predictive data modelling for different scenarios. Their applications span across sectors (Gandzeichuk 2024) such as manufacturing, healthcare, smart cities, energy, transportation, construction, and aerospace.

Within Industry 4.0 (Parrott and Warshaw 2017) digital twin technology is integral, enabling virtual replicas of physical objects, processes, organizations, or individuals. These digital twins represent unique entities and can aggregate data from multiple sources, offering a comprehensive perspective of real-world entities and their processes. For example, they can simulate and monitor power plants or entire cities, allowing comprehensive analysis and optimization. Utilizing digital twins, Industry 4.0 enhances understanding and management of intricate systems and processes.

3.8 Autonomous Robots

Robots have become increasingly versatile and are now employed across a wide range of industries and applications. There are various types of robots, such as articulated robots, humanoids, autonomous mobile robots (AMRs), automated guided vehicles (AGVs), CoBots (collaborative robots) and hybrids. AMRs function independently,

AGVs operate along predetermined paths, articulated robots replicate human arm movements, humanoids specialize in tasks centred around humans, cobots work alongside humans in collaboration, and hybrids integrate different types of robots to handle more intricate operations.

Autonomous robots (AR) "are intelligent machines capable of performing tasks in the world by themselves, without explicit human control" (Mabkhot et al. 2021) These intelligent machines operate without direct human control, enhancing efficiency and flexibility in manufacturing processes (Popović and Popović 2021). Equipped with advanced sensing technologies and artificial intelligence, they perceive their environment, deliberate actions, and autonomously execute tasks towards predefined goals. Autonomous robots are essential in converting conventional factories into intelligent and automated environments, even though there are obstacles in perception, decision-making, and striking a fair balance between autonomy and human control.

3.9 Simulations

Simulation finds extensive application in I4.0 across multiple domains. It is employed for production planning and optimization, developing digital twins, training and skill development, New product development (NPD), testing, predictive maintenance, and energy management in smart factories (Dornhöfer et al. 2020). By creating virtual models and conducting simulations, manufacturers can make informed decisions, optimize processes, reduce costs, enhance training experiences, improve supply chain efficiency, ensure product quality, optimize maintenance schedules, and drive energy efficiency (Paula Ferreira et al. 2020; Gunal 2019).

Simulation techniques are fundamental in I4.0 for analyzing, optimizing, and decision-making within complex systems. Key techniques include Discrete Event Simulation for modelling manufacturing processes, Agent-Based Modelling for simulating interactions among autonomous agents, Monte Carlo Simulation for assessing uncertainty, Continuous Simulation for modelling continuous systems, Virtual Reality Simulation for immersive experiences, and Digital Twin Simulation for predictive maintenance and real-time monitoring (Paula Ferreira et al. 2020).

3.10 Cyber Security

The transition towards digitization and interconnected systems in I4.0 brings attention to the significance of data protection. Manufacturers and consumers alike are increasingly focused on implementing policies and organization practices to safeguard computers, networks, software programs including data from unauthorized access or exploitation. Cybersecurity risks (Rani et al. 2022) which include theft of intellectual property (IP), problems with data integrity, cyber-physical damages, pose serious threats to organizations, especially those in the manufacturing sector (Ani et al. 2016). Smart factories (Peasley 2024) are not immune to cybersecurity risks, as they can be targeted by various malicious activities such as vulnerability exploitation, malware attacks, denial of service (DoS) incidents, and device hacking. The expanded attack surface of smart factories, coupled with the growth of IoT, presents manufacturers with challenges in detecting and defending against cyber threats. The consequences of these attacks can extend beyond digital realms and have severe physical implications, particularly in realm of the IIoT. Measures such as implementing secure network architectures, robust authentication, continuous monitoring, encryption, software updates, employee training, and incident response planning are essential to protect against cyber threats (Junior et al. 2021).

3.11 Cyber Physical System

Cyber-Physical Systems (CPS) are a vital component of Industry 4.0, facilitating the convergence of physical and digital technologies. By integrating sensors, actuators, and control systems, CPS enable real-time monitoring and control of physical processes, enabling data-driven decision-making and optimized production. This fusion of physical and digital systems empowers the creation of intelligent manufacturing environments, where machines and humans collaborate to produce high-quality products with increased efficiency and reduced waste (Jwo et al. 2022).

4 Industry 4.0 for Aircraft Manufacturing

The aviation sector is adopting and utilising a variety of I4.0 technologies (Ajay et al. 2023) to improve operational efficiency, safety, and passenger experience, which makes the aviation industry and I4.0 intertwined.

4.1 IoT Applications for Aircraft Manufacturing

In aircraft manufacturing, IoT technology plays a crucial role in Production monitoring (Rodrigues et al. 2022) optimizing Processes, Supply Chain Management (SCM), improving quality control, enabling predictive maintenance, and improving workplace safety. By integrating IoT devices including sensors and connected equipment, manufacturers can gather real-time data on various factors to identify bottlenecks, improve productivity, and streamline operations. IoT also enables real-time monitoring of inventory and logistics, ensuring efficient supply chain management. Predictive maintenance using IoT sensors helps prevent equipment failures and optimize maintenance schedules. Lastly, IoT devices contribute to workplace safety by monitoring worker health and environmental conditions. Overall, IoT technology offers significant benefits in enhancing efficiency and productivity in aircraft manufacturing.

4.1.1 Use Cases of IoT in Aircraft Manufacturing

Fujitsu has leveraged IoT technology to enhance production line efficiency at Mitsubishi Heavy Industries (MHI) Commercial Aircraft Divisions. This system utilizes the Fujitsu COLMINA platform (Fujitsu manufacturing Industry solution) to optimize production in a smart way and enhance processes throughout supply chain using a cloud-based service. Additionally, the implementation incorporates User Experience (UX) design principles for visualizing operational status, progress, and failures, enabling intuitive identification of issues and facilitating prompt countermeasures (IoT streamlines aircraft production 2024).

At Airbus' Saint Eloi facility, the information generated by machines and conveyors is utilized to construct a dynamic visual model known as a "digital shadow." This novel method enables accurate simulations and real-time monitoring of assembly line operations, allowing users to pinpoint and put into practice the best possible strategies for increasing operational effectiveness. Also, Airbus has implemented IoT based monitoring and control of torque levels exerted by assembly line tools. Through this technology, if excessive or insufficient torque is detected, the tool automatically halts its operation, and an instant notification is sent to the user for prompt attention (Wuggetzer 2024).

In respect of I4.0, another approach involves the utilization of flexible assembly lines and processes. Historically, aircraft programs have opted for identical fuselage diameters (e.g., Boeing B-727, B-737, B-757, or Airbus A-318, A-319, A-320, A-321) to leverage economies of scale by sharing components, jigs, and tooling. In the past, it was impractical to manufacture components from different "aircraft programs" on a single manufacturing line. Airbus has achieved this capability through the implementation of a "Mixed Model Line," enabling the production of wings for both the Airbus A-330 and A-350. This achievement necessitates digital control of the production line, typically facilitated by leveraging the IoT. This approach empowers the individual control and real-time communication of various items such as tools, machines, facilities, materials, and products (Kumar et al. 2023a; Arntz et al. 2016).

4.2 Applications of Digital Twin (DT)

Digital twins and digital threads form the basis of digital transformation, offering novel approaches to evaluate practices, processes, and product concepts within a virtual realm. A company's current system or product is virtually represented by the digital twin. encompassing machines, controls, workflows, and systems. On the other hand, the digital thread acts as a comprehensive record of a product or system throughout its entire lifecycle, from inception to retirement. Digital twin technology has made significant advancements in the aerospace industry (Singh 2023). NASA has utilized digital twins for spacecraft, while the Air Force Space and Missile Systems Center developed a DT for the Lockheed Martin Block IIR GPS satellite. Digital twins have also been used to train AI pilots and improve maintenance practices. Companies like GE and Boeing have leveraged digital twins for engine components, landing gear, and aircraft design, resulting in improved performance predictions and quality.

4.2.1 Use Cases of Digital Twins in in Aircraft Manufacturing

Boeing is prioritizing passenger safety in commercial airplanes by leveraging "Digital Twins" for seat certification testing (Pinon Fischer et al. 2022) Digital Twins enable the establishment of standard work instructions and allow for dynamic testing and compliance assessment. This approach eliminates the need for extensive physical testing and supports Boeing's digital transformation efforts. By utilizing Digital Twins, Boeing aims to ensure passenger safety while accommodating modifications to baseline seat designs.

Turkish Aerospace utilizes Iron Bird, a test environment, to validate and verify their Flight Control System (FCS) under real loads and conditions. To enhance testing accuracy and efficiency, they have implemented a Digital Twin for Iron Bird. This Digital Twin continuously updates models with real test data, allowing for real-time testing of actuator and hydraulic models. By leveraging the Iron Bird Digital Twin, Turkish Aerospace reduces the number of failed tests, improves test rig utilization, and streamlines the time required for essential testing activities (Pinon Fischer et al. 2022).

4.3 Additive Manufacturing for Aircraft Production

Additive manufacturing is transforming aerospace industry by enabling (Goyal et al. 2024b) rapid prototyping, customized component production, and lightweight structural designs. It makes it possible to design intricate internal structures and geometries that reduce weight without sacrificing strength or functionality.

Additive manufacturing also plays vital role in the production of engine components, tooling and repair parts offering on-demand manufacturing and cost efficiencies. Additive manufacturing (AM) is transforming aerospace manufacturing by optimizing designs, shortening assembly times, and improving performance, thereby spearheading innovation within the industry (Kumar et al. 2023f).

Additive manufacturing (AM) has found diverse applications in the aerospace industry, with several manufacturers actively utilizing this technology (Blakey-Milner et al. 2021).

These manufacturers include Aerojet Rocketdyne, Airbus, Boeing, GKN Aerospace, NASA, SpaceX, and Oerlikon, among others.

Aerojet Rocketdyne utilizes AM for thrust chambers, while Airbus explores AM for different components such as reflector brackets and cabin bracket connectors for the A350 aircraft.

Boeing, GKN Aerospace, NASA, SpaceX, and Oerlikon are also involved in applying AM techniques for various aerospace applications such as rocket engines, fuel tanks, turbine blades, and structural components.

These manufacturers employ different design approaches and materials, including copper alloys, titanium, and various alloys like Inconel and Hastelloy.

Technologies including, Electron beam melting (EBM), Directed energy deposition (DED) and laser powder bed fusion (L-PBF) are used to realize these AM components.

4.3.1 Use Cases of Additive Manufacturing in Aircraft Manufacturing

General Electric has successfully employed additive manufacturing techniques in its GE9X engine, by combining over 300 to just seven 3D printed components (Węgrzyn 2022; Memon 2024). The GE9X engine features a high-pressure compressor with **BLISKS** (combined bladed disks) that are created through additive manufacturing, integral casting, or welding, eliminating the need for attaching individual blades to the disk. By eliminating the requirement of attaching blades to the disk, the number of components in the compressor is significantly reduced, thus improving efficiency by up to 8%, and eliminates potential sources of crack initiation. Additionally, the use of lightweight ultra-strength materials has contributed to further part reduction and enhanced performance. GE has achieved thinner, lighter, and more efficient fan blades by utilizing advanced materials. To provide a comparison, the GE90 engine consists of 22 blades, the GENX has 18 blades, and the largest engine of the three, the GE9X, incorporates only 16 blades. The GE9X engine stands out with its advanced technology, lighter fan blades, and improved efficiency compared to its predecessors.

Boeing, a leading aerospace manufacturer, has achieved a significant milestone by successfully flight-testing of a 3D printed critical component (aluminum gearbox housing) on a Chinook helicopter (Sertoglu 2024). The flight test involved a 3D printed aluminum gearbox housing manufactured by Boeing. The usage of additive manufacturing allowed for creation of a repeatable process that met the stringent quality requirements for airworthiness of the component, offering flexibility and opening up possibilities for future innovations. By leveraging 3D printing, Boeing achieved reduced lead times, cost savings, and improved part performance, demonstrating the benefits of design freedom and optimization enabled by additive manufacturing in the aerospace industry.

4.4 Autonomous Robots for Aircraft Manufacturing

The aerospace industry is experiencing significant advancements in robotics, shaping its future and potential applications. Five key innovations are driving this change. First, automated welding is enhancing aircraft safety and manufacturing speed while reducing risks for workers. Second, robotics is automating sealing, painting, and coating processes, improving productivity and freeing up human employees for more important tasks. Third, robots are streamlining the tedious drilling and fastening process, reducing manufacturing timelines and becoming the industry standard. Fourth, automated transportation systems are increasing safety during the movement of large aircraft components within assembly facilities. Finally, robotics coupled with sensor technology is enhancing inspection accuracy and efficiency, ensuring quality and integrity in aircraft components. Looking ahead, unmanned aerial vehicles (UAVs) are anticipated to play a crucial role in transportation, relieving congestion and potentially reducing carbon emissions.

4.4.1 Use Cases of Autonomous Robots in Aircraft Manufacturing

Airbus Defence and Space

CBC factory has successfully designed, developed, and validated a full autonomous aerial robotic system (ARS) (Martínez-de Dios et al. 2018).This ARS possesses the capability to carry out intricate tasks such as logistics involving light objects and searching for missing tools (Perez-Grau et al. 2021). The system utilizes an exceptionally efficient modular architecture, incorporating all essential sensors, electronics, and processing capabilities directly within the aerial robot.

4.5 Big Data and Analytics (BDA) in Aircraft Manufacturing

BDA is playing a significant role in aircraft manufacturing in various domains such as high-precision assembly, quality assurance, process planning, predictive maintenance, and supply chain enhancement (Badea et al. 2018).Big data analytics is becoming more and more important in the aerospace industry, which has led the way in incorporating big data analytics into its manufacturing processes. This has accelerated the assembly of aircraft and significantly improved measurement accuracy (Sahoo 2022).

Significantly, the utilization of BDA has delivered substantial advantages in the production of large composite structures, fuselage skin components, and wing boxes. Additionally, big data analytics plays a pivotal role in optimizing tool life cycles, supply chain management, production forecasting, throughput, work cell design, and Product Lifecycle Management (PLM). As a result, the aerospace industry experiences more efficient and profitable operations (Sahoo 2022; Nagorny et al. 2017).

Further BDA is contributing to the design and development of new aircraft models. By analyzing data from flight simulations, wind tunnel tests, and real-world performance data, manufacturers can gain valuable insights into the aerodynamics, structural integrity, and fuel efficiency of aircraft designs. Manufacturers are able to meet strict regulatory requirements, improve performance, and optimize designs with this data-driven approach.

4.5.1 Use Cases of Big Data Analytics in Aircraft Manufacturing

Big data analytics, leveraging Machine Learning (ML) and Sparse Sensing, has been employed in Predictive Shimming for aircraft manufacturing (Manohar et al. 2018). In the aircraft Production domain, it is a common occurrence for gaps to arise between parts during the modular manufacturing, assembly, and transportation stages. Traditionally, expensive laser scans have been employed to measure these gaps, ensuring precise manufacturing of shims for gap filling. However, this process, referred to as predictive shimming, significantly extends the timelines of aircraft production. To address this issue, big data analytics (BDA) coupled with machine learning techniques have been utilized to extract patterns from historical data. This approach reduces the need for extensive data collection and processing in future aircraft production. Notably, when applied to Boeing production data, this method has demonstrated remarkable success by accurately predicting 99% of gaps in newly produced aircraft wing-to-body join scenarios using only 3% of the original measurements. As a result, this approach has the potential to greatly reduce aircraft manufacturing time, leading to significant cost savings.

5 Summary of Use Cases of Industry 4.0 Technologies in Aircraft Manufacturing

The summary of real-life applications of I4.0 in aircraft Manufacturing is presented in Table 3

6 Challenges in Adoption of Industry 4.0

Implementing Industry 4.0 in aircraft manufacturing poses a number of challenges (Sayem et al. 2022) that must be addressed. The following are the critical challenges in implementing Industry 4.0.

Table 3 Use cases for I4.0 in aircraft manufacturing

I4.0 technologies	Uses cases	References
Internet of Things (IoT)	Fujitsu has developed visualizing operational status, progress, and failures, enabling intuitive identification of issues and facilitating prompt countermeasures	Rodrigues et al. (2022), IoT streamlines aircraft production (2024)
Internet of Things (IoT)	At Airbus', the information generated by machines and conveyors is utilized to construct a dynamic visual model known as a "digital shadow"	Wuggetzer (2024)
Digital control of the production line	Airbus uses flexible assembly lines and processes to manufacture components from different aircraft programs on a single production line	Kumar et al. (2023a), Arntz et al. (2016)
Digital Twin	Air Force Space and Missile Systems Center developed a DT for the Lockheed Martin Block IIR GPS satellite Digital twins have also been used to train AI pilots and improve maintenance practices GE and Boeing have leveraged digital twins for engine components, landing gear, and aircraft design, resulting in improved performance predictions and quality	Singh (2023), Waterman (2024), Li et al. (2021), Dinter et al. 2022)
Additive manufacturing	General Electric has developed a high-pressure compressor with BLISKS (combined bladed disks) for GE9X engine through additive manufacturing Boeing has developed a aluminum gearbox housing for Chinook helicopter	Węgrzyn (2022), Memon (2024), Sertoglu (2024)
Autonomous Aerial robotic system	Airbus Defence and Space CBC factory has validated a full autonomous aerial robotic system (ARS)	Martínez-de Dios et al. (2018)
BDA for Predictive Shimming	Boeing has applied BDA to production data, for predicting gaps between parts during the modular manufacturing and assembly	Manohar et al. (2018)

6.1 Skill Gap

Implementing Industry 4.0 technologies in aircraft manufacturing necessitates the use of a highly skilled workforce (Maisiri et al. 2019; Ras et al. 2017). However, there is often a shortage of professionals with expertise in automation, artificial intelligence, data analytics, and other relevant fields. Bridging this skill gap necessitates upskilling and reskilling programs, collaboration with educational institutions, and a culture of continuous learning and adaptation.

6.2 Legacy Systems

Integration: Integrating I4.0 technologies into existing aircraft production processes will require major upgrades or replacements as these machines were not designed to capture data for analytics. It is critical to ensure seamless integration and interoperability between new and existing systems (Zawra 2019; Pessoa et al. 2018).

6.3 Data Security

I4.0 majorly depend on data collection, distribution, and analysis, making data security a critical concern. The interconnected nature of these devices or systems can result in vulnerabilities (Pereira et al. 2017). Securing sensitive aircraft manufacturing data against cyber threats necessitates implementing additional safeguards, which include robust cybersecurity practices like encryption, access controls, network segmentation, and employee awareness training (Mentsiev et al. 2020).

6.4 Interoperability and Standardization:

Adoption of I4.0 entails the unification of diverse elements, systems, and procedures within the aircraft manufacturing value chain. Nonetheless, incongruent manufacturers and suppliers might employ varying norms and methods, leading to a hurdle in compatibility (Watson et al. 2017). It is imperative to set universal standards and protocols to facilitate smooth interaction and information sharing (Albouq et al. 2022).

6.5 Regulatory Compliance

The aircraft manufacturing industry is subject to strict regulations and safety standards imposed by regulatory agencies. Agencies for regulatory compliance including European Union Aviation Safety Agency and Federal Aviation Administration the play significant roles in aircraft manufacturing processes through regulatory oversight and safety standards. Thus Implementing Industry 4.0 technologies must comply with these regulations without compromising safety (Russell et al. 2019).

Addressing these challenges will necessitate collaboration industry stakeholders, government assistance, investment in training and education, and a focus on cybersecurity and data protection. It involves upskilling the existing workforce, attracting new talent, integrating legacy systems, establishing common standards, and ensuring compliance with regulations.

7 Patent Analytics

Patent analytics leverages patent data to reveal trends and insights into innovation within specific sectors or technological domains (Patent Analytics—patent-analytics—WIPO Liferay DXP 2024). This approach offers organizations empirical insights for enhanced decision-making in research and development, innovation strategies, intellectual property (IP) monetization and licensing, collaborative research endeavors, and more. The process of patent analytics typically involves gathering data, analyzing and identifying trends, and reporting findings.

Patent analytics for Industry 4.0 technologies within the aerospace sector provides a complete overview of the innovation landscape, highlighting the areas of rapid technological advancement and strategic investment. This analysis involves examining patent filings and grants to understand the trends, key players, and technological frontiers in the integration of Industry 4.0 innovations in aerospace. Open-source database (Lens Patent) as recommend by WIPO (World Intellectual Property Organization) is utilized to conduct research. Lens Patent Database (The Lens—Free Open Patent and Scholarly Search 2023) was utilized for search due to its extensive dataset coverage across multiple patent jurisdictions such as the USPTO (US Patent Office), JPO (Japan Patent Office), EPO (European Patent Office),WIPO (World Intellectual Property Organization).,UK IPO (UK Intellectual Property Office), and more (a total of 106 jurisdictions).

Patent analytics were undertaken for the period from 2012 to 2023 using the following keywords to generate a query: (Aerospace OR Aircraft) AND (IoT OR "Internet of Things" OR "3D Printing" OR "Additive Manufacturing" OR "Digital Twin" OR "Big Data Analytics") in an open-access patent database. The results of the analytics are discussed in the following paragraphs.

Table 4 Summary of patent family

Patent records	Extended families	Simple families
"The number of patents records in the result set"	"An extended family is a collection of patent applications that stem from similar technical content patent documents."	"A simple patent family is a group of patent documents that stem from the same initial document, called the priority document."
40,503	18,275	16,947

7.1 *Patent Family*

The summary of patent family is given in Table 4

7.2 *Patent Growth*

The data on patent growth related to I4.0 technologies in aircraft manufacturing from 2012 to 2023 reveals a significant and consistent increase in the number of patents filed and published over this period. This trend underlines the fast-tracking pace of innovation and the growing importance of I4.0 technologies in this sector. The detailed analysis of patent data is covered in subsequent paragraphs.

7.2.1 Initial Growth Phase (2012–2015)

The period from 2012 to 2015 marks the initial phase of growth, with the document count starting at 65 in 2012 and experiencing a more than tenfold increase to 704 by 2015. This phase reflects the early adoption and exploration of I4.0 technologies within the aerospace domain.

7.2.2 Rapid Expansion Phase (2016–2019)

Between 2016 and 2019, the sector saw a rapid expansion in patent filings, with numbers growing from 1310 in 2016 to 4546 in 2019. This phase indicates a significant ramp-up in R&D activities, with companies and research institutions intensifying their efforts to innovate and secure intellectual property rights in key I4.0 technologies such as IoT, 3D printing and twins.

7.2.3 High Growth Phase (2020–2023)

The period from 2020 to 2023 represents a high growth phase, with the number of patents peaking at 8425 in 2023, up from 5818 in 2020 (Fig. 4). This continued growth, even in the face of major challenges such as the COVID-19 pandemic, underscores the strategic importance of I4.0 technologies in enhancing production efficiency, product innovation, and operational resilience in the aerospace sector.

7.3 Patent by Jurisdiction

The United States leads by a significant margin in the number of documents, with the count approaching 30,000. This suggests a high level of activity and possibly investment in Industry 4.0 and aerospace and aircraft sectors within the United States (Fig. 5).

The World Intellectual Property Organization (WO—WIPO) also has a substantial count, around 10,000 documents. This could indicate a global interest and participation in these sectors, as WIPO would be involved in international patent applications.

The United Kingdom and the Republic of Korea have a similar count of documents, between 2000 and 5000. This implies a moderate level of activity in these countries regarding Industry 4.0 and aerospace and aircraft innovations.

France, Russia, Hong Kong, and Taiwan have the least number of documents, all below 2000. While these countries are involved in these sectors, their activity levels, as represented by document counts, are relatively lower compared to the others listed.

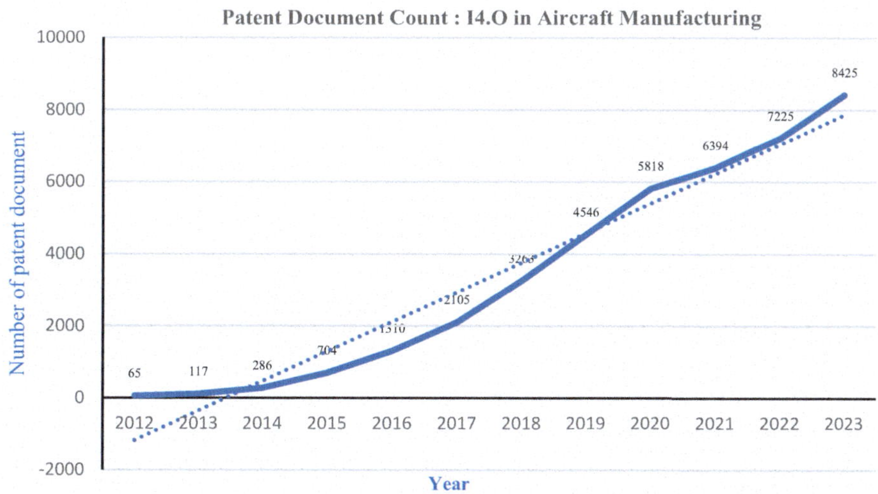

Fig. 1.4 Patents growth (2012 to 2023)

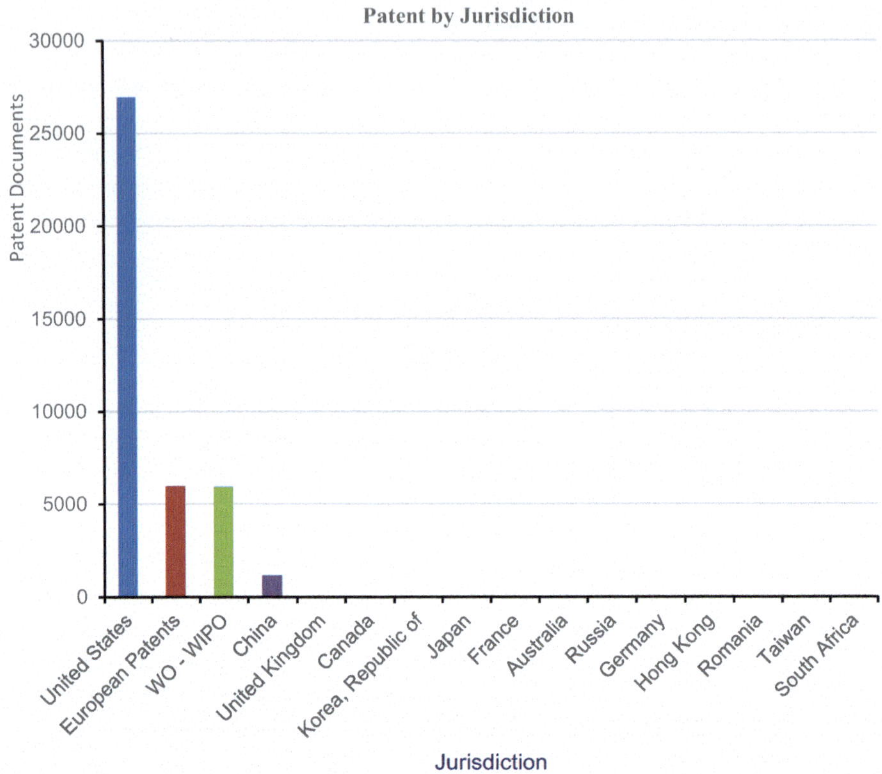

Fig. 5 Patents by Jurisdiction

There is a notable disparity between the United States and other countries. This could reflect differences in the size of the aerospace industry, the focus on Industry 4.0, the propensity to file patents, or a combination of these factors.

7.4 *Top IPC Classification*

The analyzed International Patent Classification (IPC) data underscores a diverse landscape of technological innovation, heavily influenced by Industry 4.0 paradigms across multiple sectors. The data reveals a significant emphasis on additive manufacturing (3D printing), as seen in the high number of patents related to both metallic (B33Y10/00) and non-metallic (B33Y30/00) product manufacturing processes. This trend highlights the industry's shift towards leveraging 3D printing for its versatility in creating complex, lightweight components using a wide range of materials (Fig. 6).

Furthermore, the integration of various manufacturing processes (B33Y80/00) and the focus on powder metallurgy (B22F3/105) suggest a keen interest

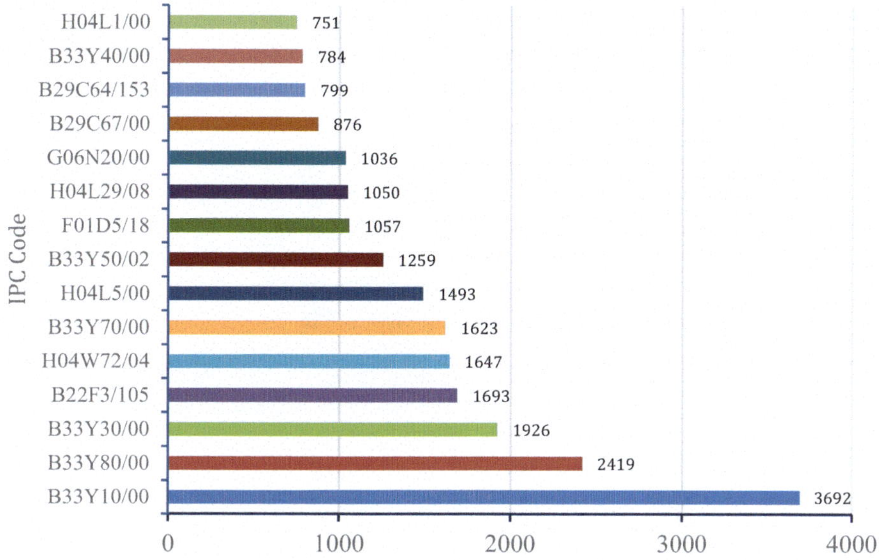

Fig. 6 Patents by IPC classification

in exploring and enhancing materials processing techniques to improve product quality and production efficiency. The importance of communication technologies is evident in the substantial number of patents related to wireless communication networks (H04W72/04) and data communication techniques (H04L5/00, H04L29/08), reflecting the critical role these technologies play in facilitating smart manufacturing environments and IoT (Internet of Things) applications. This analysis paints a picture of an industry committed to innovation and efficiency, leveraging a multidisciplinary approach that combines advancements in manufacturing technologies, materials science, communication, and artificial intelligence. The focus areas represented by the IPC data signify a transformative period where traditional manufacturing converges with digital technologies, paving the way for smarter, more sustainable, and highly efficient production processes.

7.5 Key Patent Holders

The data indicates a robust engagement in Industry 4.0 within the aerospace sector, with RTx Corporation leading with 1667 patents. Close competitors General Electric and The Boeing Company follow with 1641 and 1531 patents, respectively, underscoring their heavy investment in aerospace innovation. Qualcomm and Shanghai Langbo suggest a focus on communication technologies, essential in Industry 4.0. IBM and LG Electronics indicate a broadening of their traditional tech and electronics scope into this sector (Fig. 7).

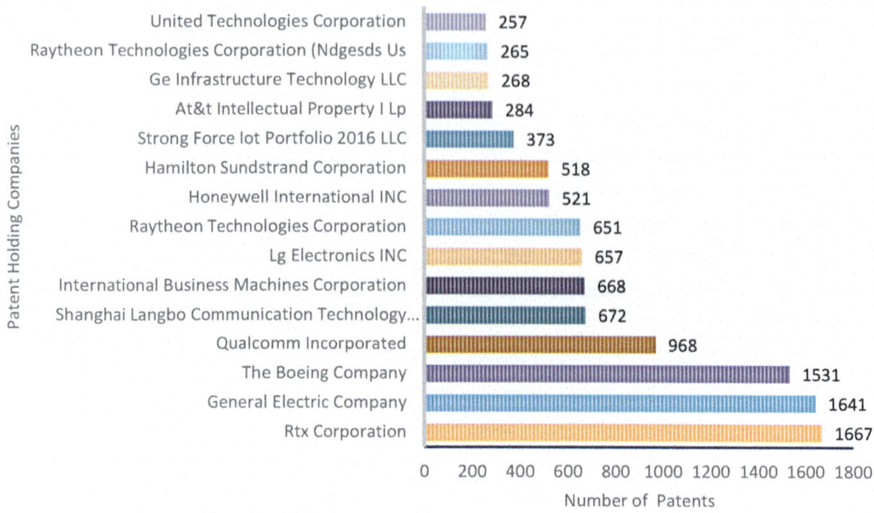

Fig. 7 Patents by top owners

7.6 *Key Observations Based Patent Analytics*

The key finding of Patent analytics are explained in brief in succeeding paragraphs.

7.6.1 Exponential Growth

The data shows an almost exponential growth in patent filings, reflecting the rapid pace of technological advancements and the increasing competition among companies to innovate and protect their inventions.

7.6.2 Strategic Importance

The consistent year-on-year increase highlights the strategic position of I4.0 technologies in aircraft manufacturing, with companies investing heavily in these areas to gain a competitive edge.

7.6.3 Technological Diversification

The broad range of keywords used in the query suggests a diversification of innovation areas, including digital manufacturing processes, data analytics, and connectivity solutions, which are pivotal for the next generation of aircraft manufacturing.

8 Future Developments

The future development in patents for Industry 4.0 and aerospace manufacturing is expected to be driven by several key trends and technologies. These include following advanced technologies:

8.1 Generative AI

Generative AI's role in design and prototyping is pivotal, offering the potential to rapidly generate multiple design variations that meet specific criteria, thus accelerating the innovation cycle. This capability extends to material development, where AI-driven simulations can discover new materials with optimal properties for aerospace applications, pushing the boundaries of what's possible in manufacturing and design.

8.2 Digital Twins and Simulation

The adoption of digital twins is increasingly becoming a trend within Industry 4.0 and the aerospace sector, including aircraft design and aeroengine simulations, highlighting its growing importance in enhancing design accuracy and operational efficiency. Latest Patents focus on Manufacturing Process Management (Aiming et al. 2021), Autonomous control of manufacturing Process (Takafumi et al. 2024) innovations that improve the accuracy and efficiency of simulations, enabling better product development and manufacturing processes.

8.3 Additive Manufacturing (3D Printing)

Innovations in additive manufacturing are likely to continue at a rapid pace, particularly in aerospace, where the demand for lightweight and complex structures is high. Latest Patent filed/granted in 2024 cover new materials ("forming the aircraft component with the combination of the SMA and the aluminum alloy" (Steele et al. 2022) processes, and machine technologies ("cold spray additive manufacturing" (Michael et al. 2019) that improve the strength, durability, and precision of 3D printed components.

8.4 Artificial Intelligence and Machine Learning

AI and machine learning are becoming integral to optimizing manufacturing processes (Chul and Ook 2024), predictive maintenance, and design. Future patents might encompass algorithms that enhance manufacturing efficiency, quality control, and the customization of aerospace components.

8.5 Edge Commuting

In the near future, companies will significantly invest in edge computing frameworks for seamless integration into production, driving Industry 4.0 towards enhanced real-time processing of data and decision-making at the edge of the network rather than cloud.

8.6 Advanced Robotics and Automation

Robotics technology will continue to advance, with patents likely focusing on autonomous robots, collaborative robots (cobots), and unmanned aerial vehicles (UAVs) that can perform complex manufacturing and assembly tasks with high precision and flexibility.

9 Discussion

The aircraft manufacturing industry is actively embracing Industry 4.0 technologies, such as additive manufacturing, robotics, Internet of Things (IoT), and advanced data analytics. While this integration presents challenges like cybersecurity risks, need for skilled workforce, and high initial costs, it unlocks transformative opportunities. Additive manufacturing enables producing complex, lightweight components using less raw material, reducing waste. Robotics enhances automation, precision, and efficiency in processes like assembly and inspection. IoT sensors collect real-time data for predictive maintenance and quality control. Data analytics optimizes design, production planning, and supply chain management.

Despite hurdles, use cases demonstrate Industry 4.0's potential: 3D printed fuel nozzles and bracketry reduce aircraft weight and fuel consumption. Collaborative robots assist humans in ergonomically demanding tasks. Digital twins virtually test aircraft performance before physical production. As companies race to patent their innovations, the patent landscape reflects fierce competition to gain competitive advantages through Industry 4.0 in this sector. Overcoming challenges while capitalizing on opportunities will shape the future competitiveness of aircraft manufacturers.

10 Conclusion

Manufacturing operations in the aviation sector have the potential to be transformed by incorporating Industry 4.0 technologies. This paper has researched into various aspects of implementing I4.0 technologies in this context. including several use cases that showcase implementation of I4.0 in aircraft manufacturing. Notably, robotics and automation offer a pathway to streamline repetitive tasks, reduce human errors, and bolster the overall efficiency of maintenance procedures.

Integrating legacy systems in an industry with diverse equipment and software, raises compatibility concerns that require careful attention. The management of high-volume data necessitates robust cybersecurity measures to safeguard sensitive information and maintain data integrity. Furthermore, there's a critical need for workforce training and upskilling to ensure the effective operation and maintenance of advanced technologies, underscoring the importance of bridging the skill gap.

The number of patents in the aerospace sector is skyrocketing, highlighting the swift advancement of Industry 4.0 technologies as a key competitive edge. This surge is both in number and in the variety of technologies, significantly influencing the direction of aircraft production.

To conclude, although integrating Industry 4.0 into aircraft maintenance brings certain obstacles, the wide range of advantages it offers makes it an attractive path for the aviation sector. By tackling these hurdles head-on and leveraging the opportunities, the industry is poised to achieve higher efficiency, safety, and productivity, resulting in better aircraft maintenance processes and superior operational performance.

References

Aggoune S, Hamadi F, Abid C et al (2024) Instabilities in the formation of single tracks during selective laser melting process. Int J Interact Des Manuf. https://doi.org/10.1007/s12008-024-01887-y

Aiming X, Can W, Na Z, Xing Z, Jingguo S, Min Z (2021) Aviation Industry cluster-oriented manufacturing process management and control system architecture. EP 4310764 A1, 26 Jul 2021 [Online]. Available: https://lens.org/074-642-599-264-976

Ajay, Singh H, Parveen, AlMangour B, Mangour A (eds) (2023) Handbook of smart manufacturing: forecasting the future of industry 4.0. CRC Press

Alabadi M, Habbal A, Wei X (2022) Industrial Internet of things: requirements, architecture, challenges, and future research directions. IEEE Access 10:66374–66400. https://doi.org/10.1109/ACCESS.2022.3185049

Albouq SS, Sen AAA, Almashf N, Yamin M, Alshanqiti A, Bahbouh NM (2022) A survey of interoperability challenges and solutions for dealing with them in IoT environment. IEEE Access 10:36416–36428. https://doi.org/10.1109/ACCESS.2022.3162219

Alouffi B, Hasnain M, Alharbi A, Alosaimi W, Alyami H, Ayaz M (2021) A systematic literature review on cloud computing security: threats and mitigation strategies. IEEE Access 9:57792–57807. https://doi.org/10.1109/ACCESS.2021.3073203

Alwada T, Al-Zitawi O, Khawaldeh S, Almasarweh M (2015) Privacy and control in mobile cloud systems. IJCA 113(1):12–15. https://doi.org/10.5120/19789-1170

Ani UPD, He H, Tiwari A (2016) Review of cybersecurity issues in industrial critical infrastructure: manufacturing in perspective. J Cyber Secur Technol 1, Dec 2016. https://doi.org/10.1080/23742917.2016.1252211

Arntz M, Gregory T, Lehmer F, Matthes B, Zierahn U (2016) Arbeitswelt 4.0—Stand der Digitalisierung in Deutschland: Dienstleister haben die Nase vorn. Institut für Arbeitsmarkt- und Berufsforschung (IAB). Nürnberg, 22/2016

Aujla GS, Prodan R, Rawat DB (2022) Big data analytics in Industry 4.0 ecosystems. Softw Pract Experience 52(3):639–641. https://doi.org/10.1002/spe.3008

Badea V, Alin Z, Boncea R (2018) Big Data in the aerospace industry. Informatica Economica 22:17–24. https://doi.org/10.12948/issn14531305/22.1.2018.02

Batista RC, Agarwal A, Gurung A, Kumar A, Altarazi F, Dogra N, HM V, Chiniwar DS, Agrawal A (2024) Topological and lattice-based AM optimization for improving the structural efficiency of robotic arms. Front Mech Eng 10:1422539

Bhatia A, Sehgal AK (2023) Additive manufacturing materials, methods and applications: a review. Mater Today Proc 81:1060–1067. https://doi.org/10.1016/j.matpr.2021.04.379

Bigliardi B, Bottani E, Casella G (2020) Enabling technologies, application areas and impact of industry 4.0: a bibliographic analysis. Procedia Manuf 42:322–326. https://doi.org/10.1016/j.promfg.2020.02.086

Blakey-Milner B et al (2021) Metal additive manufacturing in aerospace: a review. Mater Des 209:110008. https://doi.org/10.1016/j.matdes.2021.110008

Chul YS, Ook PM (2023) Apparatus and method for inspecting assembly hole of vehicle. US 2024/0020818 A1, 31 Jan 2023

Culot G, Nassimbeni G, Orzes G, Sartor M (2020) Behind the definition of Industry 4.0: analysis and open questions. Int J Prod Econ 226:107617. https://doi.org/10.1016/j.ijpe.2020.107617

Deshpande SN, Jogdand RM (2020) A survey on internet of things (IoT), industrial IoT (IIoT) and industry 4.0. IJCA 175(27):20–27. https://doi.org/10.5120/ijca2020920790

Dornhöfer M, Sack S, Zenkert J, Fathi M (2020) Simulation of smart factory processes applying multi-agent-systems—a knowledge management perspective. J Manuf Mater Process 4(3), 3, Sep 2020. https://doi.org/10.3390/jmmp4030089

Duan L, Da Xu L (2021) Data analytics in industry 4.0: a survey. Inf Syst Front. https://doi.org/10.1007/s10796-021-10190-0

Durão LFCS, Christ A, Zancul E, Anderl R, Schützer K (2017) Additive manufacturing scenarios for distributed production of spare parts. Int J Adv Manuf Technol 93(1):869–880. https://doi.org/10.1007/s00170-017-0555-z

Foidl H, Felderer M (2016) Research challenges of industry 4.0 for quality management. Apr 2016. https://doi.org/10.1007/978-3-319-32799-0_10

Gandzeichuk I (2024) Digital twin use cases and applications | softengi.com. Accessed: 06 Feb 2024. [Online]. Available: https://softengi.com/blog/use-cases-and-applications-of-digital-twin/

Goyal G, Kumar A, Sharma D (2024) 12 recent applications of rapid prototyping with 3D printing: a review. In: Kumar A, Kumar P, Sharma N, Srivastava AK (2024a) 3D printing technologies: digital manufacturing, artificial intelligence, industry 4.0. De Gruyter, Berlin, Boston, pp 245–258. https://doi.org/10.1515/9783111215112-012

Goyal G, Kumar A, Gupta A (2024b) 16 recent developments in 3D printing: a critical analysis and deep dive into innovative real-world applications. In: Kumar A, Kumar P, Sharma N, Srivastava

AK (2024) 3D printing technologies: digital manufacturing, artificial intelligence, industry 4.0. De Gruyter, Berlin, Boston, pp 335–352. https://doi.org/10.1515/9783111215112-016

Gunal MM (2019) Simulation and the fourth industrial revolution. In: Gunal MM (ed) Simulation for industry 4.0: past, present, and future in Springer Series in Advanced Manufacturing. Springer International Publishing, Cham, pp 1–17. https://doi.org/10.1007/978-3-030-04137-3_1

Hasan M (2024) Decoding digital twins: exploring the 6 main applications and their benefits. IoT Analytics. Accessed: 06 Feb 2024. [Online]. Available: https://iot-analytics.com/6-main-digital-twin-applications-and-their-benefits/

Heiselberg H (2013) Recommendations for implementing the strategic initiative INDUSTRIE 4.0. National Academy of Science and Engineering, Frankfurt. [Online]. Available: https://www.din.de/resource/blob/76902

Industrial Applications of 3D Printing: The Ultimate Guide (2023) AMFG. Accessed: 29 Jun 2023. [Online]. Available: https://amfg.ai/industrial-applications-of-3d-printing-the-ultimate-guide/

IoT streamlines aircraft production (2024) Mitsubishi Heavy Industries (MHI). Accessed: 06 Feb 2024. [Online]. Available: https://www.fujitsu.com/emeia/about/resources/case-studies/cs-2020dec-mitsubishi-heavy-industries.html

Junior AAdS, Pio JLdS, Fonseca JC, de Oliveira MA, Valadares OCdP, Silva PHSd (2021) The state of cybersecurity in smart manufacturing systems: a systematic review. Eur J Bus Manag Res 6(6):6, Dec 2021. https://doi.org/10.24018/ejbmr.2021.6.6.1173

Jwo J-S, Lee C-H, Lin C-S (2022) Data twin-driven cyber-physical factory for smart manufacturing. Sensors 22(8). https://doi.org/10.3390/s22082821

Kumar A, Rani S, Rathee S, Bhatia S (eds) (2023a) Security and risk analysis for intelligent cloud computing: methods, applications, and preventions (1st ed.). CRC Press. https://doi.org/10.1201/9781003329947

Kumar A, Kumar P, Mittal RK, Gambhir V (2023b) Materials processed by additive manufacturing techniques. In: Kumar A, Mittal RK, Haleem A (eds) Advances in additive manufacturing materials and technologies. Elsevier, pp 217–233. https://doi.org/10.1016/B978-0-323-91834-3.00014-4

Kumar A, Mittal RK, Haleem A (eds) (2023c) Advances in additive manufacturing. In: Advances in additive manufacturing, in additive manufacturing materials and technologies. Elsevier, pp i–iii. https://doi.org/10.1016/B978-0-323-91834-3.00031-4

Kumar A, Kumar P, Mittal RK, Singh H (2023d) Preprocessing and postprocessing in additive manufacturing. In: Kumar A, Mittal RK, Haleem A (Eds) Advances in additive manufacturing materials and technologies. Elsevier, pp 141–165. https://doi.org/10.1016/B978-0-323-91834-3.00005-3

Kumar A, Kumar P, Singh H, Haleem A, Mittal RK (2023e) Integration of reverse engineering with additive manufacturing. In: Kumar A, Mittal RK, Haleem A (Eds) Additive manufacturing materials and technologies. Elsevier, pp 43–65. https://doi.org/10.1016/B978-0-323-91834-3.00028-4

Kumar A, Kumar P, Srivastava AK, Goyat V (2023) Modeling, characterization, and processing of smart materials. IGI Global

Kumar A, Shrivastava VK, Kumar P, Kumar A, Gulati V (2024) Predictive and experimental analysis of forces in die-less forming using artificial intelligence techniques. In: Proceedings of the institution of mechanical engineers, Part E: journal of process mechanical engineering 0(0). https://doi.org/10.1177/09544089241235473

Kumar P, Hussain SS, Kumar A, Srivastava AK, Hussain M, Singh PK (2024a) 10 Finite element method investigation on delamination of 3D printed hybrid composites during the drilling operation. 3D Printing Technologies: Digital Manufacturing, Artificial Intelligence, Industry 4.0, 223

Kumar A, Kumar P, Sharma N, Srivastava AK (2024b) 3D printing technologies: digital manufacturing, artificial intelligence, industry 4.0. Walter de Gruyter GmbH & Co KG

Li L, Aslam S, Wileman AJ, Perinpanayagam S (2021) Digital twin in aerospace industry: a gentle introduction. IEEE Access 1–1, Dec 2021. https://doi.org/10.1109/ACCESS.2021.3136458

Lier B (2011) Connections, Information and Reality ‘ thinking about the internet of things. Systemics, Cybern Inf 9(5)

Lineberger R, Hussain A, Hanley T (2024) Aerospace and defense 4.0—Capturing the value of industry 4.0 technologies. Deloitte Insights. Accessed: 01 Feb 2024. [Online]. Available: https://www2.deloitte.com/content/dam/Deloitte/ca/Documents/energy-resources/ca-ener-aerospace-and-defense-4-aoda.pdf

Mabkhot MM et al. (2021) Mapping industry 4.0 enabling technologies into united nations sustainability development goals. Sustainability 13(5)5, Jan 2021, https://doi.org/10.3390/su13052560

Maisiri W, Darwish H, van Dyk L (2019) An investigation of industry 4.0 skills requirements. SAJIE 30(3), 3, Nov 2019, https://doi.org/10.7166/30-3-2230

Manohar K, Hogan T, Buttrick J, Banerjee AG, Kutz JN, Brunton SL (2018) Predicting shim gaps in aircraft assembly with machine learning and sparse sensing. J Manuf Syst 48:87–95. https://doi.org/10.1016/j.jmsy.2018.01.011

Marjani M et al (2017) Big IoT data analytics: architecture, opportunities, and open research challenges. IEEE Access 5:5247–5261. https://doi.org/10.1109/ACCESS.2017.2689040

Martinelli A, Mina A, Moggi M (2021) The enabling technologies of industry 4.0: examining the seeds of the fourth industrial revolution. Ind Corp Chang 30(1):161–188. https://doi.org/10.1093/icc/dtaa060

Martínez-de Dios JR, Torres-González A, Paneque JL, Fuego-García D, Ramírez JRA, Ollero A (2018) Aerial robot coworkers for autonomous localization of missing tools in manufacturing plants. In: 2018 International conference on unmanned aircraft systems (ICUAS), pp 1063–1069. https://doi.org/10.1109/ICUAS.2018.8453291

Memon DO (2024) 300 parts down to just 7: the benefits of general electric's additive manufacturing techniques. Simple Flying. Accessed: 06 Feb 2024. [Online]. Available: https://simpleflying.com/general-electric-additive-manufacturing-benefits/

Mentsiev A, Guzueva E, Magomaev T (2020) Security challenges of the Industry 4.0. J Phys Conf Ser 1515:032074. https://doi.org/10.1088/1742-6596/1515/3/032074

Michael N, Bruno ZS, Kenneth Y (2019) Systems and methods for cold spray additive manufacturing and repair with gas recovery. US 11857990 B2, 30 Aug 2019 [Online]. Available: https://lens.org/081-480-908-593-891

Mohammed CM, Zeebaree SRM (2021) Sufficient comparison among cloud computing services: IaaS, PaaS, and SaaS: a review. Int J Sci Bus 5(2):17–30. https://doi.org/10.5281/zenodo.4481415

Nagorny K, Lima-Monteiro P, Barata J, Colombo AW (2017) Big data analysis in smart manufacturing: a review. Int J Commun Netw Syst Sci 10(3), 3, Mar 2017. https://doi.org/10.4236/ijcns.2017.103003

Nain G, Pattanaik KK, Sharma GK (2022) Towards edge computing in intelligent manufacturing: past, present and future. J Manuf Syst 62:588–611. https://doi.org/10.1016/j.jmsy.2022.01.010

Oberheitmann A (2020) Industry 4.0—economic benefits and challenges, especially for small and medium-sized enterprises. In: Oberheitmann A, Heupel T, Junqing Y, Zhenlin W (eds) German and Chinese contributions to digitalization: opportunities, challenges, and impacts., in FOM-Edition. Springer Fachmedien, Wiesbaden, pp 13–22. https://doi.org/10.1007/978-3-658-29340-6_2

Osterrieder P, Budde L, Friedli T (2020) The smart factory as a key construct of industry 4.0: a systematic literature review. Int J Prod Econ 221:107476. https://doi.org/10.1016/j.ijpe.2019.08.011

Parrott A, Warshaw L (2017) Industry 4.0 and the digital twin. New York Deloitte University Press, May 2017

Patent Analytics—patent-analytics—WIPO Liferay DXP (2024) Patent-analytics. Accessed: 02 Feb 2024. [Online]. Available: https://www.wipo.int/web/patent-analytics

de Paula Ferreira W, Armellini F, De Santa-Eulalia LA (2020) Simulation in industry 4.0: a state-of-the-art review. Comput Ind Eng 149:106868, Nov 2020, https://doi.org/10.1016/j.cie.2020.106868

Peasley S (2024) Cybersecurity for smart factories in the manufacturing industry | Deloitte US. Accessed: 06 Feb 2024. [Online]. Available: https://www2.deloitte.com/us/en/pages/energy-and-resources/articles/smart-factory-cybersecurity-manufacturing-industry.html

Pereira T, Barreto L, Amaral A (2017) Network and information security challenges within industry 4.0 paradigm. Procedia Manufact 13:1253–1260. https://doi.org/10.1016/j.promfg.2017.09.047

Perez-Grau FJ et al (2021) Introducing autonomous aerial robots in industrial manufacturing. J Manuf Syst 60:312–324. https://doi.org/10.1016/j.jmsy.2021.06.008

Pessoa MAO, Pisching MA, Yao L, Junqueira F, Miyagi PE, Benatallah B (2018) Industry 4.0, how to integrate legacy devices: a cloud IoT approach. In: IECON 2018—44th annual conference of the IEEE industrial electronics society, Oct 2018, pp 2902–2907. https://doi.org/10.1109/IECON.2018.8592774

Pinon Fischer OJ et al (2022) Digital twin: reference model, realizations, and recommendations. INSIGHT 25(1):50–55. https://doi.org/10.1002/inst.12373

Popović N, Popović B (2021) Some robotics concepts for the industry 4.0 applications. Int Sci J ind 4.0 VI(4):131–134

Rani S, Tripathi K, Kumar A (2023) Machine learning aided malware detection for secure and smart manufacturing: a comprehensive analysis of the state of the art. Int J Interact Des Manuf. https://doi.org/10.1007/s12008-023-01578-0

Rani S, Tripathi K, Arora Y, Kumar A (2022) A machine learning approach to analyze cloud computing attacks. In: 2022 5th international conference on contemporary computing and informatics (IC3I), pp 22–26. https://doi.org/10.1109/IC3I56241.2022.10073468

Ras E, Wild F, Stahl C, Baudet A (2017) Bridging the skills gap of workers in industry 4.0 by human performance augmentation tools: challenges and roadmap. In: Proceedings of the 10th international conference on PErvasive technologies related to assistive environments, Island of Rhodes Greece: ACM, Jun. 2017, pp 428–432. https://doi.org/10.1145/3056540.3076192

Rodrigues D, Carvalho P, Rito Lima S, Lima E, Lopes NV (2022) An IoT platform for production monitoring in the aerospace manufacturing industry. J Cleaner Production 368:133264, Sep 2022, https://doi.org/10.1016/j.jclepro.2022.133264

Russell R et al (2019) Qualification and certification of metal additive manufactured hardware for aerospace applications. In: Froes F, Boyer R (eds) Additive manufacturing for the aerospace industry. Elsevier, pp 33–66. https://doi.org/10.1016/B978-0-12-814062-8.00003-0

Sahoo S (2022) Big data analytics in manufacturing: a bibliometric analysis of research in the field of business management. Int J Prod Res 60(22):6793–6821. https://doi.org/10.1080/00207543.2021.1919333

Sayem A, Biswas PK, Khan MMA, Romoli L, Dalle Mura M (2022) Critical barriers to Industry 4.0 adoption in manufacturing organizations and their mitigation strategies. J Manufact Mater Process 6(6), no. 6, Dec 2022. https://doi.org/10.3390/jmmp6060136

Sehrawat S, Kumar A, Prabhakar M, Nindra J (2022) The expanding domains of 3D printing pertaining to the speciality of orthodontics. Mater Today Proc 50:1611–1618. https://doi.org/10.1016/j.matpr.2021.09.124

Sertoglu K (2024) Boeing takes to the sky with Chinook's first 3D printed flight-critical part. 3D Printing Ind. Accessed: 01 Jan 2024. [Online]. Available: https://3dprintingindustry.com/news/boeing-takes-to-the-sky-with-chinooks-first-3d-printed-flight-critical-part-194134/

Sharma A, Kosasih E, Zhang J, Brintrup A, Calinescu A (2022) Digital Twins: State of the art theory and practice, challenges, and open research questions. J Ind Inf Integr 30:100383. https://doi.org/10.1016/j.jii.2022.100383

Sharma P, Singh Ghatorha K, Kang AS, Cepova L, Kumar A, Phanden RK (2024) Strategic insights in manufacturing site selection: a multi-method approach using factor rating, analytic hierarchy process, and best worst method. Front Mech Eng 10:1392543

Sharma A, Pandey H (2020) Big data and analytics in industry 4.0. In: Nayyar A and Kumar A (eds) A roadmap to industry 4.0: smart production, sharp business and sustainable development. in advances in science, technology & innovation. Springer International Publishing, Cham, pp 57–72. https://doi.org/10.1007/978-3-030-14544-6_4

Sharma LK, Ajay P, Kumar R (2023) Smart manufacturing and industry 4.0: state-of-the-art review. In: Handbook of smart manufacturing. CRC Press

Sigov A, Ratkin L, Ivanov LA, Xu LD (2022) Emerging enabling technologies for industry 4.0 and beyond. Inf Syst Front. https://doi.org/10.1007/s10796-021-10213-w

Singh S (2023) Digital twins in aerospace—A paradigm shift. Accessed: 28 Dec 2023. [Online]. Available: https://www.spsairbuz.com/story/?id=1119&h=Digital-Twins-in-Aerospace-A-Paradigm-Shift

Sony M (2018) Industry 4.0 and lean management: a proposed integration model and research propositions. Prod Manufact Res 6:416–432. https://doi.org/10.1080/21693277.2018.1540949

Srivastava AK, Kumar A, Kumar P, Gautam P, Dogra N (2023) Research progress in metal additive manufacturing: challenges and opportunities. Int J Interact Des Manuf. https://doi.org/10.1007/s12008-023-01661-6

Steele L, Ei-wardany TI, Filburn TP (2022) Shape memory alloy particle toughening of cast or additive manufactured Al—cu—mg—ag—tib2. US 11873549 B2, Mar. 18, 2022 [Online]. Available: https://lens.org/129-420-016-259-746

Stergiou C, Psannis KE, Gupta BB, Ishibashi Y (2018) Security, privacy and efficiency of sustainable cloud computing for big data & IoT. Sustain Comput Inform Syst 19:174–184. https://doi.org/10.1016/j.suscom.2018.06.003

Tadesse H, Singh B, Deresso H et al (2024) Investigation of production bottlenecks and productivity analysis in soft drink industry: a case study of East Africa Bottling Share Company. Int J Interact Des Manuf. https://doi.org/10.1007/s12008-023-01715-9

Takafumi K, Yasutaka I, Tatsuki S, Shinsuke I, Sachiko K (2023) Information management system and method for autonomous control of manufacturing process and service. US 2024/0004373 A1, 18 Sep 2023 [Online]. Available: https://lens.org/152-041-794-992-016

The Lens—Free & Open Patent and Scholarly Search (2023) The lens—Free & open patent and scholarly search. Accessed: 13 Jan 2023. [Online]. Available: https://www.lens.org/lens

Top 10 Applications & Use Cases for Digital Twins | Unity. Accessed: 06 Feb 2024. [Online]. Available: https://unity.com/solutions/digital-twin-applications-and-use-cases

Ur Rehman MH, Yaqoob I, Salah K, Imran M, Jayaraman PP, Perera C (2019) The role of big data analytics in industrial Internet of Things. Future Gener Comput Syst 99:247–259, Oct 2019. https://doi.org/10.1016/j.future.2019.04.020

Van Dinter R, Tekinerdogan B, Catal C (2022) Predictive maintenance using digital twins: a systematic literature review. Inf Softw Technol 151:107008. https://doi.org/10.1016/j.infsof.2022.107008

Waterman S (2024) Digital twinning takes flight | AFCEA international. Accessed: 06 Feb 2024. [Online]. Available: https://www.afcea.org/signal-media/digital-twinning-takes-flight

Watson V, Tellabi A, Sassmannahausen J, Lou X (2017) Interoperability and security challenges of industry 4.0, presented at the INFORMATIK 2017, Gesellschaft für Informatik, Bonn, pp 973–985. Accessed: 08 Feb 2024. [Online]. Available: https://dl.gi.de/items/1df76516-c591-47ef-8767-b6a277b0b7d8

Węgrzyn N (2022) The use of additive manufacturing for production of commercial airplane power plants components. Saf Defense 8(2), 2, Dec. 2022. https://doi.org/10.37105/sd.185

Wuggetzer I (2024) IoT: aerospace's great new connector | Airbus. IoT: Aerospace's great new connector | Airbus. Accessed: 02 Jan 2024. [Online]. Available: https://www.airbus.com/en/newsroom/stories/2019-07-iot-aerospaces-great-new-connector

Yadav AS, Kumar A, Yadav KK et al (2023) Optimization of an inventory model for deteriorating items with both selling price and time-sensitive demand and carbon emission under green technology investment. Int J Interact Des Manuf. https://doi.org/10.1007/s12008-023-01689-8

Younan M, Houssein EH, Elhoseny M (2020) Challenges and recommended technologies for the industrial internet of things: a comprehensive review. Measurements 151:107198. https://doi.org/10.1016/j.measurement.2019.107198

Zawra L (2019) Migration of legacy industrial automation systems in the context of industry 4.0-a comparative study. Feb 2019, pp 1–7. https://doi.org/10.1109/ICFIR.2019.8894776

Comparative Multi-criteria-Decision Making Approach for the Optimization of Abrasive Water Jet Machining Process Parameters Using MABAC

Bimal Tamang, Satyam Sharma, Soham Das, Spandan Guha, Ajay Kumar, and Ranjan Kumar Ghadai

Abstract Non-traditional machining (NTM) techniques have become pivotal in addressing various manufacturing challenges, ranging from handling high-strength materials to achieving precise surface finishes and minimizing production times. The NTM techniques like electro-discharge machining (EDM), abrasive water jet machining (AWJM), any many more plays vital roles in obtaining desired work-piece dimensions and smoothness. AWJM, is a versatile method used for cutting both soft and hard materials. It offers precision with minimal distortion and no heat-affected zones, making it ideal for complex designs and precise cuts. The environmental benefits of water jet cutting, such as minimal waste and no harmful byproducts, contribute to its popularity in the machining industry. multi criteria decision making (MCDM) plays a critical role in optimizing machining processes, allowing for continuous improvement and adaptation to evolving industry demands. Optimization of various machining parameters to get a desired response is a challenging task for the researchers. Recent studies showcase how (MCDM) techniques optimizes the machining parameters for specific materials like carbon nanotube-reinforced aluminum and fiber-reinforced composites, achieving improved properties and machining performance. In the current work three objective weight calculation techniques like equal weight (EW), standard deviation (STD) and entropy (ENT)

B. Tamang · S. Sharma · S. Das
Department of Mechanical Engineering, Sikkim Manipal Institute of Technology, Sikkim Manipal University, Sikkim, India

S. Guha
School of Mechanical Engineering, KIIT University, Bhubaneswar, India

A. Kumar
Department of Mechanical Engineering, School of Engineering & Technology, JECRC University, Jaipur, Rajasthan, India

R. K. Ghadai (✉)
Department of Mechanical and Industrial Engineering, Manipal Institute of Technology, Manipal Academy of Higher Education, Manipal, India
e-mail: ranjankumarbls@gmail.com

A. Kumar et al. (eds.), *Industry 4.0 Driven Manufacturing Technologies*, Springer Series in Advanced Manufacturing, https://doi.org/10.1007/978-3-031-68271-1_6

were considered while calculating the rank using multi-attributive border approximation area comparison (MABAC). From the results it is observed that exp. no 14 is the first rank for both MABAC-EW and MABAC-ENT. However, exp. no 8 is rank one for MABAC-STD. Similarly, exp. no 1 is the last rank for all three methods. From the correlation analysis it is noticed that all the three methods have strong positive correlation between each other. Thus, it can also be concluded that MABAC technique is not weight sensitive and it is showing robust results for the present AWJM of aluminium metal matrix composites.

Keywords AWJM · MCDM · MABAC · STD · ENT · Composites

1 Introduction

Nowadays, machining techniques play a crucial role in solving a range of manufacturing challenges, including machining high-strength materials, producing complex profiles, achieving superior surface finishes, attaining high precision, facilitating miniaturization, reducing waste, minimizing secondary operations, and shortening production times. Machining may be described as removing unwanted part from a product using power-driven tools. The various types of machining processes include turning, milling, facing, grinding etc. which are crucial operations that must be done on any workpiece in order to acquire a desired product of standard dimensions and smoothness. Past literatures have witnessed that the machining parameters plays critical role for achieving the desired results after the machining process. To finalize input data for any machining operation MCDM techniques are widely used by the researchers. Non-traditional machining (NTM) techniques have become pivotal in addressing various manufacturing challenges, ranging from handling high-strength materials to achieving precise surface finishes and minimizing production times. The NTM techniques like EDM, ultrasonic machining (USM), AWJM etc. play vital roles in obtaining desired workpiece dimensions and smoothness. AWJM, is a versatile method used for cutting both soft and hard materials. AWJM has been developed in the 1950s as a method for cutting soft materials using high-pressure water. Initially used for cutting softer materials like paper and rubber, AWJM gained popularity in the 1980s for its ability to cut harder materials like metals and composite. Abrasive water jet machining (AWJM) is a versatile and efficient material removal process that has gained widespread popularity in various industries due its unique benefit. AWJM is a non-traditional machining process that utilizes a high velocity stream of water mixed with abrasive particles to erode and remove material from a workpiece. AWJM can achieve high precision machining with minimal material distortion and heat affected zones. Unlike conventional machining methods such as milling and turning. Water jet technology stands out for its ability to handle a broad spectrum of materials, offering the flexibility to execute complex designs and achieve precise cuts without regard to the material's hardness. It offers precision with minimal distortion and no heat-affected zones, making it ideal for complex designs and precise cuts. The environmental benefits of water jet cutting, such as minimal waste and no harmful byproducts, contribute to its popularity in the machining industry. MCDM plays a critical role in optimizing machining processes, allowing for continuous improvement and adaptation to evolving industry demands. Optimization of various machining parameters to get a desired response is a challenging task for the researchers (Asjad and Talib 2018; Kumar et al. 2022; Vijayananth et al. 2024; Natarajan et al. 2020; Momber and Kovacevic 2012; Murthy et al. 2024; Prasad and Chaitanya 2024). This cutting method is distinguished by its gentle approach, avoiding any physical force that could lead to deformation or stress on the materials being processed. Beyond mere cutting, water jet systems boast multifunctional capabilities, enabling operations like deburring, drilling, turning, and milling, thereby enhancing manufacturing versatility (Kumar and Babu 2024). Kumar et al. investigates mechanical properties and machinability of carbon nanotube-reinforced aluminium nanocomposites, Using Taguchi's experimental design. Varying CNT weight fractions (3%, 6%, 9%) in AA7475 alloy, higher CNT concentrations exhibit improved tensile strength, compressive strength, hardness, wear, and corrosion resistance. Increased CNT necessitates higher abrasive flow rates for machining (Kumar et al. 2022). Vijayananth et al. investigated the impact of adding boron nitride (BN) in the range of 2–6 wt% over epoxy/glass fiber along with further addition of montmorillonite (MMT) through compression moulding. The experiment was conducted to investigate the impact of AWJM parameters such as transverse speed, stand-off distance, pump pressure, and filler percentage on material removal rate (MRR), surface roughness (Ra), and kerf taper (K_t). Experimental design using the Taguchi method and optimization via the CRITIC- Complex Proportional Assessment (COPRAS) technique revealed that BN addition reduced MRR, improved Ra, and decreased K_t, achieving optimal machining performance with 16.20 mm^3/min MRR, 0.29° K_t, and 3.86 μm Ra (Vijayananth et al. 2024). The absence of traditional cutting tools not only streamlines the production process by eliminating the need for frequent tool changes but also significantly reduces wear and tear on equipment. Furthermore, the high machining rates associated with water jet cutting translate into efficient production workflows and notable cost savings, underscoring its effectiveness and economic advantages in industrial applications along with environmental benefits (Abouzaid et al. 2024; Fuse et al. 2024; Kumar et al. 2018).

Decision making is the study of looking for alternate data models and selecting the best possible model using a decision maker. MCDM is an operation research theory that uses computational methods that incorporates several criteria to attain the best possible option among various alternative models to acquire the best possible outcome (Das et al. 2024). It can be used in various field from finance to engineering models. In the present study the focus has been laid over the application of various MCDM techniques for attaining the optimal machining parameters for the machining of components. Each criterion may have its own benefits and drawbacks and therefore, weightage must be given to each parameter as per their importance which plays a vital role during decision making as it affects the decision makers preference on the models. Values of parameters during machining such as feed rate, depth of cut etc. can be modified by using MCDM technique as per the requirements are attained. Satpathy et al. used EDM method on Al-SiC/SiC reinforced metal matrix composite and observed maximum MRR by keeping the input parameters as follows: High Input Current, Low Pulse on Time, High Duty Cycle and Intermediate Gap Voltage (Satpathy et al. 2017). Chenrayan et al. incorporated hybrid grey relational analysis (GRA)-principal component analysis (PCA) mathematical model to minimize the harmful effects of kerf taper and delamination caused due to AWJM process. Optimal parameters reduce kerf width and delamination by 11.72% and 33.9% respectively (Chenrayan et al. 2022). Vats et al. investigated turning operations on AISI 1040 steel with carbide inserts and coconut oil used as lubricant through minimum quantity lubrication (MQL). They employed an L9 orthogonal array for experimental design and utilized Taguchi's method combined with AHP-WASPAS to optimize process parameters. The results showed notable reductions: force by 15.95%, surface roughness by 13.92%, and power by 13.7%. (Vats et al. 2022).

The available literatures have suggested that MCDM techniques have a wide scope for selection of input process parameters in a machining process (Hosouli et al. 2024; Kumar Ghadai et al. 2023; Ziarh et al. 2024; Kalita et al. 2023; Kang et al. 2024). However, it is observed that the application of MABAC technique for the selection and optimization of input and response parameters have not been adopted ot the best of the knowledge. Therefore, the present work focuses of three main objectives (i) weight calculation of AWJM input parameters using equal weight (EW) technique, (ii) weight calculation of AWJM input parameters using standard deviation (STD) and entropy (ENT) technique, and (iii) rank calculation of AWJM process parameters using using MABAC technique.

2 Methodology

In the present case the experimental data has been taken from Gowthama et al. (2023) who have worked on the machining of Aluminium AA6026 metal matrix composites using reinforced with 20-30 μm sized Silicon Carbide (SiC) particles. The fabrication of the composite was done through stir casting process and the machining used was AWJM technique. The composition of Al AA6026 and experimental parameters (both input and response) are given in the Tables 1 and 2:

2.1 *MABAC Method*

MABAC technique is presented by Pamucar and Cirovic in 2015, is a valuable tool for decision analysis. Essentially, MABAC helps decision-makers find the best option by measuring how close alternatives are to a defined border area. This method has been adapted to work well in different fuzzy environments, like Pythagorean fuzzy sets or interval-valued intuitionistic fuzzy sets. These adaptations show that MABAC

Table 1 Composition of Al 6026

Manganese (Mn)	Iron (Fe)	Magnesium (Mg)	Silicon (Si)	Copper (Cu)	Lead (Pb)	Bismuth (Bi)	Zinc (Zn)	Chromium (Cr)	Titanium (Ti)	Other (Each)	Others (Total)	Aluminium (Al)
0.2–1	0–0.7	0.6–1.2	0.6–1.4	0.2–0.5	0–0.4	0.5–1.5	0–0.3	0–0.05	0–0.3	0–0.2	0–0.5	Balance

Table 2 Calculation of normalized matrix using equal weightage

Trial	Ra	MRR	Ka	Ra	MRR	Ka
1	3.73	9.29	1.13	− 0.0406	− 0.1178	− 0.1291
2	3.45	9.29	1.08	0.2050	− 0.1178	− 0.0009
3	3.75	9.29	1.12	− 0.0582	− 0.1178	− 0.1034
4	3.72	37.15	1.12	− 0.0319	0.0046	− 0.1034
5	3.43	37.15	1.10	0.2225	0.0046	− 0.0522
6	3.76	37.15	1.12	− 0.0670	0.0046	− 0.1034
7	3.56	83.58	1.10	0.1085	0.2087	− 0.0522
8	3.51	83.58	1.10	0.1523	0.2087	− 0.0522
9	3.61	83.58	1.10	0.0646	0.2087	− 0.0522
10	3.70	34.05	1.10	− 0.0143	− 0.0090	− 0.0522
11	3.44	34.05	1.07	0.2137	− 0.0090	0.0248
12	3.72	34.05	1.10	− 0.0319	− 0.0090	− 0.0522
13	3.71	76.61	1.08	− 0.0231	0.1780	− 0.0009
14	3.56	76.61	1.07	0.1085	0.1780	0.0248
15	3.73	76.61	1.08	− 0.0406	0.1780	− 0.0009
16	3.65	8.51	1.05	0.0295	− 0.1212	0.0760
17	3.63	8.51	1.03	0.0471	− 0.1212	0.1273
18	3.78	8.51	1.05	− 0.0845	− 0.1212	0.0760
19	3.81	69.65	1.10	− 0.1108	0.1475	− 0.0522
20	3.70	69.65	1.08	− 0.0143	0.1475	− 0.0009
21	3.75	69.65	1.10	− 0.0582	0.1475	− 0.0522
22	3.81	7.73	1.03	− 0.1108	− 0.1247	0.1273
23	3.65	7.73	1.00	0.0295	− 0.1247	0.2043
24	3.81	7.73	1.03	− 0.1108	− 0.1247	0.1273
25	3.73	30.95	1.05	− 0.0406	− 0.0226	0.0760
26	3.68	30.95	1.03	0.0032	− 0.0226	0.1273
27	3.78	30.95	1.05	− 0.0845	− 0.0226	0.0760

is versatile and can be used effectively in a wide range of decision-making situations (Liu and Cheng 2020).

For m number of alternatives starts from (X_1, X_2,..., X_m) and with n number of attributes starts from (Y_1, Y_2,..., Y_n) with weighting vector be w_j (j = 1, 2,..., n) and λ experts {d_1, d_2,..., d_λ} with weighing vector{ω_1, ω_2,..., ω_λ}.

MABAC decision-making model is presented as follows:

Step 1. To evaluate the Matrix it is constructed in the following manner

$$R = \left[X_{ij}^{\lambda}\right]_{m\times n}, i = 1, 2, \ldots, n \tag{1}$$

Thus,

$$R = \left[X_{ij}^{\lambda}\right]_{m\times n} = \begin{matrix} & \begin{matrix} Y_1 & Y_2 & \ldots & Y_n \end{matrix} \\ \begin{matrix} X_1 \\ X_2 \\ \vdots \\ X_m \end{matrix} & \begin{bmatrix} X_{11}^{\lambda} & X_{12}^{\lambda} & \ldots & X_{1n}^{\lambda} \\ X_{21}^{\lambda} & X_{22}^{\lambda} & \ldots & X_{2n}^{\lambda} \\ \vdots & \vdots & \vdots & \vdots \\ X_{m1}^{\lambda} & X_{m2}^{\lambda} & \ldots & X_{mn}^{\lambda} \end{bmatrix} \end{matrix} \quad i = 1, 2, \ldots, m, j = 1, 2, \ldots, n \quad (2)$$

where X^{λ}_{ij} (I = 1, 2, …, m, j = 1, 2, …, n) denotes the assessment in formation of alternative X_i (i = 1, 2, …, m) based on the attribute of Y_j (j = 1, 2, …, n) by decision maker d_{λ}.

Step 2. Normalization of fused results matrix, $R=[X^{\lambda}{}_{ij}]_{m\times n}$, *I = 1, 2, …, m, j=1, 2, …, n mainly depending over type of each attribute by following formula*:

For beneficial attributes:

$$N_{ij} = X_{ij}, \quad i = 1, 2, \ldots, m \text{ and } j = 1, 2, \ldots, n$$

For cost attributes:

$$N_{ij} = 1 - X_{ij}, i = 1, 2, \ldots, m \text{ and } j = 1, 2, \ldots, n$$

Step 3. Calculation of weighted normalization matrix as follows:

$$W_{ij} = w_j N_{ij} (i = 1, 2, \ldots, m \text{ and } j = 1, 2, \ldots, n) \quad (3)$$

Step 4. Calculation of Border Approximation Area (BAA) values:

$$g_j = \left(\prod_{i=1}^{m} WN_{ij}\right)^{1/m} \quad (i = 1, 2, \ldots, m, j = 1, 2, \ldots, n) \quad (4)$$

Step 5. Distance Calculation denoted by $D = [d_{ij}]_{m\times n}$ *among each alternative and BAA as*

$$d_{ij} = \begin{cases} d\left(WN_{ij}, g_j\right), \text{if} WN_{ij} > g_j \\ 0, \text{if} WN_{ij} = g_j \\ -d\left(WN_{ij}, g_j\right)\text{if} WN_{ij} < g_j \end{cases}; \quad (5)$$

where d (WN_{ij}, g_j) is the distance from WN_{ij} to g_j

If:

1. $d_{ij} > 0$, then the alternatives belong to G^{+}(UAA). (upper approximation area) (best alternate)
2. $d_{ij} = 0$, then the alternatives belong to G(BAA). (border approximation area)

3. $d_{ij} < 0$, then the alternatives belong to G^{-}(LAA). (lower approximation area) (worst alternate)

Step 6. Sum of values of all d_{ij} are calculated by the following formula:

$$S_i = \sum_{j=1}^{n} d_{ij}$$

From S_i, order of all alternatives can be derived, bigger the comprehensive evaluation result of S_i, better the result (Gowthama et al. 2023).

2.2 *Weight Allocation of Responses*

In MCDM, assigning weights to criteria is crucial for selecting process parameters. Various weight allocation techniques, such as mean/average weight, standard deviation, and entropy, can be integrated with MCDM techniques to rank alternatives effectively based on these criteria.

2.2.1 Mean Weight (MW) Method

Here, in this method equal weight is assigned to all responses (MRR, Ra, and Ka), with WMRR = 0.333, WRa = 0.333, and WKa = 0.333. It is a straightforward approach commonly used by past researchers to simplify the efforts applied for computation.

2.2.2 Standard Deviation (SDV) Method

This technique allocates weights to all criterion in an impartial manner, enhancing the accuracy of MCDM techniques and reducing personal bias in decision-making. The approach involves normalization of data in the starting decision matrix before calculating the criteria weights, ensuring fairness and objectivity in the process.

$$n''_{ij} = \frac{x_{ij} - \min(x_{ij})}{\max(x_{ij}) - \min(x_{ij})} \tag{6}$$

$$SDV_j = \sqrt{\frac{\sum_{i=1}^{m} (n''_{ij} - n''_j)^2}{m}} \quad (j = 1, 2, \ldots, n) \tag{7}$$

where n''_{ij} is the average of the normalized values for j-th criterion.

$$w_j = \frac{SDV_j}{\sum_{j=1}^{n} SDV_j} \tag{8}$$

Weights of the considered responses are estimated here using the aforementioned formulas and the experimental dataset of Table 2. $W_{MRR} = 0.40332$, $W_{Ra} = 0.3179$ and $W_{Ka} = 0.2788$.

2.2.3 Entropy (EN) Method

Here, the corresponding entropy weight function can be defined as follows:

$$e_j = -\frac{1}{\ln(m)} \sum_{i=1}^{m} n_{ij} \times \ln(n_{ij}) \tag{9}$$

For every criterion taken into consideration, the degree of diversity (d_j) is evaluated by employing Eq. (10).

$$d_j = 1 - e_j (j = 1, 2, \ldots, n) \tag{10}$$

Lastly, the following expression is used to estimate the weight for each criterion.:

$$w_j = \frac{d_j}{\sum_{j=1}^{n} d_j} \tag{11}$$

When the above formula is applied on the present dataset, the corresponding weights for MRR. Ra and Ka are achieved as $W_{MRR} = 0.3522$ and $W_{Ra} = 0.2957$, $W_{Ka} = 0.3522$.

3 Results and Discussion

The decision-making model of MABAC method was followed to find the rankings of the process parameter for equal weightage, standard deviation and entropy methods. Firstly, an evaluation matrix and its normalized matrix was calculated. The border Approximation Area (BAA) method was carried out to find the ranking order of the problem set with the parameters having the highest value of BAA having the best rankings. The BAA values are computed based on criteria Ra (Surface Roughness), MRR (Material Removal Rate), and Ka (Kerf Angle) for each trial in the experiment. The ranking for MABAC method is done through border approximation area which is calculated with formula (4) which was done for both equal weightage and standard deviation. The ranking is done in descending values of BAA with the largest value of BAA having the highest rank. It was observed that trial 14 had the highest rank for both equal weightage and entropy methods. Trial 8 had the highest ranking for standard deviation. The weights as per standard deviation were 0.3178, 0.4033, 0.2787 for Ra, MRR and Ka respectively. In the examination of input parameters, it was noted that the transverse speed (S) exerted the greatest influence on the parameter rankings, with trials 14, 8, and 7 consistently securing the highest rankings. Optimal rankings were attained when the transverse speed was set at 150 mm/min. Additionally, the weight percentage of SiC particulates (F) and the stand-off distance (D) exhibited comparatively lower impact on the rankings.

3.1 MABAC Analysis

Table 2 shows the calculation of a normalized matrix using equal weightage for criteria Ra (Surface Roughness), MRR (Material Removal Rate), and Ka (Kerf Angle). Each trial represents a different experimental run with corresponding values for Ra, MRR, and Ka. Normalization was performed to standardize the values within each column for fair comparison across criteria.

Table 3 displays the calculation of the Border Approximation Area (BAA) using a weighted normalized method based on criteria Ra (Surface Roughness), MRR (Material Removal Rate), and Ka (Kerf Angle) for each experimental trial.

- A positive BAA value indicates that the trial is closer to the border approximation area, suggesting a more desirable performance based on the criteria.
- A negative BAA value indicates that the trial is farther from the border approximation area, implying less favourable performance based on the criteria.

Table 4 presents the Border Approximation Area (BAA) values calculated using the standard deviation method for the MABAC (Multi-Attributive Border Approximation area Comparison) technique.

Table 5 presents the Border Approximation Area (BAA) values calculated using the MABAC-Entropy (M-ENT) method using equation no.6. This method utilizes entropy-based criteria weights to determine the proximity of each trial to the border approximation area in the MABAC analysis. The BAA values derived from the MABAC-Entropy (M-ENT) method provide valuable insights into the performance and proximity of experimental trials to the optimal solution in the context of decision-making and process optimization.

The ranking of all process parameters with Ra, MRR and Ka as the parameters are observed and mentioned in the Table 6.

Table 3 Calculation of Border Approximation Area (BAA) using weightage normalized method

Trial	Ra	MRR	Ka	BAA
1	0.4035	0.3402	0.3333	− 0.2875
2	0.6491	0.3402	0.4615	0.0863
3	0.3860	0.3402	0.3590	− 0.2794
4	0.4123	0.4626	0.3590	− 0.1307
5	0.6667	0.4626	0.4103	0.1750
6	0.3772	0.4626	0.3590	− 0.1658
7	0.5526	0.6667	0.4103	0.2650
8	0.5965	0.6667	0.4103	0.3089
9	0.5088	0.6667	0.4103	0.2211
10	0.4298	0.4490	0.4103	− 0.0755
11	0.6579	0.4490	0.4872	0.2295
12	0.4123	0.4490	0.4103	− 0.0930
13	0.4211	0.6360	0.4615	0.1541
14	0.5526	0.6360	0.4872	0.3113
15	0.4035	0.6360	0.4615	0.1365
16	0.4737	0.3368	0.5385	− 0.0156
17	0.4912	0.3368	0.5897	0.0532
18	0.3596	0.3368	0.5385	− 0.1297
19	0.3333	0.6054	0.4103	− 0.0155
20	0.4298	0.6054	0.4615	0.1323
21	0.3860	0.6054	0.4103	0.0371
22	0.3333	0.3333	0.5897	− 0.1081
23	0.4737	0.3333	0.6667	0.1091
24	0.3333	0.3333	0.5897	− 0.1081
25	0.4035	0.4354	0.5385	0.0128
26	0.4474	0.4354	0.5897	0.1079
27	0.3596	0.4354	0.5385	− 0.0311

3.2 Correlation of MCDM Techniques:

The correlation analysis of MABAC technique with all the three weight calculation methods are presented below Fig. 1. From the figure it is observed that all three-weight calculations such as equal weight, entropy weight and standard deviation weight combined with MABAC are showing strong correlation with each other. The results showed the robustness of the MABAC method which is not that sensitive to the various weight calculation processes.

Table 4 BAA for MABAC-standard deviation method

Trial	Ra	MRR	Ka	BAA
1	0.3848	0.4116	0.2788	– 0.2893
2	0.6191	0.4116	0.3860	0.0522
3	0.3681	0.4116	0.3002	– 0.2846
4	0.3932	0.5597	0.3002	– 0.1113
5	0.6358	0.5597	0.3431	0.1742
6	0.3597	0.5597	0.3002	– 0.1448
7	0.5270	0.8066	0.3431	0.3123
8	0.5689	0.8066	0.3431	0.3542
9	0.4852	0.8066	0.3431	0.2705
10	0.4099	0.5433	0.3431	– 0.0682
11	0.6274	0.5433	0.4075	0.2137
12	0.3932	0.5433	0.3431	– 0.0849
13	0.4015	0.7696	0.3860	0.1927
14	0.5270	0.7696	0.4075	0.3396
15	0.3848	0.7696	0.3860	0.1759
16	0.4517	0.4075	0.4504	– 0.0549
17	0.4685	0.4075	0.4932	0.0047
18	0.3430	0.4075	0.4504	– 0.1637
19	0.3179	0.7326	0.3431	0.0291
20	0.4099	0.7326	0.3860	0.1640
21	0.3681	0.7326	0.3431	0.0793
22	0.3179	0.4033	0.4932	– 0.1500
23	0.4517	0.4033	0.5576	0.0482
24	0.3179	0.4033	0.4932	– 0.1500
25	0.3848	0.5268	0.4504	– 0.0025
26	0.4266	0.5268	0.4932	0.0822
27	0.3430	0.5268	0.4504	– 0.0443

Table 5 BAA for MABAC-Entropy (M-ENT) method

Trial	Ra	MRR	Ka	BAA
1	0.3848	0.4116	0.2788	− 0.2969
2	0.6191	0.4116	0.3860	0.0564
3	0.3681	0.4116	0.3002	− 0.2854
4	0.3932	0.5597	0.3002	− 0.1327
5	0.6358	0.5597	0.3431	0.1472
6	0.3597	0.5597	0.3002	− 0.1638
7	0.5270	0.8066	0.3431	0.2616
8	0.5689	0.8066	0.3431	0.3005
9	0.4852	0.8066	0.3431	0.2227
10	0.4099	0.5433	0.3431	− 0.0773
11	0.6274	0.5433	0.4075	0.2063
12	0.3932	0.5433	0.3431	− 0.0929
13	0.4015	0.7696	0.3860	0.1667
14	0.5270	0.7696	0.4075	0.3105
15	0.3848	0.7696	0.3860	0.1511
16	0.4517	0.4075	0.4504	− 0.0215
17	0.4685	0.4075	0.4932	0.0482
18	0.3430	0.4075	0.4504	− 0.1227
19	0.3179	0.7326	0.3431	0.0024
20	0.4099	0.7326	0.3860	0.1422
21	0.3681	0.7326	0.3431	0.0491
22	0.3179	0.4033	0.4932	− 0.0955
23	0.4517	0.4033	0.5576	0.1103
24	0.3179	0.4033	0.4932	− 0.0955
25	0.3848	0.5268	0.4504	0.0204
26	0.4266	0.5268	0.4932	0.1135
27	0.3430	0.5268	0.4504	− 0.0185

Table 6 Rankings obtained through various MCDM methods

Trial	MABAC-EW	MABAC-STD	MABAC ENT
1	27	27	27
2	12	12	12
3	26	26	26
4	24	21	24
5	6	8	8
6	25	22	25
7	3	3	3
8	2	1	2
9	5	4	4
10	19	19	19
11	4	5	5
12	20	20	20
13	7	6	6
14	1	2	1
15	8	7	7
16	17	18	18
17	13	15	14
18	23	25	23
19	16	14	16
20	9	9	9
21	14	11	13
22	21	23	21
23	10	13	11
24	21	23	21
25	15	16	15
26	11	10	10
27	18	17	17

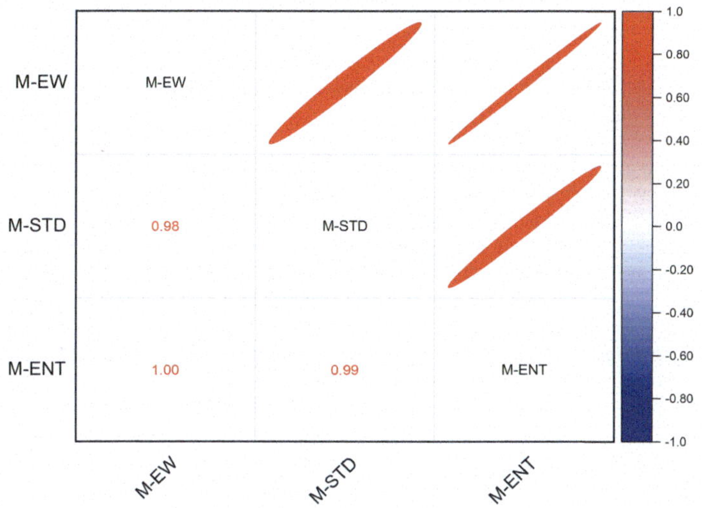

Fig. 1 Correlation analysis between MABAC-EW, MABAC-STD and MABAC-ENT

4 Conclusion

In the present case MABAC MCDM technique was used to analyse the AWJM process parameters for the machining of aluminium metal matrix composites. Three objective weight calculation techniques were considered while calculating the rank using MABAC. From the results it is observed that expt. no 14 is the first rank for both MABAC-EW and MABAC-ENT. However, expt. no 14 is rank 2 for MABAC-STD technique. Similarly, expt. no 8 is rank one for MABAC-STD and second rank for both MABAC-EW and MABAC-ENT. If we see the worst rank, then expt. no 1 is the last rank for all three methods. From the correlation analysis it is noticed that all the three methods have strong positive correlation between each other. Thus, it can also be concluded that MABAC technique is not weight sensitive and it is showing robust results for the present AWJM of aluminium metal matrix composites.

5 Future Scope of the Present Work

Future research could combine MABAC with other MCDM approaches, like TOPSIS, VIKOR, or AHP, to assess how well these strategies work in the setting of abrasive water jet machining. To improve decision-making processes, hybrid models that combine MABAC with machine learning techniques should be investigated. To verify the MABAC technique's adaptability and resilience across different manufacturing technologies, it can be used on datasets from machining methods like electrical discharge machining (EDM), laser cutting, and CNC milling.

References

Abouzaid A, Mousa S, Ibrahim AMM (2024) Effect of standoff distance and traverse speed on the cutting quality during the abrasive water jet machining (AWJM) of brass. In: Machining science and technology, pp 1–23

Asjad M, Talib F (2018) Selection of optimal machining parameters using integrated MCDM approaches. Int J Adv Oper Manage 10(2):109–129

Chenrayan V, Manivannan C, Shahapurkar K, Zewdu GA, Maniselvam N, Alarifi IM, Alblalaihid K, Tirth V, Algahtani A (2022) An experimental and empirical assessment of machining damage of hybrid glass-carbon FRP composite during abrasive water jet machining. J Market Res 19:1148–1161

Das S, Ghadai RK, Sapkota G, Guha S, Barmavatu P, Kumar KR (2024) Optimization of CNC turning parameters of copper-nickel (Cu–Ni) alloy using VIKOR, MOORA and GRA techniques. Int J Interactive Des Manuf (IJIDeM) 1–10

Fuse K, Vora J, Wakchaure K, Patel VK, Chaudhari R, Saxena KK, Bandhu D, Ramacharyulu DA (2024) Abrasive waterjet machining of titanium alloy using an integrated approach of taguchi-based passing vehicle search algorithm. Int J Interactive Des Manuf (IJIDeM) 1–15

Gowthama K, Somasheker HM, Suresha B, Singh PB, Rajini N, Mohammad F, Soleiman AA et al (2023) Characterization and optimization of abrasive water jet machining parameters of aluminium/silicon carbide composites. Mater Res Express 10(11):115505

Hosouli S, Gaikwad N, Qamar SH, Gomes J (2024) Optimizing photovoltaic thermal (PVT) collector selection: a multi-criteria decision-making (MCDM) approach for renewable energy systems. Heliyon 10(6)

Kalita K, Chakraborty S, Ghadai RK, Chakraborty S (2023) Parametric optimization of non-traditional machining processes using multi-criteria decision-making techniques: literature review and future directions. Multiscale and multidisciplinary modelling. Experiments Des 6(1):1–40

Kang D, Jaisankar R, Murugesan V, Suvitha K, Narayanamoorthy S, Omar AH, Arshad NI, Ahmadian A (2023) A novel MCDM approach to selecting a biodegradable dynamic plastic product: a probabilistic hesitant fuzzy set-based COPRAS method. J Environ Manage 340:117967

Kang YO, Yabar H, Mizunoya T, Higano Y (2024) Optimal landfill site selection using ArcGIS Multi-Criteria Decision-Making (MCDM) and Analytic Hierarchy Process (AHP) for Kinshasa City. Environ Challenges 14:100826

Kumar KR, Sreebalaji VS, Pridhar T (2018) Characterization and optimization of abrasive water jet machining parameters of aluminium/tungsten carbide composites. Measurement 117:57–66

Kumar SP, Shata AS, Kumar KP, Sharma R, Munnur H, Rinawa ML, Kumar SS (2022) Effect on abrasive water jet machining of aluminum alloy 7475 composites reinforced with CNT particles. Mater Today Proc 59:1463–1471

Kumar KN, Babu PD (2024) Improving the machining performance of polymer hybrid composite by abrasive water jet machining for precise machining. Arab J Sci Eng 1–20

Kumar Ghadai R, Chakraborty S, Kalita K (2023) On solving parametric optimization problem of an end milling process for machining of Al 1070 using MCDM techniques: a comparative analysis. In: Advances in materials and processing technologies, pp 1–23

Liu P, Cheng S (2020) An improved MABAC group decision-making method using regret theory and likelihood in probability multi-valued neutrosophic sets. Int J Inf Technol Decis Mak 19(05):1353–1387

Momber AW, Kovacevic R (2012) Principles of abrasive water jet machining. Springer Science & Business Media

Murthy BRN, Beedu R, Jayashree PK, Potti SR (2024) Study on machining quality in abrasive water jet machining of jute-polymer composite and optimization of process parameters through grey relational analysis. J Compos Sci 8(1):20

Natarajan Y, Murugesan PK, Mohan M, Khan SALA (2020) Abrasive water jet machining process: a state of art of review. J Manuf Process 49:271–322

Prasad KS, Chaitanya G (2024) Influence of abrasive water jet machining process parameters on accuracy of hole dimensions in glass fiber reinforced polymer composites. Mater Today Proc 98:135–142

Satpathy A, Tripathy S, Senapati NP, Brahma MK (2017) Optimization of EDM process parameters for AlSiC-20% SiC reinforced metal matrix composite with multi response using TOPSIS. Mater Today Proc 4(2):3043–3052

Vats P, Singh T, Dubey V, Sharma AK (2022) Optimization of machining parameters in turning of AISI 1040 steel using hybrid MCDM technique. Mater Today Proc 50:1758–1765

Vijayananth K, Pudhupalayam Muthukutti G, Keerthiveettil Ramakrishnan S, Venkatesan S, Zhou W (2024) An integrated CRITIC-COPRAS approach for multi-response optimization on AWJM of hybrid filler–reinforced polymer composite and its surface integrity. Int J Adv Manuf Technol 1–16

Ziarh GF, Kim JH, Chae ST, Kang HY, Hong C, Song JY, Chung ES (2024) Identifying the contributing sources of uncertainties in urban flood vulnerability in South Korea considering multiple GCMs, SSPs, weight determination methods, and MCDM techniques. Sustainability 16(8):3450

An Empirical Analysis of Factors Influencing Industry 4.0 Implementation in Manufacturing SMEs

Malika Aitzhanova, Dinara Dikhanbayeva, Ali Turkyilmaz, and Essam Shehab

Abstract Emerging technologies and opportunities of the fourth industrial revolution and Industry 4.0 enable companies to become more digitized. Maturity models have been developed to help organisations to assess their readiness for Industry 4.0 across various dimensions. However, there is a lack of research addressing the relationships between such dimensions and their effect on business results. This article aims to investigate critical Industry 4.0 factors used to assess and improve the digital maturity of small and medium-sized manufacturing enterprises and analyse the relationship between these critical factors and their effect on product and service performance. The research utilises structural equation modelling to test the relationships between dimensions such as strategy and organisation, workforce development, smart factory, smart processes, and smart products and services based on the data of 123 manufacturing small and medium enterprises (SMEs) from Kazakhstan. The research identifies that for successful digital transformation, the utmost attention should be placed on developing a working strategy and adapting the organisational structure to support the transition.

Keywords Industry 4.0 assessment · Critical digitisation factors · Manufacturing SMEs · Structural equation model · Emerging economies

1 Introduction

The new era of industrialisation, also known as Industry 4.0 (I4.0), symbolises a different form of human-independent, automated, and self-managing manufacturing processes. I4.0 aims to improve operational efficiency, increase productivity, and attain a higher degree of automation in production systems (Fettermann et al. 2018).

M. Aitzhanova · D. Dikhanbayeva · E. Shehab
School of Engineering and Digital Sciences, Nazarbayev University, Astana, Kazakhstan

A. Turkyilmaz (✉)
UIS School of Business and Law, University of Stavanger, 4021 Stavanger, Norway
e-mail: ali.turkyilmaz@uis.no

A. Kumar et al. (eds.), *Industry 4.0 Driven Manufacturing Technologies*, Springer Series in Advanced Manufacturing, https://doi.org/10.1007/978-3-031-68271-1_7

I4.0 transforms the approaches in supply chains by enabling intelligent and flexible organisations that can be easily adjusted to the customers' changing needs. It provides, in addition to technological development and digitisation, different organisational features and advantages such as decentralisation and service orientation (Lasi et al. 2014) and process and value chain optimisation (Cañas et al. 2022). To simplify, I4.0 may help organisations increase their overall quality, improve operational efficiency (Bayrak and Cebi 2024), and decrease costs (Kiraz et al. 2020).

Despite the opportunities that opened up with the 4th industrial revolution (Roblek et al. 2016), only around 50% of interviewed business owners believed they were ready for I4.0, according to the 2015 McKinsey poll (2015). Shifting to I4.0 may pose a challenge to all companies; however, special attention should be given to SMEs due to various issues they face, such as the absence of proper strategies, limited financial resources, lack of knowledge, qualified personnel, and other resources, as well as technology awareness (Amoozad Mahdiraji et al. 2022; Kamble et al. 2018; Turkyilmaz et al. 2020). Digital transformation in developing economies encounters many barriers, including lack of qualified personnel, especially with the required level of information and communication technology (ICT) knowledge, inefficient government policies, weak technological infrastructure, and the absence of unified digital transformation methodologies (Dikhanbayeva et al. 2020).

The complexity of the concept, lack of awareness and insufficient knowledge may lead to the oversight of some key elements and the disregard of significant indicators that are crucial for organisations during this challenging transformation (Hess et al. 2016; Adebanjo et al. 2023). Therefore, it is important for companies to prioritise critical I4.0 concepts and factors and systematically measure their performance based on these critical factors. To assist companies in developing I4.0 strategies and analysing their current digital maturity, numerous maturity models have been developed by academia and industry (Dikhanbayeva et al. 2020; Schumacher et al. 2016; Hajoary et al. 2023). However, the existing issue lies in the absence of clarity and details about the critical I4.0 dimensions, their weights, interrelations, and dependence. This lack of information leads to a knowledge gap, particularly for manufacturing SMEs, in focusing on specific dimensions to initiate the transition to Industry 4.0 (Hajoary et al. 2023; Dikhanbayeva et al. 2023). In addition, the diversity of resources about I4.0 and maturity models makes it even more challenging to identify the correct focal point. Therefore, analysis of relationships among critical I4.0 factors and their influence on the products and services could offer excellent learning material and guide the understanding of which dimension to focus to improve specific indicators, thus facilitating the I4.0 transformation (Bhatia and Kumar 2022).

Considering I4.0's objectives to transform business models and interconnections across the value chain it aims to create, this research involves an empirical analysis of the relationships between selected critical I4.0 factors and their effects on firms' products and services. The study was conducted based on data from the Kazakhstani manufacturing SMEs, which were assessed for their I4.0 readiness performance using the proposed maturity model. The research employed the Structural Equation Modelling (SEM) technique to analyse the proposed model and address the following research objectives:

- **RO1**: Understanding the critical business factors of I4.0 transformation in SMEs.
- **RO2**: Investigating the relationships between critical I4.0 dimensions and their impact on products and services.

2 Industry 4.0 Maturity Models and Critical Factors

In a competitive market, customer demands constantly change, and companies need to strive to become agile, efficient, and responsive to meet this changing demand (Sony and Naik 2020). Decisions concerning market offerings, organisational structure, technology implementation, human resources, and integral business processes are affected by the choice of business strategy and technology opportunity (Smuts et al. 2020). One of the main focuses of I4.0 is the production of *smart products and services* and the establishment of smart processes, which can be achieved through the conversion of traditional factories into smart ones (Nunes et al. 2017) This transformation is mainly dependent on various factors such as strategy, firm infrastructure and organisation, human resources and technology. These inputs are also closely connected to the supply chain, which has to be interconnected, automated, transparent, proactive, and customer-centric (Hofmann et al. 2019). Similarly, I4.0 transformation with essential information, operation systems, equipment, and installations with features of connectivity and interoperability will facilitate the establishment of autonomous systems (Santos and Martinho 2020). This connectivity then enables data collection, which affects product quality improvement (Müller et al. 2018). Thus, it is theorised that enhancement and digitisation of facilities and processes within the company will also facilitate the creation of *smarter products and services*.

Maturity models are used widely as performance management systems to assess and improve an organisation's digitisation performance across various business dimensions. They help to uncover the strengths and weaknesses in adopting digital technologies and culture in the context of Industry 4.0 (Fettermann et al. 2018; Dikhanbayeva et al. 2020; Felch et al. 2019; Ünlü et al. 2023). (Ünlü et al. 2023) conducted a systematic literature review to evaluate 22 I4.0 readiness and maturity models and assessed them for their objectivity and practical applicability in the I4.0 transition. The models, such as the Reference Architecture Model by Adolphs et al. (2015), PWC's Digital Operations Self-assessment (Geissbauer et al. 2016), Singapore Smart Industry Readiness Index (Singapore Smart Industry Readiness EDB 2018), IMPULS (Lichtblau et al. 2015), SIMMI 4.0 (Leyh et al. 2016), the Industry 4.0 Maturity Model (Schumacher et al. 2016), and the ACATECH Maturity Index (Schuh et al. 2020) provide a structured approach to assess various dimensions of digital transformation. Each model offers unique perspectives and criteria to help organisations gauge their Industry 4.0 maturity level and guide their digital strategy.

According to Sony and Naik (2020), the following themes were determined to be key components of I4.0 performance systems: organisational strategy, digitisation within the supply chain, the availability of smart products and services, employee acceptance of I4.0, and the involvement and commitment of top management. Hizam-Hanafiah et al. (2020) ‘s research based on the reviewed 30 maturity models, shows that technology, people, strategy, leadership, process, and innovation are the most common dimensions to improve the digital transformation performance (Hizam-Hanafiah et al. 2020). Similarly, (Kumar et al. 2023) identified technology, people, processes, products, and services as the most common dimensions of I4.0 and smart manufacturing (Kumar et al. 2023). Furthermore, having compared the I4.0 maturity model by Schumacher, Acatech, and Impuls models, Santos and Martinho (2020) concludes that adjustments to leadership, culture, human resources, and product and services should be considered in the roadmap of transformation to I4.0 along with operational and technological enhancements (Santos and Martinho 2020).

3 Theoretical Model

Based on the reviewed literature on the existing maturity models, this research includes an analysis of the following dimensions and their relationships in the context of manufacturing SMEs: strategy and organisation, workforce development, smart factory, smart processes, and smart products and services (Dikhanbayeva et al. 2023).

Strategy and organisation (SO): Strategic vision and the smoothness of organisational processes are crucial for every organisation’s decision-making. To manage various external and internal factors, companies must achieve strategic alignment across strategy execution, technology and competitive potential, and service level (Smuts et al. 2020). When new strategies are set and agility is applied, the organisational structure needs to be modified according to the new I4.0 strategy (Smuts et al. 2020). According to the study of Kane et al. (2015), more than 80% of companies are characterised by higher levels of digital maturity due to explicit and consistent digital strategies. Leadership support for I4.0 activities and collaboration with stakeholders regarding I4.0 projects are other key features. The faster the top management team identifies emerging technologies and opportunities, the better the company will be able to compete. This is supported by Sony (2020), who determines that strategic competitive advantage is one of the pros of I4.0. Machado et al. (2021) have also found that top management leadership and placing digital transformation at the centre of a strategy are foundation elements for organisational readiness to digital transformation. To implement initial investments into I4.0 technology and activities, including ICT and innovation, it is critical that organisational processes are well developed (Sony and Naik 2020).

Workforce development (WFD): Skilled workers are critical to manage and control integrated operations in I4.0 (Fettermann et al. 2018). Developing appropriate competencies concerning digital skills and critical thinking is required before introducing I4.0 projects (Sony and Naik 2020; Soomro et al. 2021). This can be achieved by specific investment on workforce development. Companies can face various responses to changes brought by I4.0 (Sony 2020). Proper change management to improve acceptance level is important, and it should include effective human resources strategies set by the top management (Müller et al. 2018). Therefore, the workforce development dimension investigates the criteria such as the level of digital competency of the workforce, their response to changes, and support from the organisation.

Smart factory (SF): Smart factories integrate the physical world with the digital. The five key features of smart factories are connectivity, optimisation, transparency, proactive systems, and agile flexibility (Burke et al. 2017). Being one of the main components of I4.0, the smart factory concept is responsible for a high level of automation, providing a competitive advantage, improving the quality of products and services produced, and allowing a higher customisation degree (Sony 2020). The smart factory dimension aims to assess the organisation for its automation level, communication, data exchange and integration capability within the supply chain, digitisation of enterprise data, real-time production observability, and utilisation of cloud services. The presence of such digital features in the production of products and services is crucial and can be achieved by embedding sensors and introducing IoT platforms and enhanced software programs (Dalenogare et al. 2018).

Smart processes (SP): A smart process converts raw material into a final product through I4.0 technologies (Mittal et al. 2018), which facilitate automation and innovation within organisations (Smuts et al. 2020). I4.0 calls for standardised processes within and across companies (Müller et al. 2018) and contributes to more stable and efficient operations by establishing innovative value-added processes (Gökalp et al. 2017). The digitisation of processes will help enterprises to gain a comprehensive understanding of their business processes, which is critical for efficient decision-making (Smuts et al. 2020). In this regard, the smart processes dimension is related to the standardisation of business processes, degree of automation of planning and management processes, data-driven decision-making, maintenance approach, and use of quality management and lean systems.

Smart products and services (SPS): Products and services can be considered "smart" if they poses key digital features such as data storage, communication and interaction with their environment, computation, and a high level of autonomy (Nunes et al. 2017). Integration of smart products with smart production, logistics, networks, and the Internet of Things leads to the transformation of value chains and innovative business models (Kagermann et al. 2013). Smart services within the dimension are assessed based on criteria such as the availability of online product listing and order/payment services, full traceability of service delivery, and product and service recommendation services based on customer experiences or other advanced data-driven services. Additionally, the introduction of the key elements like the use of intellectual property in product development, product customisation, frequency of product/service upgrades, customer data utilisation, and the variety of sales channels used facilitates the transition to Industry 4.0. According to Gaiardelli et al. (2021), digital technologies play a critical role for companies on their way towards service-based business models and improve information-sharing within service networks.

In order to operate and use new I4.0 technologies, employees need to possess the necessary qualifications and expertise (Müller et al. 2018). Further, as the workforce gains experience in digitisation, data usage, business processes, and lean management, the processes become more efficient and smart. For instance, Sjödin et al. (2018) state that implementing a smart factory is enhanced through developing digital skills among employees, introducing agile processes, and configuring modular technology. Hence, prior to the implementation of I4.0 technologies, it is critical that the relevant strategies be administered (Somohano-Rodríguez et al. 2020). As a result, the conceptual model (Fig. 1) for this research is structured to analyse how the digitisation of internal activities such as effective business strategies and strong organisation, smart factories, well-established smart processes, and competent workforce development affect the creation of smarter and more connected products and services for customers, either directly or indirectly.

Each arrow in the model represents a relationship (hypothesis) and direction between the factors involved. Although there may be additional relations between the I4.0 factors, the most important ones are considered based on the reviewed literature. As a result, the following hypotheses have been developed:

- Strategy and organisation for I4.0 positively impact smart products and services (H1), smart factory (H2), smart processes (H3) and workforce digital skills development (H4).
- Developing the workforce's digital skills positively affects smart processes (H5), which is expected to positively impact the smart factory (H6).
- Products and services are positively affected by the smart factory concept (H7) and smart processes (H8).

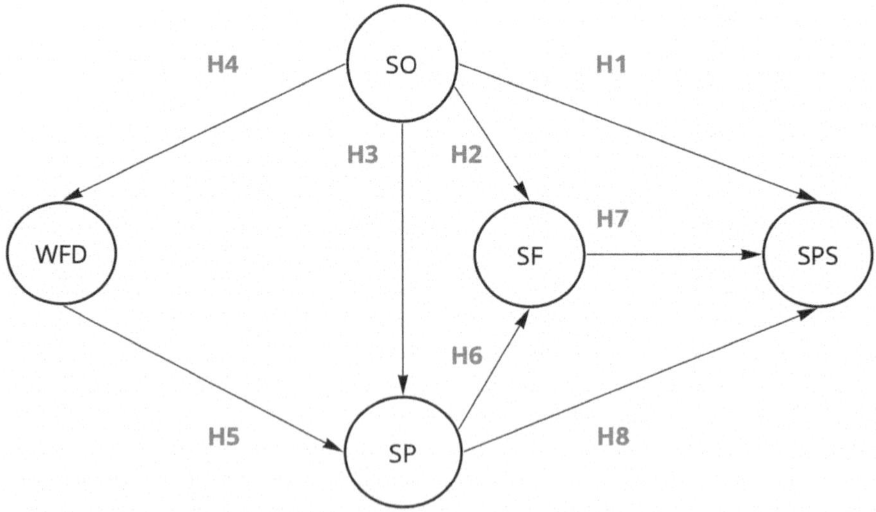

Fig. 1 Proposed structural model: the relationship between model dimensions

4 Methodology

4.1 Research Instrument

To estimate the relationships between the critical Industry 4.0 maturity factors and test the developed hypotheses, a structural equation model (SEM) was developed. The elements of the proposed SEM are the I4.0 dimensions reviewed above (i.e., strategy and organisation, workforce development, smart factory, smart processes, and smart product and services) and their corresponding indicators. The indicators were chosen as a result of extensive literature review of existing maturity models and critical I4.0 readiness measures, The model was validated with the experts in the field of digitisation and SMEs management (22 firms and researchers) to ensure that the model elements are appropriate for the SMEs of emerging economies such as Kazakhstan (Dikhanbayeva et al. 2023). Table 1 presents the main model dimensions and their corresponding sub-dimensions (observable variables) developed based on the feedback from this analysis.

The research instrument is a survey developed using the proposed model measures. It was pilot-tested with 25 SMEs for its design, wording, and correctness. The final version of the survey received the approval of the Institutional Research Ethics Committee.

Each I4.0 readiness measure in the model corresponds to a survey question designed as 1–5 scale Likert-type questions which indicate the maturity stages such as entrant (level 1), beginner (level 2), learner (level 3), integrator (level 4), and expert (level 5). Sample questions of the survey are presented in Table 2.

4.2 Data Collection

The study focused on analysing primary data collected by the research team using the developed research instrument. The survey was distributed via email to over 250 Kazakhstani manufacturing SMEs across various production fields. A total of 150 responses were collected from the manufacturing SMEs. Responses with missing data were excluded from the final sample, resulted in a final sample of 123 responses used in the study. The survey contains 40 questions including general questions about the company and its representatives, as well as questions measuring different I4.0 readiness factors.

Companies from almost all regions of the country were represented in the survey, with the majority coming from cities like Kyzylorda, Karagandy, and Petropavl. Among the respondents, 56% held senior management positions, 34% were from middle management, and 9% held junior-level positions. Table 3 provides details of the research sample, including the size and specific sector within the manufacturing industry. 54% of the companies were classified as small, with 10–50 employees, while 46% were medium-sized, employing between 50 and 200 workers. Also, the

Table 1 Main dimensions and corresponding sub-dimensions

Main dimensions/latent variable	Sub-dimensions/observable variable
Strategy and organization (SO)	SO1: I4.0 strategies SO2: business performance SO3: ICT organisation SO4: ICT financial management SO5: leadership support SO6: innovation, change management SO7: stakeholders collaboration for I4.0
Development of the workforce (WFD)	WFD1: WF digital competency WFD2: support for WF development WFD3: WF response to change
Smart processes (SP)	SP1: business process standardisation SP2: business information systems (MIS/ERP) SP3: data-driven decision-making SP4: maintenance management SP5: quality and lean management
Smart factory (SF)	SF1: automation of production processes SF2: enterprise asset and equipment upgradability SF3: level of data exchange M2M communication SF4: supply chain communication/integration SF5: enterprise data digitisation SF6: production systems observability SF7: ICT architecture SF8: cloud services
Smart product and services (SPS)	SPS1: intellectual property management SPS2: digital products SPS3: digital services SPS4: level of product customisation SPS5: upgrades in product/service systems SPS6: utilisation of customer data SPS7: sales channels

majority of companies were from the food, beverages, and tobacco production sectors (33%), followed by the textiles, leather, and apparel sectors (14%).

4.3 *Data Analysis and Results*

To achieve the research objectives, the Partial Least Squares Structural Equation Modelling (PLS-SEM) algorithm is utilised to estimate the proposed model parameters and validate the measurement model. PLS-SEM, developed by Wold (2004), combines the confirmatory factor analysis and multiple regression method to study the relationship between multiple variables. It does not have strict assumptions and perfectly works for non-normal data, small sample size, and fewer indicators for each construct (Tenenhaus et al. 2005; Henseler and Sarstedt 2013; Dijkstra and Henseler

Table 2 Sample question from the maturity survey

Question	Level 1 Entrant	Level 2 Beginner	Level 3 Learner	Level 4 Integrator	Level 5 Expert
SO1: strategy implementation level Please indicate the I4.0 strategy implementation level in your organisation	No strategy is available	The strategy is at the development stage	The strategy is formulated but not implemented	The strategy is formulated but implemented partially in some departments only	The strategy is formulated and implemented across the organisation
SF1: automation of production system Please indicate the level of automation of your enterprise production and systems processes	Not at all	Slightly automated on the level of sending the machinery and equipment data	Somewhat automated on the level of using automatic monitoring systems and devices	Moderately automated on the level of production line operations	Fully automated on the level of using a mix of industrial automation systems on the factory level

Table 3 Characteristics of the research sample

	Total number of companies: 123		
		Count	Percentage
Size	Small (from 10 to 50)	66	54
	Medium (from 50 to 200)	57	46
Manufacturing industry	Food, beverage, and tobacco	41	33
	Textiles, leather, and apparel	17	14
	Furniture	11	9
	Wood, paper, and printing	8	7
	Nonmetallic mineral	8	7
	Electric equipment, appliances, and components	7	6
	Petroleum, coal, chemicals, plastics, and rubber	6	5
	Primary metal, fabricated metal, and machinery	4	3
	Miscellaneous manufacturing	21	17

2015; Turkyilmaz et al. 2018). Due to these advantages, as well as its reliability and ability to investigate various dependencies simultaneously, the PLS-SEM technique has been selected to estimate the proposed model's parameters (Fan et al. 2016).

PLS technique performs two-phase algorithms to estimate weights, loadings, and latent variable scores (Wold 2004; Fornell and Cha 1994; Chin 1998). In the first phase, an iterative scheme of simple and/or multiple regressions is used to converge on a set of weights for the measurement variables, followed by the calculation of the latent variable scores. In the second phase, a non-iterative implementation of least squares regression is performed to estimate path coefficients, mean scores, and location parameters of the structural model.

In the developed SEM model, the latent variables are reflective, meaning the manifest variables are representative of their connected latent variable. These latent variables are defined by observable variables which correspond to indicators of the proposed I4.0 model. The results of the outer model, calculated using the SmartPLS software package (Ringle et al. 2022), are presented in Table 4.

Since the factors that are measured are closely interconnected with each other, several indicators are cross-loaded onto two or more components. In this case, indicator was assigned to a factor corresponding to a model dimension it belongs to by design of the maturity model. The only construct that has been explained by the indicators from a different dimension is Smart Processes. The results suggest that SF1 (level of automation of enterprise production and systems processes) and SF2 (upgradability extent of enterprise equipment) are more correlated with the smart processes dimension rather than the smart factory dimension. This can be explained by the fact that indicators are quite similar in their sense, and respondents might have had difficulties differentiating them.

Factor loadings are accessed in terms of their reliability. All factor loadings satisfied the criteria of reliability (> 0.708) except for WFD3 (0.640), SF4 (0.603), SPS5 (0.603), SPS6 (0.631) and SPS7 (0.653) (Hair et al. 2019). Composite reliability was used to test the internal consistency with a threshold of 0.7, and the results are consistent. Cronbach's alpha results are considered acceptable for WFD and good for the rest of the factors. It should be noted that composite reliability results are considered more reliable since the scores are weighted while Cronbach's Alpha values are unweighted (Hair et al. 2019). Convergent validity is also achieved as the results of the average variance extracted for all constructs is greater than 0.5. Discriminant validity was assessed by HTMT ratios. Estimated ratios meet the criteria of a threshold value of 0.90 as the model constructs are conceptually very similar (Hair et al. 2019; Henseler et al. 2015). The results of HTMT are presented in Table 5.

Based on the final design that meets all reliability and validity tests, the model parameters are estimated and the detailed results are presented in Fig. 2. Values on the arrows are weights and *p*-values in parentheses.

Since in PLS-SEM, the data is not assumed to have a normal distribution, a bootstrapping method is used to test the hypotheses and the structural relationships according to the model in Fig. 2. Bootstrapping employs resampling methods to calculate the significance of PLS coefficients (Garson 2016). As can be derived from Table 6, all hypotheses are confirmed with *p*-values less than 0.5, calculated using 5000 bootstrap samples.

Research hypotheses H1 and H8, which assume that smart products and services are affected by strategic planning and organisational structure of an organisation, as well as by smart processes, are confirmed. Furthermore, the smart factory dimension has a significant effect on smart products and services at a 5% level, confirming H7.

Table 4 Outer model estimation results

Construct	Indicator	Loading	Weight	Cronbach's alpha	Rho_A	Composite reliability	Average variance extracted
SO	SO1	0.784	0.170	0.911	0.913	0.929	0.652
	SO2	0.739	0.150				
	SO3	0.836	0.185				
	SO4	0.774	0.189				
	SO5	0.866	0.179				
	SO6	0.793	0.178				
	SO7	0.853	0.186				
WFD	WFD1	0.832	0.499	0.689	0.740	0.817	0.601
	WFD2	0.838	0.540				
	WFD3	0.640	0.207				
SP	SP1	0.783	0.183	0.894	0.904	0.917	0.612
	SP2	0.835	0.216				
	SP3	0.818	0.199				
	SP4	0.773	0.187				
	SP5	0.826	0.198				
	SF1	0.729	0.148				
	SF2	0.703	0.139				
SF	SF3	0.839	0.264	0.863	0.890	0.897	0.597
	SF4	0.603	0.125				
	SF5	0.789	0.202				
	SF6	0.671	0.179				
	SF7	0.829	0.257				
	SF8	0.869	0.243				
SPS	SPS1	0.760	0.235	0.833	0.851	0.874	0.501
	SPS2	0.746	0.229				
	SPS3	0.808	0.261				
	SPS5	0.603	0.157				
	SPS6	0.631	0.167				
	SPS7	0.653	0.173				

Nonetheless, the developed model predicts a substantial strong effect of SO, SF and SP on SPS with an R^2 value of 0.630. The strongest, also statistically significant, impact on SPS (0.301) is observed from SP. Similarly, SO's effect on SPS is confirmed with a path coefficient of 0.265.

Hypothesis H2 and H6 assume that SO and SP have a direct effect on the development of SF. Both SO and SP have statistically positive effects on SF with path

Table 5 Heterotrait-monotrait (HTMT) values

HTMT values					
	SF	SO	SP	SPS	WFD
SF					
SO	0.854				
SP	0.893	0.834			
SPS	0.840	0.805	0.834		
WFD	0.631	0.609	0.661	0.663	

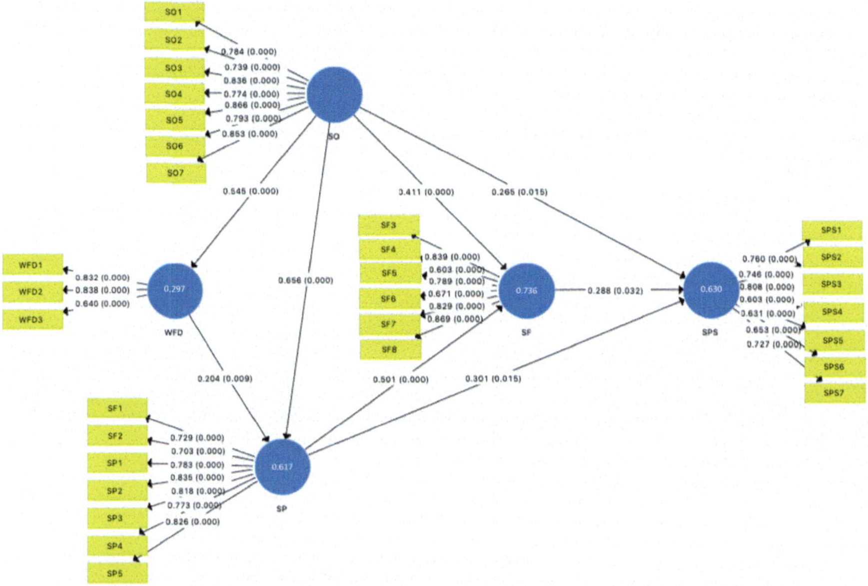

Fig. 2 Results of the developed model

Table 6 Regression results with bootstrap analysis

Hypotheses	Path	Original sample	Sample mean	Standard deviation	T stat	*P*-value	Conclusion
H1	SO → SPS	0.265	0.262	0.109	2.433	0.015	Confirmed
H2	SO → SF	0.411	0.412	0.073	5.638	0.000	Confirmed
H3	SO → SP	0.656	0.652	0.057	11.497	0.000	Confirmed
H4	SO → WFD	0.545	0.549	0.066	8.244	0.000	Confirmed
H5	WFD → SP	0.204	0.207	0.078	2.611	0.009	Confirmed
H6	SP → SF	0.501	0.502	0.080	6.277	0.000	Confirmed
H7	SF → SPS	0.288	0.295	0.134	2.147	0.032	Confirmed
H8	SP → SPS	0.301	0.297	0.124	2.424	0.015	Confirmed

coefficients of 0.411 and 0.501. The predictive power of the designed model for SF is almost 74%, which is substantial. The most significant explanation comes from the smart processes factor.

Hypothesis H3 and H5 assume that SO and WFD positively affect SP. Based on the result of the model, it can be confirmed that both SO and WFD positively affect SP. Both variables predict the SP factor with a relatively high R^2 of 0.617. It should be noted that the effect of SO on SP is the strongest (0.656) among all direct effects in the model, followed by the effect of SO on WFD (0.545).

The last hypothesis, H4, assumes that WFD is positively and significantly affected by SO. This is confirmed by the model results with a path coefficient of 0.545. However, strategy and organisation can only explain around 30% of the variation in workforce development, and the effect is weak. This can be explained by the fact that the employee development dimension is captured by other dimensions and can be additionally explained by other manifest variables and constructs.

Also, mediation effects were tested to see if other dimensions mediate the relationship between strategy and organisation and smart products and services. The results (Table 7) show that smart processes almost fully mediate the relationship between strategy and organisation and smart factory ($\beta = 0.329, p = 0.000 < 0.05$). Similarly, smart processes partially mediate the relationship between strategy and organisation and smart products and services at a 0.05 significance level ($\beta = 0.198$, $p = 0.017 < 0.05$). Moreover, it can be concluded that strategy and organisation and smart processes are related partially by workforce development ($\beta = 0.111$, $p = 0.015 < 0.05$). In addition, smart processes to some extent mediate the relationship between workforce development and smart factory ($\beta = 0.102$, $p = 0.018 < 0.05$). Finally, the indirect effect of strategy and organisation on smart factory through workforce development and smart processes is confirmed ($\beta = 0.056, p = 0.024 < 0.05$).

The relevance and importance of each construct were analysed through the Importance-Performance Map Analysis (IPMA), as it can offer great value to managers and their actions (Ringle and Sarstedt 2016). The performance values of constructs are the calculated means of latent constructs (rescaled to from 0 to 100) and the importance values with regard to the prediction power of a construct about

Table 7 Mediation effects

Specific indirect effects						
Path	Original sample	Sample mean	Standard deviation	T stat.	*P*-value	Conclusion
SP → SF → SPS	0.144	0.148	0.073	1.963	0.050	Confirmed
SO → SP → SF	0.329	0.327	0.057	5.811	0.000	Confirmed
SO → SP → SPS	0.198	0.193	0.083	2.394	0.017	Confirmed
SO → WFD → SP	0.111	0.114	0.046	2.433	0.015	Confirmed
SO → WFD → SP → SF	0.056	0.057	0.025	2.251	0.024	Confirmed
WFD → SP → SF	0.102	0.104	0.043	2.363	0.018	Confirmed

another construct either directly or indirectly and are calculated from the total effect of the relationship between the two constructs.

Figure 3 plotted on the base of Table 8 presents the importance-performance analysis map where the x-axis shows the importance of SO, SF, SP, and WFD that explain the target construct SPS. Y-axis represents the performance of SO, SF, SP, and WFD in terms of their average rescaled latent variable scores.

It can be concluded from the map that the dimension of strategy and organisation has a relatively low performance of 28.072 (below average) compared to other constructs. However, this construct's importance is relatively high among all with a total effect of 0.660 (above average), i.e., a unit increase in SO from 28.072 to 29.072 would cause an increase in the performance of SPS by 0.660 points from 35.394 to 36.394. Therefore, when the target construct of SPS needs to be improved, priority must be given to the strategy and organisation dimension followed by SP, SF, and WFD. Similarly, importance can be assessed in terms of manifest variables as well. Indicator SO7 (collaboration with stakeholders) shows the highest relative importance of 0.109 while having a relatively low performance of 24.390. Similarly, if a smart factory is chosen as a target construct, results show that strategy

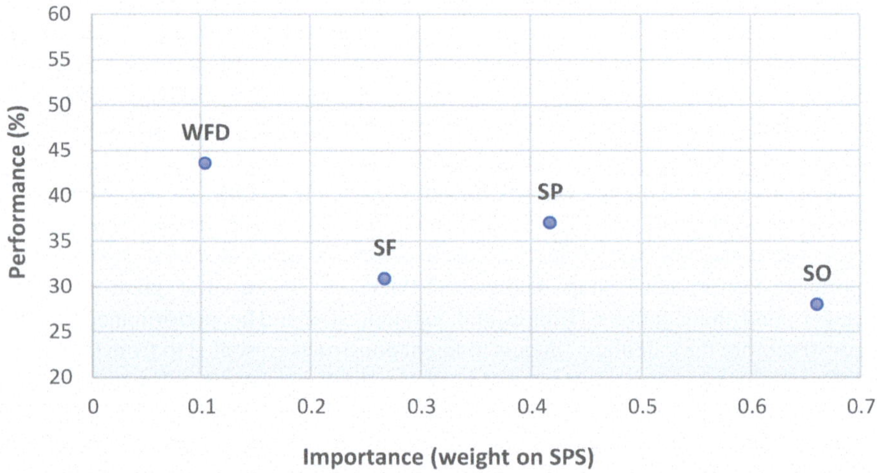

Fig. 3 IPMA for target construct of SPS

Table 8 Importance-performance analysis (IPMA)

Construct	Importance	Performance of construct
SF	0.267	30,894
SO	0.660	28,072
SP	0.417	37,053
WFD	0.104	43,603
SPS		35,394

and organisation is relatively more important than smart processes and workforce development.

5 Findings and Discussion

This study explores the relationships between the critical I4.0 maturity dimensions (strategy and organisation, workforce development, smart factory, smart processes, and smart product and services) and their measures in SMEs using the Structural Equation Modelling (SEM) technique. Therefore, it contributes to the understanding of the fit of different I4.0 measures with their generalised dimensions. The findings reveal that the measures chosen for assessing SMEs within the proposed model are adequately grouped and represent common dimensions. Some overlapping of smart factory measures with smart processes dimension has been identified and it can be explained by the fact that they can be combined under a common dimension of technology similar to the majority of maturity models (Hizam-Hanafiah et al. 2020; Mittal et al. 2018). This indicates that clear communication and explanation of each measure is required before the I4.0 readiness assessment is undertaken.

Also, the research tries to explore the relationships of the I4.0 maturity dimensions among each other and especially the effect of other dimensions on smart products and services. Thus, it contributes to understanding how I4.0 maturity dimensions affect each other directly and/or indirectly. In terms of the effect of strategy and organisation on the other four dimensions, the empirical findings confirm a direct as well as an indirect dependency of the former on the latter. Smart products and services as an outcome of an automated production are strongly affected by smart processes defined by standardised, automated, and data-driven business processes, strategic planning and organisational structure of an organisation, and smart factory defined by automated, integrated, and digitised production. Results show that almost two-thirds of the successful digitisation of smart products and services depends on how well strategy and organisation, smart factory, and smart processes dimensions are mature enough for I4.0 transformation. In relative terms, the effect of strategy and organisation on smart products and services is stronger than that of smart processes or smart factory.

The direct effect of strategy and organisation on the products and services is also indicated in the study of Simetinger and Zhang (2020), where a comparative analysis of maturity models was investigated. (Sony and Naik 2020) identifies that digitisation strategy is the most crucial component for successful Industry 4.0 implementation. The strategy should touch upon cyber-physical systems, the Internet of Things, and the Internet of Service, and it should be effectively designed to transform products and services into smart ones (Sony and Naik 2020). Furthermore, (Vrchota and Pech 2019), based on a sample of more than a thousand Czech employees, also confirms a statistically significant effect of a written strategy towards I4.0 on I4.0 implementation. Recent findings of Salume et al. (2021) based on data from more than two hundred retail companies in Brazil confirm statistically strong evidence of the greatest impact of digital strategy on digital maturity. Innovative strategic decisions unleash the possibilities of smart factory and smart processes, expanding and broadening factory and customer data utilisation, redefining the traditional relationship between supply chain members as the nature of products is changing, thus, making companies rethink almost everything of what they did internally. The main idea behind all these changes is that smart products and services enforce and bring a new competition level, where all industries are reshaping. For example, for SMEs, instead of concentrating on one discrete product, they can present a system of products and services like packages. By that, the company may extend its boundaries and compete in a broader market. Thus, the importance of strategic choices and orientation can be proved once again, as it is interdependent with the final outcome (Porter and Heppelmann 2014).

Even though a positive effect of both smart processes and strategy and organisational indicators on the smart factory dimension that encompasses the I4.0 technology measures is confirmed, the effect of the former is relatively more important. Having a clear strategy and corresponding organisational changes in small medium-sized companies is not enough for a smooth technology introduction, processes need to be upgraded first in order to achieve a well-ordered implementation of I4.0 technology in the organisation. Since human resources perform and control processes, workforce development is crucial. The findings indicate that the direct effect of strategy and organisation dimension on smart processes is more than three times higher than that of workforce development. One can expect that an improvement in strategy and organisation factors and their components will lead to substantial enhancement of processes towards smarter processes in the organisation.

Further, the results of the study also confirm the effect of strategy and organisation on the development of the workforce. Even though training employees on their adaptability to new business processes is one of the challenges that can be brought by I4.0 (Sony 2020), it is also one of the driving forces of the success of I4.0 implementation as skilled workers are required not only at the beginning of the digitisation process, but also along the whole lifecycle of the development and production of smarter products and services. In addition, the influence of the correct strategy on the development of human resources in order to achieve digital transformation has been indicated by Simetinger and Zhang (2020). The influence of top management, their leadership skills, and motivation and support are responsible for developing new skills in their personnel to adapt faster to new technologies and processes. The findings of the study of the confirmed direct and indirect effects among workforce development, smart processes, and smart factory verify the importance of people, process, and technology framework that has proved its effectiveness in change management, which is important for gradual digitisation. Similarly, in some studies as by Narula et al. (2020) the strong interdependence between three dimensions was investigated, which are organisation, technology, and personnel, in order to achieve a successful I4.0 transformation.

Overall, the results confirm the following previous qualitative studies that researched I4.0 maturity models and possible interrelations between dimensions

(Sony and Naik 2020). The results also align with the findings of Salume et al. (2021), who empirically tested the effects of key dimensions on digital maturity.

6 Conclusion and Future Work

Receiving more attention, the I4.0 concept transforms businesses, improving their performances, such as decreasing costs while improving the efficiency of the production processes. Thus, applying I4.0 was an aim of many companies; however, the lack of guidelines and tools makes the path difficult. Maturity models were developed as a helpful tool to assess the company from different perspectives or so-called dimensions such as strategy, processes, human resources, and others. To make a complex assessment, consideration of all dimensions is required; moreover, attention must be paid to the effect of one dimension on another. Despite many studies regarding those dimensions, there was an obvious lack of studies about their correlation based on empirical data in the literature.

Therefore, the research explored the relationships between the critical I4.0 dimensions and their impact on the products and services. The importance of I4.0 indicators and their corresponding dimensions used to assess the I4.0 maturity score was empirically investigated using the data collected from Kazakhstani manufacturing SMEs. The relationships between dimensions and their corresponding measures and the relationship of the dimensions to each other were tested using the PLS-SEM techniques. The validity and reliability of the proposed measurement and structural model were proved by statistical analysis; thus, RO1 and RO2 were achieved.

The importance of the study lies in its empirical investigation of the relationships between dimensions of the developed I4.0 maturity model and confirms the interrelations. Overall, it was found that the effect of the Strategy and Organization dimension is crucial for the rest of the dimensions of the model, and especially for smart products and services. As a result, at the stage of planning the transition to I4.0 and the introduction of its elements, companies should primarily take into account all indicators related to the strategy and organisation factor. As the proposed model proved, changes in strategy and organisation dimensions will bring about relatively more important enhancements to the organisation's smart products and services system than improvements in other dimensions. However, depending on their goal, managers can choose to enhance features that bring faster, more substantial results based on the interrelations studied under the structural model.

This study offers valuable insights into the dimensions and impact of I4.0 and presents several avenues for further exploration. It is important to consider the generalizability of the findings. While the sample represents diverse industry sectors, company sizes, and geographical locations, future research could expand to a broader population of manufacturing SMEs. Also, tracking the evolution of Industry 4.0 practices over time within the participant companies would allow for identifying trends and adaptations; therefore, longitudinal studies are essential. Moreover, sustainability implications within Industry 4.0 deserve attention. A similar structural model can be constructed to test the effect of I4.0 maturity assessment dimensions on organisations' environmental, social, and economic performance. And finally, as we extend our view beyond Industry 4.0, Industry 5.0 beckons, which emphasises human-centric approaches and collaboration enhanced with AI.

Acknowledgments This research has been supported by Nazarbayev University Research Grants, No: 240919FD3919 "Industry 4.0 Assessment of SMEs in Kazakhstan".

References

Adebanjo D, Laosirihongthong T, Samaranayake P, Teh P-L (2023) Key enablers of Industry 4.0 development at firm level: findings from an emerging economy. IEEE Trans Eng Manag 70(2):400–416. https://doi.org/10.1109/TEM.2020.3046764

Adolphs P, Bedenbender H, Dirzus D, Ehlich M, Epple U, Hankel M, Kärcher B et al (2015) Reference architecture model Industrie 4.0 (rami4. 0). ZVEI and VDI, Status report

Amoozad Mahdiraji H, Yaftiyan F, Abbasi-Kamardi A, Garza-Reyes JA (2022) Investigating potential interventions on disruptive impacts of Industry 4.0 technologies in circular supply chains: evidence from SMEs of an emerging economy. Comput Ind Eng 174:108753. https://doi.org/10.1016/J.CIE.2022.108753

Bayrak IT, Cebi F (2024) Procedure model for Industry 4.0 realization for operations improvement of manufacturing organizations. IEEE Trans Eng Manage 71:7901–7912. https://doi.org/10.1109/TEM.2023.3292337

Bhatia MS, Kumar S (2022) Critical success factors of Industry 4.0 in automotive manufacturing industry. IEEE Trans Eng Manag 69(5):2439–2453. https://doi.org/10.1109/TEM.2020.3017004

Burke R, Mussomeli A, Laaper S, Hartigan M, Sniderman B (2017) The smart factory: responsive, adaptive, connected manufacturing. Retrieved from https://www2.deloitte.com/content/dam/insights/us/articles/4051_The-smart-factory/DUP_The-smart-factory.pdf

Cañas H, Mula J, Campuzano-Bolarín F, Poler R (2022) A conceptual framework for smart production planning and control in Industry 4.0. Comput Ind Eng 173:108659. https://doi.org/10.1016/J.CIE.2022.108659

Chin WW (1998) The partial least squares approach to structural equation modeling

Dalenogare LS, Benitez GB, Ayala NF, Frank AG (2018) The expected contribution of Industry 4.0 technologies for industrial performance. Int J Product Econ 204:383–394. https://doi.org/10.1016/j.ijpe.2018.08.019

Dijkstra TK, Henseler J (2015) Consistent partial least squares path modeling. MIS Quart Manag Inf Syst. https://doi.org/10.25300/MISQ/2015/39.2.02

Dikhanbayeva D, Shaikholla S, Suleiman Z, Turkyilmaz A (2020) Assessment of Industry 4.0 maturity models by design principles. Sustainability (Switz) 12(23):1–22. https://doi.org/10.3390/su12239927

Dikhanbayeva D, Aitzhanova M, Lukhmanov Y, Turkyilmaz A, Shehab E, El-Thalji I (2023) Industry 4.0 readiness assessment of enterprises in Kazakhstan. In: IFIP advances in information and communication technology, vol 689. AICT. Springer Science and Business Media Deutschland GmbH, pp 297–310. https://doi.org/10.1007/978-3-031-43662-8_22

Fan Y, Chen J, Shirkey G, John R, Wu SR, Park H, Shao C (2016) Applications of structural equation modeling (SEM) in ecological studies: an updated review. Ecol Process. https://doi.org/10.1186/s13717-016-0063-3

Felch V, Asdecker B, Sucky E (2019) Maturity models in the age of Industry 4.0—do the available models correspond to the needs of business practice? In: Proceedings of the 52nd Hawaii international conference on system sciences, vol 6, pp 5165–5174. https://doi.org/10.24251/hicss.2019.620

Fettermann DC, Cavalcante CGS, de Almeida TD, Tortorella GL (2018) How does Industry 4.0 contribute to operations management? J Ind Prod Eng 35(4):255–268. https://doi.org/10.1080/21681015.2018.1462863

Fornell C, Cha J (1994) Partial least squares. In: Advanced methods of marketing research, p 407

Gaiardelli P, Pezzotta G, Rondini A, Romero D, Jarrahi F, Bertoni M, Cavalieri S et al (2021) Product-service systems evolution in the era of Industry 4.0. Serv Bus 15(1):177–207. https://doi.org/10.1007/s11628-021-00438-9

Garson GD (2016) Partial least squares: regression & structural equation models. In: Multi-label dimensionality reduction

Geissbauer R, Vedso V, Schrauf S (2016) Industry 4.0: building the digital enterprise. In: 2016 global industrial survey. Retrieved from https://www.pwc.com/gx/en/industries/industries-4.0/landing-page/industry-4.0-building-your-digital-enterprise-april-2016.pdf

Gökalp E, Şener U, Eren PE (2017) Development of an assessment model for industry 4.0: Industry 4.0-MM. In: Communications in computer and information science, vol 770, pp 128–142. https://doi.org/10.1007/978-3-319-67383-7_10

Hair JF, Risher JJ, Sarstedt M, Ringle CM (2019) When to use and how to report the results of PLS-SEM. Eur Bus Rev. https://doi.org/10.1108/EBR-11-2018-0203

Hajoary PK, Balachandra P, Garza-Reyes JA (2023) Industry 4.0 maturity and readiness assessment: an empirical validation using confirmatory composite analysis. Production Planning and Control. https://doi.org/10.1080/09537287.2023.2210545

Henseler J, Sarstedt M (2013) Goodness-of-fit indices for partial least squares path modeling. Comput Stat 28(2). https://doi.org/10.1007/s00180-012-0317-1

Henseler J, Ringle CM, Sarstedt M (2015) A new criterion for assessing discriminant validity in variance-based structural equation modeling. J Acad Mark Sci 43(1). https://doi.org/10.1007/s11747-014-0403-8

Hess T, Benlian A, Matt C, Wiesböck F (2016) Options for formulating a digital transformation strategy. MIS Quart Execut 15(2). https://doi.org/10.4324/9780429286797-7

Hizam-Hanafiah M, Soomro MA, Abdullah NL (2020) Industry 4.0 readiness models: a systematic literature review of model dimensions. Information 11(7):364. https://doi.org/10.3390/info11070364

Hofmann E, Sternberg H, Chen H, Pflaum A, Prockl G (2019) Supply chain management and Industry 4.0: conducting research in the digital age. Int J Phys Distrib Logist Manag. https://doi.org/10.1108/IJPDLM-11-2019-399

Kagermann H, Wahlster W, Helbig J (2013) Securing the future of German manufacturing industry: recommendations for implementing the strategic initiative INDUSTRIE 4.0. Final report of the Industrie 4.0 working group. https://doi.org/10.13140/RG.2.1.1205.8966

Kamble SS, Gunasekaran A, Sharma R (2018) Analysis of the driving and dependence power of barriers to adopt Industry 4.0 in Indian manufacturing industry. Comput Ind 101. https://doi.org/10.1016/j.compind.2018.06.004

Kane GC, Palmer D, Phillips AN, Kiron D, Buckley N (2015) Strategy, not technology, drives digital transformation becoming a digitally mature enterprise. Sloan Manag Rev 57181

Kiraz A, Canpolat O, Özkurt C, Taşkın H (2020) Analysis of the factors affecting the Industry 4.0 tendency with the structural equation model and an application. Comput Ind Eng 150. https://doi.org/10.1016/j.cie.2020.106911

Kumar L, Kumar A, Sharma RK, Kumar P (2023) Smart manufacturing and Industry 4.0. In: Handbook of smart manufacturing. https://doi.org/10.1201/9781003333760-1

Lasi H, Fettke P, Kemper HG, Feld T, Hoffmann M (2014) Industry 4.0. Bus Inf Syst Eng 6(4):239–242. https://doi.org/10.1007/s12599-014-0334-4

Leyh C, Bley K, Schaffer T, Forstenhausler S (2016) SIMMI 4.0—a maturity model for classifying the enterprise-wide it and software landscape focusing on Industry 4.0. In: Proceedings of the 2016 federated conference on computer science and information systems, FedCSIS 2016, vol 8, pp 1297–1302. https://doi.org/10.15439/2016F478

Lichtblau K, Stich V, Bertenrath R, Blum M, Bleider M, Millack A, Schröter M et al (2015) Impuls Industrie 4.0 readiness. VDMA, pp 1–76. https://doi.org/10.1080/03650340903302278

Machado CG, Winroth M, Almström P, Ericson Öberg A, Kurdve M, AlMashalah S (2021) Digital organisational readiness: experiences from manufacturing companies. J Manufact Technol Manag 32(9). https://doi.org/10.1108/JMTM-05-2019-0188

McKinsey (2015) How to navigate digitization of the manufacturing sector. Retrieved from https://www.mckinsey.com/~/media/McKinsey/Business%20Functions/Operations/Our%20Insights/Industry%2040%20How%20to%20navigate%20digitization%20of%20the%20manufacturing%20sector/Industry-40-How-to-navigate-digitization-of-the-manufacturing-sector.ashx

Mittal S, Romero D, Wuest T (2018) Towards a smart manufacturing maturity model for SMEs (SM3E). In: IFIP advances in information and communication technology, vol 536. https://doi.org/10.1007/978-3-319-99707-0_20

Müller JM, Kiel D, Voigt KI (2018) What drives the implementation of Industry 4.0? The role of opportunities and challenges in the context of sustainability. Sustainability (Switz) 10(1). https://doi.org/10.3390/su10010247

Narula S, Prakash S, Dwivedy M, Talwar V, Tiwari SP (2020) Industry 4.0 adoption key factors: an empirical study on manufacturing industry. J Adv Manag Res. https://doi.org/10.1108/JAMR-03-2020-0039

Nunes ML, Pereira AC, Alves AC (2017) Smart products development approaches for Industry 4.0. Proc Manufact 13:1215–1222. https://doi.org/10.1016/j.promfg.2017.09.035

Porter ME, Heppelmann JE (2014) How smart, connected products are transforming competition. Harv Bus Rev

Ringle CM, Sarstedt M (2016) Gain more insight from your PLS-SEM results. Ind Manag Data Syst 116(9). https://doi.org/10.1108/imds-10-2015-0449

Ringle CM, Wende S, Becker J-M (2022) SmartPLS 4, vol 1. http://www.smartpls.com

Roblek V, Meško M, Krapež A (2016) A complex view of Industry 4.0. Sage Open 6(2):2158244016653987. https://doi.org/10.1177/2158244016653987

Salume PK, Barbosa MW, Pinto MR, Sousa PR (2021) Key dimensions of digital maturity: a study with retail sector companies in Brazil. Rev Adm Mackenzie 22(6). https://doi.org/10.1590/1678-6971/ERAMD210071

Santos RC, Martinho JL (2020) An Industry 4.0 maturity model proposal. J Manufact Technol Manag 31(5). https://doi.org/10.1108/JMTM-09-2018-0284

Schuh GG, Anderl R, Gausemeier JJ, ten Hompel MM, Wahlster W, Ander L, Wahlster W et al (eds) (2020) Industrie 4.0 maturity index: managing the digital transformation of companies. Acatech Study

Schumacher A, Erol S, Sihn W (2016) A maturity model for assessing Industry 4.0 readiness and maturity of manufacturing enterprises. Proc CIRP 52:161–166. https://doi.org/10.1016/j.procir.2016.07.040. Retrieved from obsidian://open?vault=I4.0%20Research&file=Z-Core%2F%24%20Schumacher%20-%20A%20maturity%20model

Simetinger F, Zhang Z (2020) Deriving secondary traits of Industry 4.0: a comparative analysis of significant maturity models. Syst Res Behav Sci 37(4):663–678. https://doi.org/10.1002/sres.2708

Singapore Smart Industry Readiness EDB (2018) The smart industry readiness, p 46. Retrieved from https://www.edb.gov.sg/content/dam/edb/en/newsandevents/News/2017/advanced-manufacturing-release/Copyrighted-The-SG-Smart-Industry-Readiness-Index-Whitepaper.pdf

Sjödin DR, Parida V, Leksell M, Petrovic A (2018) Smart factory implementation and process innovation: a preliminary maturity model for leveraging digitalization in manufacturing. Res Technol Manag 61(5):22–31. https://doi.org/10.1080/08956308.2018.1471277

Smuts S, van der Merwe A, Smuts H (2020) A strategic organisational perspective of Industry 4.0: a conceptual model. In: Lecture notes in computer science (including subseries lecture notes in artificial intelligence and lecture notes in bioinformatics), vol 12066. LNCS. https://doi.org/10.1007/978-3-030-44999-5_8

Somohano-Rodríguez FM, Madrid-Guijarro A, López-Fernández JM (2020) Does Industry 4.0 really matter for SME innovation? J Small Bus Manag 00(00):1–28. https://doi.org/10.1080/00472778.2020.1780728

Sony M (2020) Pros and cons of implementing Industry 4.0 for the organizations: a review and synthesis of evidence. Product Manufact Res 8(1). https://doi.org/10.1080/21693277.2020.1781705

Sony M, Naik S (2020) Key ingredients for evaluating Industry 4.0 readiness for organizations: a literature review. Benchmarking. Emerald Group Holdings Ltd. https://doi.org/10.1108/BIJ-09-2018-0284

Soomro MA, Hizam-Hanafiah M, Abdullah NL, Ali MH, Jusoh MS (2021) Industry 4.0 readiness of technology companies: a pilot study from Malaysia. Adm Sci 11(2). https://doi.org/10.3390/admsci11020056

Tenenhaus M, Vinzi VE, Chatelin YM, Lauro C (2005) PLS path modeling. Comput Stat Data Anal 48(1):159–205. https://doi.org/10.1016/j.csda.2004.03.005

Turkyilmaz A, Temizer L, Oztekin A (2018) A causal analytic approach to student satisfaction index modeling. Ann Oper Res 263(1–2):565–585. https://doi.org/10.1007/s10479-016-2245-x

Turkyilmaz A, Dikhanbayeva D, Suleiman Z, Shaikholla S, Shehab E (2020) Industry 4.0: challenges and opportunities for Kazakhstan SMEs. Proc CIRP 96:213–218. https://doi.org/10.1016/j.procir.2021.01.077

Ünlü H, Demirörs O, Garousi V (2023) Readiness and maturity models for Industry 4.0: a systematic literature review. J Softw Evol Process. https://doi.org/10.1002/smr.2641

Vrchota J, Pech M (2019) Readiness of enterprises in Czech Republic to implement Industry 4.0: index of Industry 4.0. Appl Sci (Switz) 9(24). https://doi.org/10.3390/app9245405

Wold H (2004) Partial least squares. In: Encyclopedia of statistical sciences. https://doi.org/10.1002/0471667196.ess1914

A Cost-Minimization Approach to Production and Maintenance Planning Considering Imperfect Repairs and Human Resource Constraints

Taha-Hossein Hejazi, Donya Ramezani, and Ali Golestanian

Abstract Appropriate maintenance procedures are crucial for enhancing system availability and ensuring smooth operation. Reliability engineers and researchers have long sought optimal maintenance plans for technical systems. This research looked at various factors influencing the decision-making process regarding maintenance, including skill level, number of technicians, repair time, and the type of completed repairs, either minimal or partial (incomplete). The study also looked at time and whether a partial or minimal repair should be taken. Considering the above parameters has made the proposed model more realistically applicable. A mixed integer programming model described in this study is intended to reduce production, maintenance, and staff costs. Hence, the research gives finer data to enable one to choose the most appropriate maintenance and repair choices based on the current budget; it would be more realistic to use this model in predicting because it has more considerations when determining spare parts requirements and stock levels.

Keywords Maintenance · Optimization · Human resources planning · Partial repairs · Technician skill level · System availability

1 Introduction

Effective planning for both production and maintenance is a critical requirement for industrial entities to ensure the uninterrupted and seamless operation of their operations without any disruptions or breakdowns whatsoever. The combined improvement of production and maintenance planning has garnered significant attention from industries aiming to achieve production targets and enhance the performance

T.-H. Hejazi (✉) · D. Ramezani
Department of Industrial Engineering, College of Garmsar, Amirkabir University of Technology (Tehran Polytechnic), Tehran, Iran
e-mail: t.h.hejazi@aut.ac.ir

A. Golestanian
Department of Industrial Engineering, Sadjad University, Mashhad, Iran

A. Kumar et al. (eds.), *Industry 4.0 Driven Manufacturing Technologies*, Springer Series in Advanced Manufacturing, https://doi.org/10.1007/978-3-031-68271-1_8

and longevity of their equipment. A crucial factor to consider within this domain is the issue about the utilization of ineffective remedial actions employed alternatively rather than partially restoring a device or software system to its original operational state (Beheshti Fakher et al. 2017). Owing to the disparate expertise levels possessed by other technical personnel, this scenario manifests frequently under real-world conditions. Encompassed among such constraints is the availability and expertise proficiency of the technicians involved in the repair process, further compounding the complexity of the decision-making required. With growing area understanding in the industry due to the expansion of technical and scientific knowledge and increased standards, advancement of technologies, and better systems reliability and planning, today, research in maintenance has greatly widened to enhance efficiency. By concentrating on minimal or partial repairs and controlling human resources, which encompasses technician skill and time spent on repairs, maintenance and production costs can be reduced. This strategy can lead to lower downtimes at production lines, and companies could also utilize part-time employees who will only work when there is an issue, thus helping them save money that would otherwise be spent on extra pay when there are no problems at all. It is possible to keep production and maintenance costs as low as possible by focusing on minimal repairs or efficient human resource management in terms of technician expertise and repair time, leading to decreased production line downtime. This work advocates for an all-inclusive research objective that seeks to sift through these hurdles. It seeks to introduce difficulties arising from imperfect repairs and technician availability into the optimization modeling, thereby aiming to give it an integrated view. Such an investigation will offer some realism and a more applied structure regarding production/maintenance planning. The work we are proposing sets itself apart from existing ones by being one of the first to simultaneously consider the following perspectives:

- Taking into account the integrated production planning model, the costs of minimal and partial repairs.
- Considering the repair budget constraints.
- Compiling the salaries of high-skilled and low-skilled repair technicians.
- Considering how long-lasting repairs can be on equipment and how production plans are planned.
- It considers both the amount of time needed for minimal repairs and how it affects the time available for production.
- Investigating how the knowledge of human resources affects the time it takes to make repairs. Compiling constraints on human resource constraints.

As far as we know, previous research has not focused on simultaneously optimizing production and maintenance planning while taking into account imperfect repairs and constraints related to human resources. This research suggests a novel and unique approach to this important problem. Studies currently underway mostly look at separate issues, for instance, production planning or maintenance, without completely indicating the relationship between these decisions and real problems brought by poor repairs and the presence of competent engineers. The research presented here aims to incorporate these real-world factors into a single optimization framework, which may lead to a more holistic and realistic decision-support tool for industrial practitioners. The significant advancement that this study makes in comparison to previous studies is the inclusion of cost constraints for maintenance, repair technician availability, and the cost and impact of non-minor repairs. This integrated modeling strategy creates a minimum-cost approach that contemplates the dainty weighing of production targets against equipment reliability and human resource constraints. This novel study connects theoretical models with practical, industry-specific considerations. This research aims to provide significant input to managers in production, maintenance personnel, and decision-makers to make better data-driven judgments that will help enhance productivity levels within their firms.

The rest of this chapter is organized in the following manner: The subsequent section will examine the current literature on planning for both production and maintenance together in order to pinpoint the research deficiencies that the proposed study intends to tackle. This section investigates the present understanding of how imperfect repairs and limitations in human resources are integrated into the optimization of production-maintenance systems. Our suggested mathematical model is then outlined in Sect. 3, with the solution approach detailed in Sect. 4. Section 5 includes the presentation of Sensitivity Analyses. Lastly, Sects. 6 and 7 is dedicated to conclusions and future endeavors.

2 Review of Literature

This section presents a selective overview of the relevant literature on joint production and maintenance planning considering imperfect repairs and human resource constraints. This section aims to position our research within the existing body of knowledge, identify the gaps, and raise specific questions that need to be addressed based on our findings. We examined the studies in two key areas:

1. Production and maintenance planning in the face of imperfect repairs.
2. The production and maintenance operations are affected by limitations in human resources.

The production and maintenance operations are affected by limitations in human resources. By critically reviewing the extant literature in these domains, we seek to establish the theoretical foundations and identify the research gaps that our proposed model aims to address.

2.1 Production and Maintenance Planning with Imperfect Repairs

Organizations must ensure efficient operations and minimize disruptions by effectively coordinating production and maintenance activities. Repairs are frequently imperfect in real-world conditions, creating uncertainties and difficulties in planning. To summarize the most recent research on planning maintenance and production together under the guise of imperfect repairs, this literature review will highlight key findings, research methods, and gaps in current understanding. Viveros et al. (2023) suggested a risk analysis based on the failure probability of each component of the maintenance scheduling optimization model, which made it possible to combine various grouped PM activities. Their model attempted to compare the general behavior of the optimization model to various maintenance activity periodicities and execution times. The formulation of maintenance plans had become significantly more complicated as a result of the development of new methods to improve the realisticality of maintenance scheduling optimization models. Al-Salamah (2018) conducted research on a production-inventory system that involved an imperfect and unreliable machine, producing items that could be either perfect or imperfect and in need of rework. The study involved modeling the random time to machine failure, random repair time, and random fraction of imperfect items to derive a function for the average cost that is expected for optimizing the production lot size. Additionally, a heuristic solution approach was proposed to tackle the model's complexity. Kouedeu et al. (2011) collaborated to investigate and evaluate a manufacturing system that experienced recurring failures and subsequent repairs. The objective of their research was to reduce the adverse effects of discounted overall costs, encompassing inventory holding costs, preventive and corrective maintenance costs, and backlog costs. To achieve this, they put forth a two-tier hierarchical decision-making framework, which involved optimizing production, preventive, and corrective maintenance policies at the second level, and determining the mean time between failures at the first level. Additionally, Wu et al. (2019) proposed a two-phase opportunistic maintenance framework that integrated properties of production wait times into the decision-making process. The framework combined imperfect preventive repair with postponed maintenance during scheduled windows and opportunistic maintenance during production waits. A model for optimization was created with the goal of reducing maintenance costs and at the same time meeting a constant availability requirement. This model was solved using a genetic algorithm. In order to enhance cost-effectiveness and minimize unnecessary inspections, Rasay et al. (2022) devised a framework that incorporates three levels of warning (Signal, Alert, Alarm). By integrating maintenance clustering or opportunistic maintenance, which is recognized for its cost-effectiveness in multi-component systems with economic interdependencies, maintenance activities for such systems can be combined. The analysis and findings primarily focused on reducing costs and enhancing system availability through maintenance clustering. Further contributing to this body of work, Yang et al. (2019) investigated an innovative method for conducting preventive maintenance on a system with a single component. Their objective was to optimize the revenue generated through a performance-based contracting (PBC) arrangement. The policy consists of an imperfect maintenance phase followed by a postponed replacement phase, where the joint optimization of the inspection interval, number of inspections, and preventive replacement interval with aims to maximize the anticipated net revenue under PBC through its implementation. They applied the model to a steel converter plant case study and demonstrated the proposed policy's superior performance compared to existing maintenance policies. Cheng and Zhao (2023) presented a maintenance optimization approach that is applicable to two-component systems with dependencies, which experience degradation and imperfect repairs. They devised a random-effect imperfect repair model to accurately represent the degradation processes and maintainability of the components, considering both economic and stochastic dependencies. To address the maintenance problem, they employed the Markov decision process with an infinite horizon, and utilized the value iteration algorithm to determine the optimal solution. Extending the research on joint production and maintenance planning, Rippe and Kiesmüller (2023) conducted a study on the repair kit issue, with the objective of identifying the most efficient combination and amount of spare parts that should be carried by service technicians for on-site repairs. They incorporated the advanced demand information provided by sensor-equipped appliances, which can serve as imperfect signals of potential failures. They formulated the repair kit issue as a Markov decision process and proposed two heuristic solutions that leverage the available advanced demand information, demonstrating substantial cost savings compared to approaches that disregard this information. Pedersen et al. (2023) proposed a condition-based maintenance (CBM) strategy for a dual-component system experiencing continuous degradation and eventual hard failure. By utilizing a two-stage degradation process model, they were able to calculate the failure probability of one component and then employ a proportional hazards model to estimate the cumulative degradation of the other component. The main focus of their study was on determining the optimal preventive renewal interval and imperfect repair threshold that would minimize the long-term cost rate associated with the CBM policy.

Cheng and Zhao (2023) made an additional contribution to this line of research by presenting a maintenance optimization technique for two-component systems that are dependent on each other and are subject to degradation and imperfect repair. The economic and stochastic interdependencies between the components were considered, leading to the creation of a random-effect imperfect repair model that effectively reflects the degradation process and the maintainability of the components. Using a Markov decision process approach, the study provided structural insights into the influence of imperfect repair characteristics on optimal maintenance policies. Shabtay and Zofi (2018) (conducted a study to analyze the optimal scheduling sequence for single-machine scheduling with variable processing times and maintenance. Meanwhile, Chen et al. (2023a) investigated the impact of machine availability on a decision model aimed at reducing total delay through the utilization of genetic optimization algorithms.

Zhang et al. (2023) put forward a unified economic manufacturing quantity (EMQ) model that merges the principles of condition-based maintenance (CBM) and imperfect manufacturing processes. They developed a manufacturing process model using two indicators: a binary state represented the in-control or out-of-control condition, and the degradation of the manufacturing equipment was represented by a homogeneous Gamma process. The model accounted for potential inspection errors and the fabrication of defective products due to process degradation. The goal was to create the most efficient production lot-sizing and determine the preventive maintenance (PM) threshold to minimize the anticipated cost rate. Dehghan Shoorkand et al. (2024) have examined the integration of tactical production planning and predictive maintenance within the framework of a rolling horizon approach. They have utilized a combination of a convolutional neural network and long short-term memory (CNN-LSTM) method to analyze the sensor data and identify the optimal maintenance strategy. The effectiveness of the proposed approach, which integrates production and predictive maintenance planning, has been verified through the utilization of a benchmarking dataset. Building on these advancements, Chen et al. (2023b) introduced the concept of the "opportunity time window" (OTW) to address the maintenance planning challenges in continuous process manufacturing systems (CPMS). Due to the continuous nature of CPMS operations, maintenance can only be performed during specific time intervals. A novel maintenance optimization model was proposed by them, which takes into account three maintenance actions: doing nothing, perfect maintenance, and imperfect maintenance with epistemic uncertainty. The study developed a stochastic fuzzy flow manufacturing network (SFFMN) to evaluate machine reliability and optimize maintenance decisions by incorporating random and epistemic uncertainties in the production and degradation processes. Arani et al. (2020) conducted a study on a machine that tended to experience random breakdowns. Their objective was to address the challenges related to integrated multi-product process and maintenance planning, with the aim of minimizing both the expected corrective repairs and the production costs associated with preventive maintenance (PM). To achieve this, they developed a mixed-integer linear programming (MILP) model. The investigation of the model was prompted by the absence of the machine age effect and its potential causes. The findings revealed that the presence of the machine age effect led to a more accurate and cost-effective calculation.

Furthermore, Li et al. (2023) introduced a model for condition-based maintenance (CBM) strategy in manufacturing systems. This model considers the working schedule to find the ideal equilibrium between maintenance and production, all while considering product quality. To accurately depict real-life manufacturing scenarios, the model incorporates both imperfect and perfect maintenance actions. The researchers utilized a Markov decision process framework and devised a dual-value iteration algorithm to effectively solve the optimization problem. Hejazi et al. (2023) conducted a study to simulate the intricate relationships between maintenance strategies and their impacts on a manufacturing system. They utilized Discrete Event Simulation (DES) to develop a unique simulation model for a multi-product industry. The results of the study showed that the proposed model was able to decrease the manufacturing and maintenance costs of the system by 13%. Furthermore, the implementation of maintenance planning in the study enhanced certain aspects of the manufacturing system's efficiency.

Complementing these studies, Nobil et al. (2024) investigated an inventory-manufacturing system with strict 100% inspection protocols, a common practice in sensitive industries with precise manufacturing specifications, such as aircraft engine parts, pharmaceuticals, and electronics. They developed three models to minimize the total cost under different inventory and shortage scenarios and considered the impact of comprehensive inspection processes that can be slower than the manufacturing process. These models were especially significant for high-precision manufacturing systems in which quality control poses a bottleneck due to the demands of inspection speed. Rivera-Gómez et al. (2013) conducted a study on a single-machine manufacturing system that suffered from inaccuracies and was incapable of meeting the demands of a specific product type. The system experienced frequent deteriorations, particularly in terms of the defect rate, which continued to increase over time. To address this issue, the researchers employed simulation modeling, response surface modeling, and experiments. Through these methods, they were able to gain a better understanding of how production and overhaul policies influenced the deterioration of quality. Building upon this research, Liu et al. (2024) introduced a reliability model specifically designed for systems operating in dynamic environments with multiple dependent competing failure processes. This model took into consideration various imperfect maintenance actions performed prior to replacement, highlighting the need for decision-makers to carefully consider the trade-offs between imperfect maintenance and replacement actions.

2.2 Production and Maintenance Operations with Limitations in Human Resources

A well-known method for increasing manufacturing systems' overall performance and efficiency has been the coordination of production and maintenance schedules. However, besides the existing literature, maintenance technicians 'availability and expertise frequently overlooked the additional constraints and complexities brought on by limited human resources. Improving job satisfaction and self-motivation is achieved through multi-skilling. (Wong et al. 2013). Furthermore, skilled employees with multiple abilities can substitute for absent workers and assist in areas needing additional manpower at any time and for any length of time. These capabilities enable the expectations set forth in an additional resource-constrained project scheduling problem (RCPSP), also known as a multi-skilled resource-constrained scheduling problem (MS-RCPSP) (Maghsoudlou et al. 2021), which focuses on assigning tasks to staff members according to their skills, ability to handle workload, compensation, and other relevant factors. Bouzidi-Hassini et al. (2015) proposed a novel method for incorporating production and maintenance operations scheduling, taking into consideration the availability and expertise of human resources. A multi-agent system is employed for modeling purposes to model the production workshop and address the challenges of integrating production and maintenance activities, considering the unique constraints associated with maintenance tasks compared to production jobs. Khanizad and Montazer (2018) developed a collaborative game involving agents and tasks within an organization to allocate human resources efficiently. The model utilized fuzzification to expedite the game and enhance the chances of reaching an agreement. The findings demonstrated that organizational performance was more favorable, with increased productivity resulting from the allocation of specific human resources to tasks. Bouzidi-Hassini et al. (2015) tackled the challenge of coordinating task scheduling and preventive maintenance (PM) tasks in production workshops to minimize failures. The article presented a new approach for merging production and maintenance operations' scheduling. In updating the integrated production and maintenance schedules, the proposed method explicitly took into account the availability of human resources. Touat et al. (2017) expanded this area of study by creating remedies that specifically take into account the influence of human resource skills and availability on the simultaneous planning of production and maintenance tasks. Dui et al. (2023) introduced a maintenance metric with the goal of enhancing the efficiency of an irrigation network during drought periods. Their paper presented a new model for optimal maintenance efficiency, which facilitated the most efficient allocation of resources. This contribution addressed a significant gap in current research on maintenance strategies for irrigation systems in drought conditions. Touat et al. (2018) examined a novel scheduling challenge that takes into account production and flexible PM on a solitary machine while also factoring in human resource constraints such as availability and competence. The objective function considered both the time delay and time advancement resulting from production and maintenance activities. The problem was mathematically formulated using constraint programming (CP) principles, represented as a collection of linear constraints. This was then implemented in ILOG OPL language and solved using the precise CPLEX method for smaller instances. Additionally, a heuristic algorithm was offered by them to address larger instances of the problem. The suggested heuristic has been proven to be effective in computational experiments and is capable of finding satisfactory solutions for scenarios with up to 700 jobs within a reasonable amount of CPU time. Yu et al. (2022) examined the relationship between organizational resilience and the mediating and moderating roles of self-efficacy and self-management, respectively. By examining organizational resilience from the perspective of strategic human resource management(SHRM), this paper dissolves from previous research that examined organizational resilience from a single point of view and establishes that SHRM could help Chinese companies improve organizational resilience. Geurtsen et al. (2023) comprehensively review the literature on integrating production, maintenance, and resource scheduling. They identified this as a crucial challenge in modern manufacturing and service environments and highlighted the limited research on integrating all three scheduling problems. The review offered a classification for scheduling maintenance and resources with production and provided detailed discussions on the contributions and novelties of the existing literature, as well as identifying gaps and promising directions for future research. Levitin et al. (2023) have played a crucial role in creating models and enhancing the optimization of expected mission downtime (EMD) within a system that has limited resources. n order to successfully accomplish a mission of a specific duration, this system required replenishing the resource according to a predetermined operation and maintenance schedule (OMS).Complementing these studies, Bocewicz et al. (2023) tackled the issue of effectively allocating multi-skilled workers and sustaining the necessary skill levels during the execution of a dynamic project portfolio. The main emphasis of the paper was on ensuring that employees maintain their skills at a consistent level over time, without any decline. The aim was to develop a tool for analysis that considers the learning progress of every person, particularly employees who occasionally perform routine tasks and might forget their abilities. A new declarative programming approach was suggested, enabling the rotation of a diverse team of employees to keep their skills at the necessary level while ensuring timely completion of a specific project portfolio. The case study from the remanufacturing industry illustrated the applicability of the proposed approach to real-life companies facing the challenge of managing the competence and availability of their workforce. Birdi et al. (2008) conducted a study to examine the influence of human resources management practices on the quality of healthcare services and patient satisfaction. They employed a descriptive methodology to review and analyze existing literature on the subject. The findings of the study revealed that effective human resources management significantly impacted healthcare quality and enhanced the performance of hospital staff. The researchers also recommended the evaluation of human resources department managers' performance prior to initiating any performance development initiatives, along with continuous training and development programs for staff performance. In a separate study, Birdi et al. (2008) introduced the Operational Aircraft Line Maintenance Scheduling Problem (OALMSP) with the objective of creating a schedule that could be easily implemented by maintenance personnel. This involved assigning specific start times and resources to individual tasks, presenting a novel challenge in the field of maintenance scheduling.

The primary distinction between the present research and previous studies lies in the depth of analysis conducted on the factors taken into account, as well as the precise aim of minimizing production, maintenance, and staff costs. The current study appears to offer a more detailed and holistic approach to the joint optimization of production, maintenance, and human resource management compared to the other reviewed studies. Overall, the current work aligns well with the existing literature in terms of the overarching goals and the recognition of the importance of human resource constraints in integrated production and maintenance planning. The proposed mixed integer programming model, with its comprehensive consideration of various factors, contributes to the advancement of this research area. In Table 1, we examine some studies that focused on maintenance planning under conditions of imperfect repairs.

Despite the valuable insights provided by these studies on the effects of imperfect repairs, the coordination of production and maintenance planning, the impact of quality control procedures, the utilization of production wait times, and the maintenance policies under performance-based contracting, they have often neglected to consider the additional limitations and complexities introduced by limited human resources, including the availability and expertise of maintenance technicians. By taking into account the impact of human resource constraints, this gap in the literature presents a chance to enhance our comprehension of the joint planning of production and maintenance.

Table 1 List of some papers studying the field of imperfect repair

Study	Special solution approach	Research methods	Objective function	Mixed types of repairs	Multiple skill levels
Al-Salamah (2018)	• Imperfect, unreliable machine • Imperfect items requiring rework • Production-inventory system model • Production lot size	Stochastic modeling and optimization	Expected average cost	+	–
Wu et al. (2019)	• Framework for opportunistic maintenance in two phases • Production wait times • Steady-availability constraint	Optimization and Genetic Algorithm	Maintenance Cost	–	–
Yang et al. (2019)	• The preventive maintenance policy involves two phases • System with only one component • Performance-based contracting arrangement • Preventive replacement interval • Joint optimization	Optimization and Case Study	Expected net revenue	–	–
Rippe and Kiesmüller (2023)	• Repair kit problem • Information on demand demand • Sensor-equipped appliances	Markov decision process and Heuristics	Repair kit cost	–	–
Cheng and Zhao (2023)	• Dependent two-component systems • Random-effect imperfect repair model • Degradation process • Imperfect repair	Markov decision process and Structural insights	Cost	–	–
Zhang et al. (2023)	• Integrated EMQ model • Condition-based maintenance • Inspection errors • Fabrication of defective products	Optimization and Analytical modeling	Expected cost rate	–	–
Chen et al. (2023b)	• Opportunity time window • Random uncertainties • Epistemic uncertainties	Stochastic modeling and fuzzy optimization	Total cost	+	–

(continued)

Table 1 (continued)

Study	Special solution approach	Research methods	Objective function	Mixed types of repairs	Multiple skill levels
Li et al. (2023)	• CMB • The quality of the product • Perfect maintenance actions • Imperfect maintenance actions • Dual-value iteration algorithm	Markov decision process and Optimization algorithm	Revenue	+	–
Nobil et al. (2024)	• 100% inspection protocols • Inventory scenarios • Shortage scenarios • Inventory-manufacturing system	Optimization and Analytical modeling	Total cost	–	–
This work	• Mixed integer programming model • Type of repair (minimal or partial) • Maintenance decision-making process • Repair types • Technician skills	Mixed integer programming, Consideration of human resource constraints and repair types	Total cost	+	+

3 Problem Description and Mathematical Modeling

Repairs are not always carried out completely and permanently in the real world. Sometimes, minimal or partial repairs are performed just to prevent production from stopping, and they may not even affect the lifespan of the equipment. Additionally, the time and cost of repair may differ between a high-skilled and a low-skilled technician. This model has attempted to consider these factors.

Based on the study, the assumptions mentioned below are considered in the design of the model:

- The maintenance system consists of two types of technicians: high-skilled and low-skilled.
- The repair time of the high-skilled staff is lower than that of the low-skilled (W is the factor).
- The system has a parallel machine configuration.
- The probability distribution of the number of failures follows an exponential distribution.
- The repair time for the low-skilled technician equals the time required for a minimal repair.

3.1 Mathematical Modeling

We recall some general notions to provide a better understanding of the mathematical model. The indices, parameters, and variables of the mathematical model are defined as below:

Notations

Indices

M Number of machines
P Number of products
T Number of periods
J Number of intervals in periods

Parameters

W	High-skilled repair time factor
b_p	Unit backorder cost of product p
CMR_m	Cost of minimal repair of machine m
CPM_{mj}	Cost of jth PM level for machine m
d_{pt}	Customer demand for product p in period t
g_{pm}	Production rate of product p on machine m
h_p	Inventory holding cost of product p
L	Fixed length of periods
PMB_t	Available PM budget in period t
S_{pm}	Cost of set-up for product p on machine m
TMR_m	Time of minimal repair of machine m
θ_m	Exponential parameters for time-to-failure function (machine m)
π_{pm}	The unit processing cost of product p on machine m
CZ_t	Cost of assigning a low-skilled technician in period t
CZ'_t	Cost of assigning a high-skilled technician in period t
Delta	Minimum expected number of failures to assign a technician
V_t	The number of low-skilled technicians in period t
V'_t	The number of high-skilled technicians in period t
λ_m	The amount of reduction in the age of the machine m if minor repairs are carried out on it
CIM_{mt}	The additional cost of minor repairs in period t on machine m over minimal repairs
TIM_{mt}	The excess time of minor repairs in period t on machine m compared to minimal repairs

Variables

APT_{mt}	Available production time on machine m in period t
B_{pt}	Backorder level of product p in period t
I_{pt}	Inventory level of product p in period t
LS_{pt}	Lot-size of product p in period t

NF_{mt} Expected number of failures of machine m in period t
PM_{mtj} A binary variable representing the level of preventive maintenance of machine m in period t (if it is one, it means preventive repairs are carried out on machine m in period t at level j)
PMC_t Total preventive maintenance costs in period t
MRC_t Total minimum repair costs in the period
IMC_t Total costs of partial repairs in period t
S_{pmt} A binary variable representing setting product p in machine m in period t
TC_M The total cost of the maintenance system
TC_P The total cost of the production system
W_{mt} Age of machine m at the beginning of period t
x_{pmt} The production level of product p on machine m in period t
Y_{mt} Age of machine m at the end of period t
Z_{mt} Binary variable assigning a low-skilled technician to machine m in period t (if machine m is failed in time slot t, it will be repaired by technician k)
Z'_{mt} Binary variable assigning a high-skilled technician to machine m in period t (if machine m is failed in period t, it will be repaired by technician k)
IM_{mt} Binary variable indicating whether partial repairs or minimal repairs will be done on machine m in case of failure in period t

Equations

The total cost of the production system (TC_P) comprises the manufacturing cost, set-up cost, inventory holding costs, and backorder cost.

$$TC_P = \sum_{t \in T} \left(\sum_{p \in P} \sum_{m \in M} (x_{pmt}\pi_{pm} + S_{pmt}s_{pm}) + \sum_{p \in P} (I_{pt}h_p + B_{pt}b_p) \right) \quad (1)$$

The total costs of maintenance (TC_M) is the sum of the PM cost of repairs, minimal repairs, and partial repairs. As a result, we will have:

$$TC_M = \sum_{t \in T} (PMC_t + MRC_t + IMC_t) \quad (2)$$

The total costs of technicians, including low-skilled technicians and high-skilled technicians, are calculated below:

$$TC_T = \sum_m \sum_t Z_{mt} * CZ_t + \sum_m \sum_t Z'_{mt} * CZ'_t \quad (3)$$

Equation (4) encompasses the overall costs of the production system, maintenance expenses, and technician fees.

$$Z = TC_P + TC_M + TC_T \quad (4)$$

Equation (5) is related to the level of production.

$$x_{pmt} \le S_{pmt} g_{pm} L \forall p, \mathrm{m}, t \tag{5}$$

The total lot size of product p in period t is:

$$LS_{pt} = \sum_{m \in M} x_{pmt} \forall p, t \tag{6}$$

The equation that represents the relationship between production, inventory, backorder, and demand is as follows:

$$I_{pt} - B_{pt} = I_{pt-1} - B_{pt-1} + LS_{pt} - d_{pt} \forall p, t \tag{7}$$

Equation (8) is related to the production time available according to the repair time and the skill of the technicians.

$$APT_{mt} = \mathrm{L} - Z_{mt} NF_{mt} TMR_m - W{*}Z'_{mt}{*}NF_{mt}{*}TMR_m - IM_{mt}{*}NF_{mt}{*}TIM_{mt} \forall m, t \tag{8}$$

Equations (9) and (10) are related to the capacity and number of human resources (high-skilled and low-skilled technicians).

$$\sum_{m \in M} Z_{mt} \le V_t \quad \forall t \tag{9}$$

$$\sum_{m \in M} Z'_{mt} \le V'_t \quad \forall t \tag{10}$$

The relationship between production and available production time is shown by Eq. (11).

$$\sum_{p \in P} \frac{x_{pmt}}{g_{pm}} \le APT_{mt} \forall m, t \tag{11}$$

Equations (12) and (13) pertain to the age of the machine m at the start of the period t + 1 (based on the machine's reduced lifespan).

$$W_{m.t+1} = \left(Y_{mt} - \frac{\sum_j CPM_{mj}{*}PM_{mtj}{*}Y_{mt}}{CPM_{m1}} \right) \forall m, t \tag{12}$$

$$\sum_j PM_{mtj} = 1 \quad \forall m, t \tag{13}$$

Equation (14) is related to the possible number of failures.

$$NF_{mt} = \theta_m * (Y_{mt} - W_{mt}) \quad \forall m, t \tag{14}$$

The Eq. (15) is related to the age of the machine at the end of the period t.

$$Y_{mt} = W_{mt} + APT_{mt} - IM_{mt} * NF_{mt} * \lambda_m \quad \forall m, t \tag{15}$$

Equation (16) shows the total cost of PM.

$$PMC_t = \sum_{m \in M} \sum_j CPM_{mj} * PM_{mtj} \forall t \tag{16}$$

The relationship between the budget and the cost of PM is shown in Eq. (17).

$$PMC_t \leq PMB_t \forall t \tag{17}$$

Equation (18) shows the minimum repair cost in period t.

$$MRC_t = \sum_m NF_{mt} * CMR_m \forall t \tag{18}$$

Equation (19) examines the relationship between the possible number of breakdowns and the allocation of technicians for machine m in period t.

$$(NF_{mt} - delta)/M \leq Z_{mt} + Z'_{mt} \forall m, t \tag{19}$$

Equation (20) is related to the cost of partial repairs.

$$IMC_t = \sum_{m \in M} IM_{mt} * NF_{mt} * CIM_{mt} \forall t \tag{20}$$

$$APT_{mt}, B_{pt}, I_{pt,} LS_{pt}, NF_{mt}, MRC_t, IMC_t, TC_M, TC_P, W_{m0}, W_{mt}, x_{pmt}, Y_{mt} \geq 0 \; and \; integer \quad \forall \; m, t, j, p$$

$$Z_{mt}, Z'_{mt}, PM_{mtj}, S_{pmt}, IM_{mt} \in \{0, 1\} \forall m, t, j, p$$

The following two types of limitations have been included in the model for human resources:

1. Limiting the number of human resources
2. Human resource skill limitation

TMR_m is the minimum repair time for machine m (standard repair time), which corresponds to the repair time of a low-skilled technician, while a high-skilled technician has a repair time equal to $W * TMR_m$.

3.1.1 Linearization of the Model

Equation (8) is converted into Eq. (21) for linearization (Chen et al. 2010), and three Eqs. (22), (23), and (24) are added as constraints to the model:

$$APT_{mt} - \mathrm{L} = NF_{mt}TMR_m - W * lin1_{mt} * TMR_m \forall m, t \tag{21}$$

The model is further enhanced by incorporating the following constraints:

$$lin1_{mt} \leq M * NF_{mt} \forall m, t \tag{22}$$

$$lin1_{mt} \leq Z_{mt} \forall m, t \tag{23}$$

$$lin1_{mt} \geq NF_{mt} - M * (1 - Z_{mt}) \quad \forall m, t \tag{24}$$

Equation (12) is converted into Eq. (25) for Linearization, and Eqs. (26), (27), and (28) are added to the model.

$$W_{m.t+1} = \left(Y_{mt} - \frac{\sum_j CPM_{mj} * lin2_{mtj}}{CPM_{m1}} \right) \quad \forall m, t, j \tag{25}$$

$$lin2_{mtj} \leq Y_{mt} \quad \forall m, t, j \tag{26}$$

$$lin2_{mtj} \leq PM_{mtj} * M \quad \forall m, t, j \tag{27}$$

$$lin2_{mtj} \geq Y_{mt} - M * \left(1 - PM_{mtj}\right) \quad \forall m, t, j \tag{28}$$

In Eq. (15), the expression $IM_{mt} * NF_{mt}$ is converted into the mathematical Eq. (29) to be linearized, and Eqs. (30), (31), and (32) are added to the model.

$$IM_{mt} * NF_{mt} = lin3_{mt} \quad \forall m, t \tag{29}$$

$$lin3_{mt} \leq M * IM_{mt} \quad \forall m, t \tag{30}$$

$$lin3_{mt} \leq NF_{mt} \quad \forall m, t \tag{31}$$

$$lin3_{mt} \geq NF_{mt} - M * (1 - IM_{mt}) \quad \forall m, t \tag{32}$$

$$lin1_{mt}, lin2_{mtj} \text{ and } lin3_{mt} \geq 0,$$

4 Computational Results

Above is an optimization model utilizing Mixed-Integer Programming (MIP). Given that the problem at hand is considered a medium to large-scale problem in the real world, a numerical example was solved in small, medium, and large dimensions, and the results showed that it provides a reasonable solution with an acceptable solving time even in medium and larger dimensions. This research solved the problem using an exact approach and the GAMS version 24.8.5, 32-bit system software. The performance of the suggested model was assessed through the creation of various test problems categorized into three classes, each containing three test problems: S1-S3 in the small class (S), M1-M3 in the medium class (M), and L1-L3 in the large class (L). The structures of different instances are summarized in Table 2. The problem is solved separately in different dimensions by the CPLEX solver, and the results are displayed.

Figures 1, 2, and 3 show that As the quantity of products rises, machines, and time slots, there is a significant increase in the number of constraints, the number of variables, and the solution time. Additionally, increases in the number of machines and products substantially impact the solution time, while the number of time slots does not exert a substantial impact. As the number of time slots increases, it substantially impacts the number of constraints and variables.

5 Sensitivity Analyses

The purpose of sensitivity analysis is to determine how certain parameters impact the best solution and the adjustments to the optimal value of the objective function. The effect of altering one parameter when other parameters are regarded as fixed can be checked in this method at any time. To determine the result of the problem's output under various circumstances, several parameters likely to be effective have been set among these criteria. For this aim, an instance with two products, three machines, six-time slots with a fixed duration of 1 month, and four levels of preventive maintenance has been selected. The goal is to reduce the overall expenses, encompassing production, maintenance, and technician costs. The relevant data for the model parameters are provided in Tables A1–A5. Table A6 presents the values of the parameters for processing cost without any increase, and Table 3 presents the

Table 2 Computational results of the test problems

Dimension	Classe	# machines	# products	# time slots	# constraints	# variables	# discrete variables	CPU time	Objective function
Small	S1	2	2	6	397	307	108	0.359	306,204
	S2	3	2	6	565	433	162	0.533	329,844
	S3	3	3	7	694	568	210	0.844	598,668
Medium	M1	6	7	10	2181	2101	840	25.469	1,984,338
	M2	7	8	11	2861	2839	1155	20.109	2,311,033
	M3	8	8	12	3529	3493	1440	27.891	2,706,043
Large	L1	10	11	15	5971	6391	2700	267.453	4,655,105
	L2	11	11	16	6525	6976	2970	332.594	4,537,358
	L3	12	12	16	7777	8497	3648	855.640	524,035

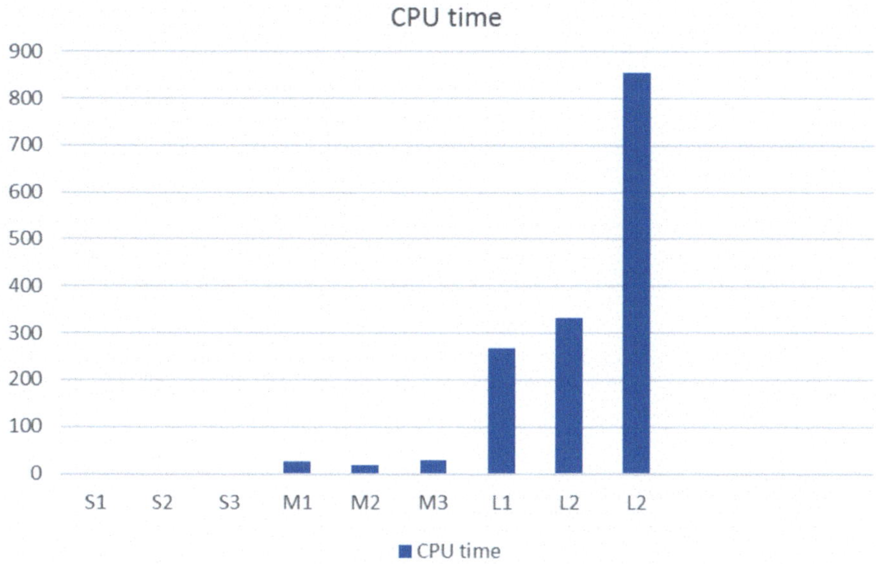

Fig. 1 Time to solve problems by GAMS software using CPLEX solver

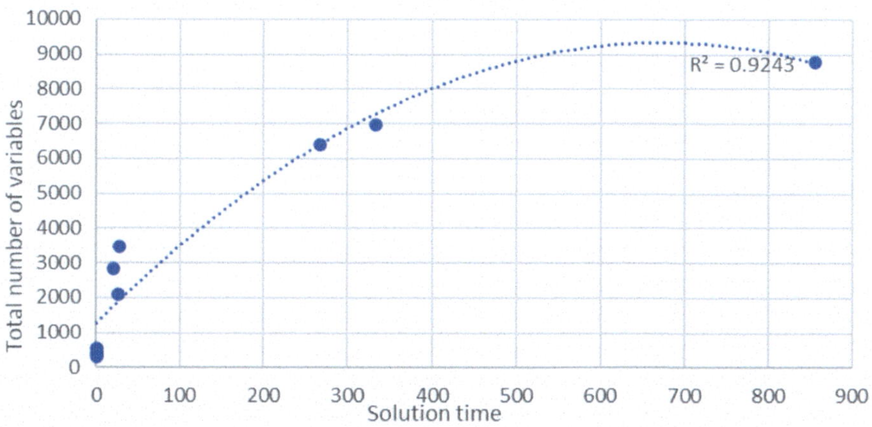

Fig. 2 Solution time trend w.r.t. the number of variables

objective function values for different percentage increases in the cost of processing product p on machine m. All the sensitive analyses are calculated based on the S2 dimension of the model.

The results show that the objective function value also increases as the processing cost increases. This indicates that the model is sensitive to changes in these parameters. At a 0% increase, the objective function value is 329,844.66. When the parameters increase by 25%, the objective function value rises to 422,218.52, representing

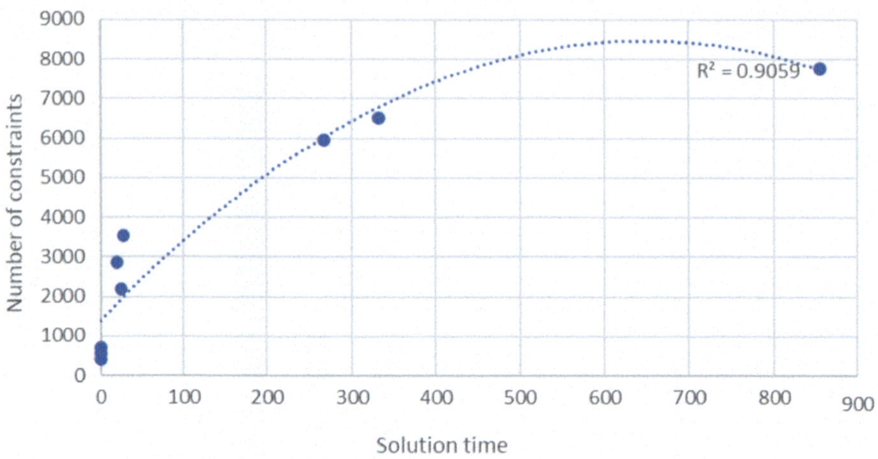

Fig. 3 Solution time trend w.r.t. the number of constraints

Table 3 Sensitivity analysis for the parameter π_{pm}

Increase in the parameter (%)	Objective function	Difference (%)
0	329,844.66	0
25	422,218.52	28
50	468,526.77	42
75	502,622.97	52

a significant impact. Further 50% and 75% increases lead to even higher objective function values of 468,526.77 and 502,622.97, respectively. The sensitivity analysis provides valuable insights into the model's behavior and the importance of accurate estimation of the unit processing cost p and the machine m (π_{pm}). Managers can use this information to make informed decisions and implement strategies to mitigate the impact of changes in these parameters on the overall system performance. Table A7 demonstrates the values of the parameters for the cost of minimal repair for machines. The results of a comparable sensitivity analysis on the minimal repair cost for machines can be found in Table 4.

Table 4 Sensitivity analysis for the parameter CMR_m

Increase in the parameter (%)	Objective function	Difference (%)
0	329,844.66	0%
25%	341,050.66	3%
50%	351,663.50	6%
75%	361,753.21	9%

The results indicate that the objective function value increases as the cost of minimal repair increases. The objective function increases to 341,050.66, which is a significant impact, as the repair cost increases by 25%. Further 50% and 75% increases result in even higher objective function values of 351,663.50 and 361,753.21, respectively. This sensitivity analysis demonstrates the model's sensitivity to changes in the cost of minimal repair for the machines.

Notably, the minimal sensitivity to technician repair cost indicates that the model can accommodate reasonable fluctuations in this parameter without significantly affecting the optimal solution. This flexibility in the decision-making process allows managers to allocate resources more effectively, as they can prioritize their efforts on the more impactful factors. Overall, the sensitivity analysis provides valuable insights into the robustness of the model to variations in the technician/operator repair cost, enabling managers to make informed decisions and optimize the system's performance.

In order to get into the details, we now analyze the impact of a 75% increase in CMR_m and CZ_t on the decision variables. The model includes four sets of decision variables:$IM_{m,t}$, $Z'_{m,t}$, $Z_{m,t}$, $X_{p,m,t}$ (current optimal values are shown in Tables 9 and 10.).

The decision variable values for two products and three machines, in the scenario of a 75% increase in the minimum repair cost of machine m (CMR_m), have been calculated as follows: the top row in the tables displays the normal case decision variable values, while the bottom row shows the values for the case with a 75% increase in the parameter. Table 5 illustrates the decision variable values for the product p's production level on machine m with an increased CMR_m in the S2 dimension.

Table 5 shows that with an increase in CMR_m, the production level has decreased, which is a logical consequence of the increase in repair costs.

In Tables 6 and 7, a comparison of the values of the decision variables $Z_{m,t}$ and $Z'_{m,t}$ in both scenarios reveals that with a 75% increase in the parameter CMR_m, or the minimum repair cost of machine m, the values of $Z'_{m,t}$ remain constant while $Z_{m,t}$ changes slightly. However, the values of the decision variable $IM_{m,t}$ do change, and during the second period of the third machine, a partial repair occurs.

In addition, in the case of a 75% increase in the parameter CZ_t in the case of two products and three machines, and by comparing the decision variables in this case with the normal case, it can be seen that the variable $X_{p,m,t}$ has not changed at all, and the increase in the cost of a low-skilled technician's wages does not affect the production level values.

Table 5 The values of the decision variable for the production level with an increase in the parameter CMR_m

X_{pmt}		t=2
p.m	2.2	641
		1500
	2.3	1000
		141

Table 6 Low-skilled technician selection with an increase in the parameter CMR_m

$Z_{m,t}$		Time slot					
		1	2	3	4	5	6
Machines	3	1	1	1	1	1	1
		1	0	1	11	1	1

Table 7 Performing partial repairs with an increase in the parameter CMR_m

$im_{m,t}$		Time slot	
		2	6
Machines	3	0	0
		1	0

With the increase in the parameter values of CZ_t, all the decision variables $Z_{m,t}$ become equal to zero, and the low-skilled technician is not used instead, according to Table 8, the values of $Z'_{m,t}$ have taken on values, and the high-skilled technician is used. Also, all the values of the decision variable $IM_{m,t}$ remain constant.

Conclusively, in this model, the inclusion of constraints concerning high-skilled and low-skilled technicians allows the model to make decisions based on the available budget, the costs of high-skilled and low-skilled technicians, and the production time available. This enables the model to determine whether a high-skilled or low-skilled technician should be assigned to repair the machine as needed. In the above model, the decision variable $IM_{m,t}$, which is related to partial repairs, takes the value of 1 in period 6 and machine 2, indicating that partial repairs were performed on machine 2 in period 6. Now, by considering the decision variable $IM_{m,t}$ as constant, such that partial repairs are performed in periods 2 and 5 on machine 1 and in periods 4 and 6 on machine 3, it is evident that the value of the objective function has risen by approximately 10% from 329,844 to 357,731. Additionally, the solution has become less coherent, as the total cost of partial repairs in the objective function increases with more partial repairs of the machines in the periods. Regarding the novelty in technician selection, if the decision variables $Z_{m,t}$ and $Z'_{m,t}$, are considered as parameters and their values are randomly set to 0 and 1; the solution will be worse than the case where the model is left free to choose between high-skilled and low-skilled technicians. To perform this, four sets of random 0 and 1 values

Table 8 Selecting low-skilled technicians with the increase in the parameter CZ_t

$Z'_{m,t}$		Time slot					
		1	2	3	4	5	6
Machines	1	1	1	1	1	1	1
	2	1	1	1	11	1	0
	3	1	1	1	1	1	1

were generated for $Z_{m,t}$ and $Z'_{m,t}$, and the objective function value for each set was obtained using the GAMS software. The average of the objective function values was then calculated and compared to the solution where the model freely decides on the technician selection, which shows that the average solution is 433,540, which is 31% worse. This contribution indicates that repairs were performed within the system in the past, but no decision was made regarding the technician's selection. In contrast, with this idea in maintenance and repair, the costs are reduced; for example, in this research, where the model freely decides on the v selection, the costs are much lower (31%) than when the model does not decide on the technician selection.

6 Conclusion

This research endeavored to enhance the model's realism by incorporating novel parameters and variables to facilitate decisions regarding the selection of repairmen and the type of repairs undertaken. By accounting for repair duration and the differentiation of repairmen by skill level, the model was brought closer to real-world circumstances. The study presents a mixed-integer programming model applicable to industrial settings to minimize costs, encompassing production, maintenance, repair, and v expenses. Aspects such as skill and number, their repair durations, the nature of repairs (minimal or partial) were examined, and the time requirements for partial or minimal repairs. The base model was originally formulated in a non-linear manner; however, in this research, the developed model incorporated three sets of non-linear constraints that were linearized, leading to a significant reduction in solution time, which was then solved using the exact solution method and the CPLEX solver.

The model's validity can be ascertained by augmenting the parameters influencing the model and scrutinizing the modeling contribution. Furthermore, by fixing or removing the associated decision variables, it can be shown that when the model is granted autonomy in selecting decision variable values, such as the choice between skilled and regular repairmen or the type of repairs, the objective function value exhibits superior performance compared to scenarios where these variable values are fixed. For instance, the decision variables $Z_{m,t}$ and $Z'_{m,t}$, whose innovative application indicates the existence of repairs within the system but without decision-making regarding repairman selection, demonstrates that this innovation in maintenance can yield cost reductions, as exemplified in this research, where the model's free decision-making on repairman selection results in significantly lower costs (31%) than when the model is not granted this decision-making capability. The flexibility inherent in decision-making empowers managers to allocate resources more judiciously, as they can now focus on the factors with the greatest impact. Moreover, the sensitivity analysis furnishes valuable insights into the model's resilience to fluctuations in the technician repair cost, equipping managers with the knowledge required to make well-informed decisions and optimize the system's overall performance.

Regarding optimizing the production system considering partial or minimal repairs and their impact on costs, it can be stated that each partial repair reduces the machine's age, allowing it to operate on the production line for a longer duration but incurring additional costs beyond the minimal repair expenses. The model, accounting for the budget, repair time, degree of failure, and the costs of minimal and partial repairs, decides to perform continuous repairs. The distinguishing feature of this model compared to previous models is its careful consideration of several crucial factors in the realm of maintenance, such as the technician's skill level, the wage cost and repair time ratio of the v, the type of repairs (partial or minimal), the time and cost of the repair types, and the number of technicians, rendering it more practically and realistically applicable in industrial settings.

7 Future Research

To further improve the model's realism and applicability, future research could focus on adding additional parameters and variables. For example, the model could be modified to incorporate other factors that influence repair decisions, such as the availability of spare parts or the need for urgent repairs. In addition, it is possible to analyze the efficiency of different repair methods, such as preventive or predictive maintenance, in order to lower expenses and enhance the system's performance. Other crucial distributions in maintenance, like the gamma and beta distributions, could be taken into account in the model rather than exclusively relying on the exponential distribution. It would also be valuable to solve the proposed model using non-linear solvers and compare it with other solution approaches to assess their efficiency and effectiveness. Expanding the model to incorporate the downtime of machines and production stoppages during repair time would provide a more comprehensive understanding of the system. Additionally, the model could consider the inventory of repair resources, equipment, and necessary parts and make decisions regarding them within the model. Taking into account uncertainty in parameters, such as demand, would contribute to a more robust model. Moreover, research could be conducted to evaluate the potential advantages of implementing the model in various industries or sectors. By applying the model to different industrial settings, researchers could assess its effectiveness and identify any limitations or areas for improvement. This would enable the model to be adapted and applied in diverse circumstances, enhancing its generalizability and practicality. Further research could explore the impact of different repair strategies, such as CBM maintenance or reliability-centered maintenance, on the performance and cost optimization of the production system. The model could be extended to incorporate multi-objective optimization, considering factors such as production throughput, maintenance costs, and equipment reliability simultaneously. Additionally, analyzing the influence of external factors, such as environmental regulations or market demand fluctuations, on the decision-making process would provide valuable insights for managers in real-world scenarios.

Appendix

Considering the different dimensions of the problem data, they are randomly generated using a uniform distribution within the ranges specified by industry experts.

The budget values PMB_t for each time are 1500 monetary units.

The cost values for assigning a low-skilled technician in period t (CZ_t) are 50 monetary units per period.

The cost values for assigning a high-skilled technician in period t (CZ_t') are 70 monetary units per period.

The value for the high-skilled repair time factor is equal to 0.4.

The number of low-skilled technicians in period t (V_t) is 100 in each period, and the number of high-skilled technicians in period t (V_t') is also 100 in each period.

The repair time coefficient for high-skilled technicians (W) is 0.4, and the parameter for the minimum expected number of failures for technician assignment (delta) is a constant value of 1. The relevant data for the model parameters are provided in Tables 9, 10, 11, 12, 13, 14, 15, 16, 17 and 18.

Table 9 Values of the parameters

Product		S_{pm}		α_{pm}		CMR_m	TMR_m	π_{pm}		ρ_m	λ_m	
		1	2	1	2			1	2			
Machine	1	40	–	0.6	–	800	0.04	6	–	1	0.3	0.847
	2	30	10	0.4	0.5	1000	0.03	8	9	1	0.1	3.812
	3	–	35	–	0.8	1200	0.025	–	10	1	0.25	6.099

Table 10 Values of the parameters (continued)

Product		g_{pm}		
		1	2	
Machine	1	2500	–	0.12
	2	1000	1500	0.14
	3	3000	–	0.15

Table 11 Additional costs of partial repairs in period t on machine m (CIM_{mt})

Machine	Time					
		1	2	3	4	5	6
	1	12	18	15	15	13	16
	2	14	16	9	15	17	16
	3	15	19	10	13	11	16

Table 12 The cost of the jth level of PM (preventive maintenance) for machine m (CPM_{mj})

Product	Preventive maintenance level				
		1	2	3	4
	1	800	400	200	0
	2	900	300	150	0
	3	500	200	100	0

Table 13 The production demand for product p in period t (d_{pt})

Product	Time						
		1	2	3	4	5	6
	1	3500	4000	1500	2500	1000	5000
	2	2500	2000	1500	1500	3500	3500

Table 14 The parameter values for processing cost without any increase in the processing cost

π_{pm}		Machines		
		1	2	3
Products	1	6	8	100
	2	100	9	10

Table 15 The parameter values for the minimal repair cost of machine m without any increase

		CMR_m	
Machines	1	800	
	2	1000	
	3	1200	

Table 16 The values of production level

$X_{p,m,t}$		Time slot					
		1	2	3	4	5	6
Product and machines	1.1	2500	2500	2500	2500	2500	2500
	1.2	0	572.66	0	0	0	0
	2.2	1500	641	1500	1500	1500	0
	2.3	1000	1000	1000	1000	1000	1000

Table 17 The values of selecting the low-skilled technician

$Z_{m,t}$		Time slot					
		1	2	3	4	5	6
Machines	1	1	1	1	1	1	1
	2	1	1	1	11	1	0
	3	1	1	1	1	1	1

Table 18 The values of performing partial repairs

$im_{m,t}$	Time slot	
Machines		
	2	1

References

Al-Salamah M (2018) Economic production quantity with the presence of imperfect quality and random machine breakdown and repair based on the artificial bee colony heuristic. Appl Math Model 63:68–83. https://doi.org/10.1016/j.apm.2018.06.034

Arani M et al (2020) Optimizing the Total Production and Maintenance Cost of an Integrated Multi-Product Process and Maintenance Planning (IPPMP) Model. Paper presented at the 2020 IEEE International Symposium on Systems Engineering (ISSE)

Beheshti Fakher H, Nourelfath M, Gendreau M (2017) A cost minimisation model for joint production and maintenance planning under quality constraints. Int J Prod Res 55(8):2163–2176. https://doi.org/10.1080/00207543.2016.1201605

Birdi K et al (2008) The impact of human resource and operational management practices on company productivity: a longitudinal study. Pers Psychol 61:467–501. https://doi.org/10.1111/j.1744-6570.2008.00136.x

Bocewicz G et al (2023) Preventive maintenance scheduling of a multi-skilled human resource-constrained project's portfolio. Eng Appl Artif Intell 119:105725. https://doi.org/10.1016/j.engappai.2022.105725

Bouzidi-Hassini S et al (2015) Considering human resource constraints for real joint production and maintenance schedules. Comput Ind Eng 90:197–211. https://doi.org/10.1016/j.cie.2015.08.013

Chen K et al (2023a) Single-machine scheduling with autonomous and induced learning to minimize total weighted number of tardy jobs. Eur J Oper Res 309(1):24–34. https://doi.org/10.1016/j.ejor.2023.01.028

Chen Z et al (2023b) Opportunistic maintenance optimization of continuous process manufacturing systems considering imperfect maintenance with epistemic uncertainty. J Manuf Syst 71:406–420. https://doi.org/10.1016/j.jmsy.2023.10.001

Chen D-S, Batson RG, Dang Y (2010) Applied integer programming: modeling and solution. Wiley, Hoboken

Cheng W, Zhao X (2023) Maintenance optimization for dependent two-component degrading systems subject to imperfect repair. Reliab Eng Syst Saf 240:109581. https://doi.org/10.1016/j.ress.2023.109581

Dehghan Shoorkand H, Nourelfath M, Hajji A (2024) A hybrid CNN-LSTM model for joint optimization of production and imperfect predictive maintenance planning. Reliab Eng Syst Saf 241:109707. https://doi.org/10.1016/j.ress.2023.109707

Dui H et al (2023) Performance-based maintenance analysis and resource allocation in irrigation networks. Reliab Eng Syst Saf 230:108910. https://doi.org/10.1016/j.ress.2022.108910

Geurtsen M et al (2023) Production, maintenance and resource scheduling: a review. Eur J Oper Res 305(2):501–529. https://doi.org/10.1016/j.ejor.2022.03.045

Hejazi T-H, Hekmatnia B, Soltanzadeh M (2023) A novel approach for planning imperfect preventive maintenance in manufacturing systems by a simulation-optimization approach. J Syst Thinking Pract 2(3):1–20. https://doi.org/10.22067/jstinp.2023.84099.1072

Khanizad R, Montazer G (2018) Optimal allocation of human resources based on operational performance of organizational units using fuzzy game theory. Cogent Eng 5(1):1466382. https://doi.org/10.1080/23311916.2018.1466382

Kouedeu A, Jean-Pierre K, Songmene V (2011) Production, preventive and corrective maintenance planning in manufacturing systems under imperfect repairs

Levitin G, Xing L, Dai Y (2023) Minimum downtime operation and maintenance scheduling for resource-constrained system. Reliab Eng Syst Saf 238:109465. https://doi.org/10.1016/j.ress.2023.109465

Li S et al (2023) A novel maintenance strategy for manufacturing system considering working schedule and imperfect maintenance. Comput Ind Eng 185:109656. https://doi.org/10.1016/j.cie.2023.109656

Liu P, Wang G, Tan Z-H (2024) Calendar-time-based and age-based maintenance policies with different repair assumptions. Appl Math Model 129:592–611. https://doi.org/10.1016/j.apm.2024.02.013

Maghsoudlou H, Afshar-Nadjafi B, Niaki STA (2021) A framework for preemptive multi-skilled project scheduling problem with time-of-use energy tariffs. Energy Syst 12(2):431–458. https://doi.org/10.1007/s12667-019-00374-8

Nobil AH et al (2024) Economic production quantity models for an imperfect manufacturing system with strict inspection. Ain Shams Eng J 15(5):102714. https://doi.org/10.1016/j.asej.2024.102714

Pedersen TI, Liu X, Vatn J (2023) Maintenance optimization of a system subject to two-stage degradation, hard failure, and imperfect repair. Reliab Eng Syst Saf 237:109313. https://doi.org/10.1016/j.ress.2023.109313

Rasay H, Fallahnezahd MS, Bazeli S (2022) Clustering of condition-based maintenance activities with imperfect maintenance and predication signals. IUST 33(4):1–19. https://doi.org/10.22068/ijiepr.33.4.11

Rippe C, Kiesmüller GP (2023) The repair kit problem with imperfect advance demand information. Eur J Oper Res 304(2):558–576. https://doi.org/10.1016/j.ejor.2022.04.019

Rivera-Gómez H, Gharbi A, Kenné JP (2013) Joint production and major maintenance planning policy of a manufacturing system with deteriorating quality. Int J Prod Econ 146(2):575–587. https://doi.org/10.1016/j.ijpe.2013.08.006

Shabtay D, Zofi M (2018) Single machine scheduling with controllable processing times and an unavailability period to minimize the makespan. Int J Prod Econ 198:191–200. https://doi.org/10.1016/j.ijpe.2017.12.025

Touat M et al (2017) A hybridization of genetic algorithms and fuzzy logic for the single-machine scheduling with flexible maintenance problem under human resource constraints. Appl Soft Comput 59:556–573. https://doi.org/10.1016/j.asoc.2017.05.058

Touat M, Tayeb FB-S, Benhamou B (2018) An effective heuristic for the single-machine scheduling problem with flexible maintenance under human resource constraints. Proc Comput Sci 126:1395–1404. https://doi.org/10.1016/j.procs.2018.08.091

Viveros P et al (2023) Extended framework for preventive maintenance planning: risk and behaviour analysis of a proposed optimization model. Complexity 2023:2701439. https://doi.org/10.1155/2023/2701439

Wong CS, Chan FTS, Chung SH (2013) A joint production scheduling approach considering multiple resources and preventive maintenance tasks. Int J Prod Res 51(3):883–896. https://doi.org/10.1080/00207543.2012.677070

Wu T et al (2019) Proactive maintenance scheduling in consideration of imperfect repairs and production wait time. J Manuf Syst 53:183–194. https://doi.org/10.1016/j.jmsy.2019.09.011

Yang L et al (2019) A two-phase preventive maintenance policy considering imperfect repair and postponed replacement. Eur J Oper Res 274(3):966–977. https://doi.org/10.1016/j.ejor.2018.10.049

Yu J et al (2022) A study of the impact of strategic human resource management on organizational resilience. Behav Sci (Basel) 12(12). https://doi.org/10.3390/bs12120508

Zhang N et al (2023) Optimal production lot-sizing and condition-based maintenance policy considering imperfect manufacturing process and inspection errors. Comput Ind Eng 177:108929. https://doi.org/10.1016/j.cie.2022.108929

Environment-Friendly Practices for Integrating Green Business with Green Supply Chain Management: Industry 4.0 Perspectives and Beyond

B. K. Mawandiya, Aditya Choksi, Anjaney Lokapur, Ajay Kumar, M. A. Makhesana, K. M. Patel, and P. R. C. Gopal

Abstract In any organization, Supply Chain Management (SCM) plays a vital role in improving business performance. With the scarcity of resources and the increase in the population and demand of goods, adapting and shifting to Green Supply Chain Management (GSCM) practices is the need of the hour. SCM integrated with Green Business (GB) operations encompasses most of the firms' activities. This review paper covers the current research work carried out in GB and GSCM for Industry 4.0 perspectives and beyond. It is highlighted that how environment-friendly practices empower the organization to gain a competitive advantage. The conceptual framework provided shows the link between all the practices in a linear flow and highlights the importance of each intermediate step.

Keywords Industry 4.0 · Sustainability · Supply chain management · Green supply chain management · Green business · Closed-loop supply chain

1 Introduction

Green Supply Chain Management (GSCM) has its roots in the green revolution that engulfed the whole manufacturing and service industry globally in the late 1980s-early 1990s. The main driving force behind this movement was the birth

B. K. Mawandiya · A. Choksi · A. Lokapur · M. A. Makhesana (✉) · K. M. Patel
Mechanical Engineering Department, Institute of Technology, Nirma University, Ahmedabad 382481, India
e-mail: mayur.makhesana@nirmauni.ac.in

A. Kumar
Department of Mechanical Engineering, School of Engineering and Technology, JECRC University, Jaipur, Rajasthan 303905, India

P. R. C. Gopal
Department of School of Management, National Institute of Technology, Warangal, Telangana 506004, India

A. Kumar et al. (eds.), *Industry 4.0 Driven Manufacturing Technologies*, Springer Series in Advanced Manufacturing, https://doi.org/10.1007/978-3-031-68271-1_9

of an idea to develop environmentally conscious business operations that can be financially and ecologically profitable. Supply chain (SC) was not an integral part of the company's operations back in the 1980s. However, gradually, with the increase in market complexity and the emergence of new firms, each firm saw the need to revamp their business model. And SC played a central role in most companies' new business strategies. The business leaders viewed greening the whole supply chain as an essential step. This was motivated by both external and internal factors. Introducing environment-friendly features in various supply chain practises was a task that involved a great deal of innovation. GSCM practices involve inspection of the suppliers' environmental performance and the firm's internal practices. GSCM encompasses the whole supply chain and finds ways to keep a check on the environmental activities of other members. So, each practise has an indirect or direct impact on the environment and they need to be monitored accordingly (Darnall et al. 2008). The execution of this process depends on the company's core competencies and the employees' technical know-how.

The primary driving forces behind the transition to GSCM were the rise in consumer awareness and the modifications made to government regulations, which required businesses to adapt their business practices and adopt more environmentally friendly ones. Following the idea's dissemination throughout many industries, it is witnessed that the fusion of traditional practices with end-of-life product management encompasses re-manufacturing and reverse logistics. This partially addressed the customers' increased awareness of environmental issues and improved customer service. It was a positive move both ecologically and financially (Li et al. 2019). In an already competitive industry, companies were searching for ways to reduce their operational costs and look for more efficient methods to reduce waste and increase the productivity and ultimately the firm's profitability. Many new companies revolutionized the whole industry by emphasizing supply chain at the turn of the century. With the growing popularity of supply chain and customer demand for sustainable products, this was the perfect opportunity for the firms to move towards GSCM. Green business (GB) consists of two words, "Green" and "Business". This idea originated when two concepts, i.e. green and business, were combined. The main purpose behind this was to create an environment-friendly approach to business operations to generate economic and ecological value and drive the company forward. Adopting green corporate practices and policies would boost customer satisfaction, save the environment, encourage the use of renewable energy sources, improve material recyclability, and lessen the spread of harmful substances (Karagülle 2012). A green firm is dedicated to implementing the principles of environmental sustainability in its operations and reducing the adverse effects of its operations on the environment. GB's ideas came into business leaders' minds after a widespread surge of ecological conservation amongst the consumers (Čekanavičius et al. 2014). Furthermore, the hierarchy was compelled by government policies to modify their working procedures. Green Marketing, Green Finance, and Green Information Systems make up GB. Each of these elements contributes to the overall goal of introducing environmentally friendly practices within a company. Through the application of GB, businesses hoped to achieve their main goal of increasing revenues in addition to environmental conservation. The cost-saving element entered the picture later on as businesses noticed a decrease in the use of raw materials, the utilisation of alternative energy sources, and product reuse (Sharma et al. 2010). Implementing GB by firms early on had a transformative effect on the whole industry. Others followed this trend and gradually adopted it into their operations. So, ecologically it had a massive impact and increased customer goodwill toward the firms. The companies were able to meet legal requirements, obtain government tax incentives, and by incorporating consumers' and managerial concerns about the environment, they achieved superior business advantage and enhanced firm reputation (Sharma et al. 2010).

The review paper has aimed to interlink and incorporate the GB practices with the GSCM practices to highlight the possibilities of firms leveraging this concept to gain higher revenue share, fulfill their environmental responsibilities and become more profitable. For many years, supply chain management was viewed as a part of the operations department rather than a way to increase profitability and customer satisfaction. Nevertheless, gradually many large companies started to view it as a critical part of the business model. Many large companies long term success/failure has been directly linked to their supply chain model. With increased green awareness amongst consumers, legal compliances, and other external pressure, GSCM came into existence. GSCM comprises five practices: Design, Procurement, Logistics, Manufacturing and Product Recovery. These practices cover the significant portion of the operations department in any firm. But for any successful GB model, coordination and interlinking between various departments is a must to generate profit and at the same time promote sustainable development. A business value chain interconnecting the various departments to efficiently produce any good at the lowest possible cost and provide the highest possible value to the customer by delivering it to the customers at the correct time is integral for the firm's success in the foreseeable future. GSCM cannot function without involvement of GB practises. Finance, marketing and information systems all add decision-making capacity, SC design inputs and viability of the venture's long-term profits. In short they help to maximize the ability of the firm to extract value from the supply chain.

Figure 1 shows a conceptual flowchart which aims to provide a flow to the research literature we have consolidated in our article. This flowchart presents a product's journey from idea generation to post use value recovery. The very first step is GB. Before launching any new product, the company always prefers to get information from their target market and get a grip on the customers' needs and expectations. Such a customer-centric approach enables the company to develop and introduce products that strongly conform to the customers' demands and requirements. With the help of green marketing, they can publicize their green philosophy in the market, collect data through surveys, and then use demand analytics on that data to predict and forecast rough estimates of future demand for that product. Once they have analysed the potential of the new project, the company moves on the next step, Green Financing. This practise involves checking the financial feasibility of any green project and making any necessary funding arrangements through green loans, bonds, stocks and through collaboration with other members of the supply chain. Once the capital budgeting of the project has been done and approved, we move on

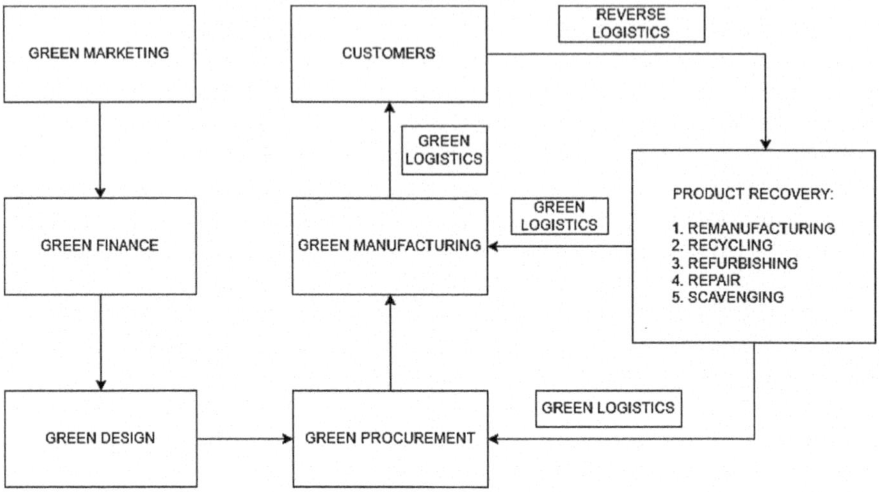

Fig. 1 Conceptual flowchart

from GB to GSCM. The very first GSCM practise, Green Design involves product development, prototype design, life-cycle analysis of the product and the product's technical requirements which the manufacturer needs to adhere to. At this stage, the designer has to develop a product with considerable value left in it after its usable life is over. After the product's design is ready, the purchasing department procures necessary ecologically sustainable raw materials for the product's manufacturing. Next up, we move from Green Procurement to Green Logistics, where the raw materials are transported in an eco-friendly way from the supplier to the manufacturer. In Green Logistics, the mode of transportation, type of packaging and the route they use are considered. The next practise is Green Manufacturing where the product is manufactured sustainably, which doesn't impact the environment negatively. By using various production techniques, ecologically sustainable lubricants, raw materials and energy-efficient machines, this practise aims to decrease the input and the extra losses. After the product is made, it is again transported from the manufacturer to its end destination- the customers. After the product has been used and discarded, it is returned via reverse logistics to the manufacturer, where product recovery occurs.

2 Green Marketing

Green Marketing (GM) is considered as a holistically integrated approach that re-evaluates how firms achieve corporate objectives continuously and meets customer demands while minimizing harm to the environment (Polonsky and Rosenberger 2001). Numerous industries have adopted eco-friendly practices as a result of the negligent disposal of industrial waste and the escalation of global warming.

Green marketing was created as a result. Environmental issues can be resolved via marketing strategies. The marketing team's duty in a market situation is to satisfy human demands by utilising resources as efficiently as possible (Peattie 1999). The marketing process provides important information required to define product concepts and design. Marketing inputs are required to create product designs and concepts that lessen the environmental impact of industrial activities. The manufacturer can create and produce environmentally sustainable items by considering those factors (Gustavo et al. 2021; Rex and Baumann 2007). Figure 2 shows a visual representation of the number of paper published in various journals for Green Marketing as referred in this review paper.

The marketing practices can lead to the creation of a green market by communicating with customers and enhancing their environmental sustainability knowledge, as well as educating them on the environmental benefits of sustainable products (Rex and Baumann 2007). Internal and external pressure influence the greening process the most. Customer requirements, competition greening initiatives, and Supplier requests to change input are among the internal pressures, while cost and philosophy are among the external pressures (Polonsky and Rosenberger 2001). GM can improve the efficiency by planning so that the products are produced without waste instead of handling it, by eliminating the concept of waste (Kumar 2020). Consumers are concerned about whether products are eco-friendly and whether businesses participate in green marketing campaigns. Growing awareness makes consumers more conscious of what they buy, resulting in little to no environmental damage (Tsai et al. 2020). They would rather select green products than traditional ones. As a result, GM significantly impacts how people choose which products to purchase. GM is a way to include environmentally friendly product packaging, design, and manufacture. This further improves a company's reputation with customers and motivates them to support these initiatives by purchasing (Dangelico and Vocalelli 2017). Further,

Fig. 2 Data visualization for Green Marketing

customers are encouraged to have a repurchase intention when their demands are satisfied, which leads to market expansion, increased market share, and increased profitability for the firm (Tsai et al. 2020). The author also pointed out that customer satisfaction is the most valuable outcome of marketing activity and is driven by marketing orientation.

Customers choose to purchase non-green products even though they know the advantages of green products. This is because, although green products have greater starting costs than non-green ones, they have lower long-term costs. Appropriate marketing strategies can address this. Without an effective marketing strategy, creating environmentally friendly goods and services would not result in environmental sustainability (Dangelico and Vocalelli 2017). Green advertising has significantly impacted customers' emotional processing and has sparked their desire to buy green products (Zubair et al. 2020). Olsen et al. (2014) suggested that negative message framing is more effective than positive message framing in research on green product communication. Also, it was pointed out that consumers trust negative information more than positive information. Thus, while advertising items, corporations should emphasize how non-green products deplete the environment rather than how their product would preserve the environment since this will increase sales. GM's performance is based on reducing the environmental effect of its operations during the purchase, manufacture, sale, and disposal of raw materials (Tsai et al. 2020). In order to achieve optimal results for marketing activities, the campaigns should link customers' emotions to environmental concerns (Chou et al. 2020).

There are numerous benefits to implementing genetically modified agriculture (GM) methods, including adhering to international environmental trends, increasing the value of products, strengthening competitive advantages in the market, strengthening company image, and investigating green options and possibilities (Chou et al. 2020; Chen and Yang 2019). Many companies have begun to adopt genetically modified organisms (GM) as a market expansion strategy in response to fierce market competition and growing consumer awareness of environmental protection. Many businesses have embraced green marketing strategies to be competitive in today's highly competitive marketplaces. These strategies encourage consumers to purchase eco-friendly products, foster brand loyalty, boost sales and market share, and maintain sustainable operations (Tsai et al. 2020; Dangelico and Vocalelli 2017). Green marketing improves firm performance and competitiveness by utilising sustainable supply chains (Gustavo et al. 2021). The focus is on stumbling through corporate social responsibility and getting involved in politics by implementing GM practices. This has been proven to have a significant beneficial impact on boosting the image of businesses and distinguishing them from competitors in customers' minds. This has further improved the operational performance of profit of the organization (Amores-Salvadó et al. 2014).

3 Green Finance

The term "green finance" describes how financial institutions actively encourage funding for projects that save energy and protect the environment. It is crucial for modifying the industrial structure and promoting economic expansion." Financial instruments such as green bonds, green credit, and green stock index are categorised under green finance (GF) (Wang et al. 2021a). Green bonds are fixed income instruments that finance projects promoting sustainable development and eco-friendly practises. Green credit is credit offered by the supplier/vendor to the manufacturer by extending the proposed payment duration to decrease the firm's financial burden. It incentivizes the manufacturer to produce green products. The green stock index is a special financial index that measures the performance of the green firms. This indirectly affects the likelihood of private investment and the firm's equity capital generation. So, firms are driven to perform better to sustain their green stock index to attract potential investment.

Figure 3 shows the distribution of papers in the domain of Green Finance based on the first author's country as used in this paper. Besides project financing using green instruments, green financing deals with carbon finance and green project risk management. Green financing also considers the financial analysis of a new project to check its feasibility after carrying out a demand forecast on marketing and sales numbers. This is typically done during the infancy stages of a new product development. Manufacturers generally consider internal and external financing. External financing is done by parties from outside the supply chain like bank loans and financing using green instruments (Wang et al. 2021b). Manufacturers traditionally consider bank loans as their major source of capital for a new project. But banks demand collateral and have shown an uninterested attitude towards green projects in the past considering the high risk involved. Bank loans with high interest rates and collateral directly affect the wholesale rate of their products, affecting the market share and reducing customer pull. This increases the inventory and related costs for the firm and drastically reduces the inventory turnover which affects the company massively. Though going downstream in the supply chain, we can observe that it is relatively easier for the retailers selling these green products to loan money from banks (Cao et al. 2020). To minimise risk and obtain a proper balance of capital and credit to finance their operations, manufacturers prefer hybrid financing, which is a combination of internal and external financing. The main source of internal finance for manufacturers is trade credit from other parties or SC members. A prepaid ratio that the supplier establishes during the agreement stage with the manufacturers is used to quantify trade credit. It is dependent on the profit margins of the members and the degree of risk aversion of the supplier. In order to ensure payment from the manufacturer, the supplier tries to take on a risk-hedging role in the financing. The manufacturer's bank loan failure risk is mitigated to a greater extent when the pre-paid ratio is lower (Yu et al. 2021). Businesses frequently use short-term finance to complete their procurement tasks when strapped for funds. Contracts for trade credit are frequently utilised for short-term funding. It manifests as a discount on the

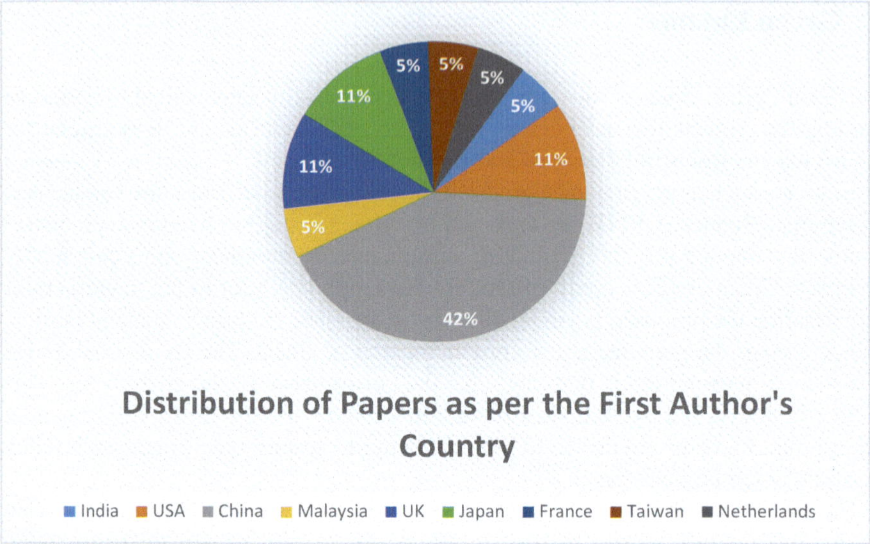

Fig. 3 Data visualization for Green Finance

manufacturer's wholesale price for early payment. The buyer can make interest-free payments for a predefined period of time. In addition to motivating the buyer to pay early to receive the discount, it improves the supplier–buyer relationship (Kouvelis and Zhao 2018).

The company with a greater power structure than its upstream partners could obtain lower loan pricing and less restrictive loan terms. When the bank uses cutting-edge technology, it could charge a greater service fee to lower the supply chain's financing risk (Wang et al. 2021b). For retailers, a major part of financing is used for refilling inventory. Their financial portfolio consists of a large part of trade credit with small bank loans (Yang and Birge 2018). From the government's perspective, green financing policies include carbon tax, government subsidy, green bond, and green investment policies (Wang et al. 2021a). Governments charge high taxes to manufacturers and provide benefits for recycling/re-manufacturing industries using money-raising activities to promote ecologically sustainable activities (Sheu 2011). Fixed carbon emission quotas are allotted to companies by the government. They often sell their surplus carbon emission quotas to gain profit or buy some from the carbon trading market to fulfil their requirements. They risk getting fined or restrictions on their production capacity if the firm does not implement such measures. Carbon allowances allocation rules like grandfathering rules and benchmarking rule are in practice. Benchmarking rules rely on average carbon emissions in the industry. This quote is directly proportional to the production quantity of the firm. At the same time, the rule allocates the quota based on the historical records of the firm (Wang et al. 2021a).

An excellent tool for encouraging manufacturing companies to switch to more environmentally friendly methods is a carbon tax or cap on emissions. The carbon emission cap is useful for reducing greenhouse gas emissions since it may charge fossil fuel consumption more heavily and incentivize businesses to utilise cleaner fuels and produce more environmentally friendly goods. The government can encourage environmentally friendly development by offering subsidies to businesses that employ green technology and development and raising the "greenness" of their products. This programme has the potential to reduce carbon emissions as well. It is a means of encouraging the advancement of green technology (Wang et al. 2021b; Cao et al. 2020). Private enterprises face bigger constraints on R&D activities than public firms because of a bias from state-owned banks who are partial towards public industries. Government can set up regional green development funds to support private firms' environmental activities and promote the development of eco-friendly industries. Ownership discrimination is observed in China, where public firms prefer private firms due to the dominance of large public banks (Yu et al. 2021). The government can develop credit guarantee schemes to make the banks ready to finance green projects. They can share and absorb some of the risk associated with the financing. It also removes the need for collateral as the guarantee itself is collateral. The government can later be compensated by the spillover effect where there is an increase in prosperity in the region due to the new green project. They can then collect taxes as their initial investment returns (Taghizadeh-Hesary and Yoshino 2019). A survey was conducted in developed economies across the globe regarding the relationship between pollution and green financing. The results showed that demand for green finance investment also increases with increased pollution. The result showed the inverse relationship between green finance and carbon emissions in selected states (Saeed Meo and Karim 2021).

4 Green Design

Green Design (GD) is defined as a method for determining if items and processes are compatible with environmental well-being without sacrificing function or quality by evaluating them (Navinchandra 1990). The work provided some of the initial literature for GD. It emphasised the need for eco-friendly design to decrease product waste's environmental effect. Green product design incorporates both ecological and economic considerations while creating innovative and valuable goods. Figure 4 shows the data visualization for the number of papers published in different journals on Green Design against their year of publication, as referred to in this paper.

The green product demand is increasing due to consumer desire for green product design (Zhang et al. 2015). The aims of green product design, also known as Design for Environment (DfE), are to limit the number of harmful emissions while minimizing the usage of non-renewable resources. Design for disassembly (DfD) and design for recycling (DfR) strategies, as well as life cycle evaluations, are some of the methods for creating a green and profitable supply chain (Kuo et al. 2001; Hoek

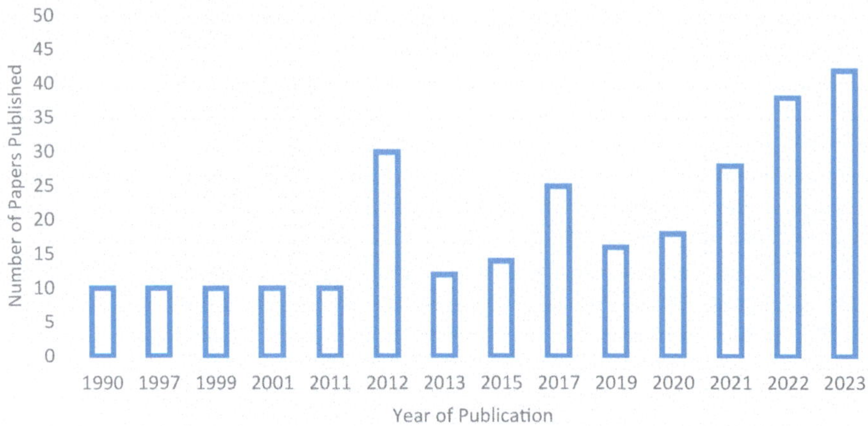

Fig. 4 Data visualization for Green Design

1999). Generally, product recovery is given very little attention as it is a labour-intensive job. Mostly the products are designed to serve the large population. But, the benefits of Green Design can be understood better when return products are reprocessed for reuse of materials. This is either done in the form of subassemblies or the whole product. To create a system for recovering products, Taleb and Gupta (1997) developed algorithms. Based on their research, "core algorithms" and "allocation algorithms" are the scheduling systems that would help reduce waste.

As far as the design modification doesn't negatively impact other objectives for product design, such as functionality and cost, it's acceptable. Green design and reverse logistics are cost-cutting strategies (Green et al. 2012). The former, however, appears more promising because it employs an eco-sustainable strategy advantageous to the environment and the economy. One way to determine how a product is valued is to assess its residual value. Time-sensitive handling is given to those with a high residual value, while economical treatment is given to those with a low residual value (Eltayeb et al. 2011; Hazen et al. 2012). The marginally expensive green product design may increase the recycling benefits and residual values of used goods (Li et al. 2021a). Atasu and Subramanian (2012) examined that the specified objectives on the manufacturer's product design, profit, and customer surplus are easier to encourage the manufacturer to apply green product design through individual responsibility mode. The research findings of Li et al. (2021b) presented that the mandated collection target does not always decrease the quantity of new product or manufacturer's profit. However, it increases them depending upon total collection and disposal cost.

5 Green Procurement

An organisation that participates in green programmes and whose internal operations and processes are environmentally sustainable can be considered a chain of companies that practise green supply chain management (GSCM). This can help a manufacturer adopt sustainable practices and potentially save resources. The process of considering environmental, social, ethical, and economic issues is known as sustainable purchasing. Its goal is to manage the organization's external resources in a way that guarantees the supply of all goods, services, (Wilding et al. 2012).

Figure 5 shows the distribution of the number of papers published in the domain of green procurement per year of publication as used in this review paper. Green purchasing is a practice which involves environmentally conscious procurement/ sourcing of raw materials/intermediate goods by the manufacturers to reduce wastage and hazards to the environment without compromising the quality of the final output. The term "environmental purchasing" refers to the purchasing process's participation in supply chain management initiatives that support resource reduction, recycling, and reuse (Carter et al. 2000). More firms and companies get recognition by adopting green practices and providing their customers with sustainably produced goods/ products. This incentivizes other firms to move towards green practices and increases sustainable development. Until the early 1990s, price, quality and delivery were the predominant factors which dictated purchasing policies, supplier selection and evaluation processes. But with the increase in popularity of green purchasing, the manufacturers became more selective in choosing their suppliers. Thus a new term was coined, Green supplier selection (GSS). This led to the greening of the supplier, and it affected the upstream supply chain.

In order to identify the best supplier who complies with the company's environmental criteria, green supplier selection must match its supplier selection procedure

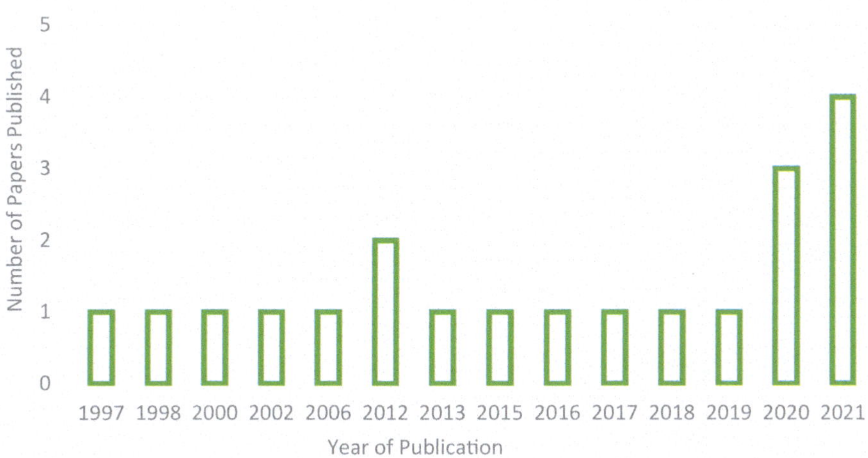

Fig. 5 Data visualization for Green Procurement

with those standards (Sousa Jabbour et al. 2015). The supplier was able to take advantage of this and leverage its market share by offering their strategic client green raw materials, which completely altered the nature of the supplier–buyer relationship. This improves both parties' economic and environmental performance (Igarashi et al. 2013; Zhu and Sarkis 2006). When assessing a supplier's degree of sustainability, manufacturing businesses typically consider the supplier's operations, distribution method, packaging, recycling capacity, and material choices. Suppliers can provide eco-friendly packaging and products to build a strong environmental relationships with buyers (Zhu and Sarkis 2006). Procurement is a boundary function in the supply chain as it links external and internal practices. Therefore, it plays a much more important role in enhancing the greenness of the supply chain by setting internal environmental standards for the upstream members of SC and making use of the environmental expertise of the supplier to its advantage by getting more customers and increasing their market share and gain the advantage over their competitors (Preuss 2002). Purchasing augments the effectiveness of a source reduction strategy in ways such as:

1. Reducing the volume of purchased materials/items which are hazardous to the environment.
2. Increasing the purchase of recycled or reused items at a more economical rate which promotes sustainable production.
3. To check the type of packaging used by the suppliers (Min and Galle 1997).

The sourcing managers must consider the disposition of raw materials and components and the life cycle duration of the goods. Cost reduction can be facilitated by establishing environmental regulations and standards by the firm, which are difficult to imitate by the competitors and provides the firm with an advantage. Firms can benefit economically by reducing material wastage (Green et al. 2012). It was found that green purchasing has a positive link with economic performance but no real relation with the firm's environmental performance. It is observed that the party's position in the supply chain plays a massive role on its ability to influence other parties both upstream and downstream in adopting green and sustainable practices. Generally, the "main driver" is an end retailer who exerts a strong purchasing power on the upstream members (Green et al. 1998). Manufacturers can use the R&D department of their supplier to get an advantage in innovation. It is observed that it is not feasible for manufacturing firms to invest large firms in R&D and innovation and green product development and reap its benefits. Therefore, outsourcing key innovation activities to specialists has become an everyday activity (Bag 2017).

6 Green Logistics

The process of planning, executing, and overseeing the economical and effective transportation and storage of completed goods, in-process inventories, raw materials, and related data from the point of origin to the point of consumption in order to meet customer demands is known as logistics, according to the American Council of Logistics Management. There was a need to look into a more sustainable concept of logistics with the issue of ever-increasing carbon emissions from various modes of transport. Thus came the need for "green logistics" (GL). Material handling, waste management, packaging, and transportation are the main areas of focus for supply chain management techniques and tactics that lessen the energy and environmental impact of freight distribution (Seroka-Stolka 2014).

Figure 6 shows the distribution of the papers used as a reference in the domain of Green Logistics based on the journal in which they were published. Another idea connected to green logistics is reverse logistics (RL). After reaching the end of their life cycle, the items or products return from the consumer to the source, where some value is recovered, and they are recycled or utilised again in related applications. This is more environmentally friendly because less garbage is produced. Reclaiming money from returned goods helps RL compensate for losses and offset increased expenses. Whereas in GL, the concern is transportation network, packaging and warehousing. Market demand and environmental concerns compel companies to green their logistics services (Kumar 2015). Green shipping techniques, also known as resource conservation and waste reduction in cargo handling and distribution, are environmental management strategies used by shipping companies. Feeder boats are used between two ports in place of more traditional means of transportation, such as trucks, in order to save energy, cut emissions, and transport more goods. For the company, this is a financially and environmentally advantageous choice (Lai et al. 2011).

Moreover, companies can prevent harm to the environment by considering the implications of the logistics on nature and then designing logistics networks. For the German 3PL industry, evidence of a positive association is discovered between adopting environmental practices and economic success, consistent with the findings

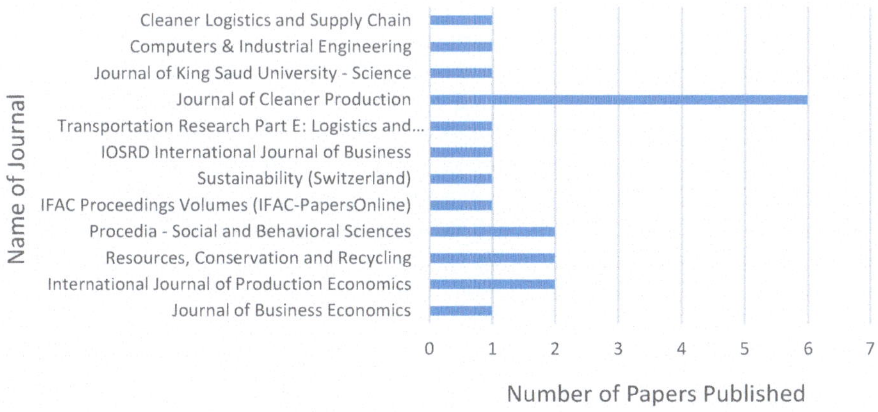

Fig. 6 Data visualization for Green Logistics

of other studies that revealed a similar relationship (Maas et al. 2018). For economically and environmentally efficient transportation, the manufacturer company must make decisions about the mode of transportation and the design of an ideal logistics network. Whether to employ outside logistics companies to manage the department's operations or maintain it in-house is another choice that needs to be made. Here, the company can assess the 3PLs' green practices and compare the two possibilities using economic analysis. Small and medium-level firms decide to use 3PLs as these service providers already have a well set-up network. They first consolidate goods going to a particular destination and then transport them together, reducing their overall cost.

Moreover, they take care of warehousing, inventory management and pick up/drop off, which relieves the burden off such firms and they get to use the expertise of 3PLs, providing them with a competitive advantage. Sheu et al. (2005) worked on an integrated logistics model where they combined logistics and reverse logistics to evaluate the firm's economic performance. They considered green practices and calculated the various costs and revenue with the help of a mathematical model. The Numerical analysis results show that, in comparison to the current supply chain operational performance, the chain-based aggregate net profits of a certain notebook computer manufacturer can be increased by 21.1% by using the suggested integrated logistics operational model. Yang et al. (2009) tried to establish a conclusive relationship between logistics services and the firm performance of shipping firms. Resource allocation was also used as an intermediate variable between logistics, firm performance, and innovation capability. To gain the advantage, container shipping service firms need to use their resources efficiently to improve the logistics service. Green logistics are about win–win relationships on economic and environmental performance and major points that motivate a firm to green their logistics are the reduced costs for all parties involved, judicious use of the resources, and increase in market revenue and share and improvement in customer relationship and service. Use of renewable energy as fuel for various modes of transportation reduces both carbon emissions and the operational costs of logistics. Warehousing is another critical component in the design of an efficient logistics network. It is a bridge between the buyer and the vendor. Properly placed warehouses diminish the need for carrying high volumes of goods for longer distances, benefiting the firm economically. In a logistics network, proper positioning of warehouses must be done to reduce the lead time and to fulfil the customers' demands on time. It was concluded from the analysis that the sustainable approach towards logistics combined with information technology, network optimization and green standardization benefits the firm financially. Prioritizing the consolidation of shipments improves the efficient use of the capacity of the vehicles and reduces fuel costs and harmful carbon emissions. Another way is to choose another mode of transportation that causes less pollution.

7 Green Manufacturing

Green manufacturing can be defined as using sustainable processes to minimize wastage and use of hazardous materials to produce environment-friendly products. Green manufacturing came into the picture when there were demands from consumers to get products made from sustainable activities. Manufacturing firms are traditionally one of the primary sources of pollution and material wastage (Kumar et al. 2023a, b; Kumar and Mittal 2023; Kumar et al. 2023). By incorporating green manufacturing practices, the firm can improve its standing in the market where consumers are environmentally conscious. It can be a way to reduce costs and maximize profits if implemented correctly. Green practices are driven by government rules and regulations and the consumers' environmental consciousness. Green process innovation enhances enterprise competitiveness by improving performance. Proper implementation of green manufacturing in practice is very important in the long term for a firm.

Figure 7 shows the distribution of papers in the domain of Green Manufacturing based on their type as referred in this review paper. Chen et al. (2006) defined "green core competence" as the efficiency levels of the company/firm to implement environmental policies in practice. Green core competence enhances the optimal usage of energy resources by bringing out the best in their employees and integrating external and internal knowledge to create a unique green path which gives them competitive advantage over their competitors (Kumar et al. 2023b; Chen et al. 2006). Higher green core competence shows that the firm can adapt rapidly to the ever-changing demands of the consumers and thus enhance their operational performance. The manufacturing enterprises gained profit through the commercialization of green technology and generated additional revenue by reusing and recycling the waste into marketable products (Dangelico and Pontrandolfo 2015). Environmentally Responsible Manufacturing (ERM) is an economically driven approach of reducing waste during disposal and manufacturing of products (Curkovic and Landeros 2000). The firm's size directly impacts the company's positive performance after green practices are incorporated in their operations (Li et al. 2021b). They also found that green practices have a positive implication on the green core competence of the firm. Proper process planning, factory layout and maintenance of the machines also play an important role in energy-efficient manufacturing. The firm must check the use of cleaning and lubricating fluids during manufacture. They can use other substitutes like water-based solvents or mechanical cleaning methods to reduce the use of toxic and hazardous lubricants in manufacturing processes (Curkovic 2003). Zhu and Sarkis (2004) studied the relationship between GSCM practises and the operational performance of the enterprise. They considered JIT and TQM as moderating factors between them. They analysed data using statistical tools and tried to validate the hypothesis. In the end, they found a positive relationship between the GSCM practices and the firm's economic growth. JIT (Just in Time) and TQM (Total Quality Management) are operations management and quality control techniques used by

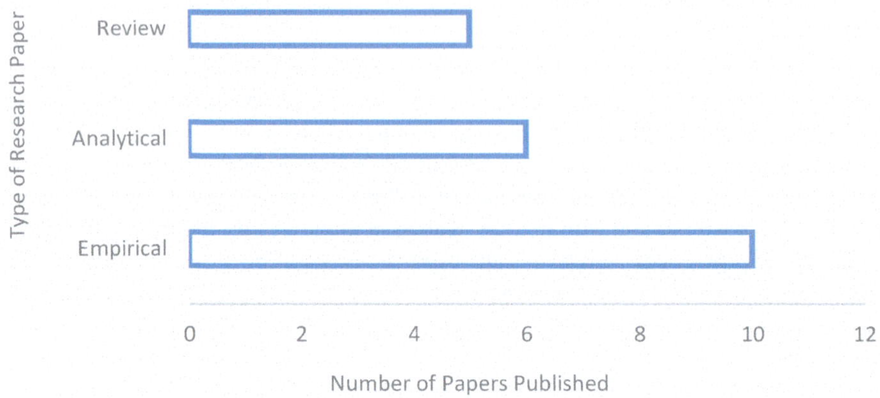

Fig. 7 Data visualization for Green Manufacturing

manufacturing firms to reduce wastage of raw materials, inventory and defective goods, respectively which helps minimize the wastage of various inputs.

A system model for green manufacturing was proposed. The model captures the current levels of greenness, tools needed for green transformation and the necessary policies to be incorporated to sustain that level of greenness (Deif 2011). With the help of an industrial case study, the economic benefits of green practices were exhibited. The study projected cost savings of around 46,000$ annually by adopting green manufacturing practices. This structured approach to the model is a very efficient way of analyzing the various stages of the adoption of green practices by a manufacturing firm. The study covered cost savings in water, energy, waste disposal, and sustainable use of raw material and emissions.

8 Product Recovery

Product Recovery ensures the best way to repossess an item and retain its economic value as much as possible. It aims to reduce the environmental wastage of items. By adopting the product recovery process, one not only caters to the scarcity of resources but also reduces the volume of landfills. Our paper lists 5 significant product recovery practices widely used with a special emphasis on Re-manufacturing.

8.1 Re-manufacturing

With the development of globalized markets, there has been a significant increase in the depletion of natural resources. As a result, the focus has shifted from manufacturing an entire new product to introducing the re-manufactured product in the market. Recovery from used products is emphasized as it forms the raw materials needed to re-manufacture the product. Re-manufacturing is collecting, processing, and reselling discarded items as new ones in marketplaces (Zheng et al. 2021). Remanufacturing helps companies minimise their manufacturing expenses directly related to materials. Remanufacturing is where corporate social responsibility and lucrative business intersect. Remanufacturing is regarded as the most practical and efficient method of conserving natural resources in terms of energy and materials among the several available recovery strategies (Nie et al. 2021; Giuntini and Gaudette 2003). Remanufacturing has been emphasised as one of the best ways to prolong the useful life of commodities within the economy while maintaining the intrinsic worth of materials and resources. Remanufactured goods can save the industries between 30 and 40% of manufacturing costs when compared to fresh goods creation (Steinhiper and Nagel 2017). The engine re-manufacturing activities save about 50% cost, 60% energies and 70% virgin raw materials with no solid waste (Liao et al. 2021). Re-manufacturing saves energy and protects the environment by providing an efficient method of recycling resources. It also saves money and gives you an exact response time when you get the demand (Kaya 2010). The concept can be extended to numerous industries like electronics, automobiles, cosmetics, and appeals. For cosmetic industries, re-manufacturing can be viable as it has shorter production time and lower costs (Reimann 2016). Customers value a new product more than a re-manufactured one. Thus, it is necessary to quote sufficiently lesser prices for re-manufactured goods, for customers to buy them (Kaya 2010). Manufacturers can utilize re-manufacturing as a marketing technique to maintain their market share by generating price difference in a competitive market (Heese et al. 2005). The competing manufacturer can outweigh the sale of new products by introducing low priced re-manufactured products. Only when the fixed cost of re-manufacturing the commodity is lower is re-manufacturing a viable option. It is cost effective since it recovers the residual value from old items. In a competition between new and re-manufactured goods, a firm can attract the 'low end customers' by providing the optimal pricing for re-manufactured goods, and thereby increasing the overall profit (Ovchinnikov 2011). The core or used products are the most fundamental raw materials in re-manufacturing. Additionally, obtaining them is a vital task since it aids in fulfilling market needs while continuing to operate the firm (Chakraborty et al. 2021). Therefore, to increase the economic benefits of the process, it is necessary to collect/obtain quality cores. The pricing-related decisions for re-manufactured goods are determined by various factors like quality of used products, quality of core customer valuation of re-manufactured products and market type (Kumar and Ramachandran 2016). Government intervention can pave the way to re-manufacturing. In case of high fixed cost, the government subsidizing the fixed cost can help the manufacturer make more profit. In order to promote the practice of re-manufacturing, correct and valid data is needed indicating the benefits over the recycling process, to avoid the dissipation of materials.

8.2 Recycling

Recycling is the process by which waste material is converted into new materials. According to Mobley et al., it presents opportunities for job creation and better social behaviour. It suggested the customers positively view recycled products. This leads to customers' willingness-to-pay (WTP) to protect the environment. The paper also pointed the various risk perception associated with recycled products. To mitigate the risks, the recycled products should be given proper guarantee/warranty and should be subjected to proper branding.

8.3 Refurbishment

Refurbishment refers to selling previously returned items by the manufacturer in the market. The customers might not previously use them but may contain some defects which are repaired by the manufacturer and released in the market. Refurbished goods can offer an environmentally favourable alternative because their lifetime can be extended, reducing their negative influence on the environment (Wallner et al. 2022). We may cut down on the energy required to dispose of used goods by designing durable products and maintaining and repairing them, as well as by reusing, remanufacturing, refurbishing, and recycling them. It is acknowledged that life cycle assessment (LCA) is a suitable technique for estimating environmental impacts. Iteratively, it evaluates the environmental effects of processes and products over the course of their life cycles (Potr et al. 2021).

8.4 Repair

Repairing failed parts can vastly decrease the safety stock quantities companies must keep when they face uncertain demand. Repairing parts and returning them after re-assembly can stretch the product's life and add eco-friendly value to it (Behfard et al. 2018). Repairing the damaged parts can also reduce the need to source new parts and optimize the number of items flowing in the supply chain. Replacement is the method used to repair capital goods, whereby a malfunctioning component is removed from the system and replaced with a working one. So, we can replace the damaged part instead of discarding the whole component (Basten et al. 2011).

8.5 Scavenging

Scavenging is the process of gathering garbage from businesses or disposal sites and redistributing it to businesses that might profit from recycling or reuse inside the circular ecosystem (Ghisellini et al. 2016). Scavengers and manufacturers may have a stronger relationship than recycling efforts since it allows producers to employ less expensive materials, reduce their capacity for collection, and comply with regulations. In circular ecosystems, scavengers play a crucial role in maintaining resource sustainability and continual availability, which in turn supports the stability of the firms.

A two-tier supply chain has been optimised under carbon tax/cost policy and carbon cap-and-trade policy for random demand. A potential hybridization of these two policies has also been offered, along with comparisons between them. The demand is thought to be arbitrary in nature. Two distinct models of mixed integer nonlinear programming was created and resolved while considering two distinct carbon policies. These models were expected to assist organisations in identifying the best order amount, reorder point, and number of shipments under the majority of carbon policies that are currently in use (Ghosh 2021; Ghosh et al. 2020). In one of the similar approaches, it was demonstrated how carbon emission concerns could be incorporated into operational decision-making concerning production, inventory management, and procurement using widely-used and rather simple models. It also demonstrated how conventional models can be adjusted to facilitate decision-making that considers cost and carbon footprint by linking different choice variables to carbon emission characteristics. The effects of these parameters' values were analyzedon costs and emissions and the parameters of regulatory emission control policies. Using the models, it was proposed to address the need for carbon reduction through operational changes in lieu of or in addition to expensive investments in carbon-reducing technologies (Benjaafar et al. 2013; Ghosh et al. 2021). Another significant method for developing a time-dependent demand and variable holding cost inventory model is proposing a two-storage production inventory model, where demand is contingent on both price and time (Aarya et al. 2022). The best pricing and replenishment strategies were recommended by the study. The manufacturing industries were reported to benefit from using fuzzy membership functions, goal programming, and weighted decision-making with varying degrees of confidence in handling multi-objective transportation problems. It increases efficiency by integrating the confidence level and weights together (Joshi et al. 2023). The growth of smart manufacturing techniques and supply chain management strongly emphasised the value of green supply chain management. The possibility of green supply chain management playing a significant role in Industry 4.0 logistics and inventory was highlighted (Kumar and Sharma 2023; Kumar et al. 2023c, d).

9 Conclusion and Summary

In this paper, the existing literature related to the Green Supply chain and green business are reviewed and elucidated on various green business and green supply chain practises which can be incorporated by the manufacturing firms to improve their business performance and promote sustainable business operations. In current work, an effort has been made to integrate various GB and green supply chain practises and establish a link between them to show their interdependency. The following important summary can be drawn from the work:

1. It aims to highlight that to become a fully green organization, all its functions must go through greening practices. Any organization must align its company goals and objectives with sustainability. Only then can the green practices be practically implemented and benefit the firm's economic growth.
2. Being a green organisation requires a significant investment of time and money. On the other side, it gives the company a competitive advantage and long-term financial stability. Academicians, researchers, and practitioners will benefit from our classifications of green practices as they will provide a more thorough understanding of the integrated green supply chain—green business model.
3. As shown in the conceptual framework, the customer is an integral part of the supply chain loop as they are the primary source of upstream cash flow. It was also evident that the customers' demands and their overall mindset towards the term "green", facilitates the adoption of green practices by the firm. The organization can collect valuable data for launching new green products with sophisticated green marketing techniques that analyze consumer psychology and purchasing behaviour.
4. With various green finance instruments, the company can stabilize both short-term and long-term financial health by managing risk and providing benefits to its supply chain partners. This process is conducive to increasing supply chain surpluses, ultimately the common objective of any firm involved. In green design, we observe that inculcating an environment-friendly approach at the design stage provides better utilization of resources in later processes. One of the key components of green design is taking the product's life cycle into account and extracting value from discarded goods. The practical use of green procurement is contingent upon the core competences of the organisation and the attitude of purchasing managers to prioritise the use of environmentally friendly resources. It is intimately related to both green manufacturing practices and green design.
5. In green manufacturing, basic research has been carried out on incorporating alternate energy-saving processes and selecting eco-friendly lubricants to prevent environmental damage and reduce the costs. Our work observed that by combining these features with the existing advanced production planning and inventory management techniques like Kaizen, JIT and better forecasting methods, we can create a streamlined manufacturing system, reduce waste generation, and provide better quality to the customers.

10 Future Scope of Work

There is plenty of scope in green logistics with wide-ranging fields like warehousing, transportation and inventory and the dearth of documentation of industry practices in the current research work. Reverse logistics and re-manufacturing are essential features that systematically expedite the reverse flow of goods from the customers towards the manufacturers. It allows for proper waste disposal and extraction of spare parts from the waste, which can be further utilized in manufacturing recycled products.

There is further scope for the study of the impact of re-manufacturing on the costs and the subsequent benefits to the firm. So, as we can see, there are plenty of topics on which further research can be carried out. Literature on the role of data analytics and the use of mathematical models to simulate the performance of green practises in industry hasn't been covered extensively in this paper. So, there is scope for using alternative methodology and real-life data to give the research work some practical context.

Acknowledgements Declaration of Conflicting Interest The Author(s) declare(s) that there is no conflict of interest.

References

Aarya DD, Rajoria YK, Gupta N, Raghav YS, Rathee R, Boadh R, Kumar A (2022) Selling price, time dependent demand and variable holding cost inventory model with two storage facilities. Mater Today Proc 56:245–251. https://doi.org/10.1016/j.matpr.2022.01.111

Amores-Salvadó J, Castro GMD, Navas-López JE (2014) Green corporate image: moderating the connection between environmental product innovation and firm performance. J Clean Prod 83:356–365. https://doi.org/10.1016/j.jclepro.2014.07.059

Atasu A, Subramanian R (2012) Extended producer responsibility for e-waste: individual or collective producer responsibility? Prod Oper Manag 21:1042–1059. https://doi.org/10.1111/j.1937-5956.2012.01327.x

Bag S (2017) Role of green procurement in driving sustainable innovation in supplier networks: some exploratory empirical results. Jindal J Bus Res 6:155–170. https://doi.org/10.1177/2278682117727208

Basten RJI, Heijden MC, Van Der Schutten JMJ (2011) A minimum cost flow model for level of repair analysis. Int J Prod Econ 133:233–242. https://doi.org/10.1016/j.ijpe.2010.03.025

Behfard S, Al Hanbali A, Heijden MC, Van Der Zijm WHM (2018) Last time buy and repair decisions for fast moving parts. Int J Prod Econ 197:158–173. https://doi.org/10.1016/j.ijpe.2017.12.012

Benjaafar S, Li Y, Daskin M (2013) Carbon footprint and the management of supply chains: insights from simple models. IEEE Trans Autom Sci Eng 10(1):99–116. https://doi.org/10.1109/TASE.2012.2203304

Cao K, Xu B, He Y, Xu Q (2020) Optimal carbon reduction level and ordering quantity under financial constraints. Int Trans Oper Res 27:2270–2293. https://doi.org/10.1111/itor.12606

Carter CR, Kale R, Grimm CM (2000) Environmental purchasing and firm performance: an empirical investigation. Transp Res Part E Logist Transp Rev 36:219–228. https://doi.org/10.1016/S1366-5545(99)00034-4
Čekanavičius L, Bazytė R, Dičmonaitė A (2014) Green business: challenges and practices. Ekonomika 93:74–88. https://doi.org/10.15388/ekon.2014.0.3021
Chakraborty K, Mukherjee K, Mondal S, Mitra S (2021) A systematic literature review and bibliometric analysis based on pricing related decisions in re-manufacturing. J Clean Prod 310:127265. https://doi.org/10.1016/j.jclepro.2021.127265
Chen HC, Yang CH (2019) Applying a multiple criteria decision-making approach to establishing green marketing audit criteria. J Clean Prod 210:256–265. https://doi.org/10.1016/j.jclepro.2018.10.327
Chen YS, Lai SB, Wen CT (2006) The influence of green innovation performance on corporate advantage in Taiwan. J Bus Ethics 67:331–339. https://doi.org/10.1007/s10551-006-9025-5
Chou SF, Horng JS, Sam Liu CH, Lin JY (2020) Identifying the critical factors of customer behavior: an integration perspective of marketing strategy and components of attitudes. J Retail Consum Serv 55:102113. https://doi.org/10.1016/j.jretconser.2020.102113
Curkovic S (2003) Environmentally responsible manufacturing: the development and validation of a measurement model. Eur J Oper Res 146:130–155. https://doi.org/10.1016/S0377-2217(02)00182-0
Curkovic S, Landeros R (2000) An environmental Baldrige? Am J Bus 15:63–76. https://doi.org/10.1108/19355181200000012
Dangelico RM, Pontrandolfo P (2015) Being "Green and Competitive": the impact of environmental actions and collaborations on firm performance. Bus Strateg Environ 24:413–430. https://doi.org/10.1002/bse.1828
Dangelico RM, Vocalelli D (2017) "Green Marketing": an analysis of definitions, strategy steps, and tools through a systematic review of the literature. J Clean Prod 165:1263–1279. https://doi.org/10.1016/j.jclepro.2017.07.184
Darnall N, Jolley GJ, Handfield R (2008) Environmental management systems and green supply chain management: complements for sustainability? Bus Strat Env 17:30–45. https://doi.org/10.1002/bse
de Sousa Jabbour ABL, Frascareli FCDO, Jabbour CJC (2015) Green supply chain management and firms' performance: understanding potential relationships and the role of green sourcing and some other green practices. Resour Conserv Recycl 104:366–374. https://doi.org/10.1016/j.resconrec.2015.07.017
Deif AM (2011) A system model for green manufacturing. J Clean Prod 19:1553–1559. https://doi.org/10.1016/j.jclepro.2011.05.022
Eltayeb TK, Zailani S, Ramayah T (2011) Green supply chain initiatives among certified companies in Malaysia and environmental sustainability: investigating the outcomes. Resour Conserv Recycl 55:495–506. https://doi.org/10.1016/j.resconrec.2010.09.003
Ghisellini P, Cialani C, Ulgiati S (2016) A review on circular economy: the expected transition to a balanced interplay of environmental and economic systems. J Clean Prod 114:11–32. https://doi.org/10.1016/j.jclepro.2015.09.007
Ghosh A (2021) Optimisation of a production-inventory model under two different carbon policies and proposal of a hybrid carbon policy under random demand. Int J Sustain Eng 14(3):280–292. https://doi.org/10.1080/19397038.2020.1800858
Ghosh A, Jha JK, Sarmah SP (2020) An integrated supply chain with uncertain demand and random defect rate under carbon cap-and-trade policy. Int J Ind Eng Theor Appl Pract 27(2):209–228. https://doi.org/10.23055/ijietap.2020.27.2.3973
Ghosh A, Kumar N, Choudhary P (2021) Proposal of "carbon-lockdown" policy in the light of environmental silver lining of COVID19 lockdowns. MDIM Bus Rev II(I):36–48
Giuntini R, Gaudette K (2003) Re-manufacturing: the next great opportunity. Bus Horiz 41–48

Green K, Morton B, New S (1998) Green purchasing and supply policies: do they improve companies' environmental performance? Supply Chain Manag 3:89–95. https://doi.org/10.1108/13598549810215405

Green KW, Zelbst PJ, Meacham J, Bhadauria VS (2012) Green supply chain management practices: impact on performance. Supply Chain Manag 17:290–305. https://doi.org/10.1108/13598541211227126

Gustavo JU, Trento LR, de Souza M, Pereira GM, Lopes de Sousa Jabbour AB, Ndubisi NO, ChiappettaJabbour CJ, Borchardt M, Zvirtes L (2021) Green marketing in supermarkets: conventional and digitized marketing alternatives to reduce waste. J Clean Prod 296. https://doi.org/10.1016/j.jclepro.2021.126531

Hazen BT, Hall DJ, Hanna JB (2012) Reverse logistics disposition decision-making: developing a decision framework via content analysis. Int J Phys Distrib Logist Manag 42:244–274. https://doi.org/10.1108/09600031211225954

Heese HS, Cattani K, Ferrer G, Gilland W, Roth AV (2005) Competitive advantage through take-back of used products. Eur J Oper Res 164:143–157. https://doi.org/10.1016/j.ejor.2003.11.008

Igarashi M, De Boer L, Fet AM (2013) What is required for greener supplier selection? A literature review and conceptual model development. J Purch Supply Manag 19:247–263. https://doi.org/10.1016/j.pursup.2013.06.001

Joshi VD, Agarwal P, Kumar A (2023) Fuzzy transportation planning: a goal programming tactic for navigating uncertainty and multi-objective decision making. Int J Interactive Des Manuf 1–29. https://doi.org/10.1007/s12008-023-01634-9

Karagülle AÖ (2012) Green business for sustainable development and competitiveness: an overview of Turkish logistics industry. Proc Soc Behav Sci 41:456–460. https://doi.org/10.1016/j.sbspro.2012.04.055

Kaya O (2010) Incentive and production decisions for re-manufacturing operations. Eur J Oper Res 201:442–453. https://doi.org/10.1016/j.ejor.2009.03.007

Kouvelis P, Zhao W (2018) Who should finance the supply chain? Impact of credit ratings on supply chain decisions. Manuf Serv Oper Manag 20:19–35. https://doi.org/10.1287/msom.2017.0669

Kumar A (2015) Green logistics for sustainable development: an analytical review. IOSRD Int J Bus 1:7–13

Kumar R, Ramachandran P (2016) Revenue management in re-manufacturing: perspectives, review of current literature and research directions. Int J Prod Res 54:2185–2201. https://doi.org/10.1080/00207543.2016.1141255

Kumar V (2020) Organizational sustainability and green business practices—the way forward. Mater Today Proc. https://doi.org/10.1016/j.matpr.2020.09.791

Kumar A, Mittal RK, Haleem A (2023) Advances in additive manufacturing artificial intelligence, nature-inspired, and biomanufacturing. Elsevier. https://doi.org/10.1016/C2020-0-03877-6

Kumar L, Sharma RK (2023) Smart manufacturing and industry 4.0: state-of-the-art review. In: Handbook of smart manufacturing, pp 1–28. https://doi.org/10.1201/9781003333760-1

Kumar A, Kumar P, Mittal RK, Singh H (2023) Printing file formats for additive manufacturing technologies. In: Kumar A, Mittal RK, Haleem A (eds) Additive manufacturing materials and technologies, advances in additive manufacturing. Elsevier, pp 87–102. ISBN 9780323918343. https://doi.org/10.1016/B978-0-323-91834-3.00006-5

Kumar R, Kumar V, Kumar A (2023) A review on effect of computer-aided machining parameters in incremental sheet forming. In: Handbook of flexible and smart sheet forming techniques: industry 4.0 approaches, pp 29–58

Kumar A, Singh H, Kumar P, AlMangour B (2023) Handbook of smart manufacturing: forecasting the future of industry 4.0. CRC Press. https://doi.org/10.1201/9781003333760

Kumar A, Mittal RK, Goel R (2023) Waste recovery and management: an approach toward sustainable development goals. CRC Press. https://doi.org/10.1201/9781003359784

Kumar P, Kumar RK, Mittal HS (2023) Integration of reverse engineering with additive manufacturing. In: Advances in additive manufacturing. https://doi.org/10.1016/B978-0-323-91834-3.00028-4

Kuo TC, Huang SH, Zhang HC (2001) Concurrent engineering and DFMA/DFX in the development of automotive components. Proc CIRP 41:241–260

Lai KH, Lun VYH, Wong CWY, Cheng TCE (2011) Green shipping practices in the shipping industry: conceptualization, adoption, and implications. Resour Conserv Recycl 55:631–638. https://doi.org/10.1016/j.resconrec.2010.12.004

Li G, Shao S, Zhang L (2019) Green supply chain behavior and business performance: evidence from China. Technol Forecast Soc Change 144:445–455. https://doi.org/10.1016/j.techfore.2017.12.014

Li F, Xu X, Li Z, Du P, Ye J (2021b) Can low-carbon technological innovation truly improve enterprise performance? The case of Chinese manufacturing companies. J Clean Prod 293:125949. https://doi.org/10.1016/j.jclepro.2021.125949

Li B, Wang Y, Wang Z (2021) Managing a closed-loop supply chain with take-back legislation and consumer preference for green design. J Clean Prod 282. https://doi.org/10.1016/j.jclepro.2020.124481

Liao H, Li C, Nie Y, Tan J, Liu K (2021) Environmental efficiency assessment for re-manufacture of end of life machine and multi-objective optimization under carbon trading mechanism. J Clean Prod 308:127168. https://doi.org/10.1016/j.jclepro.2021.127168

Maas S, Schuster T, Hartmann E (2018) Stakeholder pressures, environmental practice adoption and economic performance in the German third-party logistics industry—a contingency perspective. J Bus Econ 88:167–201. https://doi.org/10.1007/s11573-017-0872-6

Min H, Galle WP (1997) Green purchasing strategies: trends and implications. Int J Purch Mater Manag 33:10–17. https://doi.org/10.1111/j.1745-493x.1997.tb00026.x

Navinchandra D (1990) Steps toward environmentally compatible. Carnegie-Mellon Univ Pittsburgh Pa Robot. Inst No CMU-RI

Nie J, Liu J, Yuan H, Jin M (2021) Economic and environmental impacts of competitive re-manufacturing under government financial intervention. Comput Ind Eng 159:107473. https://doi.org/10.1016/j.cie.2021.107473

Olsen MC, Slotegraaf RJ, Chandukala SR (2014) Green claims and message frames: how green new products change brand attitude. J Mark 78:119–137. https://doi.org/10.1509/jm.13.0387

Ovchinnikov A (2011) Revenue and cost management for re-manufactured products. Prod Oper Manag 20:824–840. https://doi.org/10.1111/j.1937-5956.2010.01214.x

Peattie K (1999) Trappings versus substance in the greening of marketing planning. J Strateg Mark 7:131–148. https://doi.org/10.1080/096525499346486

Polonsky MJ, Rosenberger PJ (2001) Reevaluating green marketing: a strategic approach. Bus Horiz 44:21–30. https://doi.org/10.1016/S0007-6813(01)80057-4

Potr T, Jordan S, Ruschi M, Saade M (2021) An LCA methodolody for assesing the environmental impacts of building components before and after refurbishment, vol 327. https://doi.org/10.1016/j.jclepro.2021.129527

Preuss L (2002) Green light for greener supply. Bus Ethics A Eur Rev 11:308–317. https://doi.org/10.1111/1467-8608.00290

Reimann M (2016) Accurate response with refurbished consumer returns. Decis Sci 47:31–59. https://doi.org/10.1111/deci.12150

Rex E, Baumann H (2007) Beyond ecolabels: what green marketing can learn from conventional marketing. J Clean Prod 15:567–576. https://doi.org/10.1016/j.jclepro.2006.05.013

Saeed Meo M, Karim MZA (2021) The role of green finance in reducing CO_2 emissions: an empirical analysis. Borsa Istanbul Rev. https://doi.org/10.1016/j.bir.2021.03.002

Seroka-Stolka O (2014) The development of green logistics for implementation sustainable development strategy in companies. Proc Soc Behav Sci 151:302–309. https://doi.org/10.1016/j.sbspro.2014.10.028

Sharma A, Iyer GR, Mehrotra A, Krishnan R (2010) Sustainability and business-to-business marketing: a framework and implications. Ind Mark Manag 39:330–341. https://doi.org/10.1016/j.indmarman.2008.11.005

Sheu JB (2011) Bargaining framework for competitive green supply chains under governmental financial intervention. Transp Res Part E Logist Transp Rev 47:573–592. https://doi.org/10.1016/j.tre.2010.12.006

Sheu JB, Chou YH, Hu CC (2005) An integrated logistics operational model for green-supply chain management. Transp Res Part E Logist Transp Rev 41:287–313. https://doi.org/10.1016/j.tre.2004.07.001

Steinhiper R, Nagel A (2017) New opportunities and incentives for re-manufacturing by 2020's car service trends. Proc CIRP 61:183–188. https://doi.org/10.1016/j.procir.2016.11.233

Taghizadeh-Hesary F, Yoshino N (2019) The way to induce private participation in green finance and investment. Financ Res Lett 31:98–103. https://doi.org/10.1016/j.frl.2019.04.016

Taleb KN, Gupta SM (1997) Disassembly of multiple product structures. Comput Ind Eng 32:949–961. https://doi.org/10.1016/s0360-8352(97)00023-5

Tsai PH, Lin GY, Zheng YL, Chen YC, Chen PZ, Su ZC (2020) Exploring the effect of Starbucks' green marketing on consumers' purchase decisions from consumers' perspective. J Retail ConsumServ 56. https://doi.org/10.1016/j.jretconser.2020.102162

Van Hoek RI (1999) From reversed logistics to green supply chains. Supply Chain Manag 4:129–134. https://doi.org/10.1108/13598549910279576

Wallner TS, Magnier L, Mugge R (2022) Resources, conservation & recycling do consumers mind contamination by previous users ? A choice-based conjoint analysis to explore strategies that improve consumers' choice for refurbished products. Resour Conserv Recycl 177:105998. https://doi.org/10.1016/j.resconrec.2021.105998

Wang M, Li X, Wang S (2021b) Discovering research trends and opportunities of green finance and energy policy: a data-driven scientometric analysis. Energy Policy 154:112295. https://doi.org/10.1016/j.enpol.2021.112295

Wang M, Zhao R, Li B (2021) Impact of financing models and carbon allowance allocation rules in a supply chain. J Clean Prod 302. https://doi.org/10.1016/j.jclepro.2021.126794

Wilding R, Wagner B, Miemczyk J, Johnsen TE, Macquet M (2012) Sustainable purchasing and supply management: a structured literature review of definitions and measures at the dyad, chain and network levels. Supply Chain Manag Int J 17:478–496. https://doi.org/10.1108/13598541211258564

Yang SA, Birge JR (2018) Trade credit, risk sharing, and inventory financing portfolios. Manage Sci 64:3667–3689. https://doi.org/10.1287/mnsc.2017.2799

Yang CC, Marlow PB, Lu CS (2009) Assessing resources, logistics service capabilities, innovation capabilities and the performance of container shipping services in Taiwan. Int J Prod Econ 122:4–20. https://doi.org/10.1016/j.ijpe.2009.03.016

Yu CH, Wu X, Zhang D, Chen S, Zhao J (2021) Demand for green finance: resolving financing constraints on green innovation in China. Energy Policy 153:112255. https://doi.org/10.1016/j.enpol.2021.112255

Zhang L, Wang J, You J (2015) Consumer environmental awareness and channel coordination with two substitutable products. Eur J Oper Res 241:63–73. https://doi.org/10.1016/j.ejor.2014.07.043

Zheng M, Shi X, Xia T, Qin W, Pan E (2021) Production and pricing decisions for new and remanufactured products with customer prejudice and accurate response. ComputInd Eng 157. https://doi.org/10.1016/j.cie.2021.107308

Zhu Q, Sarkis J (2004) Relationships between operational practices and performance among early adopters of green supply chain management practices in Chinese manufacturing enterprises. J Oper Manag 22:265–289. https://doi.org/10.1016/j.jom.2004.01.005

Zhu Q, Sarkis J (2006) An inter-sectoral comparison of green supply chain management in China: drivers and practices. J Clean Prod 14:472–486. https://doi.org/10.1016/j.jclepro.2005.01.003

Zubair M, Wang X, Iqbal S, Awais M, Wang R (2020) Attentional and emotional brain response to message framing in context of green marketing. Heliyon 6. https://doi.org/10.1016/j.heliyon.2020.e04912

Barrier Analysis for the Sustainable Business Practice of a Textile and Apparel Industry in Fiji Using an ISM Approach

D. Seenivasagam, R. Ananthanarayanan, A. K. Jeevanantham, and P. Kumar

Abstract The textile and apparel (TA) industries in Fiji are one of the pillars of the manufacturing sector bearing the nation's GDP. Yet these industries face some unique challenges as Fiji is a geographically isolated developing country. The present study pertains to barriers identification and framework modulation for the TA industry to adopt sustainable business practices (SBP). Thirteen barriers were identified through a literature survey, of which ten barriers were considered most relevant to Fiji's TA industries using the experts' opinion through surveys. The barriers' inter-relationship and the framework were established using the interpretive structural modelling (ISM) approach. The results reveal that the lack of skill across the different levels of the organization is the most influential barrier demanding the maximum attention of the decision and policymakers for achieving SBP, while the lack of knowledge in reducing the waste from fabric is the least effective.

Keywords Sustainable business practices · Barriers · Textile and apparel industry · Interpretive structural modelling

---

D. Seenivasagam
Management Department, College of Business, Hospitality and Tourism Studies, Fiji National University, Nasinu, Fiji

R. Ananthanarayanan (✉) · P. Kumar
Mechanical Engineering, School of Information Technology, Engineering, Mathematics and Physics, The University of the South Pacific, Suva, Fiji
e-mail: r.ananthanarayanan@usp.ac.fj

A. K. Jeevanantham
Mechanical Engineering Department, School of Mechanical Engineering, Vellore Institute of Technology, Vellore 632014, India

A. Kumar et al. (eds.), *Industry 4.0 Driven Manufacturing Technologies*, Springer Series in Advanced Manufacturing, https://doi.org/10.1007/978-3-031-68271-1_10

1 Introduction

Sustainable business practices (SBP) and pollution prevention strategies have historically focused on primary processes in the textile and apparel industries (Yang et al. 2023). With small and medium-sized enterprises (SMEs) being accountable for about 70% of global pollution (Caldera et al. 2018), there is an urgency for this sector to adopt SBP at the secondary level. SBP are essential for the livelihood of an organization, impacting the lives of its staff, workers, and owners. Any disruption or imbalance in the sector can have severe consequences for the social dimension of sustainability, particularly in Fiji, where a substantial portion of the population is directly or indirectly involved in and dependent on the TA industries (Radinibaravi 2024). Given the significant role that the TA industries play in Fiji's economy, any negative impact on this sector would ripple through the community, affecting the well-being and financial stability of countless individuals and families. Therefore, maintaining sustainability in the TA industry is crucial for ensuring social and economic stability in the region.

Sustainability from the environmental aspect of the TA industry draws immense attention due to the chemical involved in the dying process as part of the primary process. It is observed that the TA industry ranks among the leading contributors to environmental pollution globally (Chen et al. 2017). However, this context does not directly apply to Fiji's TA industry, which primarily centers on tailoring, clothing, and predominantly readymade garment production. Unlike the primary processing stages in the textile industry, which involve converting raw materials into yarn and then fabric, Fiji's TA industry predominantly focuses on secondary processing specifically, the transformation of fabric into apparel. Therefore, the concerns typically associated with the primary stages of textile production are not pertinent to Fiji's TA industry, given its focus on fabric-to-apparel processes.

In Fiji, the TA industry was founded in the year 1982 and has since undergone several ups and downs, contributing sustainability to the country's economy (Narayan 2001). Statistics show that there has been continuous gradual growth in this sector in the recent years (Radinibaravi 2024). This growth is due to the overall country's infrastructural development and government policies that pave channels for industries to grow (FijiSun 2021). Further, it is noted that most of the TA industries in Fiji fall under the SMEs category, where the majority of the businesses operate under family/home businesses, therefore, competition among them is high from the local to the export market.

Being a geographically isolated developing nation, Fiji heavily depends on importing raw materials like fabrics, tools, and machinery. Furthermore, for skilled personnel, industries often rely on expatriate expertise, presenting distinctive challenges compared to similarly developing countries with better geographical connectivity. In these countries, supply chains are seamless, technological advancements occur without interruption, market competition is not confined to localized areas, and human resources are readily available. Consequently, many TA industries in Fiji encounter numerous challenges, ranging from basic survival to achieving sustained growth. This underscores the central objective of this research, which aims to address the barriers hindering the attainment of SBP in the industry.

By identifying and addressing the barriers within the TA industry, it becomes transparent where resources and policies need to be focused to achieve SBP. This process also supports the transition to lean manufacturing within the context of Industry 4.0. In this case industry study, the MCDM (multiple-criteria decision-making) technique, specifically Interpretive Structural Modeling (ISM), is crucial as it aids in pinpointing these barriers and developing a conceptual framework to identify the most significant ones. These barriers are identified through comprehensive literature surveys and further examined through expert opinions and questionnaire surveys to assess their relevance in the TA industry in Fiji. It is essential to systematically establish the interrelationship of these barriers and develop a framework using the ISM approach within the case industry. This method brings to light the most dominant barriers that the TA industry must overcome to achieve SBP. The thorough analysis and strategic framework development ultimately pave the way for more informed decision-making and effective policy formulation, fostering sustainable growth and operational efficiency in the TA industry.

2 Literature Research

2.1 Sustainable Business Practice (SBP)

Sustainability, within the framework of the triple bottom line (TBL), is described in terms of three key pillars: social, environmental, and financial aspects (Gimenez et al. 2012). However, depending on the application and the nature of the research, several authors extended the TBL with other factors such as productivity and technology (Sharma et al. 2019). Caldera et al. (2018) argued there are no concrete models or measurement indexes (MIs) for SMEs, which can act as guidelines to the SBP. They attempted to establish the MIs in consultation with Australian senior industry experts, and as a result, nine SBP characteristics were derived with respect to environmental stewardship, process excellence, and sustainable-oriented culture (Caldera et al. 2018).

There are several tools and techniques available to establish sustainable development (SD) in manufacturing industries. For instance, lean principles are applied phenomenally in many case studies to support SD (Caldera et al. 2019). As such, lean alone is mostly used to optimize or reduce resource utilization in improving productivity and process efficiency; however, lean in combination with green has evolved as a 'lean and green' approach that is commonly applied for SBP.

2.2 Small and Medium-Sized Enterprises (SMEs) and Their Importance to SBP

The scale and the definition of SMEs differ from country to country. For example, the US Small Business Administration states that SMEs should generally have fewer than 500 employees within 12 months (Liberto 2023), while the European Commission defines SMEs by the number of employees and the financial assets that should consist of less than 250 employees and 50 million euros, respectively (European Commission 2020). However, in both countries invariably SMEs occupy 99% of all businesses. Almost the same scenarios can be observed in the UK and Australia (Caldera et al. 2018). Whereas, in Fiji, SMEs provide 60% of employment, and accounts for 12% of GDP. According to the Reserve Bank of Fiji, an enterprise that consists of a total asset between 100 and 500 k FJD, and employee numbers between 21 and 50 can be considered as SMEs (Reserve Bank of Fiji 2020). For most of these countries, SMEs are considered as primary drivers of their economies and job creation. It is noted that while SMEs are dominating the world economy, they are also responsible for about 70% of global industrial waste pollution and account for about 13% of global final energy consumption annually (Herce et al. 2023). Therefore, it is evident that SMEs shoulder a considerable responsibility to implement SBP. SMEs have various advantages over large multinational corporations (MNCs) when it comes to swiftly adopting to changes. SMEs often have simpler organizational structures, with owners directly involved in operations. This suggests that SMEs prioritize people over systems, contrasting with the approach of MNCs. As a result, SMEs can promptly address market demands by leveraging their innovative capacities to meet customer needs. This same principle should similarly guide the implementation of SBP within SMEs.

2.3 Textile and Apparel (TA) Industry in SBP

Chen and Burns (2006) conducted a case study to evaluate the environmental impact of various textile materials production processes with the objective to reduce the adverse environmental impact of the case industry. The study concluded that a sustainable solution could be attained only when the consumer, government, and producer come on the same platform to work toward environmental sustainability, otherwise, it is difficult and complex to find a viable solution that can work for the businesses. Neto et al. (2019), reported that the extant literature reveals cleaner production processes improve economic and environmental benefits through innovative technologies and strategies, however, the benefits of the cleaner production processes are not linked with sustainable development goals (SDG). Therefore, a framework was proposed to evaluate resource allocation for process improvement in the textile industry owing to financial and environmental impacts using environmental strategies in relation to the SDGs as the textile industries have the highest consumption levels for water and soil pollution. Chen et al. (2017) attempted to address this issue by proposing a tool that not only measures water consumption and pollution at the process level but also determines the undiscoverable disadvantages of using technology. Concisely, the literature review reveals that TA industries are one of the sectors that draw immense attention to SBP.

2.4 Multi-criteria Decision Approach in SBP of TA Industry

Al Sawaf and Karaca (2018) conducted a survey to seek different stakeholders' opinions on the sustainability of common textile wastewater technologies in Turkey. Analytic hierarchy process (AHP) technique was utilized to determine the most sustainable wastewater treatment technology in light of economic, environmental, social, and technical aspects. In a similar context, Zhu et al. (2011), studied various factors that could hinder apparel industries from stepping into eco-friendly production. The grey-based decision-making trial and evaluation laboratory (DEMATEL) methodology was employed to understand the hierarchical relationship between the factors. The findings indicate that addressing the market demand for eco-friendly products and enhancing human resource capabilities are the primary areas of concern.

2.5 Barriers to SBP in the TA Industry

The literature review unveiled numerous barriers within the TA industries as documented in journal articles in their respective regional contexts. These articles, along with their identified barriers, are presented in Table 1. Through the literature survey, a total of thirteen barriers were identified, from which common barriers were further deduced to the ten most relevant barriers. Expert opinions were solicited through surveys to refine these barrier selections. These barriers, along with their descriptions and corresponding sustainability indices encompassing economic, environmental, and social factors are presented in Table 2.

2.6 Gap Analysis and Research Importance

There exists a notable lack of research concerning the secondary processing aspect of the TA industry, particularly in applying MCDM for sustainability purposes. Hence, this current study employs the ISM approach within Fiji's TA industries to discern and establish interrelationships among various barriers based on expert opinions. Additionally, ISM is chosen for its straightforwardness, making it comprehensible to stakeholders, and for its ability to develop a hierarchical model conducive to the implementation process.

Table 1 Barriers identification

No. of barriers identified	List of barriers	Source
12	Absence of a reward system for suppliers, complexity in designing green processes and systems, inadequate enforcement of legislation, lack of commitment from top management, insufficient eco-literacy and training, scarcity of green suppliers, insufficient guidance and support from regulatory authorities, lack of economic benefits, insufficient consumer support and encouragement, high costs for implementation and maintenance, lack of trust and environmental partnerships among supply chain partners, and the lack of effective environmental measures	Majumdar and Sinha (2019)
8	Inadequate reverse logistics strategies, uncertainty in the quality, quantity and timing of returned used products, lack of technology, insufficient management participation and initiative, ineffective production planning and control, improper distribution channels, fear of product cannibalization, ambiguous government policies and regulations, lack of expertise, flawed pricing strategies, lack of management foresight, poor consumer acceptance, unwillingness to return products, and negligence towards environmental concerns	Singhal et al. (2018)
12	Insufficient knowledge on process development, lack of understanding of customer needs, inadequate internal encouragement programs, lack of criteria for supplier selection, lack of involvement operational staff in planning decisions, failure to adapt to new occupational health and safety policies, lack of supportive policies and institutional uncertainty, lack of collaboration for implementation, absence of internal environmental initiatives, shortage of trained personnel and training initiatives, resistance to adopting new practices, and lack of encouragement from international customers	Awan et al. (2018)
7	Inadequate infrastructure, insufficient governmental policies, complex supply chain, low-level integration, limited foreign investment, poor quality of raw materials, skill shortages, demonetization, lack of product quality control, high production costs, lack of a positive brand image, negative environmental and societal impacts, declining exports, and low productivity	Gardas et al. (2018)
11	Commitment from top management, technological opportunities, competitive pressure, development capabilities, new product development, supply chain coordination, long-term strategic goals, information sharing, research and development, continuous improvement, and financial performance	Dewangan et al. (2015)

Table 2 List of barriers

No.	Barrier name	Description	Sustainability index	Sources
1	Lack of waste minimization or elimination	Uncontrollable/unnecessary wastages in fabrics while converting into apparel. It also indicates waste in raw materials, processed materials, underutilized equipment, and human resources	Economic, environment and social	Gardas et al. (2018)
2	Lack of floor or operational workers	Operational level workers are one of the pillars of the organization, either enough operators are not available, or even if they are available, they do not possess enough skills to operate the machines safely	Economic and social	Serageldin (1996)
3	Skill shortage	Skilled workers refer to intermediate or supervisory level workers who could have proficiency in looking after the operational management independently and possess an ability to understand and implement operations management tools and techniques	Economic and social	Caldera et al. (2019)
4	Raw materials quality	Raw materials are imported, whose quality is determined at the receiving end rather than at the supplier end. Sometimes, poor-quality goods have to be rejected and/or processed, which results in wasting time and other resources	Economic and environment	Gardas et al. (2018)
5	Lack of government policies	Skill and workforce development-related policies to cater to the need of industry's survivability. Policies - related to the easing of the customs service. Strict norms that govern or guide environment and social exploitation	Economic, environment and social	Singhal et al. (2018)
6	High implementation cost	Includes initial investment in the men (professionals), materials (new machines), and methods (scientific and managerial techniques)	Economic, and social	Sharma et al. (2019), Majumdar and Sinha (2019)

(continued)

Table 2 (continued)

No.	Barrier name	Description	Sustainability index	Sources
7	Supply chain or logistics complications	Usually, the time required to procure raw materials or equipment is not definite. Thus, it highly varies as the industries rely on imports of those items. Besides, the items are sometimes stuck for customs clearance and may take a bit longer than usual	Economic, and social	FijiSun (2021), Dewangan et al. (2015)
8	Fear of failure	The fear of failing to implement new strategies could result in a substantial monetary loss for the company	Economic	Narayan (2001), Majumdar and Sinha (2019)
9	Lack of quality assurance system	Product value is highly regarded with product quality, which in turn depends on the quality control or assurance process to govern the quality of products and avoid waste of resources	Economic, environment, and social	Singhal et al. (2018), Gardas et al. (2018)
10	Low productivity	This can be attributed to various factors, including extended machine downtime, inexperienced workers, and outdated machinery	Economic, environment, and social	Sharma et al. (2019)

In Fiji, the TA industries rank as the third most growing sector within manufacturing industry, making consistent contributions to the nation's GDP growth. These industries predominantly engage in apparel production from fabrics, with many organizations catering to local demands while others prioritize exports, primarily focusing on Oceania regions/countries. Despite governmental investments in infrastructural development and policy initiatives aimed at facilitating growth in both local and export markets, the challenges encountered by TA industries have yet to undergo systematic analysis. Although certain facts are transparent, considering the raw material and the supply being entirely dependent on import, it cannot be classified as the primary reason for not accommodating SBP; proper identification of barriers to SBP is inevitable. Understanding the interrelationship between barriers and identifying the most influential barrier can lead stakeholders to change or revise their planning and strategies. Hence, this research work is a foundational attempt for TA industries in Fiji to implement SBP.

3 Research Methodology

The identification of various barriers was done from the literature and research team consultation, which summed up to 13 different barriers. Alternatively, the research team was formulated that consisted of academic and industry experts, and research wards. A questionnaire was prepared consisting of all 13 barriers with their descriptions and a Likert's scale provided against each barrier to assess the relevancy of the barriers against the implementation of SBP. The survey results were consulted with the research team to deduce the most relevant barriers, which are shown in Table 2. The identified barriers are subjected to ISM analysis in the context of the case industry. The methodology applied for this research work is portrayed in Fig. 1.

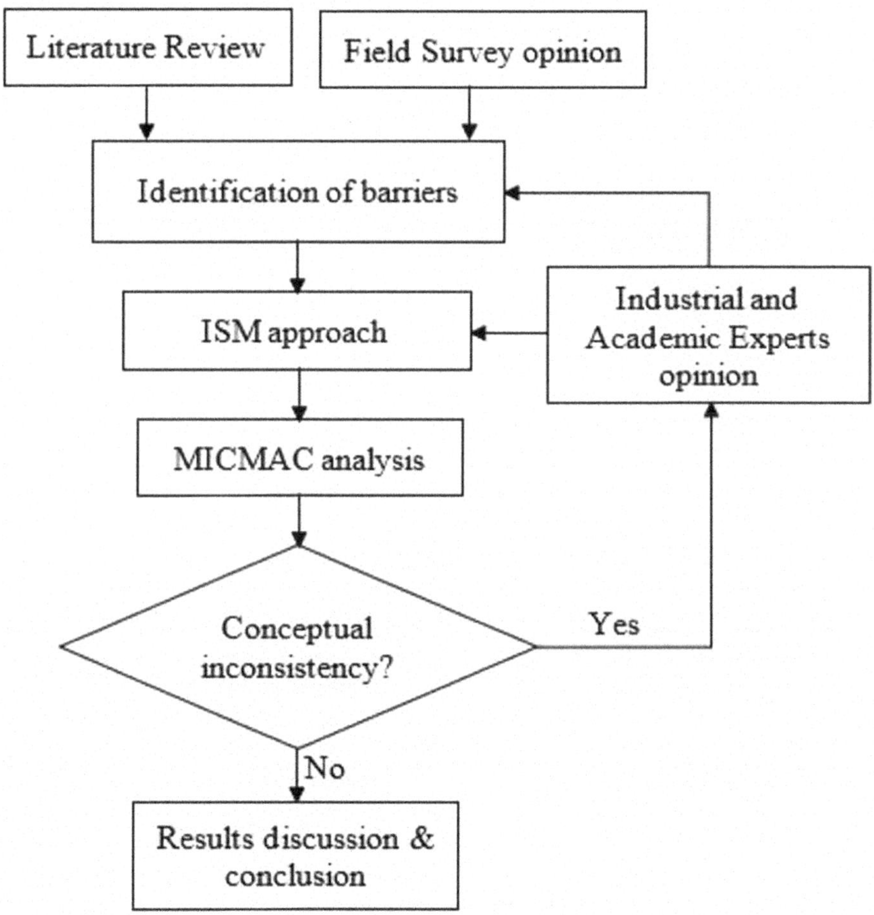

Fig. 1 Research methodology

3.1 ISM Approach

ISM is a MCDM technique that relies heavily on experts' opinions to determine the relationship between the variables. The ISM methods are detailed by Abuzeinab et al. (2017). The first step in the ISM methodology is to list all the relevant barriers and identify the contextual relationship between them, followed by the development of the structural self-interaction matrix (SSIM) with the information gathered from the field survey and expert committee. The reachability matrix is developed from the SSIM with barrier transitivity. The transitivity is done with the basic assumption that if barrier A is related to barrier B and barrier B is related to barrier C, then barrier A can be also related to barrier C. The barriers are then partitioned into different levels forming a hierarchy of the influencing barriers. A digraph is drawn from the relationship obtained in the reachability matrix, and the transitivity is delinked, which is then converted into an ISM model by replacing the barrier nodes with statements. The developed model is subjected to conceptual inconsistency, and the modifications are amended if necessary. The process by which the model was obtained is shown in Fig. 1.

4 Methods Application

4.1 Data Collection

From the literature research, thirteen relevant barriers have been identified, of which ten are shortlisted, as shown in Table 2. Questionnaires were developed consisting of all 13 barriers with description and Likert's scale. It is noted that the survey was conducted only in the eastern and western region (Suva, Nadi, and Lautoka) of Viti Levu as nearly 90% of the industries are located in this region and is considered as the industrial hub of Fiji. The survey is then subjected to the research team's opinion, and it was decided to eliminate the three different barriers as it respectively received less than 15% of confidence from the respondents, these were lack of infrastructural development by the government, lack of top management commitment, and lack of consideration of the negative impact on the environment and society.

The SSIM (Table 3) was developed to determine the interrelationship between barriers denoted by the symbol, V, A, X, and O. For instance, as per Table 3, lack of waste minimization does not influence the complex supply chain, denoted by the letter 'O'. In another case, the lack of human resources directly controls the low productivity; therefore, the symbol 'V' is placed, while skill shortages and lack of quality control of the products influence each other hence indicated by the letter 'X'. The high implementation cost is controlled by complex supply, resulting in the usage of the symbol 'A'. Similarly, all the pairwise comparisons are made using the expert committee.

Table 3 SSIM

Barriers	B 2	B 3	B 4	B 5	B 6	B 7	B 8	B 9	B 10
B1	O	O	A	A	O	O	A	A	O
B2		V	O	O	O	O	X	V	V
B3			O	O	O	O	A	X	X
B4				O	X	X	A	A	V
B5					O	O	O	O	O
B6						A	O	O	O
B7							A	O	A
B8								X	X
B9									X

4.2 Reachability Matrix and Level Partitions

Qualitative data in SSIM is converted into quantitative data as shown in Table 4. The entry of 'V' is replaced by one in cell (i, j), while cell (j, i) receives zero. In the case of 'A', cell (i, j) becomes zero while cell (j, i) receives one. Similarly, for the 'X' and 'O' symbols, both cells receive one and zero respectively. The final reachability matrix is obtained by introducing transitivity (Table 5) explained elsewhere (Singhal et al. 2018), and shown in Table 6.

The level partition works with the identification of the reachability set, antecedent set, and intersection set from the final reachability matrix. If the reachability set and intersection set barriers are identical, then it is partitioned to level one. The intersection set barriers are identified from the barriers that are common in the intersection and antecedent set. The levelled barriers are removed from the sets, and the next iterations are carried out until all the barriers are allocated to various levels. In the

Table 4 Initial reachability matrix

Barriers	B1	B2	B3	B4	B5	B6	B7	B8	B9	B10
B1	1	0	0	0	0	0	0	0	0	0
B2	0	1	1	0	0	0	0	1	1	1
B3	0	0	1	0	0	1	1	0	1	1
B4	1	0	0	1	0	1	1	0	0	1
B5	1	0	0	0	1	0	0	0	0	0
B6	0	0	0	1	0	1	0	0	0	0
B7	0	0	0	1	0	1	1	0	0	0
B8	1	1	1	1	0	0	1	1	1	1
B9	1	0	1	1	0	0	0	1	1	1
B10	0	0	1	0	0	0	1	1	1	1

Table 5 Determining the transitivity

Barriers	B1	B2	B3	B4	B5	B6	B7	B8	B9	B10
B1	1	0	0	0	0	0	0	0	0	0
B2	1*	1	1	1*	0	1*	1*	1	1	1
B3	1*	1*	1	1*	0	1	1	1*	1	1
B4	1	0	0	1	0	1	1	1*	1*	1
B5	1	0	0	0	1	0	0	0	0	0
B6	1*	0	0	1	0	1	1*	0	0	1*
B7	1*	0	0	1	0	1	1	0	0	1*
B8	1	1	1	1	0	1*	1	1	1	1
B9	1	1*	1	1	0	1*	1*	1	1	1
B10	1*	1*	1	1*	0	1*	1	1	1	1

Table 6 Final reachability matrix

Barriers	B1	B2	B3	B4	B5	B6	B7	B8	B9	B10	Driving
B1	1	0	0	0	0	0	0	0	0	0	**1**
B2	1	1	1	1	0	1	1	1	1	1	**9**
B3	1	1	1	1	0	1	1	1	1	1	**9**
B4	1	0	0	1	0	1	1	1	1	1	**7**
B5	1	0	0	0	1	0	0	0	0	0	**2**
B6	1	0	0	1	0	1	1	0	0	1	**5**
B7	1	0	0	1	0	1	1	0	0	1	**5**
B8	1	1	1	1	0	1	1	1	1	1	**9**
B9	1	1	1	1	0	1	1	1	1	1	**9**
B10	1	1	1	1	0	1	1	1	1	1	**9**
Dependence	**10**	**5**	**5**	**8**	**1**	**8**	**8**	**6**	**6**	**8**	

present work, three iterations are performed; accordingly, three levels are partitioned, as shown in Table 7.

4.3 *ISM Model and MICMAC Analysis*

The barriers that secured level 1 while partitioning is placed in the top position in the ISM model. Likewise, subsequently levelled barriers are vertically positioned one after the other downwards. There are only three levels in the present investigation, as shown in Fig. 2. The MICMAC (Cross-Impact Matrix Multiplication Applied to the Classification analysis) is a tool that expresses the distribution of the barriers

Table 7 Level partition

Barriers	Reachability set	Antecedent set	Intersection	Level
B1	1	1, 2, 3, 4, 5, 6, 7, 8, 9, 10	1	I
B2	2, 3, 8, 9	2, 3, 8, 9	2, 3, 8, 9	III
B3	2, 3, 8, 9	2, 3, 8, 9	2, 3, 8, 9	III
B4	4, 6, 7, 8, 9, 10	2, 3, 4, 6, 7, 8, 9, 10	4, 6, 7, 8, 9, 10	II
B5	5	5	5	II
B6	4, 6, 7, 10	2, 3, 4, 6, 7, 8, 9,10	4, 6, 7, 10	II
B7	4, 6, 7, 10	2, 3, 4, 6, 7, 8, 9, 10	4, 6, 7, 10	II
B8	2, 3, 8, 9	2, 3, 8, 9	2, 3, 8, 9	III
B9	2, 3, 8, 9	2, 3, 8, 9	2, 3, 8, 9	III

graphically in four different quadrants (autonomous, dependent, linkage, and independent—as per Fig. 3 anticlockwise starting from the lower-left corner), with the primary objective to analyze the driving and dependence power of the barriers. The MICMAC chart is constructed using Table 6, from the total score obtained for the driving column and the dependence row.

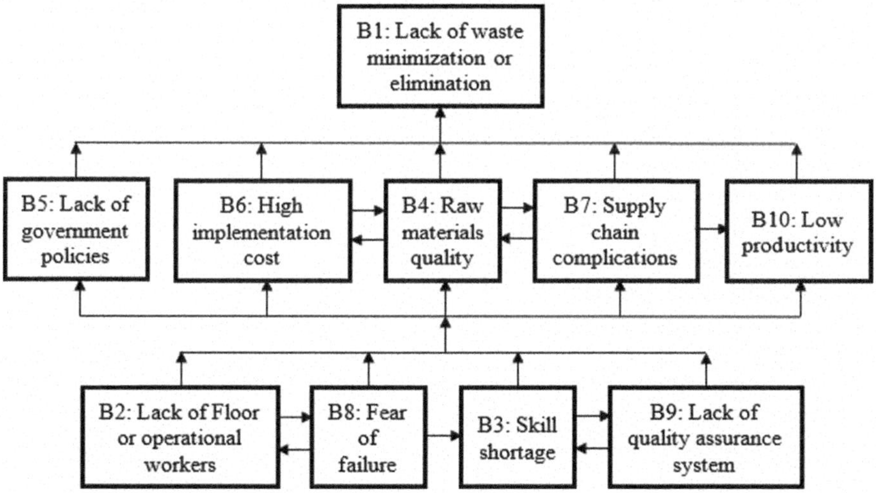

Fig. 2 ISM model

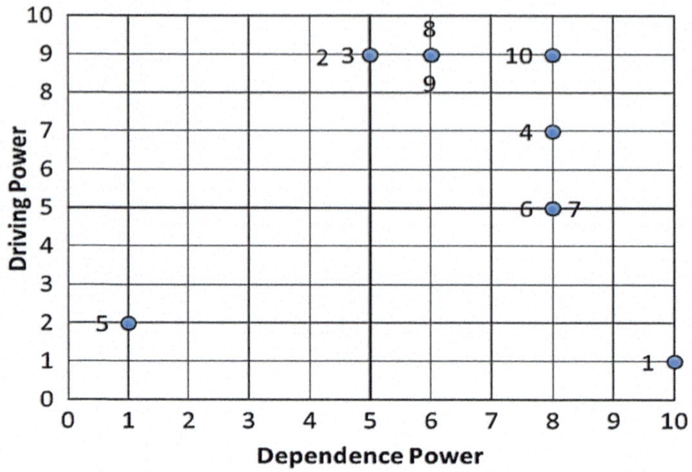

Fig. 3 MICMAC analysis

5 Results Discussion

SMEs' sustainable business operations play a vital role in bolstering the nation's GDP. Among Fiji's manufacturing sectors, the TA industries stand out as key contributors to GDP growth. Achieving sustainable operations hinges on a multitude of factors, with government policies emerging as pivotal influencers. Particularly crucial is the role of government policies in fostering skill development and workforce readiness to meet industry demands, as well as in formulating and implementing environmental and social policies to responsibly manage resources. It can be observed from Fig. 3 that B5 (government policies) is in the autonomous quadrant, which reflects that it has low dependence but little higher driving power. Observing the ISM model (Fig. 2), it can further be understood that B5 eventually governs B1. This indicates that B5 comes to importance provided if at all the third-level barriers are addressed. The B1 was observed to have almost nil driving power but high dependence power as per Fig. 3; this is the reason B1 is in the first level of the ISM model and is driven by second and third-level barriers.

Intriguingly, the remaining eight barriers are situated in the third quadrant of the MICMAC analysis, which discloses that each barrier has both driving and dependence power. This means that the eight barriers have the potential to influence and get influenced by the other barriers. However, careful observation of this quadrant reveals that the barriers are 'dispersed' within the quadrant, which indicates some significance; for example, the B6 and B7 have high dependence power and moderate driving power, whereas the B2 and B3 possess high driving power and moderate dependence power; the same scenario is almost applicable for B8 and B9 too. As a result, the barriers B2, B8, B3 and B9 are positioned in the ISM model (Fig. 2) at the third level as it got relatively higher driving power. This indicates the importance of addressing these barriers as it drives the remaining barriers.

The field survey and the experts' discussion reveal that, in general, the TA industries and, for this investigation, the case industry is willing to invest in SBP as it engenders the success of the business and places it on the path of continuous improvement. Indeed, industries possess the capacity to invest, but fear of failure (B8) hinders their interest. This is mainly due to a lack of skilled professionals who can confidently introduce the 'lean and green approach' or any similar approach. Besides, even if the industry occasionally obtains such skilled professionals from foreign countries, the lack of floor or operational-level workers to implement is a major setback. The combined barriers (B8, B2 & B3) not just affect the quality assurance system (B9) but also the implementation cost (B6) in terms of men, material and/or equipment and its subsequent operation and maintenance. So, these combined barriers eventually affect the quality of incoming material (B4), which eventually affects productivity (B10).

On the contrary, these combined barriers directly influence the fear of high implementation costs, owing to any new equipment bought and whose operation and maintenance heavily rely on the local personnel, as the expatriate professionals' longevity is uncertain. On the other hand, industries in this sector heavily depend on imported raw materials to sustain the TA businesses. Given Fiji's relative isolation, there are often delays in the arrival of goods. Occasionally, these goods may remain held up at customs offices for extended periods, unbeknownst to employers. Subsequently, it complicates the supply chain activities (B7), which indeed affects productivity. It is noted that the quality of raw materials tremendously affects productivity and product quality.

Based on the findings, the study reveals that the Ministry of Education collaborates with tertiary education and training institutions to prioritize the development of human resource capacity and capability. Simultaneously, government policies should be geared towards enhancing supply chain activities for SMEs. Moreover, outdated governmental policies need to be revamped to ensure that SMEs can achieve economic, environmental, and social benefits, thereby setting them on a sustainable business trajectory. It is evident from the ISM model in Fig. 2 that for most of the barriers in the lower level (B2, B3, B8, and B9) involve human intervention. And as part of Industry 4.0, digitalization is inevitable, this means that as part of solving the lower-level barriers, a digital transformation would be necessary in the TA manufacturing industry that would help in resolving the dominant barrier (B1). Siemens (2019) discussed three key imperatives for successful digital transformation, one of which is agile models (Mateo 2020), which was also proposed by Jabbari and Rosemann (2022) as a mechanism for achieving modularity in manufacturing in Industry 4.0. They describe agile model as models that are designed to adapt and update their structures and behaviours in response to unexpected changes. They are iterative, incremental, self-organizing, and flexible, allowing their configuration and actions to adjust based on the situation. Agile models enable cost-effective responses to unpredictable requirement changes and support the rapid and responsive development of systems tailored to evolving needs and circumstances (Jabbari and Rosemann 2022).

These models facilitate real-time interaction between physical and digital spaces, akin to agile digital twins in Industry 4.0 (Aheleroff et al. 2021).

6 Conclusion and Scope for Future Work

This study contributes to SBP by offering a comprehensive and contextual understanding of the inter-relationships among barriers, drawing from both the literature and senior decision-makers' perspectives. The investigation highlights that the success of SMEs, particularly in Fiji's textile and apparel (TA) industry, heavily relies on human resources (B2), especially skilled and professional labour (B3). This dependency creates a fear of failure (B8) in undertaking advanced or high-cost initiatives involving people, materials, methods, and machinery. Among the ten significant barriers identified, these three are particularly influential, subsequently affecting the other barriers. Government legislation (B5) stands alone, not significantly impacted by other barriers but crucial in paving the way for sustainable industry growth. Other barriers, such as lack of quality control (B9), high implementation costs (B6), quality of raw materials (B4), complex supply chains (B7), and low productivity (B10), are moderately important. These barriers can be overcome by systematically addressing the fundamental factors to ensure SBP.

The present work primarily depends on the case industry; thus, the decision made at several stages of the research approach has the influence of case industry experts in making the ISM model. The resultant model, however, is qualitative and immensely meaningful to the case industry but may not be for others. An alternative method suggests that the research approach can extend with the inclusion of several TA industries; accordingly, the number of barriers can also be expanded to develop a robust framework for wide application. In addition, a quantitative analysis can be developed from real data using structural equation modelling that can be compared with the qualitative model to get a promising framework to follow SBP in TA industries. Further, addressing the lower-level barriers by means of digital transformation in the industry would solve the need for employing more skilled labour and floor workers. By employing agile manufacturing model in the TA industry would result in a dynamic business model with lean principles.

References

Abuzeinab A, Arif M, Qadri MA (2017) Barriers to MNEs green business models in the UK construction sector: an ISM analysis. J Clean Prod 160:27–37

Aheleroff S, Xu X, Zhong RY, Lu Y (2021) Digital twin as a service (DTaaS) in Industry 4.0: an architecture reference model. Adv Eng Inform 47

Awan U, Kraslawski A, Huiskonen J (2018) Understanding influential factors on implementing social sustainability practices in manufacturing firms: an interpretive structural modelling (ISM) analysis. Procedia Manuf 17:1039–1048

Caldera H, Desha C, Dawes L (2018) Exploring the characteristics of sustainable business practice in small and medium-sized enterprises: experiences from the Australian manufacturing industry. J Clean Prod 177:338–349

Caldera H, Desha C, Dawes L (2019) Evaluating the enablers and barriers for successful implementation of sustainable business practice in 'lean' SMEs. J Clean Prod 218:575–590

Chen H-L, Burns LD (2006) Environmental analysis of textile products. Cloth Text Res J 24(3):248–261

Chen L, Wang L, Wu X, Ding X (2017) A process-level water conservation and pollution control performance evaluation tool of cleaner production technology in textile industry. J Clean Prod 143:1137–1143

de Oliveira Neto G, Correia JMF, Silva PC, de Oliveira Sanches AG, Lucato WC (2019) Cleaner Production in the textile industry and its relationship to sustainable development goals. J Cleaner Prod 228:1514–1525

European Commission (2020) SME policy in the EU, 16 Nov 2020. [Online]. Available: https://www.eu2020.de/eu2020-en/news/article/looking-back-looking-ahead-sme/2416916. Accessed 29 May 2024

Dewangan DK, Agrawal R, Sharma V (2015) Enablers for competitiveness of indian manufacturing sector: an ISM-Fuzzy MICMAC analysis. Procedia Soc Behav Sci 189:416–432

FijiSun (2021) Business boom in clothing & textile sector, 1 Feb 2021. [Online]. Available: https://fijisun.com.fj/2021/02/01/business-boom-in-clothing-textile-sector/. Accessed 29 May 2024

Gardas BB, Raut RD, Narkhede B (2018) Modelling the challenges to sustainability in the textile and apparel (T&A) sector: A Delphi-DEMATEL approach. Sustain Prod Consum 15:96–108

Gimenez C, Sierra V, Rodon J (2012) Sustainable operations: their impact on the triple bottom line. Int J Prod Econ 140(1):149–159

Herce C, Biele E, Martini C, Salvio M, Toro C, Brandl G, Lackner P, Reuter S (2023) A methodology to characterize energy consumption in small and medium-sized enterprises at national level in European countries. Clean Technol Environ Policy 26:93–108

Jabbari AM, Rosemann M (2022) Modelling 4.0. In: Conceptual modelling in the "Digital First" era—a joint AIS SIGSAND/SIGPrag workshop

Liberto D (2023) Small and midsize enterprise (SME) defined: types around the world, 14 Aug 2023. [Online]. Available: https://www.investopedia.com/terms/s/smallandmidsizeenterprises.asp. Accessed 29 May 2024

Majumdar A, Sinha SK (2019) Analyzing the barriers of green textile supply chain management in Southeast Asia using interpretive structural modeling. Sustain Prod Consum 17:176–187

Mateo FW (2020) Industry 4.0 and the emergence of new business models. In: 18th LACCEI international multi-conference for engineering, education, and technology, Buenos Aires

Narayan PK (2001) Fiji's garment industry: an economic analysis. J Econ Soc Policy 6(1)

Radinibaravi M (2024) Reviving Fiji's garment industry, 12 Jan 2024. [Online]. Available: https://www.fijitimes.com.fj/reviving-fijis-garment-industry/. Accessed 29 May 2024

Reserve Bank of Fiji (2020) Press Release No. 13—Reserve Bank of Fiji Offers Financial Lifeline to Fijian Businesses, 28 April 2020. [Online]. Available: https://www.rbf.gov.fj/press-release-no-13-reserve-bank-of-fiji-offers-financial-lifeline-to-fijian-businesses/#:~:text=In%20Fiji%2C%20an%20SME%20is,are%20eligible%20for%20RBF%20relief.. [Accessed 29 May 2024].

Sawaf MBA, Karaca F (2018) Different stakeholders' opinions toward the sustainability of common textile wastewater treatment technologies in Turkey: a Case study Istanbul province. Sustain Cities Soc 42:194–205

Serageldin I (1996) Sustainability as opportunity and the problem of social capital. Brown J World Affairs 3(3):187–203

Sharma JK, Rajeshkannan A, Sharma AA, Seenivasagam DR (2019) Some aspects on the sustainable process design in a timber mill using the design for manufacturability/sustainability. Int J Sustain Eng 13(3)

Singhal D, Tripathy S, Jena SK, Nayak KK, Dash A (2018) Interpretive structural modelling (ISM) of obstacles hindering the remanufacturing practices in India. Procedia Manuf 20:452–457
Yang Y, Yang X, Xiao Z, Liu Z (2023) Digitalization and environmental performance: an empirical analysis of Chinese textile and apparel industry. J Clean Prod 382
Zhu Q, Sarkis J, Geng Y (2011) Barriers to environmentally-friendly clothing production among Chinese apparel companies. Asian Bus Manag 10(3):425–452

Strategic Design Optimization of Cutting Tools for Enhanced Manufacturing Efficiency

Abhishek Agarwal, Parveen Kumar, Ajay Kumar, and Ranjan Kumar Ghadai

Abstract The efficient utilization of manufacturing technology with respect to cutting tool design for the intended improvement of its structural characteristics is deemed critical. This study uses Finite Element Analysis (FEA) and the response surface approach to determine the effect of the design parameters of the cutting tool on its structural performance. The ANSYS simulation package is applied to capture valuable information from the explicit dynamic analysis with the help of which the critical stress zones and the chipping zones are obtained. The Box-Behnken optimization is used for the optimization of the Design of Experiments (DOE) and response surface which helps in deciding the design parameters that yield the best results. To critically evaluate the findings of this study, perceptions of 2D linearized curves and 3D response surface plots show that the base length has a positive impact on equivalent-stress and equivalent-elastic-strain while total deformation is most affected by the base angle. As evident from the findings of this study, more attention needs to be paid to the relationship between the design characteristics of cutting tools and their structural behaviors. Thus, it is possible to improve the tool performance, decrease the wear, and achieve the effective manufacturing costs. This research makes a contribution to the development of knowledge in the area of Cutting tools design and optimization and provides suggestions for enhancing technical processes and decreasing production expenses.

A. Agarwal
Department of Mechanical Engineering, College of Science and Technology, Royal University of Bhutan, 21101 Phuentsholing, Bhutan
e-mail: agarwala.cst@rub.edu.bt

P. Kumar
Department of Mechanical Engineering, Rawal Institute of Engineering and Technology, Faridabad, Haryana 121004, India

A. Kumar (✉)
Department of Mechanical Engineering, School of Engineering and Technology, JECRC University, Jaipur, Rajasthan, India
e-mail: ajay.kumar30886@gmail.com

R. K. Ghadai
Department of Mechanical and Industrial Engineering, Manipal Institute of Technology, Manipal Academy of Higher Education, Manipal, India

A. Kumar et al. (eds.), *Industry 4.0 Driven Manufacturing Technologies*, Springer Series in Advanced Manufacturing, https://doi.org/10.1007/978-3-031-68271-1_11

Keywords Cutting tools design · Turning · FEA · ANSYS · Machining

1 Introduction

Turning operation is one of the most common and efficient methods of machining with a broad application in the production of both metallic and non-metallic products for various industries. This process is crucial in designing and manufacturing precise parts that conform to an industry's demanding specifications to guarantee that gears operate effectively in different mechanical systems (Ghule et al. 2024; Gangwar et al. 2024). Turning is a machining process that entails the use of a cutting tool to transform a revolves workpiece to attain an improved diameter and finish by successively eliminating material as shown in Fig. 1. The traditional turning process has been characterized by the fact that both external and internal steaks of the rotary axes are obtainable. This versatility is made possible through the use of the lathe which is one of the most basic, yet perhaps one of the most flexible tools in classical machinery.

The lathe functions based on the input workpiece that is revolved at a precise speed while the tool cutting edge that could either be mounted as translating on a carriage, turret or tailstock performs the machining. Cross slide or X-axis moves in the direction transverse to the part's axis while the carriage or turret moves along the bed-ways parallel to the part's axis or Z-axis. Workpiece holding methods are in many forms and they include the use of chucks, collets, face plates, or mandrels. On the same note, another technique entails placing the workpiece between two tapered centers and then rotating the spindle. This indicates that workpieces holding on this lathe can be done in several ways, thus confirming the all-around use of the lathe when it comes to turning operations.

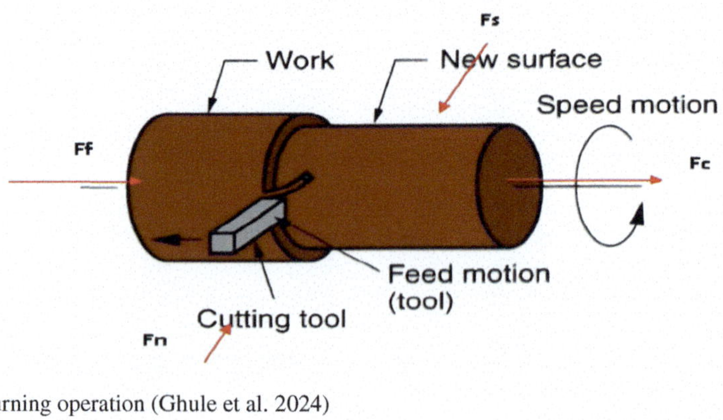

Fig. 1 Turning operation (Ghule et al. 2024)

A significant factor of turning operations is the shape of the cutting tool, which consists of cutting edges, corners, flanks, as well as the face or blade surface. The geometry or shape of these tools is dependent on the characteristics of the work material and the characteristics of the tool material (Kumar et al. 2024). Right-hand and left-hand cut tools are types of single-point cutting tools that are mostly used in turning operations where the input is oriented in a particular manner. Pertinent properties, as stipulated in the international standard nomenclature include rake, relief, and end angles which are significant as they determine the tool's functionality and productivity. The tool's contouring feature is most often designed in order to allow for the most effective interaction between the tool and the workpiece with regard to the most efficient material removal rates while at the same time trying not to compromise on tool life as well as the integrity of the workpiece (Wu et al. 2024). For example, the rake angle affects the cutting forces and chip formation while the relief angle helps in avoiding rubbing and the subsequent heating. The end angles are helpful in enhancing the stability in usage and capacity for efficient cutting in the machines, guaranteeing the tool's performance in high-precision cutting. Turning operation is one of the most significant operations of today's manufacturing that allows the production of parts used in numerous applications through accurate control of the manufacturing process. Thus, the interaction between the functional elements of a lathe and the geometry of the cutting tool indicates the complex nature of this process, and the subsequent development of machining technology, and advances in the creation of high-quality industrial products (Palaniappan and Subramaniam 2024). Many industries rely on turning operation as one of the cornerstones of precision manufacturing, which constantly transforms with new technologies and surface coatings on cutting tools. The literature review has found out that modern scientific advancements have paid significant attention to the role of surface texturing in the improvement of cutting tools. This literature review aims to identify some of the important studies that address different aspects concerning tool texturing in detail, as well as developments made on new techniques and their impact on the machining process.

1.1 Texturing Techniques and Their Impact

Sasi et al. (2017) explored the use of laser surface texturing on high-speed steel cutting tools, employing a pulsed Nd: The YAG laser to put small, micro–Contemporary CVD techniques such as using laser [YAG] to put micro–textures on the rake face of the tool. Their investigative work was exclusively concerned with enhancing some of the laser characteristics like the fluence and the wavelength in order to achieve circular and equally dimensioned impressions. From the results carried out on the textured tools, it was observed that a considerable decrease in thrust and cutting forces during the dry machining of Al7075-T6 aerospace alloy. This research also reveals that laser texturing may be used as a technique to increase the tool's performance and overcome issues based on friction while cutting metals. Machado et al. (2021) did a detailed literature review on the recent technologies in surface texturing for cutting tools with a focus on tribological improvement. Several geometries of texture, materials, and machining conditions were in the range of their assessment and focused mainly on laser ablation. The findings showed that the dimple geometries and the linear channel textures attained better machining performances on uncoated WC–Co tools for titanium alloys as well as hardened steel. The studies revealed that using textured tools can minimize cutting forces and also prolong the life span of the cutting tool particularly under conditions of dry machining.

1.2 Advanced Texturing Patterns

Sugihara and Enomoto (2017), also conducted a study on the orthogonal cutting process using new cutting tools with dimple-shaped patterns at the rake face. Via apparent comparisons of dimple and groove textures under various cutting circumstances, he proved that the dimple textures were superior to the groove textures, especially in high-lubrication conditions. The task of achieving superior texture patterns was recognized as one of the major difficulties, and it was stated that further experimental studies should be carried out in order to define the best solutions. As concerning the study of Ahmed et al. (2020), the work discussed the material used for the experiment, which was AISI 304 stainless steel, a material that is generally difficult to machine because of the formation of built-up edge (BUE).

1.3 Dual Texture Geometries

To address the high friction that arises during dry machining of the Titanium alloys, Siju et al. (2021) developed new geometries of grooves and dimples in the tool-chip contact area. The studies conducted by them showed that the use of dual-textured tools helped to cut down on the cutting and thrust forces to much lower levels than would be experienced by using nontextured tools. Due to the dual textures that were used, it was possible to minimize workpiece adhesion and formation of BUE, and hence the nails used promoted smooth chip formation and longer tool life.

1.4 Tool Wear and Machinability

High-pressure coolant was used in the study conducted by Hoier et. al. (2017) to establish tool wear mechanisms in uncoated WC–Co tools when machining Alloy 718. EDX and SEM were used to investigate the worn tool surfaces in their study. From the experiment, sawdust indicated that high-pressure coolant jets heightened the natural deterioration of the Co-binder and how it affected tool wear and workpiece-precipitate bonding. Such findings are valuable for comprehending tool wear patterns and enhancing the tool's durability under the high-stress conditions of machining. Kawasegi et al. (2009) employed femtosecond laser ablation to cut tools with nano- and micro-morphological characteristics. In their turning tests with aluminium alloys, they also pointed out that textures perpendicular to the direction of the chips effectively meant low efficient cutting forces hence giving support to the texture orientation. Based on the above observations, it is realized that there is a potential to enhance machinability through micro-textured tools since they reduce friction and extend tool life when high cutting velocities are applied.

1.5 Micro-grooved PCD Tools

In a work conducted by Su et al. (2017), micro-grooved PCD tools were used in the cutting tests and the results of the dry_machining of Ti6Al4V. According to the results, both micro-grooved PCD tools had achieved better tribological characteristics than untextured PCD tools. Such properties comprised superior anti-adhesion properties, a reduced mean coefficient of friction, and beneficial force on the cutting tools. Other important parts of the research included micro-pin texture production on the rake face of PCD tools through using a Fiber laser; this gave insight into how the micro-pin texture affects the Tribal behaviors and formation of Titanium carbide (TiC) during the Machining process. Consequently, the significance of micro-texturing was outlined in terms of increasing the performance and longevity of tools, which is vital to proceed with the creation of innovative tools.

1.6 Texturing for Composite Machining

Arulkirubakaran et al. (2019) employed the machinability investigation relating to linear and areal textures in CNC tungsten carbide tools when machining Al-Cu/TiB_2 composites. They concluded that the least cutting forces, specific energy and wear on the tool are achieved when linear textures are erected perpendicular to the chip flow direction. This work is highly relevant to the topic of using textured tools in machining composite material thus stressing the role of texture direction and geometry in enhancing the machining outcomes. As described in the previous research, major progress has been made in the improvement of surface texturing on cutting tools. The above-mentioned investigations indicate that, from laser ablation to the use of innovative dual textures, texture tools can be an effective means of enhancing the rate of machining, reducing tool wear, and thereby developing a better surface finish. In a future study, design parameters should be emphasized as various texture patterns are available in the market; therefore, future investigations must identify the finest texture designs and novel materials and conditions for texturing to promote the utilization of textured cutting tools in different industries. However, significant research has been published on the surface texturing of cutting tools, and some critical issues are still unresolved regarding the specific effects of the parameters of texture design on the properties of cutting tools.

The literature review revealed that most works done in this area encompass the methodology of analyzing the percentage enhancement in the tool's performance characteristics like the cutting forces and tool wear. Nevertheless, there is a dearth of knowledge about how particular texture parameters like geometry, size, orientation, and distribution can influence the mechanics' behavior and wear of cutting tools when subjected to dynamic loads. It is also necessary to investigate how the indicated parameters depend on operational variables such as cutting speed, feed rate, and machining environment with the help of complex simulation tools. Based on this rationale, this research hypothesizes that POS of the design parameters of cutting tool textures through systematic can highly improve the structural characteristics and machining performance. Through FEA and RSM procedures for finding better texture profiles that decrease the size of the area that creates stress concentration and improve the wear characteristic, the tool's durability and machining capability will be improved. Based on the information mentioned in the methodology, a few of the generalized research questions that have been framed are-

1. High-speed machining means the efficiency of making holes of necessary sizes and there is the need to determine how the geometry, size, and orientation of texture in the making affects the strength of the cutting tools and its response.
2. While designing a texture for a tool, which of the parameters goes to create the longest tool life and the smallest cutting forces?
3. How does the range of variations in the matching conditions (e.g. In what manner is the texture of the surfaces influenced by certain cutting factors like speed, feed rate, and type of lubricant that affect the tool?
4. Is it possible that information obtained from enriched simulation methods such as the explicit dynamic analysis yields the outside world application performance of textured cutting tools?

The key research question that was addressed in this study is the lack of an adequately general mathematical model that establishes the maximum characteristics of the textile structures of cutting tools and their parameters. Literature review contributing to the awareness of the benefits of surface texturing is insufficient with regard to the systematic identification of the optimal textures, while many factors and linkages between the cutting tool geometry and machining parameters remain unexamined as shown in Table 1.

This gap hindered the progression of cutting tools that would be lively and adequate to cater to the present highly accurate manufacturing demand. This means that in this research the cutting tool textures were optimized in a way that had not been attempted before hence the novelty; The two methodologies employed included FEA and Response Surface (RSM) Methodology. Thus, with the aid of these progressive methods, the objectives of the present study are to put forward an optimal and accurate methodological framework for the design of enhanced cutting tools with preferable

Table 1 The research contributed to enhancing the understanding of design optimization of cutting tools for enhanced manufacturing efficiency

Literature	Texturing techniques and their impact	Dual texture geometries	Tool wear and machinability	Advanced texture approach	Texturing for composite machining	Systematic identification of the optimal textures
Sasi et al. (2017)	✓					
Machado et al. (2021)	✓		✓			
Sugihara and Enomoto (2017)	✓			✓		
Ahmed et al. (2020)	✓					
Siju et al. (2021)		✓				
Hoier et. al. (2017)			✓			
Kawasegi et al. (2009)			✓			
Su et al. (2017)				✓		
Arulkirubakaran et al. (2019)		✓			✓	
This study	✓	✓	✓	✓	✓	✓

structures. ANSYS: This explicitly employs dynamic analysis, which could help to determine the nature of the tool under real-life machining situations Box-Behnken optimization: This helped determine the various configurations of texture, which were favorable for use. Besides, such an approach helps not only to fill the gap in research under consideration but also to open the prospects for creating cutting tools with improved performance and durability characteristics. This research aims to narrow the existing knowledge gap that exists with regard to the enhancement of cutting tool textures using state-of-the-art simulations and optimization methods. In addition to adding to the existing literature on the subject, the findings of this study will be beneficial in the design and production of high-performance cutting tools for the industry.

2 Methodology

The approach to the fulfillment of this research lies in the systematic analysis of the influence of cutting tool design parameters on its structural characteristics using such tools as the CAD models, FEA, and RSM (Agarwal and Mthembu 2023; Letsatsi and

Agarwal 2022). This systematizes the approach and guarantees the comprehensiveness of the analysis, including the objectives formulated in the research and the gaps in the literature. Initially, a cutting tool and workpiece are modeled in the ANSYS environment using the ANSYS Design Modeler (Agarwal et al. 2022; Molwane et al. 2020). The tool material employed in this study is High Carbon High Chromium (HCHC) steel which is known to possess certain desirable properties that are suitable in cutting processes. When fully heat treated, HCHC steel has a Rockwell hardness of 58–62 HRC; as a result, the ideal edge angle could be preserved longer than other varieties that necessitate weekly regrinding. Its chromium content is between 12 and 14% of the composition which leads to the formation of chromium carbides which increases the tools wearing ability particularly when cutting hard and abrasive materials. Thus, HCHC steel has high hardness, but at the same time, it has reasonably high toughness, which allows it to deflect impacts and shocks during cutting and prevent chipping and breakage. It also has a relatively small change in dimensions during heat treatment, and thus is likely to have good dimensional stability required in cutting tasks. HCHC steel's high compressive strength also means that the cutting edge is not deformed under high forces thereby enabling efficient material removal. These properties make HCHC steel to be used in cutting tools which need wear resistance, and hardness, which also needs to be combined with the ability to resist fatigue as well as great compressive strength. The modeling phase entails the articulation of details of the workpiece and the cutting tool in terms of modeling, and extrusion of the models as shown in Fig. 2.

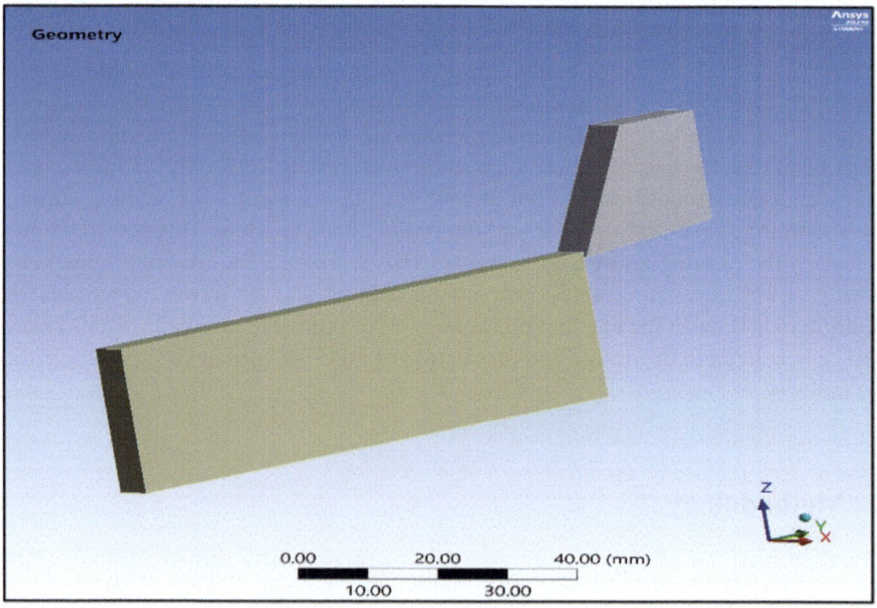

Fig. 2 Designed model of cutting tool and workpiece

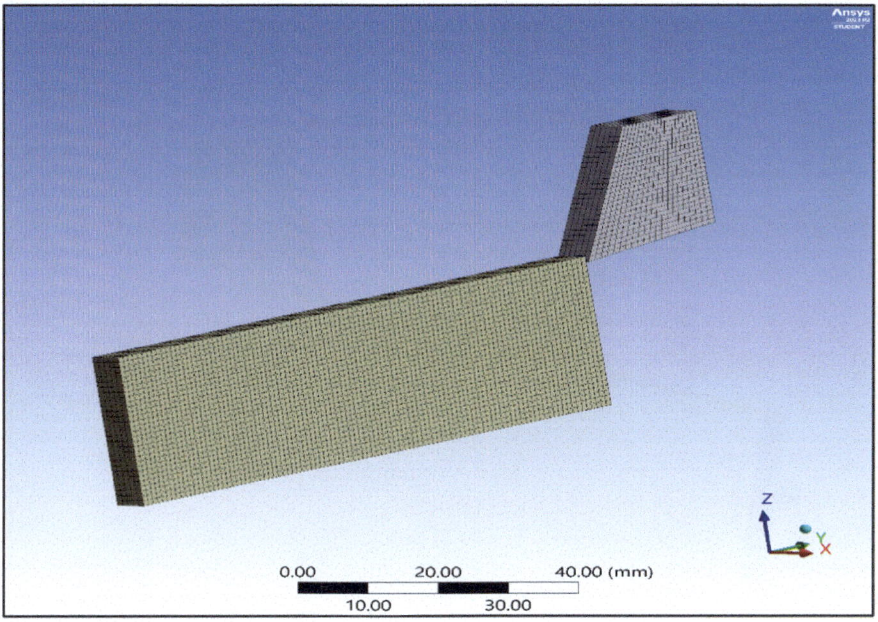

Fig. 3 Meshed model of cutting tool and workpiece

After the generation of CAD models, the next process that is performed is discretization in which the model is discretized for finite element analysis. Since the model is topologically accurate hexahedral element meshing has been used to achieve accurate and efficient simulations. The parameters that influence the meshing are the growth rate, normal inflation, and fine sizing (Agarwal et al. 2024). Hence, after the completion of the meshing process, the model is brought to a meshed model with features like the number of elements—18,276 Number of nodes—22,944 The above process is illustrated in Fig. 3.

After meshing, BCs (Boundary conditions) are imposed on the model in order to support explicit dynamic analysis. Most of the analyses are set up with boundary conditions where the front bottom and rear faces of the piece get constrained to limit the movements during the analysis. The single-point cutting tool as shown in Fig. 4 undergoes a velocity of 140 m/s on the axial direction.

Following the BC application, the other aspects of simulation include the selection of the solver and the specification of the end time for the simulation process. The settings of the solver are time-related and contain an end time equal to 0.001 s, a maximum energy cycling of 10^7, and the time step safeguard factor set at 0.9. These settings are important for the reliable simulation, as shown in Fig. 5.

The simulation is then performed to iteratively predict the dynamic behavior and interaction between the cutting-tool/nose and the workpiece, and it takes 75,009 cycles to complete the simulation process. The conclusion reached from the end simulation process is then calculated and applied in the determination of the stability

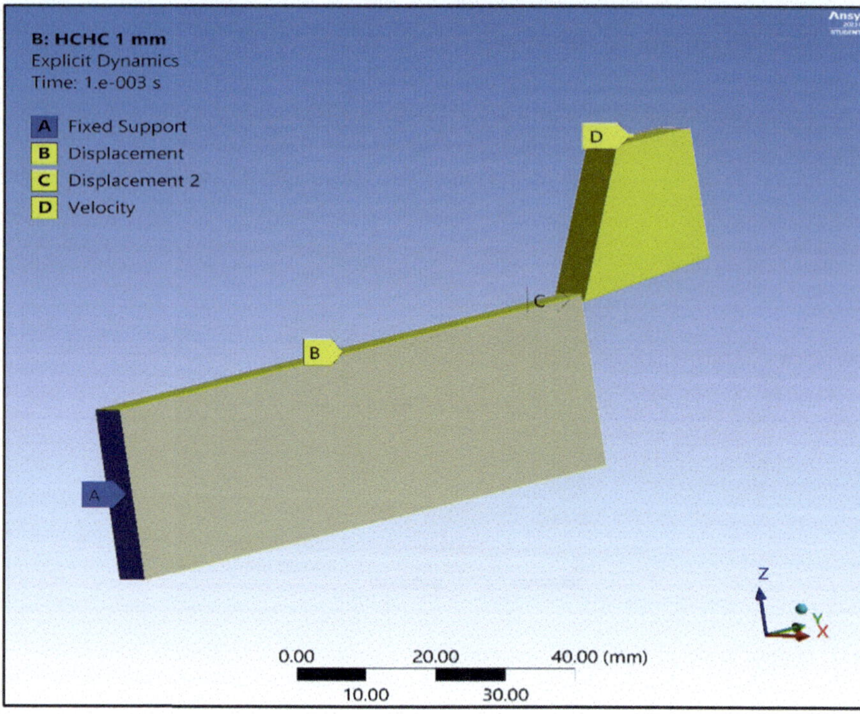

Fig. 4 Applied boundary conditions

Step Controls	
Resume From Cycle	0
Maximum Number of Cycles	1e+07
End Time	1.e-003 s
Maximum Energy Error	100.
Reference Energy Cycle	1e+07
Initial Time Step	Program Controlled
Minimum Time Step	Program Controlled
Maximum Time Step	Program Controlled
Time Step Safety Factor	0.9
Characteristic Dimension	Diagonals
Automatic Mass Scaling	No

Fig. 5 Solver setting definition

as well as the performance of the cutting tool to the test conditions. Subsequently, the DOE process is complied with by applying the Box Behnken optimization strategy. This is familiar by altering the levels of the independent variables of design and then assessing the effects on the tool (Singh and Al Mangour 2023). The incorporation of Taguchi's RSM assists in the development of mathematical models and, in addition to this, helps to determine the most suitable levels of the parameters for enhancing the structure and function of the cutting tool. Therefore, the implementation of the suggested strategy in design for manufacturing is expected to enhance the systematic assessment and optimization of design parameters of cutting tools using CAD modeling, FEA, and optimization algorithms. This process will be of immense help in enhancing the efficiency of cutting tools and increasing their life span—an area of great concern in the current manufacturing industries of the respective fields.

3 Results and Discussion

In the case of the dynamic analysis, from the deformation plot at various time steps, a lot of insight can be gathered about the machining process. The final deformation is presented in the last part of the plot labelled in Fig. 6; this demonstrates the maximum amount of chipping that occurs at the end of the simulation is in excess of 875. 55 mm.

Such a significant change in the material speaks to the extent of the cutting process and underlines the significance of using correct cutting tool parameters to lessen the amount of wear. Machining-induced equivalent-elastic-strain is recorded at various time intervals thus depicting the material behavior under stress. First of all, elastic deformation is observed, as presented in Fig. 7 at 1.2838 mm/mm max.

During machining, the workpiece material undergoes elastic and plastic deformation. Elastic-strain occurs first as the material deforms elastically before yielding and undergoing plastic deformation. Understanding the elastic-strain helps in predicting the onset of plastic deformation, which is crucial for accurate chip formation. Proper control over this transition can lead to more efficient material removal and better surface finish. The magnitude and nature of elastic-strain affects the stress distribution on the cutting tool. High elastic-strain can lead to increased tool wear, reducing the tool's operational life. Central to this change is the inherent crystal characteristic that between 0.0002 s and plastic deformation, a material's deformation mechanism transforms. The amount as well as the distribution of the elastic-strain depends on the chip formation and the tool wear. The findings reveal that the magnitude of the elastic-strain reading obtained at the tool-work interface relates to a higher degree of tool wear as depicted in Figs. 8, 9 and 10 with maximum elastic-strain values of 1. 5617 mm/mm, 1. 715 mm/mm, and 2. 4009 mm/mm, respectively.

After 0.00035 s of simulation (Fig. 8), the maximum elastic-strain observed is 1.5617 mm/mm at the work tool interface zone as represented by red-colored zones. The results obtained matches closely with the results in the literature (Soni 2019) which validates our findings. After 0.00055 s of simulation (Fig. 8), the maximum elastic-strain observed is 1.715 mm/mm at the work tool interface zone as represented

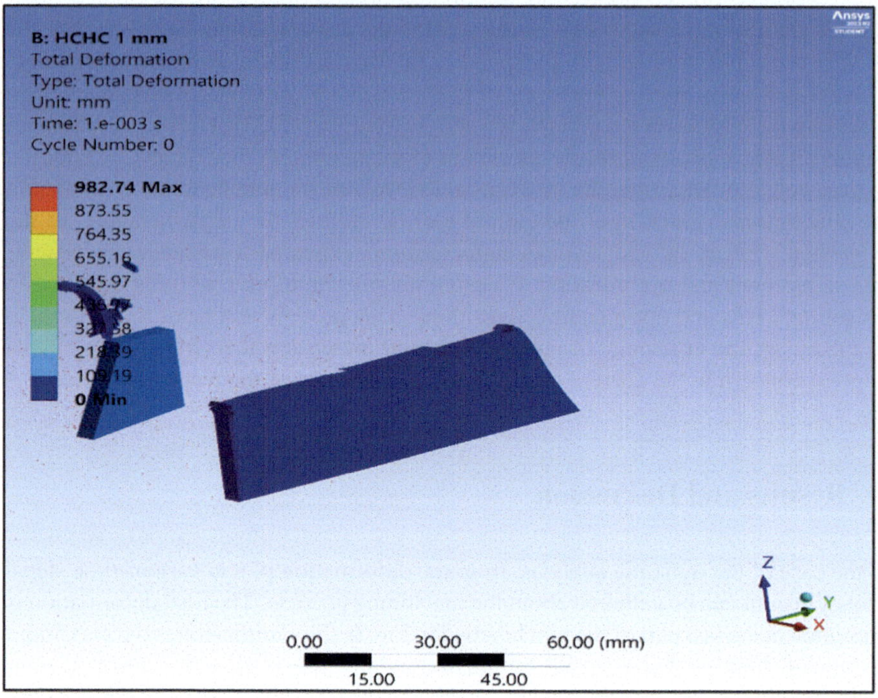

Fig. 6 Deformation at the end of the machining operation

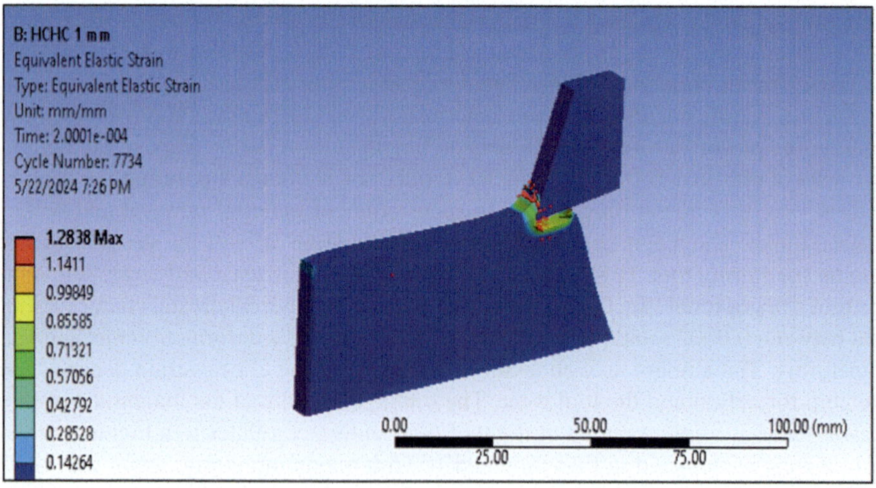

Fig. 7 Equivalent-elastic-strain at 0.0002 s

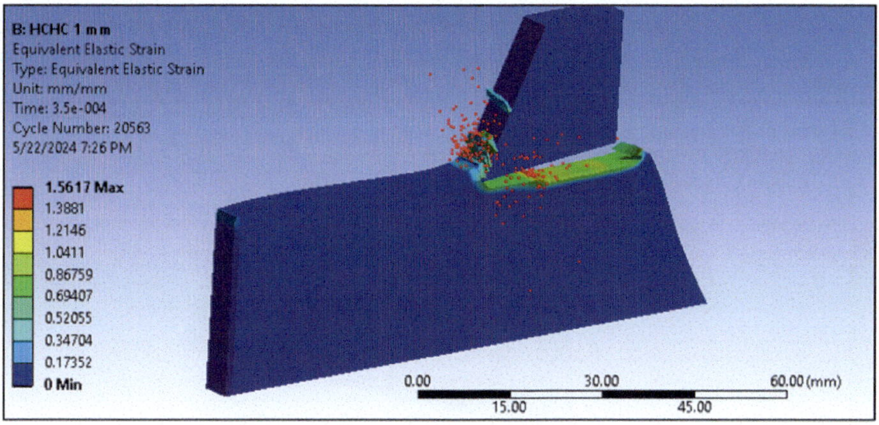

Fig. 8 Equivalent-elastic-strain at 0.00035 s

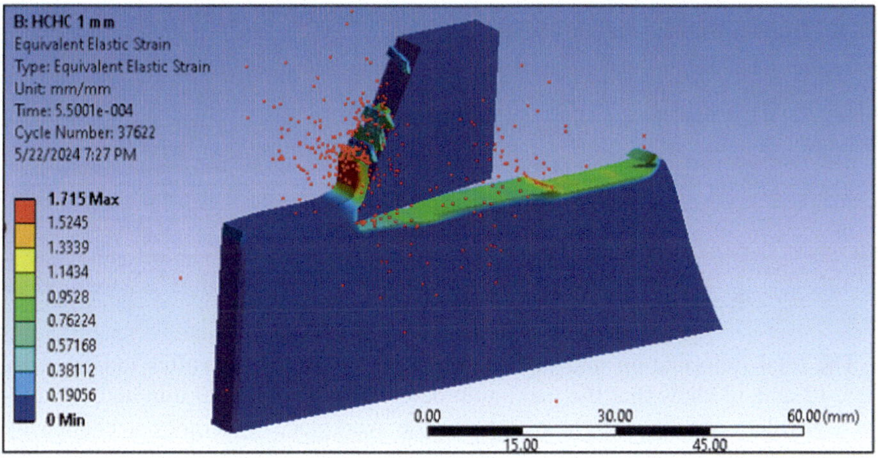

Fig. 9 Equivalent-elastic-strain at 0.00055 s

by red colored zones while at the completion of the machining process, the maximum elastic-strain obtained is 2.4009 mm/mm as shown in Fig. 10. To check the quality of the mesh and to verify the correctness of the simulation a grid independency study was carried out and the details of it are reported in Table 2.

The findings of the paper indicate that our implementation yielded constant elastic strain regardless of the number of elements in the model, thus verifying the effectiveness of the meshing technique. The second analysis which is the static analysis was carried out to find out the maximum force. From the explicit dynamic analysis, the maximum force acting on the cutting-tool is determined. The line pressure load of 675.95 N/mm is applied on the cutting tool edge as shown in Fig. 11.

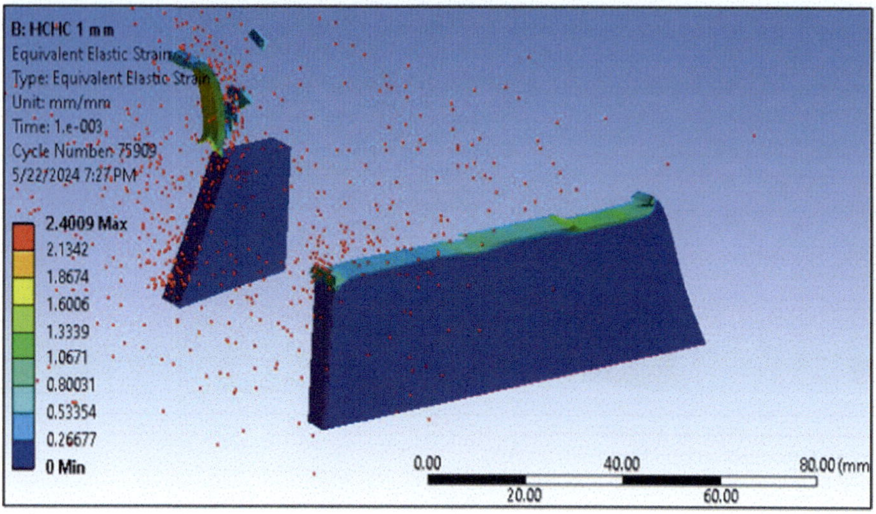

Fig. 10 Equivalent-elastic-strain at 0.001 s

Table 2 Grid Independence Evaluation

Elements	Equivalent-elastic-strain (mm/mm)
17,885	1.5391
17,980	1.5589
18,110	1.5616
18,276	1.5617

The total deformation and equivalent-stress variations thereafter indicated by Figs. 12 and 13 show that the maximum deformation is 0.12889 mm and substantial equivalent stress, in light of which, the workpiece undergoes a change by experiencing the state of plastic deformation.

The equivalent-stress is significantly high which causes plastic deformation of the metal work-piece. The equivalent-stress experienced during machining has a direct impact on the surface integrity of the finished part. High equivalent-stress levels can induce residual stresses, which affect the mechanical properties of the part, such as its fatigue life, corrosion resistance, and dimensional stability. From the results of the static analysis described in the preceding section, the equivalent-elastic-strain is obtained as depicted in Fig. 14 with the value of 0. 5512 mm/mm for the tool-work interface, which leads to the prospects of tool wear and the need to work on the optimization of this factor. The higher elastic-strain causes tool wear, reducing the tool's operational life.

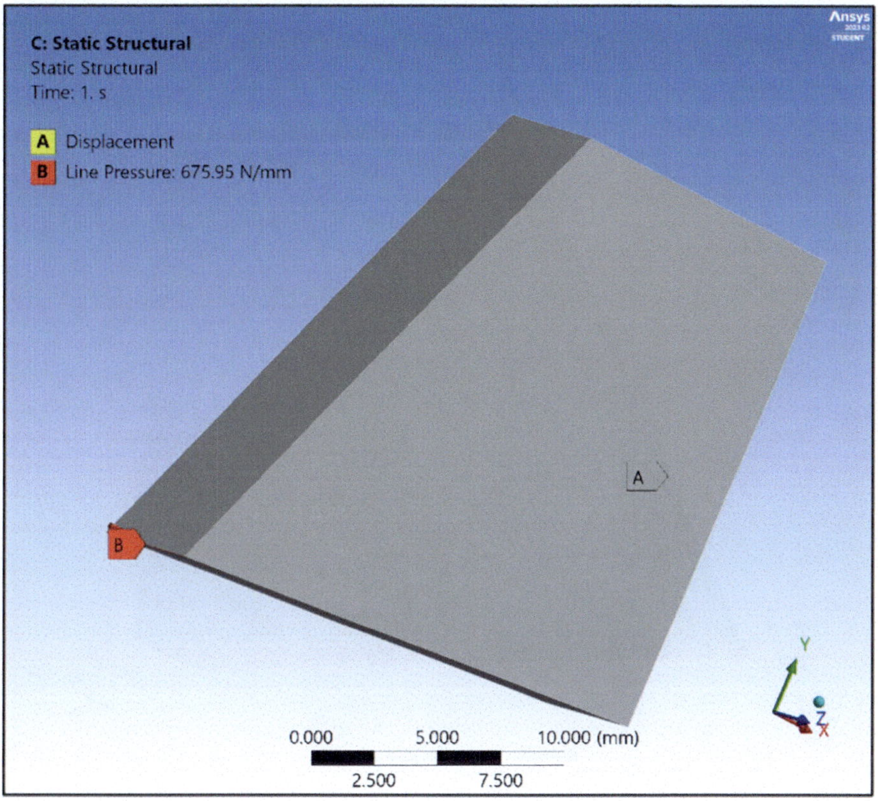

Fig. 11 Loading conditions for static analysis

As the elastic-strain can be controlled the performance of the tool can be improved, lessening downtime and manufacturing expenses. The optimization process using the Box-Behnken design provided a DOE chart as established in Fig. 15 to determine the best design parameters.

For design point number four, the cutting tool was optimized with a base angle of 4.5 degrees and a base length of 27. 5 mm they reached the minimum value of 10,882 MPa of equivalent stress. The optimization process resulted in a goodness of fit curve as depicted below in Fig. 16: Observed values are depicted by the red and blue boxes, while the expected values are exemplified by the straight line that indicates that there is almost no variation between these two sets of values.

The goodness of fit curve is shown in Fig. 16 and sensitivity plots in Figs. 17–22 wherein the changes in base angle and base length affect the equivalent stress, total strain, and total deformation. Figure 17 illustrates the increase in the equivalent stress with an increased base angle as a linear function of the base length.

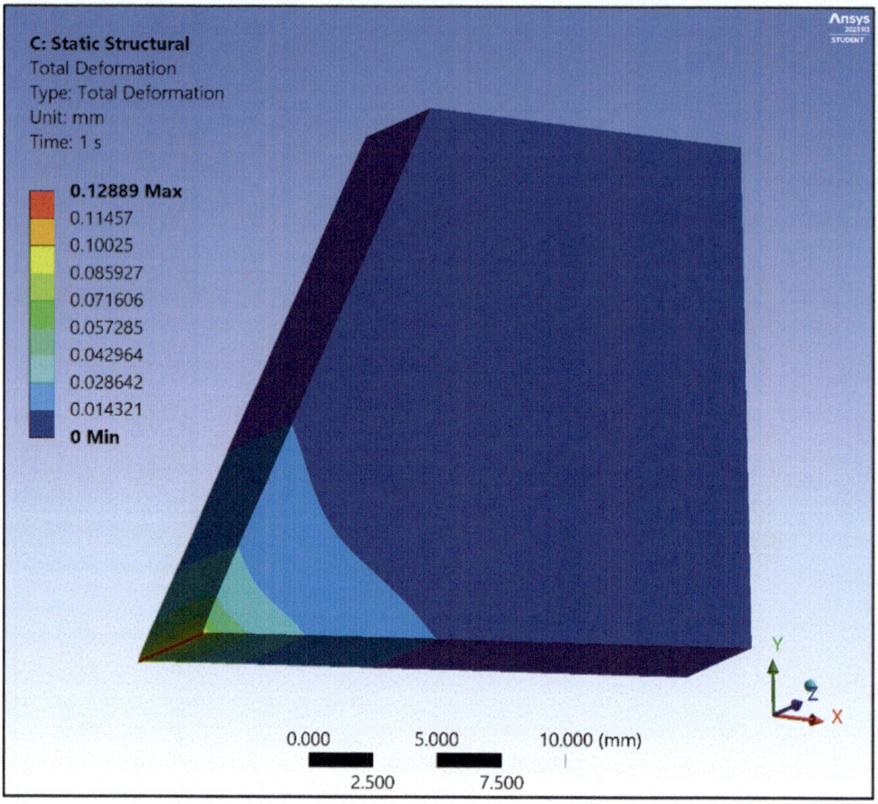

Fig. 12 Total deformation plot from static analysis results

On the other hand, based on the Analysis of the result as shown in Fig. 18, there is a linear descending trend of the equivalent stress when the base length is increased. 5 mm. The obtained results indicate that the peak equivalent stress is reached with a base length of 22. The lowest equivalent stress on the other hand is found to be at a base length of 27. 5 mm.

Figure 19 also shows the trend of equivalent-elastic-strain versus the base angle, and this figure clearly indicates that the value of the strain increases as the base angle.

Likewise, Fig. 20 displays the correlation between equivalent-elastic-strain and base length where on average, the strain declines with the increment of the base length. Consequently, the highest equivalent-elastic-strain was equal to 0. This is achieved at a base length of 22 and the L/mM ratio of 0578/1 is obtained. 5 mm, while the minimum strain is obtained at 27. 5 mm.

Figure 21 indicates the relationship between base angle and total deformation, and it is evident that total deformation increases with the base angle. Figure 22 is also a figure showing the curve of total deformation increasing with the base length. Thus, the total amount of deformation is maximum when the length of the base is

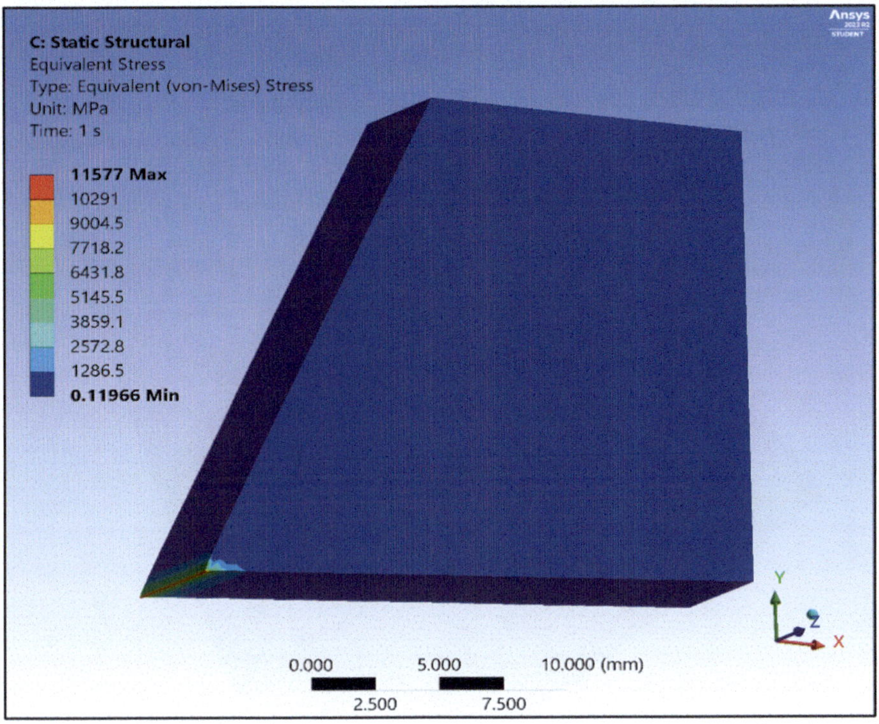

Fig. 13 Equivalent-stress plot from static analysis results

27. The maximum deformation is at 32 The minimum deformation is 27 mm, and the middle deformation is at 5 mm. 5 mm, suggesting that the deformation traits are subject to a lot of influences and interdependencies of the parameters.

From the 3D shape of RSM plots of stress, strain, and deformation as shown in Figs. 23, 24 and 25, it is possible to visualize the changes.

These plots show that in a bid to reduce the equivalent-stress as well as the elastic strain, the base length should be between 25.5 and 27.5 mm. As for the geometry, the working length should not be more than 5 mm, and the base angle should be no less than 4.5° to 5.5°. Likewise, minimum deformation is realized when he adopts a base angle of 4.5° to 4.8° and a base length of 22.5–24.5 mm. The sensitivity plot in Fig. 26 depicts how the base angle and the base length affect the life and sharps of the cutting tool.

The base length has been found to have a higher sensitivity percentage with respect to equivalent-stress and elastic-strain meaning the more the base length, the more the stress and the strain. On the other hand, the base angle significantly influences the total deformation more and hence is very critical when it comes to structuring the cutting tool. The detailed analysis and optimization of characteristics of cutting tools also give useful information on how to enhance the life of the cutting-tool. Thus, by regulating these factors, it is possible to effectively select appropriate cutting

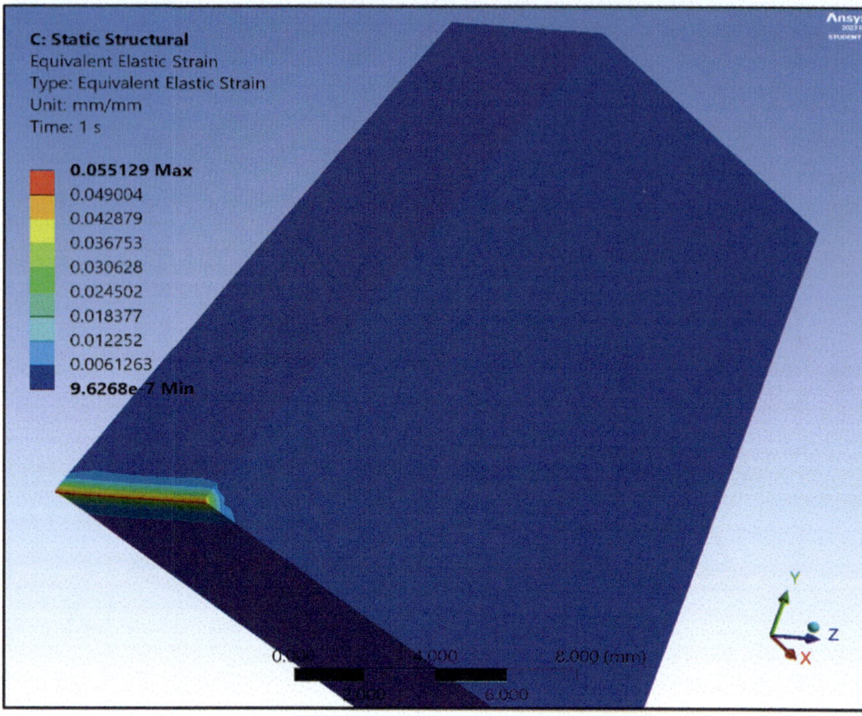

Fig. 14 Equivalent-elastic-strain plot from static analysis results

Table of Outline A2: Design Points of Design of Experiments

	A	B	C	D	E	F
1	Name	P4 - base_angle (degree)	P5 - base_length (mm)	P1 - Equivalent Stress Maximum (MPa)	P2 - Equivalent Elastic Strain Maximum (mm mm ^-1)	P6 - Total Deformation Maximum (mm)
2	1 DP 0	5	25	11577	0.055129	0.12889
3	2	4.5	22.5	12120	0.057714	0.12154
4	3	5.5	22.5	12189	0.058041	0.12895
5	4	4.5	27.5	10882	0.051817	0.12853
6	5	5.5	27.5	11054	0.052638	0.13665

Fig. 15 DOE chart

conditions minimize the material removal rate as well as the wear of cutting tools and improve the quality of the machined-parts. Besides contributing to the existing literature on cutting tool mechanics, this research will be interesting and useful for representatives of numerous industries who may want to improve relevant machining processes.

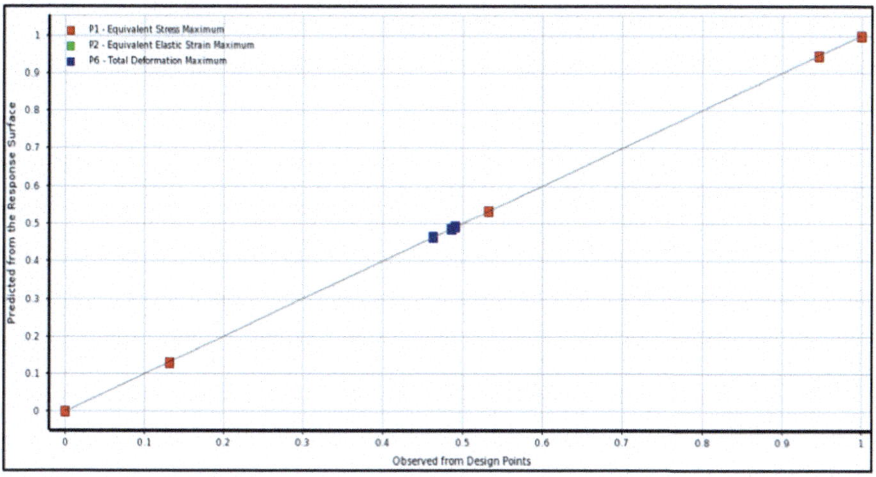

Fig. 16 Goodness of fit curve

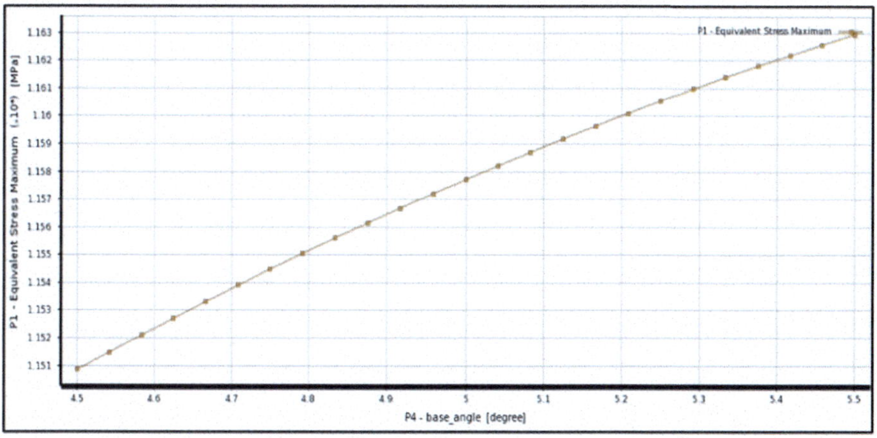

Fig. 17 Equivalent-stress versus base angle

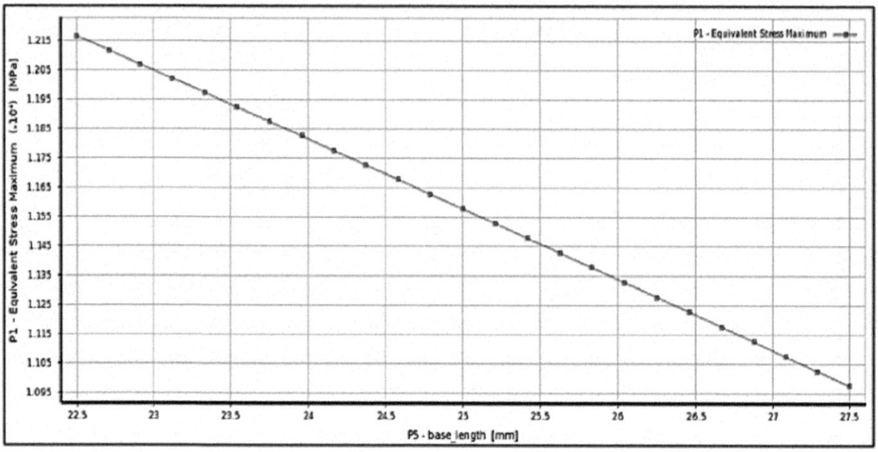

Fig. 18 Equivalent-stress versus base length

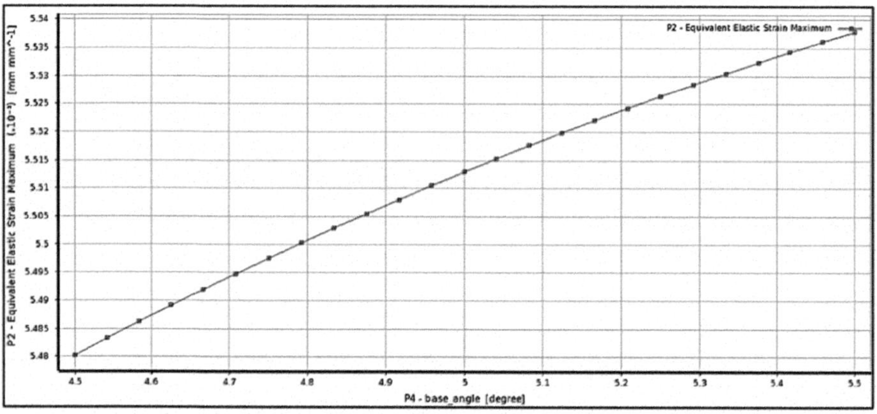

Fig. 19 Equivalent-elastic-strain versus base angle

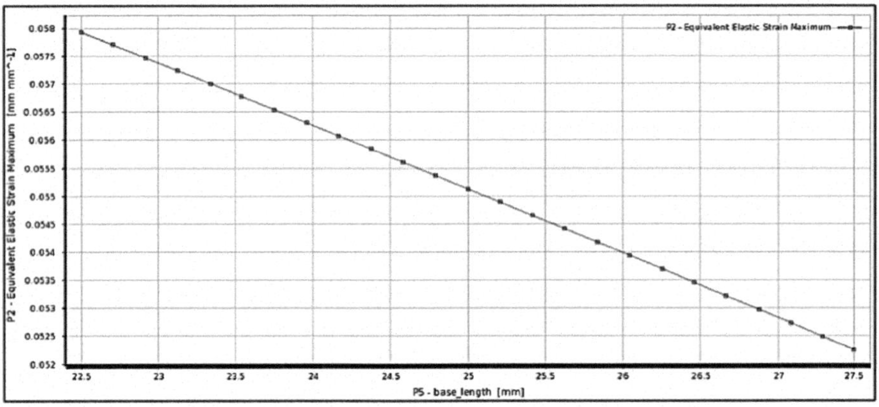

Fig. 20 Equivalent-elastic-strain versus base length

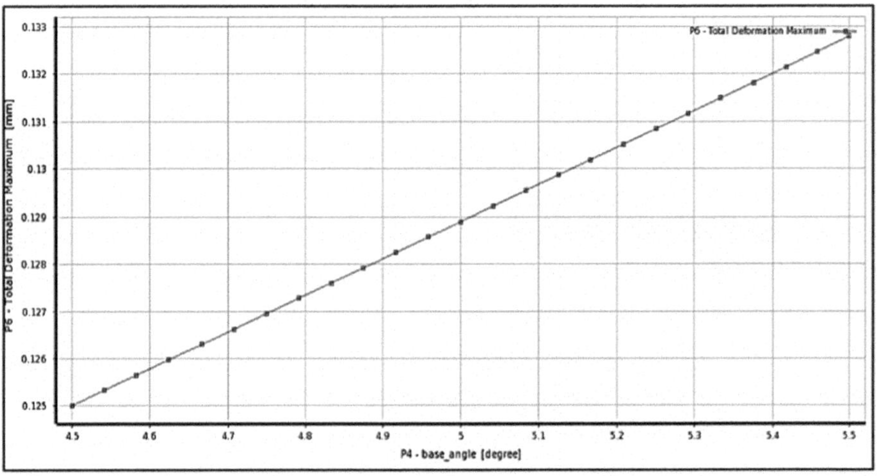

Fig. 21 Total deformation versus base angle

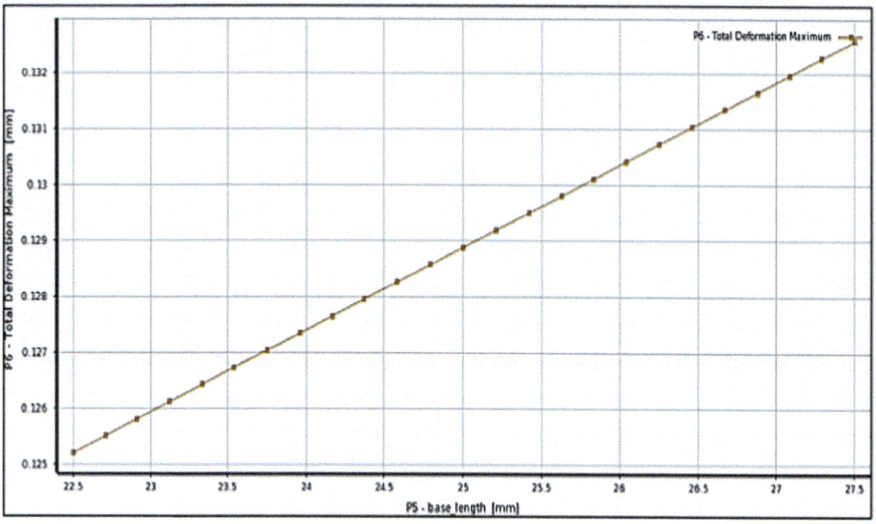

Fig. 22 Total deformation versus base length

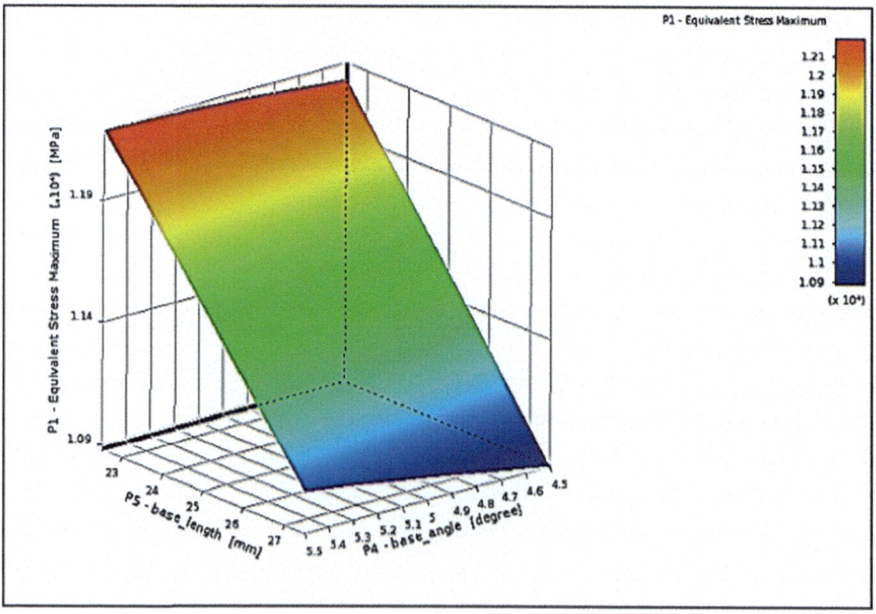

Fig. 23 3D RSM plot of equivalent stress

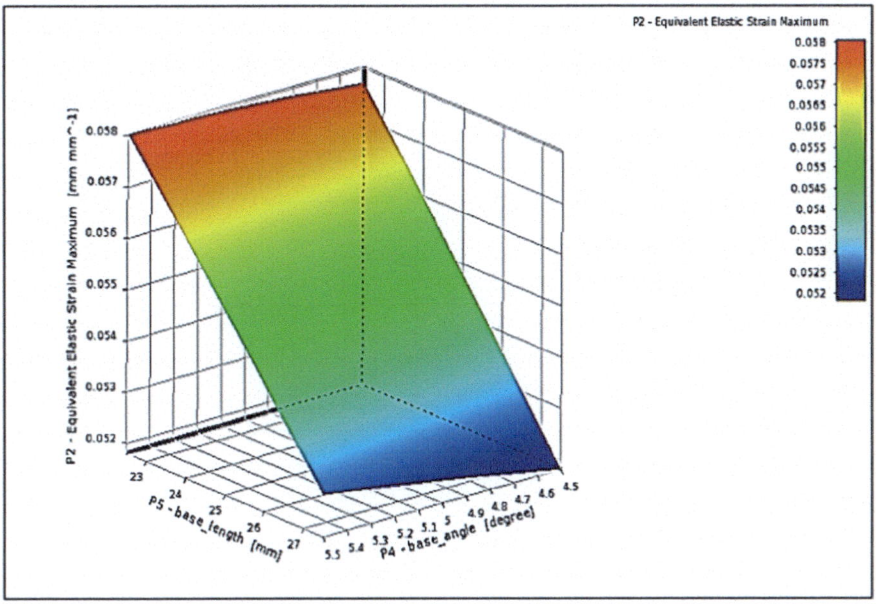

Fig. 24 3D RSM plot of equivalent elastic strain

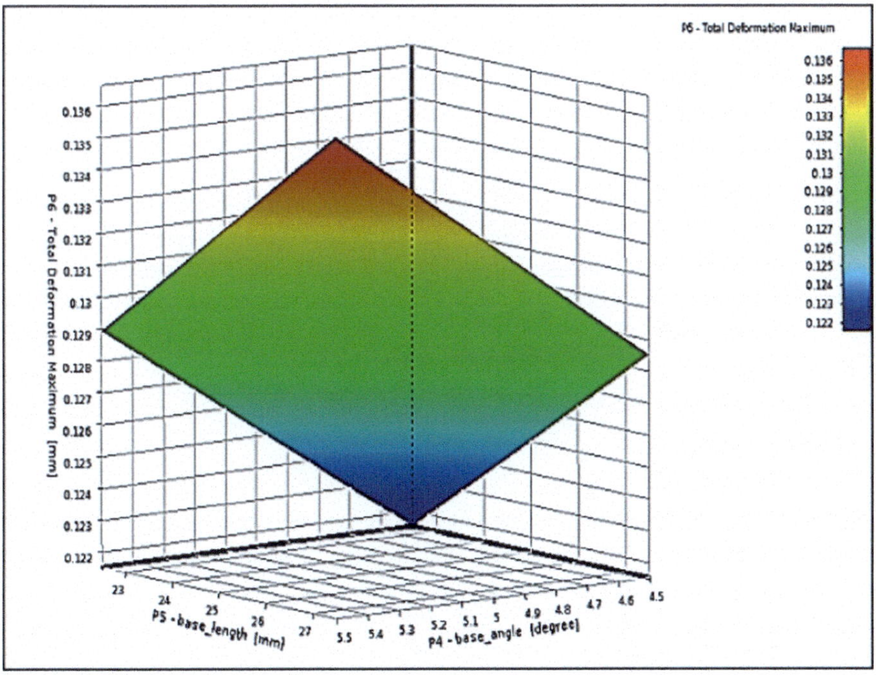

Fig. 25 3D RSM plot of total deformation

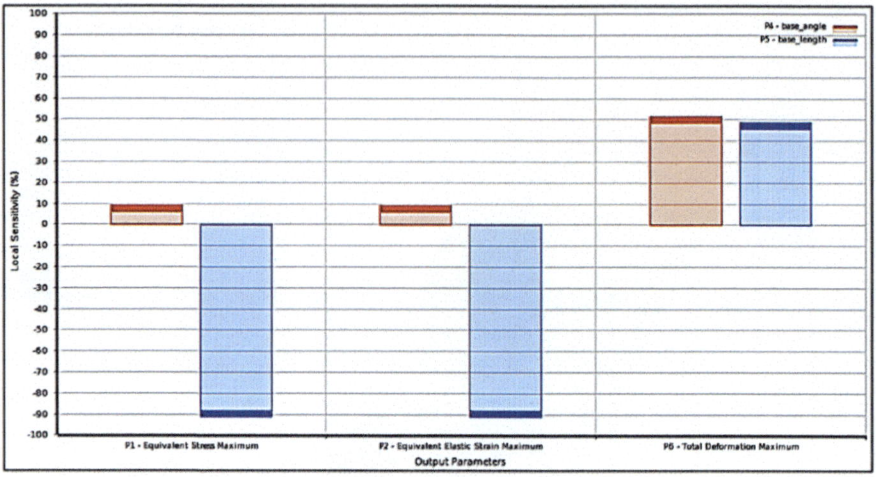

Fig. 26 Sensitivity plot

4 Conclusion

This study analyzed the design parameters of cutting tools for the improvement of structure and performance characteristics through FEA and RSM. The research can be considered successful in proving that the improvement of the base angle and base length of the cutting tool can lead to a much-needed increase and positive change in the tool's equivalent stress, strain, and deformation. The results give a highly affirmative nod to the hypothesis that time for optimization of systematic parameters of the cutting tool can bring about improved structural features along with improved machining aptitude. The study also answered the following key research questions: The variations of texture geometry significantly affected the tools' performance; the configuration to minimize stress concentrations and maximize wear resistance was determined; the effect of different machining conditions interacting with the textured surface was evaluated.

5 Future Scope

Future studies can consider the investigation of more design factors and the relationships between the factors This work can also consider proving the design parameters from the optimization process through experimentation. In conclusion, this research provides a framework for coming up with improved designs of the cutting tools and formulation of better strategies for optimizing the cutting tools that are of paramount importance when it comes to improving the machining processes and eventually reducing the cost of manufacturing products.

References

Ghule GS, Sanap S, Chinchanikar S, Cep R, Kumar A, Bhave SY, Kumar R, Altarazi F (2024) Investigation of conventional and ultrasonic vibration-assisted turning of hardened steel using a coated carbide tool. Front Mech Eng 10:1391315. https://doi.org/10.3389/fmech.2024.1391315

Gangwar S, Mondal SC, Kumar A et al (2024) Performance analysis and optimization of machining parameters using coated tungsten carbide cutting tool developed by novel S3P coating method. Int J Interact Des Manuf. https://doi.org/10.1007/s12008-024-01852-9

Kumar A, Kumar P, Sharma N, Srivastava AK (eds) (2024) 3D Printing technologies: digital manufacturing, artificial intelligence, Industry 4.0. Walter de Gruyter GmbH & Co KG. https://doi.org/10.1515/9783111215112

Wu H, Zhang X, Zhu L, Ren M, Rahman M (2024) Parallel tool servo turning of microstructured surfaces. CIRP Ann. https://doi.org/10.1016/j.cirp.2024.04.092

Palaniappan T, Subramaniam P (2024) Investigation in optimization of process parameters in turning of mild steel using response surface methodology and modified deep neural network. Mater Today Commun 38:108425. https://doi.org/10.1016/j.mtcomm.2024.108425

Sasi R, Kanmani Subbu S, Palani IA (2017) Performance of laser surface textured high speed steel cutting tool in machining of Al7075-T6 aerospace alloy. Surf Coat Technol 313:337–346. https://doi.org/10.1016/j.surfcoat.2017.01.118

Machado AR, da Silva LRR, de Souza FCR, Davis R, Pereira LC, Sales WF, de Rossi W, Ezugwu EO (2021) State of the art of tool texturing in machining. J Mater Process Technol 293:117096. https://doi.org/10.1016/j.jmatprotec.2021.117096

Sugihara T, Enomoto T (2017) Performance of cutting tools with dimple textured surfaces: a comparative study of different texture patterns. Precis Eng 49:52–60. https://doi.org/10.1016/j.precisioneng.2017.01.009

Ahmed YS, Paiva JM, Arif AFM, Amorim FL, Torres RD, Veldhuis SC (2020) The effect of laser micro-scale textured tools on the tool-chip interface performance and surface integrity during austenitic stainless-steel turning. Appl Surf Sci 510:145455. https://doi.org/10.1016/j.apsusc.2020.145455

Siju AS, Gajrani KK, Joshi SS (2021) Dual textured carbide tools for dry machining of titanium alloys. Int J Refract Metal Hard Mater 94:105403. https://doi.org/10.1016/j.ijrmhm.2020.105403

Hoier P, Klement U, Tamil Alagan N, Beno T, Wretland A (2017) Flank wear characteristics of WC-Co tools when turning Alloy 718 with high-pressure coolant supply. J Manuf Process 30:116–123. https://doi.org/10.1016/j.jmapro.2017.09.017

Kawasegi N, Sugimori H, Morimoto H, Morita N, Hori I (2009) Development of cutting tools with microscale and nanoscale textures to improve frictional behavior. Precis Eng 33:248–254. https://doi.org/10.1016/j.precisioneng.2008.07.005

Su Y, Li Z, Li L, Wang J, Gao H, Wang G (2017) Cutting performance of micro-textured polycrystalline diamond tool in dry cutting. J Manuf Process 27:1–7. https://doi.org/10.1016/j.jmapro.2017.03.013

Arulkirubakaran D, Senthilkumar V, Chilamwar VL, Senthil P (2019) Performance of surface textured tools during machining of Al-Cu/TiB_2 composite. Measurement 137:636–646. https://doi.org/10.1016/j.measurement.2019.02.013

Agarwal A, Mthembu L (2023) FE structural analysis and experimental investigation of HMV Chassis. Presented at the (conference). https://doi.org/10.1007/978-981-19-6945-4_70

Letsatsi MT, Agarwal A (2022). Study the effects of dimensional parameter using free vibrational modal analysis of composite laminate. Presented at the (conference). https://doi.org/10.1007/978-981-19-0244-4_83

Agarwal A, Letsatsi MT, Pitso I (2022) Response surface optimization of heat sink used in electronic cooling applications. Presented at the (conference). https://doi.org/10.1007/978-981-19-0244-4_13

Molwane OB, Agarwal A, Marumo R (2020) Industrial computational analysis of aerodynamic characteristics of delta-shaped aircraft. In: Kumar PA, Dirgantara T, Krishna PV (eds) Advances in lightweight materials and structures. Springer, Singapore, pp 761–770. https://doi.org/10.1007/978-981-15-7827-4_77

Agarwal A, Cavicchioli Batista R, Tashi (2024) Crashworthiness evaluation of electric vehicle battery packs using honeycomb structures and explicit dynamic analysis. E3S Web Conf 519:04010. https://doi.org/10.1051/e3sconf/202451904010

Singh H, Al Mangour B (eds) (2023) Handbook of smart manufacturing: forecasting the future of Industry 4.0. CRC Press, Boca Raton. https://doi.org/10.1201/9781003333760

Soni R (2019) Explicit dynamic analysis of single point cutting tool using ANSYS. Int J Res Appl Sci Eng Technol 7:478–488. https://doi.org/10.22214/ijraset.2019.11078

Exploring the Challenges of Integrating Lean Green Practices in Industry 4.0 Manufacturing Frameworks: An Empirical Study

Rajesh Kumar, Rajender Kumar, and Ashwini Kumar

Abstract In the era of Industry 4.0, the Lean Green approach stands as a beacon of environmental sustainability and waste reduction within manufacturing processes. This study delves into its implementation dynamics within the Indian manufacturing landscape. Employing a comprehensive methodology, a widespread questionnaire survey was conducted across companies to gauge and rank barriers hindering seamless integration. The analytical framework, structured around a three-tier hierarchy diagram, leveraged the Analytical Hierarchy Process (AHP), a renowned Multi-Criteria Decision Making (MCDM) technique. Through this systematic approach, insights were gained into pivotal barriers, with employee motivation emerging as the linchpin for driving Lean Green initiatives. Lack of motivation surfaced as the most salient barrier, significantly impacting operational efficacy. These findings underscore the critical importance of addressing motivational factors to propel sustainable manufacturing practices amidst Industry 4.0 advancements.

Keywords Lean Green concept · Analytical hierarchy process (Multi criteria decision making technique) · Manufacturing

1 Introduction

The genesis of the Lean Green paradigm can be traced back to the 1990s, marking a pivotal moment in the evolution of sustainable manufacturing methodologies within the framework of Industry 4.0-driven manufacturing technologies (Alsadi et al.

R. Kumar (✉) · R. Kumar · A. Kumar (✉)
Department of Mechanical Engineering, School of Engineering and Technology, Manav Rachna International Institute of Research and Studies, Faridabad, Haryana 121003, India
e-mail: rajeshybm14@gmail.com

A. Kumar
e-mail: ashwinikumar.set@mriu.edu.in

R. Kumar
e-mail: rajender.fet@mriu.edu.in

A. Kumar et al. (eds.), *Industry 4.0 Driven Manufacturing Technologies*, Springer Series in Advanced Manufacturing, https://doi.org/10.1007/978-3-031-68271-1_12

2023). This holistic approach, as elucidated by represents a systematic, economically motivated strategy aimed at orchestrating waste elimination across the entire spectrum of material and product lifecycle processes, encompassing manufacturing, design, and disposal (Badri et al. 1995).At the heart of the Lean Green framework lie initiatives such as green production planning and the integration of cutting-edge production technologies, which are fundamental to contemporary manufacturing systems (Batista et al. 2024). These initiatives underscore a commitment to resource optimization and energy efficiency, thereby fostering environmental sustainability while concurrently bolstering economic resilience (Burande et al. 2024).

Furthermore, the adoption of Lean Green strategies not only addresses safety concerns but also mitigates health risks inherent in industrial operations, as underscored by Campos and Vazquez-Brust (2016). A fundamental tenet of Lean Green is the relentless pursuit of waste reduction, targeting environmental as well as process-related inefficiencies. Through the prioritization of continuous improvement initiatives, as advocated by Lean Green endeavors to optimize both temporal and financial metrics, thereby cultivating a culture of efficiency and sustainability within manufacturing ecosystems (Carvalho et al. 2011).

The symbiotic relationship between lean principles and green practices underscores their shared objective of waste minimization and environmental stewardship. Deif (2011) emphasize the interconnectedness of lean and green methodologies in addressing production waste and mitigating the adverse impacts of pollution on the environment, thereby fostering a more sustainable and ecologically balanced industrial landscape. Within the domain of production management, the fusion of diverse paradigms, including lean and green principles, serves to augment operational efficiency and fortify the overarching quality assurance framework (Ding et al. 2023). Advocates for the synergistic alignment of lean and green practices, contending that such convergence not only streamlines production processes but also enhances the value proposition of the quality management system, thereby nurturing a culture of excellence and sustainability. The intricate interplay between lean and green principles is visually depicted in Fig. 1, offering a comprehensive portrayal of the multifaceted Lean-Green concept and its ramifications for sustainable manufacturing practices within the context of Industry 4.0 (Dolci et al. 2024).

The implementation of a lean green approach plays a pivotal role in mitigating environmental effects and minimizing waste, thereby fostering a paradigm shift towards enhanced productivity. By adhering to the standards of green manufacturing, companies can harness the power of green innovation to not only reduce costs but also gain a competitive edge in the market (Handfield et al. 1997). The ethos of the lean green approach extends beyond mere efficiency gains; it encompasses a holistic approach aimed at delivering superior value by elevating product quality and refining services. They studied on industry 4.0 combined with lean and agile manufacturing approach a frame work proposed to linking their relationship and find that cost reduced, increase the productivity and increase the overall performance (Gallo et al. 2021).

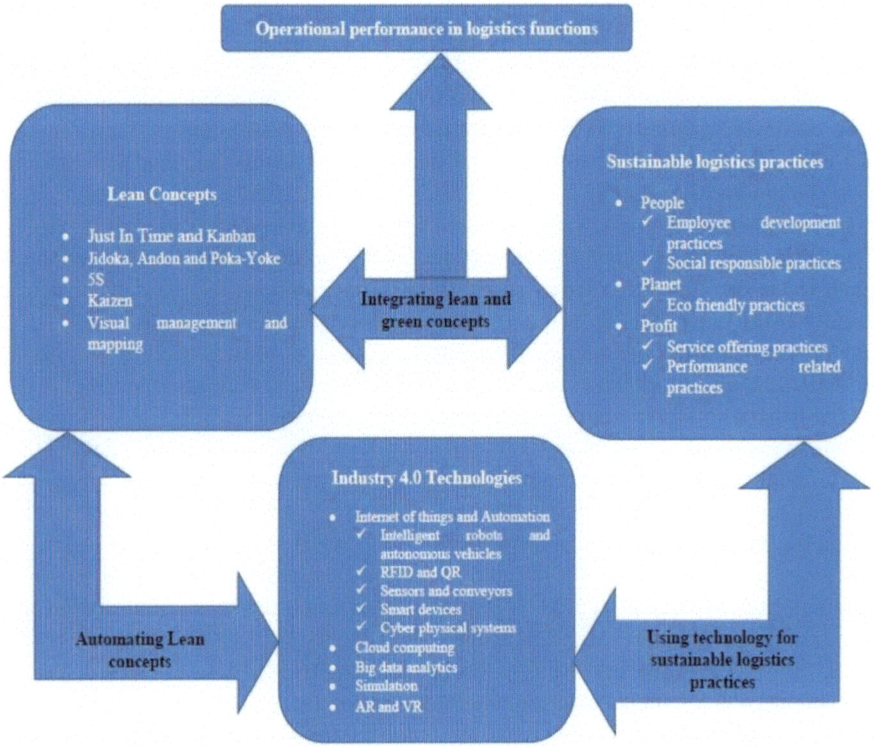

Fig. 1 Lean-Green concept in detail

Moreover, it targets the reduction of both manufacturing and environmental wastes, thereby aligning with sustainable business practices.

The multifaceted benefits of the lean green approach are underscored by its positive impact on various performance indicators, including the cost-effectiveness and quality of manufactured products. Through meticulous implementation, it facilitates the generation of tangible value while fortifying the operational prowess of manufacturing entities this study delves into the realm of manufacturing companies, meticulously scrutinizing the barriers impeding the seamless integration of the lean green approach (Gangwar et al. 2024). By evaluating the justification behind these barriers and identifying the pivotal obstacles hindering the systematic adoption of this concept, the research seeks to illuminate pathways towards sustained performance enhancement and environmental stewardship.

2 Literature Review

They studied on industry 4.0 model used with various technologies like cloud, IoT, AI to use in manufacturing company for improving the sustainability. They studied on Combined used of lean approach 4.0 and industry 4.0 to find that benefits and Gaps (Gupta and Jain 2013).

Posited an intricate framework delineating the six fundamental attributes pivotal to comprehensively assessing the green quotient of any product. These attributes encapsulate the entire lifecycle of a product, from its inception to its eventual disposal, encompassing aspects such as manufacturing processes, developmental endeavors, sales and distribution mechanisms, packaging considerations, recycling initiatives, and maintenance protocols (Jovane et al. 2003). The crux of their argument lies in the assertion that a holistic evaluation of a product's environmental impact necessitates an exhaustive analysis across these six dimensions. Moreover, their research delved into the formulation of a robust green product evaluation index system, which serves as a foundational tool for assessing the eco-friendliness of various design projects. This evaluation system, meticulously crafted by amalgamates 37 distinct performance indicators, each meticulously calibrated to capture nuanced aspects of environmental sustainability (Kaswan et al. 2023). By integrating this index system into the evaluative framework, stakeholders can effectively gauge the ecological footprint of their endeavors and strategize accordingly. Building upon this foundation, expanded upon the concept of green manufacturing, elucidating its multifaceted dimensions (Kumar et al. 2023).They studied on 3D technology with additive manufacturing is used to removing the extra material from the final product and convert into desire shape of the product with accurate size and dimensions. 3D is used for complex shape and various type of product Design. 3D technology very good utilizing for manufacturing company. It is used in artificial intelligence (AI) and Industry 4.0 both is better for manufacturing company. It used in very beneficial in aerospace, automotive, life science, and defense, which will have the biggest impact.

Their conceptualization encompasses not only the production processes themselves but also extends to encompass usage patterns, product recovery mechanisms, packaging considerations, transportation logistics, and waste management strategies. By adopting a comprehensive outlook that encapsulates the entire value chain, underscore the importance of adopting a holistic approach towards green manufacturing. Furthermore, introduced a paradigm-shifting perspective by emphasizing the pivotal role of stakeholder engagement and designer integration in fostering environmentally conscious design practices (Kumar et al. 2021). Their framework posits that by fostering synergy between stakeholders and designers, organizations can cultivate an ethos of environmental stewardship that permeates every facet of the design process. This integration engenders a hierarchical framework wherein environmental considerations are seamlessly interwoven into the fabric of product design, thereby facilitating the emergence of more sustainable solutions.

Additionally, elucidated the concept of lean green manufacturing, advocating for the adoption of ecosystem-centric design methodologies. Central to their thesis is the notion that by embracing lean principles and incorporating environmental considerations into the very fabric of organizational strategy, firms can foster a symbiotic relationship with their ecological surroundings (Kumar et al. 2024). By delineating internal environmental strategies and identifying key drivers within the external environment, provide a roadmap for organizations seeking to navigate the complex terrain of sustainable manufacturing. Questionnaire survey conducted of 167 manufacturing company and they design frame work of lean and green combined approach is beneficial for sustainability and beneficial for environment, increase the organizational performance. They promote the industry for using combined used of industry 4.0 with economy circular (Kumar and Kumar 2023).

Figure 2, as depicted in their work, serves as a visual taxonomy delineating the various dimensions of the green manufacturing approach. They studied on used of industry technology 4.0 with lean tool is beneficial for an organization. Combined used of lean and industry 14.0 improve the production and quality of a manufacturing company (Madu et al. 2002).

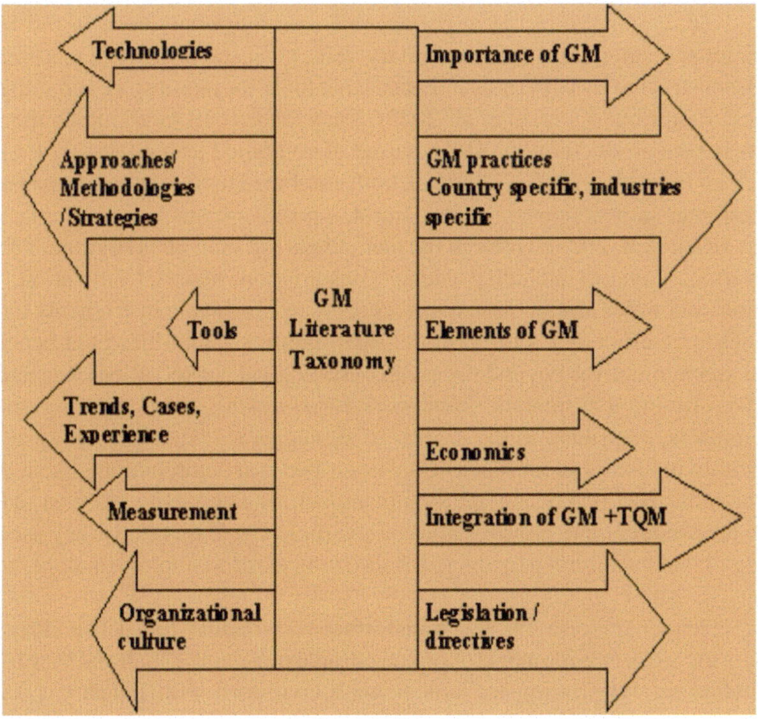

Fig. 2 Taxonomy of Green manufacturing approach

Through this illustrative representation, readers are provided with a comprehensive overview of the interconnected components comprising the green manufacturing paradigm. From production processes to logistical considerations, waste management strategies to stakeholder engagement initiatives, Fig. 2 serves as a roadmap for organizations seeking to embark on the journey towards environmental sustainability. They studied on to make a conceptual framework of ancombined link of GLSS and industry 4.0 technology to increase the performance of a manufacturing company (Martín-Gómez et al. 2024).

Delve into the intricate dynamics of green manufacturing and sustainable development within the context of business models (Mazur and GołaśH. 2011). Their comprehensive analysis underscores the pivotal role of environmental design, coupled with the integration of cutting-edge manufacturing technologies, in fostering sustainable business practices. By elucidating the symbiotic relationship between environmental consciousness and industrial advancement, highlight the potential for synergistic growth wherein ecological concerns and economic imperatives converge (Narula et al. 2023). Building upon this foundation, assert that the full realization of the lean and green concept hinges upon a profound understanding of the transformative journey required to recalibrate traditional supply chains (Ojha 2023). Central to this paradigm shift is the imperative to navigate and mitigate the myriad barriers inherent in the adoption of lean and green methodologies. Their argument underscores the need for a nuanced comprehension of the consequences and complexities involved, emphasizing the importance of fostering confidence in the face of organizational change. In a complementary vein, posit the utility of environmental matrices constructed by experts as indispensable tools for monitoring and mitigating ecological footprints (Qureshi et al. 2023). They studied on to eliminate the waste and increase the productivity in SME to used of combined lean 4.0 technology with supply chain industry 14.0.They find that both combined used to increase the productivity, customer satisfactions, process control, a positive impact on a manufacturing company (Rani et al. 2023).Through the meticulous tracking of environmental waste across various domains including energy consumption, material utilization, waste generation, and water usage, these matrices serve as invaluable instruments for optimizing resource efficiency. Furthermore, shed light on the multifaceted benefits of lean and green initiatives beyond the realm of ecological conservation (Rao and Holt 2005). By offering a systematic framework for assessing and addressing environmental impacts, contribute to the arsenal of strategies aimed at promoting sustainability within industrial ecosystems. They described that combination of lean supply chain and industry 4.0 is beneficial for a manufacturing company like cost reduced, increase the production to improve the overall efficiency. In this study they used ISM modeling to find the interdependencies. A direction made to implementation the both approach (Rossini et al. 2024).

Their research underscores the transformative potential of such endeavors in enhancing the work environment for organizational personnel (Shah and Ward 2003). They studied on 4D printing technology with combined with Industry 4.0 in the context of Indian manufacturing. They described on challenge and beneficial for a manufacturing company (Sharma et al. 2024). After using 4d printing technology with 4.0 industries they found that good impact of environment, social, increase the productivity of the manufacturing company (Singh and Rahman 2021).

By streamlining operations, reducing hazardous exposures, and fostering safer workplace conditions, lean and green practices not only mitigate environmental risks but also bolster employee well-being and productivity (Tadesse et al. 2024). They findings underscore the holistic nature of sustainability initiatives, wherein ecological stewardship intertwines with socio-economic considerations to engender lasting organizational prosperity (Taghaboni-Dutta et al. 2010). Figure 3 serves as a visual encapsulation of the lean and green performance indicators elucidated throughout the discourse. Through graphical representation, it offers a succinct yet comprehensive overview of the key metrics and benchmarks essential for gauging the efficacy and impact of lean and green endeavors (Tege and Kumar 2024). By providing a visual roadmap for stakeholders, Fig. 3 facilitates informed decision-making and strategic planning, underscoring the importance of data-driven approaches in fostering sustainable business practices (Worley and Doolen 2006).They studied on combined effect of industry 4.0 and industry 5.0 technology to increases the social and environment sustainability of a manufacturing company (Yadav et al. 2023). A frame work proposed industry 5.0 to enable the new technology and human skill to used to increase the performance of a manufacturing company (Zailani et al. 2012).

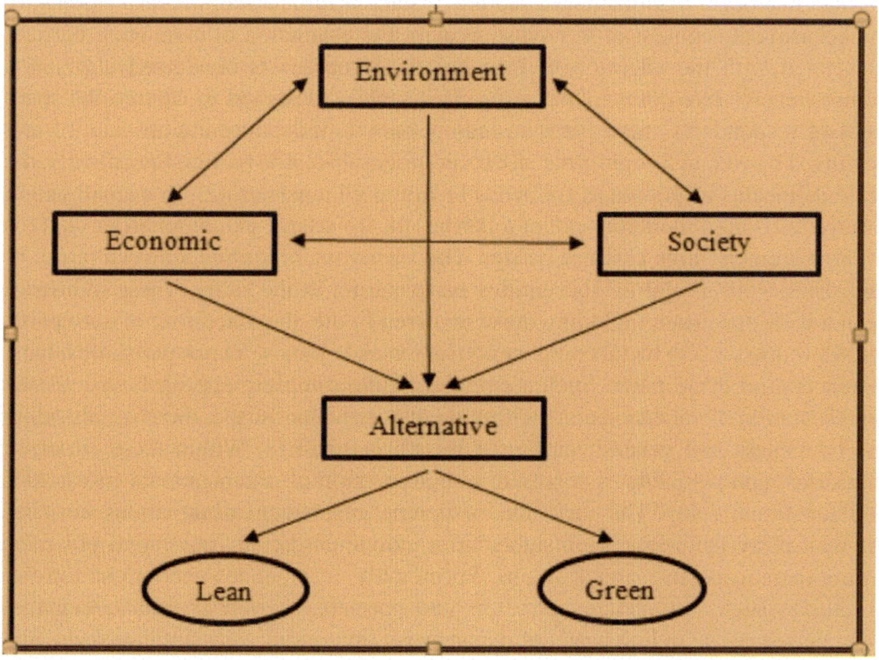

Fig. 3 Lean–Green performance indicators

They studied on industry 4.0 (14.0) connected with digital accelerators to increase the manufacturing growth. They used analysis model DEMATEL (Decision Making Triàl and Evaluation Laboratory) combination of 18 factors and find together approach. They find the challenge and barrier for used in small industry (Zhou et al. August 2008).

3 Research Framework and Methodology

A comprehensive questionnaire survey was conducted, employing a convenient sampling approach due to its practicality and accessibility. The survey tapped into the extensive database compiled by the Directorate of Indian Industry and the Confederation of Industry, ensuring a robust representation of industrial sectors. Additionally, the Northern Indian Industrial Directory was leveraged to enrich the survey dataset with diverse inputs. Furthermore, a snowball sampling technique was implemented, where subsequent industries were identified through referrals from preceding ones, thereby broadening the scope and depth of the research. For a visual depiction of the methodology employed in this study, please refer to Fig. 4.The questionnaire employed for this study comprises two distinct sections. The primary section encompasses crucial details pertaining to the company under investigation, the specific product line being manufactured, and the contact information of the respondents. In the subsequent segment of the questionnaire, the evaluation of significant barriers associated with the adoption of lean green approaches is conducted utilizing a comprehensive five-point Likert scale. This scale is designed to capture the spectrum of respondents' perceptions, ranging from 1, indicating the absence of any perceived barrier, to 5, signifying an exceedingly substantial barrier. Specifically, the scale delineates responses as follows: (1) Not at all a barrier, (2) To a small extent a barrier, (3) To a moderate extent a barrier, (4) To a large extent a barrier, and (5) To an extremely large extent a barrier. The survey encompasses a diverse array of industries, with a total of 160 entities participating in the study. These industries span a wide spectrum, including those involved in the manufacturing of auto parts, multi-products, sheet metal components, billets and blooms, tractor parts, rods, bars, fasteners, and cycle parts. Such a comprehensive sampling approach ensures the representation of various sectors within the industrial landscape, thereby enhancing the robustness and generalizability of the study findings. Within these surveyed industries, and respondents represent a cross-section of organizational hierarchies and functional roles. The inclusion of diverse respondent designations enriches the dataset by incorporating insights from individuals occupying varied positions within their respective organizations. Specifically, respondents encompass top-tier executives such as managing directors and partners, as well as middle management personnel including heads of departments, managers, assistant managers, and a cadre of technical professionals such as senior engineers and engineers spanning different departments. By encompassing a wide array of industries and incorporating insights from individuals across various organizational roles, the questionnaire aims

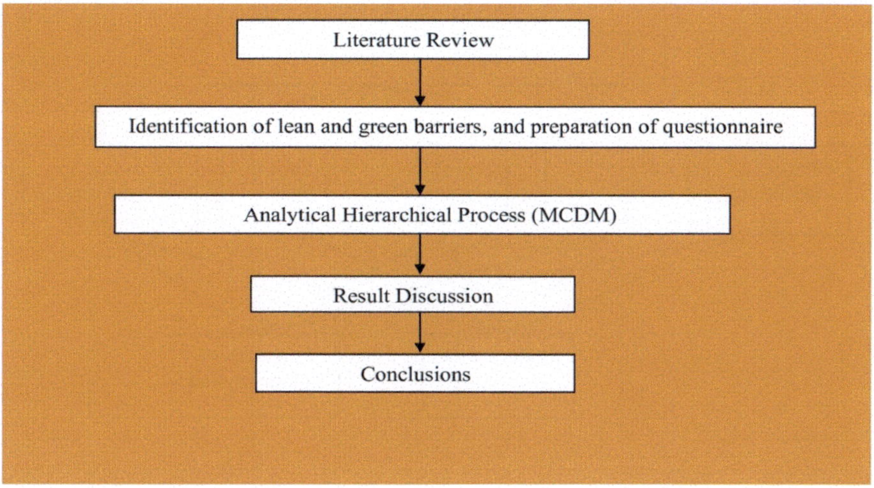

Fig. 4 Research methodology

to provide a comprehensive understanding of the barriers encountered in the adoption of lean green approaches within the industrial landscape. This holistic approach ensures that the findings derived from the study are not only robust but also reflective of the multifaceted challenges faced by organizations striving to embrace sustainable and efficient operational practices.

4 Results and Discussion

4.1 Analytical Hierarchy Process (MCDM Technique)

- Step 1: Preparation of Model for AHP application

To substantiate the significance of various lean green (LG) barriers in influencing performance interruptions, it becomes imperative to evaluate and rank these barriers while also quantifying their respective contributions. In this pursuit, Analytic Hierarchy Process (AHP) methodology emerges as a valuable tool. At the first level of hierarchy, the focus lies on elucidating the roles played by LG barriers. This involves understanding their individual impacts and how they collectively shape performance outcomes within manufacturing systems. Moving to the second level, the emphasis shifts towards delineating the practical application of these barriers within the intricate processes of current manufacturing systems. Here, the nuanced ways in which these barriers manifest and interact with operational workflows come into play, impacting the occurrence and mitigation of performance interruptions.

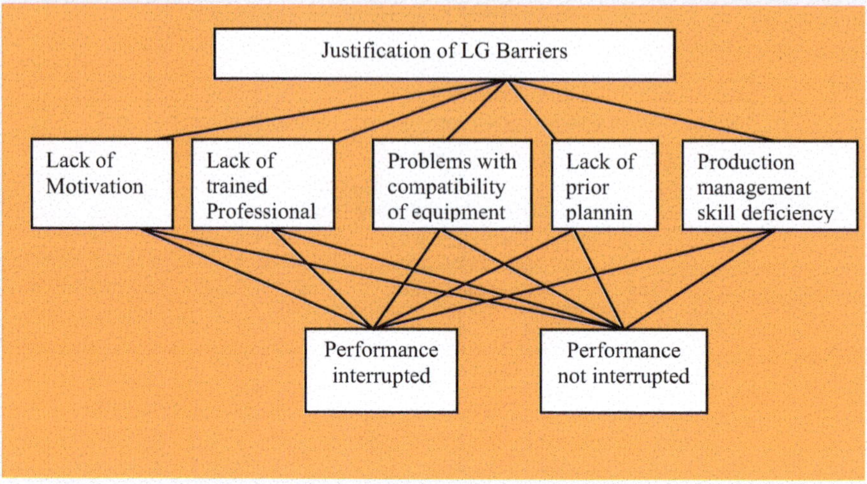

Fig. 5 Model for justification of LG practices

Finally, at the third level, the discourse expands to encompass the diverse performance outcomes engendered by the interplay of LG barriers. These outcomes span a spectrum from interrupted performance to the absence of improvement, shedding light on the multifaceted nature of performance dynamics within manufacturing contexts. Through this comprehensive hierarchical approach, AHP enables a nuanced understanding of how LG barriers influence performance, thereby informing targeted strategies for improvement and optimization. Figure 5 show the model prepared for the application for AHP.

- Step 2: Degree of preference

The assessment of the significance of different barriers has been conducted utilizing a comprehensive 9-point scale, ranging from low to high. When the barrier on the right side exhibits a greater contribution compared to the left, its importance is retained as is. However, if the barrier on the left side is deemed more significant than the right, it is represented through the reciprocal of the scale. Each barrier has been evaluated according to respondent perceptions, employing a multi-criteria decision-making framework. Specifically, the Analytical Hierarchy Process (AHP) has been employed on a substantial sample size to validate and rank these barriers systematically. This methodological approach ensures a thorough and justified assessment of the barriers' relative importance.

- Step 3: Pair-wise comparison of different sub-objectives

The importance of *i*th sub-objective is compared with *j*th sub-objectives is calculated. The Pair-wise comparison matrix for the sub-objectives is shown in Table 1.

Table 1 Pair –Wise comparison of different sub-objectives

Barriers	1	2	3	4	5
Lack of motivation **(1)**	1	10.25	9	8	6.66
Lack of trained professionals **(2)**	10.25	1	4.83	3.25	7.5
Problems with compatibility of equipment **(3)**	9	4.83	1	3.91	10
Lack of prior planning **(4)**	8	3.25	3.91	1	2
Production management skill deficiency **(5)**	6.66	7.5	10	2	1
Total	34.91	26.83	16.77	18.16	27.16

Thus, the approximate priority weight (W_1, W_2, W_j) for each attribute is obtained as shown in Table 2.

$$W_j = 1/n \times \sum_{i=1}^{n} a_{ij} \tag{1}$$

The values of consistency test are given in Table 3.

- Step 4: Priority weights for alternatives with respect to attribute

The analysis of how LG barriers affect performance interruption delves into the evaluation of various alternatives based on specific attributes. Table 4 provides a comprehensive overview of the decision-making process, illustrating the percentage contribution of these attributes towards the final determination. Through this methodical examination, the significance of LG barriers in influencing performance interruptions becomes clearer, shedding light on their role in shaping outcomes.

The Justification index is calculated by multiplying priority weight by attribute weight and taking summation of all attributes.

Table 2 Normalized matrix of sub-objectives

	1	2	3	4	5	Weight
1	0.029	0.382	0.537	0.441	0.245	**0.327**
2	0.294	0.037	0.288	0.179	0.276	**0.215**
3	0.258	0.18	0.059	0.215	0.368	**0.216**
4	0.229	0.121	0.233	0.055	0.074	**0.142**
5	0.191	0.279	0.596	0.11	0.037	**0.243**

Table 3 Consistency test

Maximum eigen value	C.I	R.I	C.R
9.808	0.1335	1.53	0.087

Table 4 Decision index table

		Justification	Not justification	Priority weight
1	Performance interrupted	1	3.57	0.649
	Performance not interrupted	0.315	1	0.228
2	Performance interrupted	1	3.69	0.778
	Performance not interrupted	0.284	1	0.217
3	Performance interrupted	1	2.47	0.695
	Performance not interrupted	0.47	1	0.303
4	Performance interrupted	1	2.312	0.708
	Performance not interrupted	0.37	1	0.286
5	Performance interrupted	1	1.343	0.610
	Performance not interrupted	0.538	1	0.380

$$\begin{aligned}\text{Justification Index of Performance Interrupted} &= 1 - 0.232 * 0.327 \\ &+ 0.217 * 0.215 + 0.303 * 0.216 \\ &+ 0.286 * 0.142 + 0.380 * 0.243 = 0.6790 \end{aligned} \tag{2}$$

5 Conclusions and Limitations

In the realm of Industry 4.0-driven manufacturing technologies, the conclusions drawn from this study highlight the critical importance of motivation, or its absence, as the primary barrier obstructing engagement in lean green initiatives. This insight is reinforced by a series of challenges stemming from deficiencies in production management skills, issues with equipment compatibility, a shortage of trained professionals, and inadequate prior planning, all of which significantly impede the seamless adoption of lean green practices within manufacturing processes.

Interestingly, while meticulous pre-planning traditionally holds sway in orchestrating production activities, its significance appears somewhat diminished in the context of implementing lean green methodologies. This observation prompts a reconsideration of traditional managerial approaches, signaling the need for innovative strategies tailored to the dynamics of Industry 4.0 environments.

Moreover, the provision of comprehensive training programs tailored to industry professionals emerges as a linchpin in guiding organizations toward embracing the lean green paradigm. Empowering and motivating the workforce with the requisite skills and knowledge emerges as crucial for successfully integrating environmentally conscious practices into operational frameworks, highlighting the intrinsic link between human factors and organizational sustainability initiatives in the Industry 4.0 landscape.

Quantitatively, the cumulative impact of these barriers on overall performance disruption is substantial, accounting for a significant 67.90% of the observed variance. However, it is essential to acknowledge methodological limitations inherent in the study design, particularly the reliance on a single respondent per company. Future research endeavors could employ robust case study methodologies to assess the efficacy of identified barriers across diverse organizational contexts, thereby enhancing the generalizability of findings.

Furthermore, the deliberate selection of a limited number of barriers is aimed at mitigating sample fluctuations and reducing variability attributable to interdependencies among identified impediments. Yet, the intricate interplay between these barriers presents an opportunity for further exploration. Employing interpretive structural equation modeling could help delineate distinct categories such as autonomous, dependent, and independent barriers, shedding light on the nuanced relationships between these impediments. By doing so, organizations can develop more targeted interventions to strengthen their resilience against impediments to lean green adoption, fostering sustainable operational paradigms conducive to long-term ecological stewardship and competitive advantage within the Industry 4.0 landscape.

6 Future Scope

- Exploring Motivation Dynamics: Delve deeper into understanding the intricacies of motivation within the context of lean green adoption. Conducting qualitative studies or surveys to uncover specific factors that influence motivation levels among employees and management could provide actionable insights for fostering a culture of sustainability.
- Innovative Managerial Approaches: Investigate novel managerial strategies tailored to Industry 4.0 environments that go beyond traditional pre-planning methods. This could involve exploring agile management techniques or integrating artificial intelligence and data analytics to optimize lean green practices and mitigate barriers.
- Enhancing Training Programs: Develop and assess the effectiveness of comprehensive training programs designed to equip industry professionals with the necessary skills and knowledge for lean green implementation. This could involve partnerships with educational institutions or specialized training providers to ensure relevance and effectiveness.
- Case Study Methodologies: Expand the scope of research by conducting in-depth case studies across diverse organizational contexts. By examining real-world implementations of lean green initiatives, researchers can gain a deeper understanding of contextual factors and identify best practices for overcoming barriers.

- Interplay between Barriers: Further investigate the complex interrelationships between different barriers to lean green adoption. Utilizing advanced modeling techniques such as interpretive structural equation modeling can help elucidate the hierarchical structure of barriers and prioritize interventions accordingly.
- Long-term Ecological Stewardship: Explore the long-term ecological and economic impacts of lean green practices within Industry 4.0 manufacturing. This could involve longitudinal studies to track sustainability metrics and assess the sustainability performance of organizations over time.
- Competitive Advantage and Market Positioning: Analyze the link between lean green adoption and competitive advantage within the Industry 4.0 landscape. Research could focus on identifying how organizations leveraging sustainable practices gain market share, attract investment, and enhance brand reputation.

References

Alsadi J, Antony J, Mezher T, Jayaraman R, Maalouf M (2023) Lean and Industry 4.0: A bibliometric analysis, opportunities for future research directions. Qual Manage J 30(1):41–63

Badri MA, Davis D, Davis D (1995) A study of measuring the critical factors of quality management. Int J Qual Reliab Manage 12(2):36–53

Batista RC, Agarwal A, Gurung A, Kumar A, Altarazi F, Dogra N, M, Vishwanatha HM, Chiniwar DS, Agrawal A (2024) Topological and lattice-based AM optimization for improving the structural efficiency of robotic arms. Front Mech Eng 10. https://doi.org/10.3389/fmech.2024.1422539

Burande DV, Kalita K, Gupta R, Kumar A, Chohan JS, Kumar D (2024) Machine learning metamodels for thermo-mechanical analysis of friction stir welding. IJIDEM. https://doi.org/10.1007/s12008-024-01871-6

Campos LMS, Vazquez-Brust DA (2016) Lean and green synergies in supply chain management. Supply Chain Manag 21(5):627–641

Carvalho H, Duarte S, Cruz Machado V (2011) Lean, agile, resilient and Green: divergencies and synergies. Int J Lean Six Sigma 2(2):151–179

Deif AM (2011) A system model for green manufacturing. J Clean Prod 19(2):1553–1559

Ding B, Ferras Hernandez X, Agell Jane N (2023) Combining lean and agile manufacturing competitive advantages through Industry 4.0 technologies: an integrative approach. Prod Plann Control 34(5):442–458

Dolci V, Bigliardi B, Petroni A, Pini B, Filippelli S, Tagliente L (2024) Integrating Industry 4.0 and circular economy: a conceptual framework for sustainable manufacturing. Procedia Comput Sci 232:1711–1720

Handfield R, Walton S, Seegers L, Melnyk S (1997) Green value chain practices in the furniture industry. J Oper Manage 12(5):38–53

Gallo T, Cagnetti C, Silvestri C, Ruggieri A (2021) Industry 4.0 tools in lean production: a systematic literature review. Procedia Comput Sci 180:394–403

Gangwar S, Mondal SC, Kumar A, Ghadai RK (2024) Performance analysis and optimization of machining parameters using coated tungsten carbide cutting tool developed by novel S3P coating method. Int J Interact Design Manuf (IJIDeM). https://doi.org/10.1007/s12008-024-01852-9

Gupta S, Jain SK (2013) A literature review of Lean manufacturing. Int J Manage Sci Eng Manage 8(4):241–249

Jovane F, Koren Y, Boer N (2003) Present and future of flexible automation: towards new paradigms. CIRP Ann 52(2):543–547

Kaswan MS, Rathi R, Cross J, Garza-Reyes JA, Antony J, Yadav V (2023) Integrating Green Lean six sigma and Industry 4.0: a conceptual framework. J Manuf Technol Manag 34(1):87–121. https://doi.org/10.1108/JMTM-03-2022-0115

Kumar A, Singh H, Kumar P, Al Mangour B (2023) Handbook of smart manufacturing. CRC Press eBooks. https://doi.org/10.1201/9781003333760

Kumar A, Kumar V, Modgil V, Kumar A (2021) Stochastic petri nets modelling for performance assessment of a manufacturing unit. Mater Today Proc 56(2214–7853):215–1219. https://doi.org/10.1016/j.matpr.2022.01.073

Kumar A, Kumar P, Sharma N, Srivastava A (2024) 3D printing technologies: digital manufacturing, artificial intelligence, Industry 4.0. De Gruyter, Berlin, Boston. https://doi.org/10.1515/9783111215112

Kumar R, Kumar R (2023) Impact of Lean and Green strategies on manufacturing industry of India—an empirical investigation. Scope 13(3):1413–1429

Madu CN, Kuei C, Madu IE (2002) A hierarchic metric approach for integration of green issues in manufacturing: a paper recycling application. J Environ Manage 64(3):261–272

Martín-Gómez AM, Agote-Garrido A, Lama-Ruiz JR (2024) A framework for sustainable manufacturing: integrating Industry 4.0 technologies with Industry 5.0 values. Sustainability 16(4):1364. https://doi.org/10.3390/su16041364

Mazur A, Gołaś H (2011) Application of pro-quality methods and tools for production process improvement. In: Borkowski S, Kryke M (eds) Improvement of production processes. Trnava, pp 99–114

Narula S, Puppala H, Kumar A, Luthra S, Dwivedy M, Prakash S, Talwar V (2023) Are Industry 4.0 technologies enablers of lean? Evidence from manufacturing industries. Int J Lean Six Sigma 14(1):115–138

Ojha R (2023) Lean in Industry 4.0 is accelerating manufacturing excellence—a DEMATEL analysis. The TQM J 35(3):597–614

Qureshi KM, Mewada BG, Kaur S, Qureshi MRNM (2023) Assessing lean 4.0 for Industry 4.0 readiness using PLS-SEM towards sustainable manufacturing supply chain. Sustainability 15(5):3950

Rani S, Tripathi K, Kumar A (2023) Machine learning aided malware detection for secure and smart manufacturing: a comprehensive analysis of the state of the art. Int J Interact Des Manuf. ISSN: 1955-2505. https://doi.org/10.1007/s12008-023-01578-0

Rao P, Holt D (2005) Do Green supply chains lead to competitiveness and economic performance? Int J Oper Prod Manag 25(9):898–916

Rossini M, Ahmadi A, Staudacher AP (2024) Integration of Lean supply chain and Industry 4.0. Procedia Computer Science 232:1673–1682

Shah R, Ward PT (2003) Lean manufacturing: context, practice bundles, and performance. J Oper Manag 21(2):129–149

Sharma P, Singh Ghatorha K, Kang AS, Cepova L, Kumar A, Phanden RK (2024) Strategic insights in manufacturing site selection: a multi-method approach using factor rating, analytic hierarchy process, and best worst method. Front Mech Eng 10. https://doi.org/10.3389/fmech.2024.1392543

Singh AP, Rahman Z (2021) Integrating corporate sustainability and sustainable development goals: towards a multi-stakeholder framework. Cogent Bus Manage 8(1). https://doi.org/10.1080/23311975.2021.1985686

Tadesse H, Singh B, Deresso H, Lemma S, Singh GK, Srivastava AK, Dogra N, Kumar A (2024) Investigation of production bottlenecks and productivity analysis in soft drink industry: a case study of East Africa Bottling Share Company. Int J Interact Design Manuf (IJIDeM)

Taghaboni-Dutta F, Trappey AJC, Trappey CV (2010) An XML based supply chain integration hub for green product lifecycle management. Expert Syst Appl 37(11):7319–7328

Tege S, Kumar P (2024) 7 review of 4D printing and materials enabling Industry 4.0 for implementation in manufacturing: an Indian context. In: 3D printing technologies: digital manufacturing, artificial intelligence, Industry 4.0, p 143.

Worley J, Doolen TL (2006) The role of communication and management support in a lean manufacturing implementation. Manag Decis 44(2):228–245

Yadav AS, Kumar A, Yadav KK, Rathee S (2023) Optimization of an inventory model for deteriorating items with both selling price and time-sensitive demand and carbon emission under green technology investment. IJIDEM. https://doi.org/10.1007/s12008-023-01689-8

Zailani SHM, Eltyeb TK, Hsu CC, Tan KC (2012) The impact of external institutional drivers and internal strategy on environmental performance. Int J Oper Prod Manag 32(6):721–745

Zhou X, Zhang QS, Zhang M, Li X (2008) Research on evaluation and development of green product design project in manufacturing industry. In: International conference on wireless communications, networking and mobile computing, Crete, Greece, 6–8 Aug 2008, pp 1–5

Robotic Arm 3D Printing: Technological Advancements and Applications

Sourabh Anand, Gayatri, Manoj Kumar Satyarthi, Pushpendra S. Bharti, Ajay Kumar, and Sarita Rathee

Abstract Robotic Arm 3D Printing (RA3DP), which combines the flexibility of 3D printing with the accuracy of robotic arms, is a significant turning point in the history of manufacturing technology. This review offers a thorough investigation of RA3DP, an innovative robotics and additive manufacturing intersection. Highlighting RA3DP's transformative potential in manufacturing, the paper navigates through the dynamic landscape of RA3DP, unveiling technological advancements and a variety of applications. The study covers the built-in benefits of RA3DP, such as more design flexibility, less need for supports, and the ability to print at almost any angle. Sustainability and materials are important, and RA3DP embraces pellet or chip forms, bringing in affordable and ecologically friendly solutions. Leading businesses in the industry are highlighted in the paper, including KUKA, CEAD, ABB Robotics, and others. Their noteworthy projects and contributions are highlighted. As RA3DP keeps pushing the limits of additive manufacturing, this review acts as a thorough manual for the technological advancements and uses that are driving this rapidly evolving industry.

Keywords Robotic Arm 3D Printing · Large scale 3D printing · Robot manipulator · Additive manufacturing · Metal additive manufacturing · Pallet extrusion

---

S. Anand · Gayatri · M. K. Satyarthi · P. S. Bharti
University School of Information, Communication and Technology, Guru Gobind Singh Indraprastha University Sec 16C, Dwarka, New Delhi, India

A. Kumar (✉)
Department of Mechanical Engineering, School of Engineering and Technology, JECRC University, Jaipur, Rajasthan, India
e-mail: ajay.kumar30886@gmail.com

S. Rathee
Department of Electronics and Communication Engineering, School of Engineering and Technology, JECRC University, Jaipur, India

A. Kumar et al. (eds.), *Industry 4.0 Driven Manufacturing Technologies*, Springer Series in Advanced Manufacturing, https://doi.org/10.1007/978-3-031-68271-1_13

1 Introduction

Over the years, the manufacturing industry has undergone a significant transformation driven by the unwavering quest for efficiency and innovation (Ruffa 2008; Iansiti and Lakhani 2020). The innovative idea of additive manufacturing, which has completely changed how we design and make objects, is at the centre of this evolution, summarized in Table 1 (Gao et al. 2015; Ajay et al. 2023; Singh et al. 2023). Additive manufacturing has had an interesting historical journey, from its modest beginnings as a novel idea to its current status as a major player in the manufacturing arena (Gibson et al. 2021; Mittal et al. 2022). One of the most important chapters in the evolution of additive manufacturing opens with the introduction of Robotic Arm 3D Printing (RA3DP) (Praveena et al. 2022; Attaran 2017). This review seeks to provide a thorough understanding of RA3DP's transformative potential in modern manufacturing by navigating through its technological advancements, historical roots, and diverse applications.

The idea of layer-by-layer material deposition first emerged in the 1980s, which is when additive manufacturing's history began, (Beaman et al. 2020; Kumar et al. 2023;

Table 1 Overview of RA3DP's transformative potential in modern manufacturing

Elements	Description
Transformation of Manufacturing (Ruffa 2008; Iansiti and Lakhani 2020)	Driven by efficiency and innovation, the manufacturing industry has evolved significantly through additive manufacturing
Historical Development of Additive Manufacturing (Beaman et al. 2020; Kumar et al. 2023, 2024b)	Began in the 1980s with layer-by-layer material deposition; crucial technologies include selective laser sintering and stereolithography
Introduction of RA3DP (Praveena et al. 2022; Attaran 2017)	Represents a major advancement in additive manufacturing integrates a 3D printer head with a multi-axis robotic arm, overcoming fixed-axis limitations
Technological Advancements in RA3DP	Provides unparalleled design freedom and flexibility, enabling the creation of intricate and complex geometries from multiple angles
Applications of RA3DP	Utilized in various sectors such as industrial manufacturing, architecture, art, and sculpture, demonstrating its wide-ranging impact
Advantages of RA3DP	Offers enhanced design flexibility, reduces the need for supports, and allows printing at nearly any angle
Sustainability and Material Evolution	Transition from filament rolls to pellet or chip forms enhances sustainability and economic efficiency
Industry Pioneers and Contributions	Highlights significant contributions and projects from leading companies like KUKA, CEAD, and ABB Robotics

Abdulhameed et al. 2019; Laguna et al. 2021; Goyal et al. 2024a) visualized through Fig. 1. Early innovations paved the way for methods like selective laser sintering and stereolithography, which in turn led to a paradigmatic change in manufacturing practices (Praveena et al. 2022; Laguna et al. 2021; Jiménez et al. 2019; Mahmood et al. 2022). These fundamental technologies changed the face of manufacturing as they opened the door to a wide range of applications, such as customised part production and rapid prototyping (Ian Gibson 2015; Rani et al. 2023; Aggoune et al. 2024; Kumar et al. 2024; Kumar et al. 2023a). The background of additive manufacturing's past evolution sheds light on the dynamic development that ultimately resulted in the combination of robotics and 3D printing (Ian Gibson 2015; Fidan et al. 2023; Butt 2020; Parmar et al. 2022). The recent development of Robotic Arm 3D Printing represents a major turning point in the journey of additive manufacturing (Fidan et al. 2023; Praveena et al. 2022; Pires et al. 2021). By combining an extrusion-capable 3D printer head with a multi-axis robotic arm, RA3DP surpasses the limitations of fixed-axis 3D printers (Batista et al. 2024; Kumar et al. 2024a; Goyal et al. 2024b). This combination of 3D printing and robotics opens up new possibilities with unmatched design freedom and flexibility. The recent past of RA3DP is becoming more and more important as its uses spread throughout a variety of sectors, from industrial manufacturing and architecture to art and sculpture (Bhardwaj et al. 2023; Sehrawat et al. 2022a; Kumar et al. 2023b).

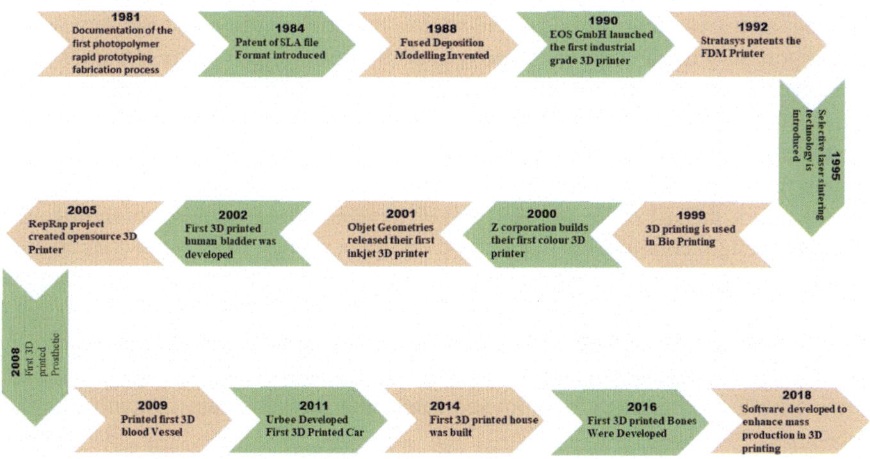

Fig. 1 Historical background of 3D printing

1.1 Working Principle and Process of Robotic Arm 3D Printing

RA3DP uses a multi-axis robotic arm with an end effector—like a laser head or extruder—as part of its working mechanism and deposit material Layer by layer, according to a pre-programmed computer design (Tang et al. 2024) initially using Computer-Aided Design (CAD) software, a digital three-dimensional model is created. The robotic arm's movements are successfully directed by a very accurate toolpath that is produced after the model is sliced into horizontal layers using slicing software. (Kumar et al. 2023a) For the best possible accuracy and quality of the finished printed product, important factors including print speed, temperature, and material flow must be optimized during the preprocessing stage. The particular material qualities and the goals of the printing process will determine which of these factors are used.

During the printing phase, the robotic arm is carefully aligned with the toolpath, making it easier to deposit material and layer the object. This particular attribute facilitates the creation of exceedingly elaborate designs and intricate geometries that are typically unachievable through traditional manufacturing methods. The incorporation of post-processing techniques is frequently required to enhance the mechanical properties and improve the surface quality of the printed object. The subsequent procedures may encompass surface finishing methodologies such as smoothing or finishing, heat treatments aimed at enhancing material rigidity, and the removal of any support structures employed throughout the printing procedure. (Kumar et al. 2023c) The RA3DP system exhibits an extensive range of materials processing capabilities, enabling the handling of various materials such as thermoplastics, composites, metals, and concrete. The exceptional versatility exhibited by RA3DP greatly expands the range of applications within diverse industries. In the aerospace and automotive sectors, the utilization of RA3DP is observed for the production of lightweight and high-strength components that exhibit complex shapes. Within the construction industry, the utilization of this technology facilitates the development of wide-ranging architectural components that possess customized designs. The capacity to manipulate a diverse range of materials and create complex structures with high precision. RA3DP as a revolutionary technology within the domain of additive manufacturing, showing fresh opportunities for innovation and enhanced efficiency in manufacturing procedures.

As we begin this thorough analysis, we will examine all of RA3DP's benefits, such as its capacity to increase design freedom and get around conventional constraints like the requirement for supports (Ishak and Larochelle 2017; Arleo et al. 2021; Luu et al. 2021). The past progression of materials in 3D printing from rolls of filament to the use of pellet or chip forms will be carefully examined in terms of how it affects sustainability and economy. In addition, we will explore the past paths of industry pioneers and cutting-edge technologies, illuminating their contributions to the domain and the noteworthy initiatives that have influenced the development of RA3DP. Exploring the historical perspectives on early applications reveals the remarkable impact of

RA3DP in architecture, large-scale prototyping, industrial manufacturing, and the creation of artistic sculptures (Sehrawat et al. 2022b; Mostafavi 2016; Pan et al. 2021; Srivastava 2023). Exploring the challenges encountered during the historical development of RA3DP and the evolution of solutions over time will offer valuable insights into the maturation of this technology. Finally, we will consider the future directions and emerging trends in RA3DP, taking into account historical contexts to predict the path of this ever-evolving field. This review takes readers on a historical journey, exploring the development of additive manufacturing and ultimately reaching the breakthrough of Robotic Arm 3D Printing. Through the integration of historical narratives, technological advancements, and diverse applications, our goal is to offer a comprehensive view of the transformative journey of RA3DP in modern manufacturing.

2 Advantages of Robotic Arm 3D Printing

The development of RA3DP brings about a significant change in additive manufacturing, providing numerous benefits that redefine the potential of modern production methods. RA3DP offers an exceptional level of design freedom, setting it apart from other options. With the integration of a multi-axis robotic arm, RA3DP breaks free from the limitations of traditional 3D printers (Sehrawat et al. 2022a; İpekçi and Ekici 2024; Anand and Satyarthi 2024). This innovative technology allows for the printing of intricate and complex geometries from a wide range of angles (Shembekar et al. 2019; Kontovourkis and Tryfonoas 2020). The flexibility of this method goes beyond the constraints of traditional printing techniques, enabling the production of structures with curves and shapes that were previously considered difficult or even impossible to achieve (Kumar et al. 2023c; Krčma and Paloušek 2022) as shown in Fig. 2a, b.

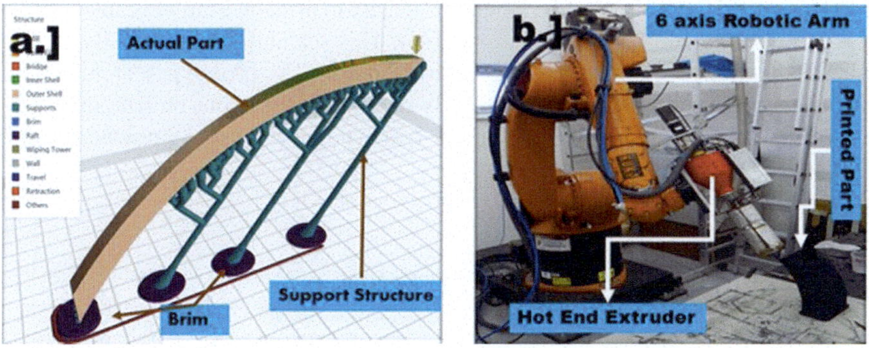

Fig. 2 **a** Support structure required for fixed axis 3D printer, **b** support Structure not required for RA3DP (Krčma and Paloušek 2022)

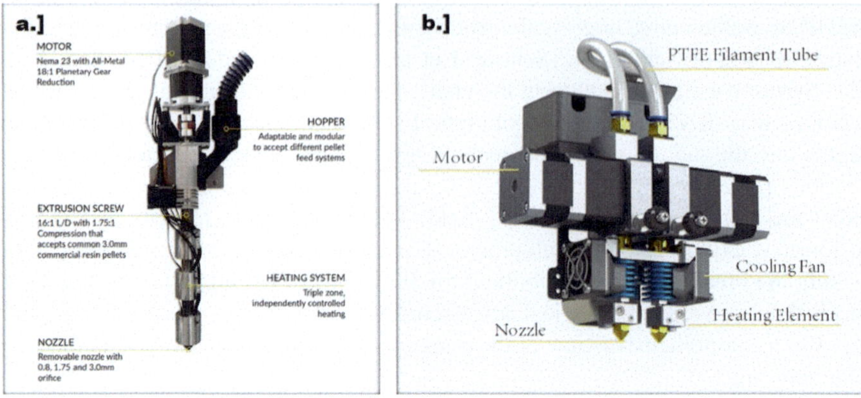

Fig. 3 **a** Pellet extrusion printing head for RA3DP (Little 2020), **b** filament extruder head for fixed axis 3D printer

Furthermore, RA3DP eliminates the requirement for conventional support structures during the printing process. This development is quite noteworthy as the lack of supports not only allows for more creative designs but also leads to substantial savings on material costs. The inherent benefits of RA3DP-printed parts make them highly valuable for large-scale projects, including architectural elements, sculptures, and industrial prototypes (Goyal et al. 2024b; Sehrawat et al. 2022b; Budig 2014; Kumar et al. 2023d). These projects require careful attention to cost-efficiency and design intricacy. RA3DP distinguishes itself in material utilization by rejecting the traditional use of filament rolls (as in Fig. 3b) and instead opting for polymer materials in pellet or chip form (Fig. 3a). This shift not only enables the use of a wider variety of materials, such as plastics, metals, clay, and concrete, but also promotes the adoption of cost-effective and sustainable practices. Utilizing polymer pellets, a common practice in industries such as injection molding, is more cost-effective compared to using traditional filament rolls, resulting in higher material throughput. In addition, RA3DP possesses the inherent capacity to print on a larger scale, with the ability to reach dimensions of 30 m or greater. This scalability is especially beneficial for industries engaged in the manufacturing of extensive prototypes, architectural components, and aerospace applications. By removing limitations on size, manufacturers can now pursue ambitious projects that were previously impractical using traditional 3D printing techniques.

3 Limitation of Robotic Arm 3D Printing

Despite it has many benefits, robotic arm 3D printing (RA3DP) has a number of limitations that might affect its efficiency and broad use. The high initial expense of acquiring and setting up robotic arms which might be unaffordable for small and medium-sized businesses is one of the main drawbacks. Furthermore, operating without professional assistance is difficult since multi-axis movement programming and control need certain knowledge and abilities. Inaccuracies in the finished output may also result from calibration problems and the physical restrictions of the robotic arm's range of motion. Furthermore, material compatibility, while wide, still presents difficulties; some materials may not be appropriate for RA3DP if they need particular conditions (such as high temperatures or exact environmental controls). Requirements for post-processing, like surface polishing and support structure removal, prolong the process and perhaps lower efficiency. Finally, the incorporation of RA3DP into current production processes might be difficult and need for major adjustments and alterations to the present systems and procedures. These restrictions emphasize the need of continuous research and development to solve these issues and improve the capabilities of RA3DP technology.

4 Leading Companies and Technologies in Robotic Arm 3D Printing

The adoption of RA3DP has experienced a significant increase, propelled by forward-thinking companies utilizing state-of-the-art technologies. The individuals mentioned below are leading figures who are significantly influencing the development of RA3DP with their innovative technological progress and pioneering contributions.

4.1 ABB Robotics

ABB Robotics is a significant contributor to industrial automation, enhancing the flexibility of RA3DP with the use of advanced robots like the IRB 7600. These robots, renowned for their accurate wrist movements, can be used with ABB's RobotStudio software, providing customized solutions for 3D printing with low production volume and a wide variety of materials (Smith 2019). ABB's dedication to additive manufacturing is apparent through its creation of the 3D Printing PowerPac and its collaborations with extruder manufacturers.

4.2 CEAD

CEAD's AM Flexbot is an innovative and groundbreaking solution for large-scale additive manufacturing. This modular system guarantees precision and a large printing table through the incorporation of Siemens' Sinumerik 840D controller. CEAD demonstrates its dedication to adaptability by ensuring compatibility with a wide range of materials, such as high-temperature plastics (https://ceadgroup.com/portfolio-items/3d-printing-with-cead-robot-extruder-in-largest-innovation-centre-in-the-netherlands/). This positions the company as a leading player in the RA3DP industry.

4.3 Kuka

KUKA's industrial robots, such as the KR Quantec and KR Cybertech Nano, demonstrate exceptional performance in the field of metal additive manufacturing. KUKA's ProLMD system is a laser metal deposition system that plays a significant role in the company's progress in hybrid metal additive manufacturing (Mascellino 2020). The shelf-mounted robots are distinguished by their efficiency and rapidity, as they optimize both space and resources.

4.4 Bloom Robotics

Bloom Robotics specializes in the field of pellet extrusion, specifically focusing on the ADE25 pellet extruder that is specifically designed to be used with ABB robotic arms (https://www.bloom-robotics.com/ade25h-extruder/). This technology, which has the ability to process a wide range of plastics and composites, improves the efficiency of using materials. The AR System Cell by Bloom Robotics guarantees efficient printing with minimal support material, making it a favoured option for applications such as custom piping solutions.

4.5 Branch Technology

Branch Technology is an innovative company that is revolutionizing 3D printing methods with its cutting-edge Cellular Fabrication (C-Fab) technology. This novel technique employs a polymer matrix infused with composite material, yielding structures that are both lightweight and resilient. Branch Technology's commitment to sustainability is apparent in its significantly reduced material consumption, which is 20 times less than that of traditional methods (Sher 2021) (Table 2).

Table 2 Mapping the landscape of robotic arm 3D printing innovations

Company	Key technologies	Noteworthy products/systems	Specializations
ABB Robotics (Smith 2019)	Flexible wrist movements High accuracy	IRB 7600 robotic arm RobotStudio 3D Printing PowerPac	Low-volume, high-mix 3D printing Collaborations with extruder manufacturers I
CEAD (https://ceadgroup.com/portfolio-items/3d-printing-with-cead-robot-extruder-in-largest-innovation-centre-in-the-netherlands/)	Modular system Sinumerik 840D controller	AM Flexbot Compatibility with high-temperature plastics	Large-scale additive manufacturing Versatility in material usage
KUKA (Mascellino 2020)	ProLMD system for metal additive manufacturing	KR Quantec, KR cybertech Nano Robots	Industrial robots for metal 3D printing Efficiency and speed
Branch Technology (Sher 2021)	Cellular Fabrication (C-Fab) technology	3D printed freeform polymer matrix	Sustainable 3D printing Lightweight and durable structures
Bloom Robotics (https://www.bloom-robotics.com/ade25h-extruder/)	ADE25 pellet extruder for ABB robotic arms	ADE25 Pellet Extruder	Pellet extrusion technology Efficient printing with minimal support material

5 Applications of RA3DP

Robotic Arm 3D Printing (RA3DP) has become a powerful and influential technology, demonstrating a wide range of uses that challenge traditional manufacturing methods. Utilizing the extensive range of motion of robotic arms, this technology excels in various large-scale projects, including Mold-making, creating prototypes on a large scale, crafting intricate artistic sculptures, producing architectural elements, designing custom furniture (Fig. 4), and even constructing metallic bridges (Fig. 5) and rockets (Fig. 6) (https://ceadgroup.com/the-use-of-robotic-arms-for-3d-printing-versus-for-assembly/; Kauppila 2023; Ranjit 2020). RA3DP revolutionizes 3D printing by offering unprecedented design flexibility, allowing the creation of intricate, curved shapes from any perspective.

RA3DP specializes in Mold-making, providing a versatile and effective method that is particularly adept at creating both large-scale prototypes as shown in Fig. 7 and intricate designs. The technology's capacity to convert intricate digital designs into tangible and intricate forms is advantageous for artistic sculptures. RA3DP has made significant progress in the domains of architectural elements and custom-made furniture. The printer's ability to accommodate larger print sizes enables the production of significant structural components with exceptional precision.

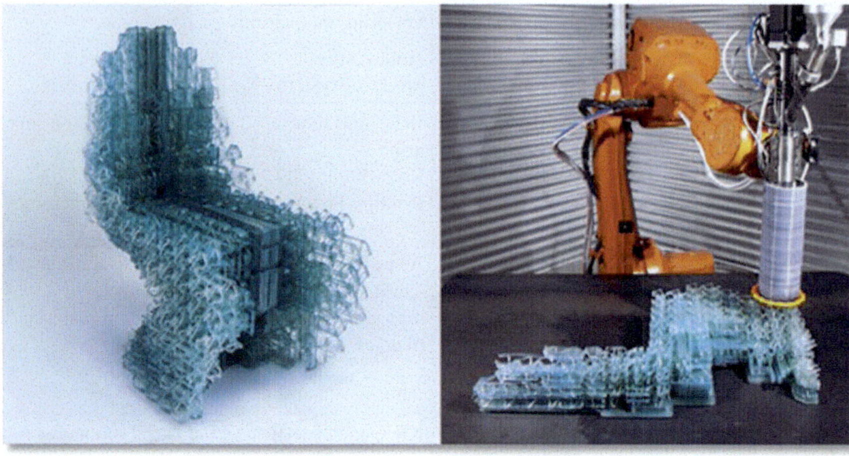

Fig. 4 RAD3P Printed Voxel Chair (Ranjit 2020)

Fig. 5 RA3DP printed bridge (Block 2018)

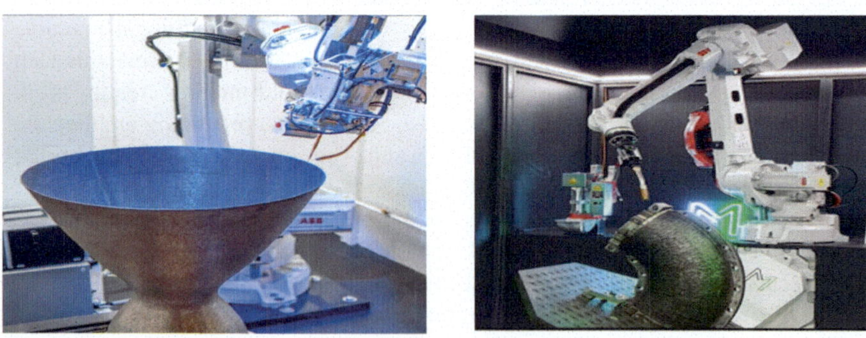

Fig. 6 RA3DP printed metallic functional part (Saunders 2023)

Fig. 7 RA3DP printed large structure (Scott 2018)

In addition, RA3DP's ability to work with various materials goes beyond the use of traditional filament rolls. It typically employs polymer materials in the form of pellets or chips, providing cost-efficient solutions. RA3DP demonstrates versatility by not being limited to plastics, but also encompassing metals, clay, and concrete, thus showcasing its adaptability across a wide range of materials.

The elimination of the requirement for supports in printed components enhances the efficiency of RA3DP, resulting in reduced material expenses and increased design flexibility. To overcome challenges associated with overhanging designs, innovative solutions entail repositioning the building platform to generate self-supporting structures. The technology's distinctive attribute of dispensing with the need for slicing software, due to the utilization of specialized 3D printing software that employs multi-axis toolpaths, simplifies the workflow, albeit necessitating precise programming instructions. Although RA3DP systems are typically designed for customization by the user, companies such as Evo3D are currently striving to facilitate their implementation by offering comprehensive packages that consist of robotic arms, extruder systems, software, as well as essential training and support. The impact of the technology goes beyond traditional 3D printing, especially in industrial manufacturing. The incorporation of RA3DP into current processes holds promise for improved efficiency and the emergence of innovative production opportunities. The ongoing development of RA3DP is expected to lead to an expansion of its practical uses in various industries, resulting in a significant and revolutionary change in the field of additive manufacturing.

6 Challenges and Future Perspectives

The field of RA3DP is rich with challenges, covering an intricate terrain that involves skilled navigation by researchers and practitioners. A major difficulty arises from the sophisticated programming required to synchronize the movements of the robotic arm and the 3D printing head. RA3DP, unlike conventional 3D printers that have fixed axes, needs precise programming to coordinate the motion of the robotic arm with the material deposition. Even the smallest miscalculation or programming error can lead to collisions between the robotic arm and the printed part, highlighting the significance of creating completely reliable programming methodologies. Another challenge arises from the customizable nature of numerous RA3DP systems. Although this feature enables customization and flexibility, it presents an impediment for potential users who might not have the necessary skills to set up and calibrate the system's components, thus restricting their ability to use it. To tackle this challenge, it is necessary to simplify the integration process, possibly by creating more accessible and ready-to-use RA3DP solutions. The wide range of materials available in RA3DP is advantageous, but it also presents difficulties. The shift from traditional filament rolls to pellet or chip forms brings about complications.

in material manipulation, encompassing challenges pertaining to feeding, melting, and extrusion. Ensuring a consistent and high-quality flow of materials is of utmost importance, particularly when working with a variety of substances, including polymers, metals, and concrete. Research in this field should prioritize the improvement of extrusion mechanisms and the creation of versatile systems that can efficiently handle a wide range of materials. Support structures pose an ongoing difficulty in 3D printing, and RA3DP is no different. Although RA3DP can minimize the requirement for supports in various scenarios, intricate geometries and overhangs still require careful considerations. Advanced software algorithms can help overcome these challenges by reorienting the build platform and optimizing support structures. This allows for more design freedom without compromising the structural integrity. RA3DP has a promising outlook for the future, as continuous research and development efforts are expected to overcome existing challenges. An area of investigation involves improving the level of automation in programming by utilizing artificial intelligence and machine learning algorithms. These technologies have the potential to gain knowledge from previous successful printing experiences, gradually enhancing the accuracy and effectiveness of RA3DP systems. Standardization and the establishment of industry-wide protocols are crucial for promoting wider acceptance. With the increasing prevalence of RA3DP, it is important to establish standards for programming languages, materials, and system components. This will enable interoperability and collaboration across various platforms. The process of standardization will enhance the accessibility of RA3DP, thereby facilitating its adoption across various industries and by a wider user base. The incorporation of sustainable practices into RA3DP processes is a major area of future emphasis. Investigating the utilization of recycled materials, maximizing energy efficiency, and reducing waste production are in line with worldwide endeavours to promote environmentally sustainable manufacturing. Researchers and industry leaders should cooperate to create RA3DP solutions that not only demonstrate exceptional performance but also comply with environmentally sustainable practices. The process parameters have a big impact on how well RA3DP works. The key parameters that are subject to variation are the extrusion temperature, print speed, layer height, and bed temperature. These parameters are contingent upon the specific material that is being printed as mentioned in Table 3. In order to achieve optimal melting and flow characteristics of various materials, such as thermoplastics, metals, or composites, it is essential to make necessary adjustments to the extrusion temperature as discussed in Table 3. The build time of a product can be influenced by the print speed, which in turn can have an impact on the precision and surface finish of the final product. While higher speeds may expedite production, they can potentially compromise the overall quality. The resolution and smoothness of the printed object are contingent upon the layer height, whereby thinner layers yield more intricate details, albeit at the expense of extended print duration. The temperature of the bed plays a critical role in facilitating optimal adhesion of the initial layer and mitigating warping, especially in the case of thermoplastics. The determination of overall print quality and structural integrity of printed objects is influenced by various additional parameters. These parameters include the inert gas flow rate for metals, pump pressure for concrete, and cooling rate for thermoplastics. The calibration and optimization of these parameters play a crucial role in attaining prints of superior quality and streamlining production processes for enhanced efficiency.

7 Conclusion

In conclusion, the investigation of RA3DP demonstrates a technological boundary that has swiftly progressed and broadened its uses. The historical context examines the combination of 3D printing and robotic arms, laying the foundation for the revolutionary development observed in recent times. The advancements in Rapid Additive 3D Printing have introduced a new era characterized by enhanced design flexibility and expanded manufacturing capabilities. The synergy between a 3D printer head and a multi-axis robotic arm allows for the creation of expansive prototypes, intricate sculptures, awe-inspiring architectural structures, and practical furniture, demonstrating the adaptability of this groundbreaking method. The distinctive characteristics of RA3DP, such as its capacity to print from different perspectives and its limited dependence on support structures, enable designers to investigate intricate geometries and expand the limits of creativity. The utilization of pellet or chip formats of various substances, such as polymers, metals, clay, and concrete, indicates a transition towards sustainability and cost-efficiency. The popularity of RA3DP in industrial manufacturing settings is due to its ability to effectively utilize recycled materials and its faster printing speeds achieved through pellet extrusion. RA3DP's ability to work with various materials and its potential for recycling make it an important player

Table 3 Summary of process parameters for various materials used in Robotic Arm 3D Printing

Material	Extrusion temperature/ laser power	Print speed	Layer height	Bed temperature	Other parameters
Metal (Chen et al. 2022)	Laser power 200–500 W	200–600 mm/s	20–100 microns	Not applicable	Inert gas flow:10–20 L/min
Thermoplastic (Anand and Satyarthi 2023)	PLA: 190–22 °C	30–100 mm/s	100–300 microns	PLA:50–70 °C	Cooling rate: Adjusted Fan Speed; ABS Bed temp 90–110 °C
Concrete (Carneau et al. 2022)	Mix composition dependent	50–200 mm/s	10–30 mm	Ambient or controlled	Pump pressure: 1–5 bar; curing conditions: Humidity and temperature
Composites (Tian et al. 2016)	Carbon fiber PLA: 200–220 °C	30–60 mm/s	100–300 microns	60–90 °C (depending on polymer matrix)	Nozzle diameter 0.4 mm or more
Ceramics (Kim et al. 2019)	Binder Jetting: Binder specific	20–40 mm/s	20–50 microns	Not applicable	Sintering temperature: > 1200 °C
Elastomers (Bruère et al. 2023)	TPU: 210–240 °C	20–40 mm/s	100–300 microns	40–60 °C (If required)	None
Bio materials (Huber et al. 2022)	PCL: 80–100 °C	10–30 mm/s	50–200 microns	Sterile environment	Must be printed in sterile conditions; Consider biocompatibility

in the effort to promote environmentally friendly manufacturing methods. An analysis of prominent companies and technologies in RA3DP reveals the current state of the industry, with industry leaders such as ABB Robotics, CEAD, KUKA, and Bloom Robotics taking the lead. These companies provide modular systems, adaptable solutions, and ready-to-use options, which help to make RA3DP more accessible in different industries. The utilization of specialized software, such as ABB's RobotStudio and Adaxis's AdaOne, simplifies the programming and simulation procedures, thereby enhancing the accessibility of RA3DP. RA3DP has versatile applications in architecture, manufacturing, and art, demonstrating its flexibility in various domains. RA3DP's influence is evident in the works of companies such as Branch Technology, Aectual, and Nagami Design, where they have contributed to the creation of both expansive architectural elements and customized furniture. RA3DP's capacity to produce tailored designs with accuracy and effectiveness establishes it as a versatile instrument for both prototyping and final production. Although RA3DP shows potential, it is not devoid of obstacles. Researchers and practitioners must address the challenges posed by the intricacies of programming, complexities of system integration, and nuances of material handling. The future prospects for RA3DP depend on progress in automation, standardization, and sustainability measures. The utilization of artificial intelligence and machine learning has the capacity to simplify programming, while the implementation of standardization initiatives can improve compatibility and increase the acceptance of RA3DP. Robotic Arm 3D Printing is a powerful and revolutionary technology in the field of additive manufacturing. The technological progress, combined with a wide range of uses, highlights its capacity to revolutionize the process of designing, manufacturing, and incorporating objects into different sectors. With ongoing research and development, RA3DP is positioned to become a fundamental element of future manufacturing methods, providing a combination of accuracy, personalization, and environmental consciousness.

References

3D Printing with CEAD Robot Extruder in Largest Innovation Centre in the Netherlands (2021). Available from: https://ceadgroup.com/portfolio-items/3d-printing-with-cead-robot-extruder-in-largest-innovation-centre-in-the-netherlands/

Abdulhameed O et al (2019) Additive manufacturing: challenges, trends, and applications. Adv Mech Eng 11(2):1687814018822880

Aggoune S, Hamadi F, Abid C et al (2024) Instabilities in the formation of single tracks during selective laser melting process. Int J Interact Des Manuf. https://doi.org/10.1007/s12008-024-01887-y

Ajay P, Ahmad S, Sharma J, Gambhir V (eds) (2023) Handbook of sustainable materials: modelling, characterization, and optimization, 1st edn. CRC Press. https://doi.org/10.1201/9781003297772

Anand S, Satyarthi MK (2023) Parametric optimization of fused filament fabrication process. In: Advances in mechanical and energy technology. Springer Nature Singapore, Singapore

Anand S, Satyarthi M (2024) Predictive modeling and optimization of tensile and flexural strength in FDM 3D printing using decision trees and Bayesian optimization. J Polym Compos 11:203–214

Arleo L et al (2021) I-support soft arm for assistance tasks: a new manufacturing approach based on 3D printing and characterization, vol 6, pp 243–256

Attaran M (2017) The rise of 3-D printing: The advantages of additive manufacturing over traditional manufacturing. Bus Horizons 60(5):677–688

Batista RC, Agarwal A, Gurung A, Kumar A, Altarazi F, Dogra N, Chiniwar DS, Agrawal A (2024) Topological and lattice-based AM optimization for improving the structural efficiency of robotic arms. Front Mech Eng 10:1422539. https://doi.org/10.3389/fmech.2024.1422539

Beaman J et al (2020) Additive manufacturing review: early past to current practice. J Manuf Sci Eng 142(11):110812

Bhardwaj A, Bhatnagar A, Kumar A (2023) Current trends of application of additive manufacturing in oral healthcare system, in a book: advances in additive manufacturing artificial intelligence. In: Nature-inspired, and biomanufacturing. Elsevier, Amsterdam, pp 479–491

Bliemert N. ADE25(h) extruder. Available from: https://www.bloom-robotics.com/ade25h-extruder/

Block I (2018) Robots complete span of 3D-printed bridge for Amsterdam canal. [17 Apr 2018]. Available from: https://www.dezeen.com/2018/04/17/mx3d-3d-printed-bridge-joris-laarman-arup-amsterdam-netherlands/

Bruère V et al (2023) The influence of printing parameters on the mechanical properties of 3D printed TPU-based elastomers. Progr Additive Manuf 8(4):693–701

Budig M et al (2014) Design of robotic fabricated high rises: integrating robotic fabrication in a design studio. In: Robotic fabrication in architecture, art and design, pp 111–129

Butt JJD (2020) Exploring the interrelationship between additive manufacturing and Industry 4.0. Designs 4(2):13

Carneau P et al (2022) Layer pressing in concrete extrusion-based 3D-printing: experiments and analysis. Cem Concr Res 155:106741
CEAD|Large Scale Additive Manufacturing Logo the Use of Robotic Arms for 3D Printing Versus for Assembly. Available from: https://ceadgroup.com/the-use-of-robotic-arms-for-3d-printing-versus-for-assembly/
Chen S-G et al (2022) Review on residual stresses in metal additive manufacturing: formation mechanisms, parameter dependencies, prediction and control approaches. J Market Res 17:2950–2974
Fidan I et al (2023) Recent inventions in additive manufacturing: holistic review. Inventions 8(4):103
Gao W et al (2015) The status, challenges, and future of additive manufacturing in engineering. Comput Aided Des 69:65–89
Gibson I et al (2021) Additive manufacturing technologies, vol 17. Springer, Heidelberg
Goyal G, Kumar A, Gupta A (2024) 16 recent developments in 3D printing: a critical analysis and deep dive into innovative real-world applications. In: 3D printing technologies: digital manufacturing, artificial intelligence, industry 4.0, vol 335
Goyal G, Kumar A, Sharma D (2024) 12 recent applications of rapid prototyping with 3D printing: a review. In: Kumar A, Kumar P, Sharma N, Srivastava AK (eds) 3D printing technologies: digital manufacturing, artificial intelligence, industry 4.0. De Gruyter, Berlin, Boston, pp 245–258. https://doi.org/10.1515/9783111215112-012
Huber F et al (2022) Influence of 3D printing parameters on the mechanical stability of PCL scaffolds and the proliferation behavior of bone cells. Materials 15:2091. https://doi.org/10.3390/ma15062091. https://doi.org/10.3390/ma15062091
Ian Gibson IG (2015) Additive manufacturing technologies 3D printing, rapid prototyping, and direct digital manufacturing. Springer, Heidelberg
Iansiti M, Lakhani KR (2020) Competing in the age of AI: strategy and leadership when algorithms and networks run the world. Harvard Business Review Press
İpekçi A, Ekici B (2024) An innovative composite elbow manufacturing method with 6-axis robotic additive manufacturing for fabrication of complex composite structures. J Thermoplastic Compos Mater 37(1):402–425
Ishak I, Larochelle P (2017) Robot arm platform for additive manufacturing: 3D lattice structures. In: 30th Florida conference on recent advances in robotics
Jiménez M et al (2019) Additive manufacturing technologies: an overview about 3D printing methods and future prospects
Kauppila I (2023) Robotic arm 3D printing—the ultimate guide. [2023 22 May 2023]. Available from: https://all3dp.com/1/robotic-arm-3d-printing-platforms-software/
Kim NP, Cho D, Zielewski M (2019) Optimization of 3D printing parameters of Screw Type Extrusion (STE) for ceramics using the Taguchi method. Ceramics Int 45(2, Part A):2351–2360
Kontovourkis O, Tryfonos G (2020) Robotic 3D clay printing of prefabricated non-conventional wall components based on a parametric-integrated design. Autom Constr 110:103005
Krčma M, Paloušek D (2022) Comparison of the effects of multiaxis printing strategies on large-scale 3D printed surface quality, accuracy, and strength. Int J Adv Manuf Technol 119(11):7109–7120
Kumar A, Kumar P, Srivastava AK, Goyat V (2023) Modeling, characterization, and processing of smart materials. In: Advances in chemical and materials engineering book series. https://doi.org/10.4018/978-1-6684-9224-6
Kumar P, Hussain SS, Kumar A, Srivastava AK, Hussain M, Singh PK (2024) 10 finite element method investigation on delamination of 3D printed hybrid composites during the drilling operation. In: 3D printing technologies: digital manufacturing, artificial intelligence, industry 4.0, p 223
Kumar A, Kumar P, Sharma N, Srivastava AK (2024) 3D printing technologies: digital manufacturing, artificial intelligence, industry 4.0. Walter de Gruyter GmbH & Co KG
Kumar A, Shrivastava VK, Kumar P, Kumar A, Gulati V (2024) Predictive and experimental analysis of forces in die-less forming using artificial intelligence techniques. In: Proceedings of the

institution of mechanical engineers, part e: journal of process mechanical engineering, vol 0, issue 0. https://doi.org/10.1177/09544089241235473
Kumar A et al (2023) Chapter 6—Printing file formats for additive manufacturing technologies. In: Kumar A, Mittal RK, Haleem A (eds) Advances in additive manufacturing. Elsevier, pp 87–102
Kumar A et al (2023) Chapter 4—Integration of reverse engineering with additive manufacturing. In: Kumar A, Mittal RK, Haleem A (eds) Advances in additive manufacturing. Elsevier, pp 43–65
Kumar A et al (2023) Chapter 9—Preprocessing and postprocessing in additive manufacturing. In: Kumar A, Mittal RK, Haleem A (eds) Advances in additive manufacturing. Elsevier, pp 141–165
Kumar A et al (2023) Chapter 12—Materials processed by additive manufacturing techniques. In: Kumar A, Mittal RK, Haleem A (eds) Advances in additive manufacturing. Elsevier, pp 217–233
Laguna O et al (2021) A review on additive manufacturing and materials for catalytic applications: milestones, key concepts, advances and perspectives. Mater Des 208:109927
Little H (2020) Gigabot X: recycling plastic via pellet 3D printing. Sustainable Technologies/Future Energy 2020. Available from: https://contest.techbriefs.com/2020/entries/sustainable-technologies-future-energy/10136
Luu QK, La HM (2021) A 3-dimensional printing system using an industrial robotic arm. In: 2021 IEEE/SICE international symposium on system integration (SII), IEEE
Mahmood A et al (2022) On the evolution of additive manufacturing (3D/4D printing) technologies: materials, applications, and challenges. Polymers 14(21):4698
Mascellino A (2020) KUKA releases ProLMD project for advancements in hybrid production with laser technology [cited 2020 16 June 2020]. Available from: https://control.com/news/kuka-releases-prolmd-project-for-advancements-in-hybrid-production-with-laser-technology/
Mittal RK, Haleem A, Kumar A (2022) Advances in additive manufacturing: artificial intelligence, nature-inspired, and biomanufacturing. Elsevier
Mostafavi S, Bier H (2016) Materially informed design to robotic production: a robotic 3D printing system for informed material deposition. In: Robotic fabrication in architecture, art and design, pp 338–349
Pan Y et al (2021) 3D printing in construction: state of the art and applications. Int J Adv Manuf Technol 115(5–6):1329–1348
Parmar H et al (2022) Advanced robotics and additive manufacturing of composites: towards a new era in Industry 4.0. Mater Manuf Process 37(5):483–517
Pires JN et al (2021) The role of robotics in additive manufacturing: review of the AM processes and introduction of an intelligent system. Ind Robot Int J Robot Res Appl 49(2):311–331
Praveena BA et al (2022) A comprehensive review of emerging additive manufacturing (3D printing technology): methods, materials, applications, challenges, trends and future potential. Mater Today Proc 52: 1309–1313
Praveena B et al (2022) A comprehensive review of emerging additive manufacturing (3D printing technology): methods, materials, applications, challenges, trends and future potential. Mater Tpday Proc 52:1309–1313
Rani S, Tripathi K, Kumar A (2023) Machine learning aided malware detection for secure and smart manufacturing: a comprehensive analysis of the state of the art. Int J Interact Des Manuf. https://doi.org/10.1007/s12008-023-01578-0
Ranjit J (2020) Bartlett researchers create robot-built voxel chair using new 3D printing software. [12 Nov 2020]. Available from: https://parametric-architecture.com/voxel-chair-by-bartlett-researchers/
Ruffa SA (2008) Going lean: how the best companies apply lean manufacturing principles to shatter uncertainty, drive innovation, and maximize profits. AMACOM
Saunders S (2023) The state of metal 3D printing: ADDiTEC, Nikon SLM solutions, additive industries. [19 Dec 2023]. Available from: https://3dprint.com/305520/metal-3d-printing-formnext-2023-additec-nikon-slm-solutions-additive-industries/

Scott C (2018) Branch technology and Techmer PM unveil a 3D printed band shell and hanging gardens. [7 Aug 2018]. Available from: https://3dprint.com/221462/branch-technology-techmer-pm/

Sehrawat S, Kumar A, Grover S, Dogra N, Nindra J, Rathee S, Dahiya M, Kumar A (2022a) Study of 3D scanning technologies and scanners in orthodontics. Mater Today Proc 56:186–193

Sehrawat S, Kumar A, Prabhakar M, Nindra J (2022b) The expanding domains of 3D printing pertaining to the speciality of orthodontics. Mater Today Proc 50:1611–1618

Shembekar AV et al (2019) Generating robot trajectories for conformal three-dimensional printing using nonplanar layers. J Comput Inf Sci Eng 19(3):031011

Sher D (2021) Branch technology's C-Fab process used for giant 3D printed facade and exterior wall panels. 2021 17 Feb 2021]. Available from: https://www.voxelmatters.com/branch-technologys-c-fab-process-used-for-giant-3d-printed-facade/

Singh H, Parveen, AlMangour B (eds) (2023) Handbook of smart manufacturing: forecasting the future of industry 4.0, 1st edn. CRC Press. https://doi.org/10.1201/9781003333760

Smith R (2019) ABB enables 3D printing via RobotStudio® for faster digital manufacturing [cited 2019 2019-12-16]. Available from: https://new.abb.com/news/detail/52810/abb-enables-3d-printing-via-robotstudior-for-faster-digital-manufacturing

Srivastava AK et al (2023) Research progress in metal additive manufacturing: challenges and opportunities, pp 1–17

Tang P et al (2024) A review of multi-axis additive manufacturing: potential, opportunity and challenge, p 104075

Tian X et al (2016) Interface and performance of 3D printed continuous carbon fiber reinforced PLA composites. Compos A Appl Sci Manuf 88:198–205

Elephant Swarm Water Search Algorithm-Based Optimization of a Laser Beam Machining Process

Sunny Diyaley, Shankar Chakraborty, Ajay Kumar, and Kanak Kalita

Abstract Laser beam machining (LBM) is a popular non-conventional material removal process. It is observed that quality of cut in LBM process performance can only be achieved through proper tuning of its different input parameters. It is thus vital to explore the effects of LBM process parameters on the quality of cut. In this paper, a newly developed metaheuristic algorithm, i.e. elephant swarm water search algorithm (ESWSA), inspired by the behavior of social elephants, is employed to determine the optimal combination of gas pressure (P_a), pulse width (W_p), pulse frequency (f_p) and cutting speed (S_c) during Nd:YAG laser-based straight profile cutting of thin aluminium alloy sheet. During multi-objective optimization, minimum values of kerf taper (KT) and average surface roughness (Ra) are achieved as 0.294° and 0.133 μm respectively at a parametric combination of $P_a = 7.437$ kg/cm^2, $W_p = 1.6$ ms, $f_p = 8$ Hz and $S_c = 6$ mm/min while providing equal importance to both the responses. It is noticed that for both single and multi-objective optimization problems, ESWSA supersedes its peers, like genetic algorithm (GA), particle swarm optimization (PSO), ant colony optimization (ACO), artificial bee colony (ABC) and differential evolution (DE) with respect to accuracy and deviation of the derived solutions, and computational effort. Application of ESWSA achieves 8.41, 2.00,

S. Diyaley
Department of Mechanical Engineering, Sikkim Manipal Institute of Technology, Sikkim Manipal University, Majitar, Sikkim, India

S. Chakraborty
Department of Production Engineering, Jadavpur University, Kolkata, West Bengal, India

A. Kumar (✉)
Department of Mechanical Engineering, School of Engineering and Technology, JECRC University, Jaipur, Rajasthan, India
e-mail: ajay.kumar30886@gmail.com

K. Kalita
Department of Mechanical Engineering, Vel Tech Rangarajan Dr. Sagunthala R&D Institute of Science and Technology, Avadi, India

University Centre for Research and Development, Chandigarh University, Mohali 140413, India

K. Kalita
e-mail: drkanakkalita@veltech.edu.in

A. Kumar et al. (eds.), *Industry 4.0 Driven Manufacturing Technologies*, Springer Series in Advanced Manufacturing, https://doi.org/10.1007/978-3-031-68271-1_14

14.53, 27.76 and 2.32%; and 44.81, 55.52, 18.90, 29.63 and 60.42% improvements respectively for KT and *Ra* against ABC, ACO, PSO, DE and GA techniques when both the responses are equally preferred.

Keywords LBM · ESWSA · Optimization · Parameter · Response

Abbreviations

d_{en}	Entry Hole Diameter
d_{ex}	Exit Hole Diameter
C_{en}	Circularity of Entry Hole
C_{ex}	Circularity of Exit Hole
f_p	Pulse Frequency
I	Lamp Current
P_a	Gas Pressure
P_{laser}	Laser Power
R_a	Average Surface Roughness
S_c	Cutting Speed
W_p	Pulse Width
N	Number of iterations
pop	Population Size
p_c	Crossover Probability
SF_{lb}	Lower Bound of Scaling Factor
SF_{ub}	Upper Bound of Scaling Factor
p_m	Mutation Probability
ABC	Artificial Bee Colony
ACO	Ant Colony Optimization
ANFIS	Adaptive Neuro-Fuzzy Inference System
AR	Aspect Ratio
CHOA	Chimp Optimization Algorithm
CSA	Cuckoo Search Algorithm
DE	Differential Evolution
ESWSA	Elephant Swarm Water Search Algorithm
FA	Firefly Algorithm
GA	Genetic Algorithm
GWO	Grey Wolf Optimizer
HAZ	Heat Affected Zone
KD	Kerf Deviation
KT	Kerf Taper
KW	Kerf Width
LBM	Laser Beam Machining
LSTM	Long Short-Term Memory
MRR	Material Removal Rate

PSO	Particle Swarm Optimization
RSM	Response Surface Methodology
TLBO	Teaching-Learning-Based Optimization
WOA	Whale Optimization Algorithm.

1 Introduction

Continuous demand for machining of various hard-to-cut advanced engineering materials and accessibility of moder day high-power lasers have led to the growing popularity of laser beam machining (LBM) process. It is a thermal energy-based material removal process being extensively employed in modern-day manufacturing industries to fulfill the requirements of higher flexibility, increased productivity, reduced processing cost, improved product quality, low maintenance cost due to minimum wear and tear, adaptability to automation, elimination of finishing operations, capability of processing both conductive and non-conductive work materials and green machining environment (Sun and Brandt 2013). In LBM process, a high power pulsed laser is directed at a specific location on the workpiece to be cut (Dubey and Yadava 2008a, 2008b). The energy beam is absorbed into the surface of the work material and the energy of laser is subsequently converted into heat, helping in melting and vaporization of the work material. The molten material is blown away from the machining zone with the help of a pressurized gas. The LBM can provide a wide range of machining operations, like cutting, drilling, turning, grooving, scribing, marking etc. Among different types of laser sources employed in manufacturing industries, CO_2 and Nd:YAG have been mostly established to cut varieties of materials, including ferrous and non-ferrous metals, ceramics, stone, polycarbonate, polyethylene etc. Although the general principles of both these laser types are almost similar, CO_2 laser has found more applicability in industries due to its lower cost, higher resolution, faster processing, capability to machine almost all types of work materials etc. (Vasiga and Channankaiah 2015). Despite several advantages, LBM processes have some limitations, like higher initial investment, requirement of skilled operator, higher energy consumption and formation of taper while machining thicker work materials.

It has been noticed that in LBM processes, excellent quality of cut with respect to lower kerf taper (KT), kerf width (KW), kerf deviation (KD), average surface roughness (*Ra*), heat affected zone (HAZ); and higher material removal rate (MRR) can only be achieved through proper adjustment and tuning of various input parameters, like laser power (P_{laser}), lamp current (I), pulse width (W_p), pulse frequency (f_p), cutting speed (S_c), gas pressure (P_a) etc. For effective utilization and exploring the fullest potential of an LBM process, it is always recommended to determine the optimal settings of these input parameters leading to attainment of the preferred quality characteristics of the machined components (Bakhtiyari et al. 2021; Singh et al. 2022). Any deviation from the optimal parametric combination may lead to increased energy consumption and poorer product quality (Madić et al. 2022). Thus, it is required to employ a mathematically sound global optimization tool to single out the best parametric intermix of the LBM process. In this paper, the application of an almost unexplored metaheuristic algorithm in the form of elephant swarm water search algorithm (ESWSA) is proposed to identify the optimal settings of W_p, f_p, S_c and P_a during Nd:YAG-based straight profile cutting of thin aluminium alloy sheet. A unique combination of these LBM process parameters would result in simultaneous minimization of both KT and *Ra* with minimum computational effort.

This paper is organized as follows: Sect. 2 provides a concise literature review of the applications of different metaheuristics for solving parametric optimization problems of LBM processes. The detail of ESWSA is presented in Sect. 3, while both the single and multi-objective optimization problems are solved in Sect. 4 using ESWSA. Results and discussions are presented in Sect. 5 and conclusions are drawn in Sect. 6. The future scopes are discussed in Sect. 7.

2 Literature Review

Finding the optimal process parameter combination of LBM plays pivotal role with respect to attainment of the most desired quality characteristics of the machined components. To resolve this issue, the past researchers applied different metaheuristic algorithms, like genetic algorithm (GA), particle swarm optimization (PSO), artificial bee colony (ABC), teaching–learning-based optimization (TLBO), firefly algorithm (FA), grey wolf optimizer (GWO), cuckoo search algorithm (CSA), whale optimization algorithm (WOA), chimp optimization algorithm (CHOA) etc. Table 1 provides a concise summary of the various studies using different metaheuristics for finding the optimal process parameter of LBM processes.

Goswami and Chakraborty (2015) solved both single and multi-objective optimization problems of an LBM process using fireworks algorithm and CSA, and noticed that although both the considered algorithms would provide almost similar solutions, CSA would excel over fireworks algorithm with respect to computational time and cost. Considering P_{laser}, S_c, focal position, P_a and working distance as the input parameters, and KT and KW as the responses, Hiwale and Basavarajappa (2020) applied two swarm-based metaheuristics, i.e. black hole and krill herd algorithms for optimization of an LBM process while machining Hastelloy C-276 material. Based on a comparative analysis, it was concluded that krill herd algorithm would provide better results than black hole algorithm for the considered process. Rajamani et al. (2021) endeavored to maximize MRR, and minimize KT and *Ra* by integrating an adaptive neuro-fuzzy inference system (ANFIS) with WOA. It was noticed that an optimal combination of LBM parameters as $P_a = 3$ Bar, $S_c = 319.8$ mm/min, pulse energy = 5.93 J and stand-off distance = 2.97 mm would lead to attainment of the corresponding MRR, KT and *Ra* values as 236.98 mg/min, 1.135° and 1.109 μm respectively. Najjar et al. (2022) integrated long short-term memory (LSTM) and CHOA for solving the parametric optimization problem of an LBM process. The

Table 1 Applications of different metaheuristics for optimization of LBM processes

Author(s)	LBM process parameters	Response(s)	Metaheuristic(s)
Phipon and Pradhan (2012)	P_a, W_p, f_p, S_c	KT, *Ra*	GA
Madić and Radovanović (2013)	P_{laser}, S_c, P_a, focus position	*Ra*	CSA
Mukherjee et al. (2013)	I, f_p, P_a, W_p, S_c	HAZ, taper, upper deviation, lower deviation, depth deviation	ABC
Goswami and Chakraborty (2015)	I, f_p, P_a, workpiece thickness, P_{laser}, wait time, W_p	Hole circularity at exit, hole taper, MRR	Fireworks algorithm, CSA
Chatterjee et al. (2017)	Flushing pressure, P_{laser}, f_p	Spatter area	Harmony search
Gautam and Mishra (2019)	I, W_p, f_p, P_a, S_c	KW, KD, KT	TLBO
Sibalija et al. (2019)	P_a, beam focus position, P_{laser}, S_c	KD, KT, micro-hardness, grate, *Ra*, roughness root mean square, roughness peak-to-valley	PSO
Gautam and Mishra (2020)	I, f_p, W_p, P_a, S_c	KW, KD, KT	FA
Hiwale and Basavarajappa (2020)	P_{laser}, S_c, P_a, working distance, focal position	KW, KT	Black hole algorithm, krill herd algorithm
Madić et al. (2020)	S_c, P_{laser}, P_a	MRR	Exhaustive iterative search algorithm
Sibalija et al. (2021)	f_p, pulse duration	d_{en}, d_{ex}, C_{en}, C_{ex}, AR, taper, spatter	PSO, SA, GA
Elsheikh et al. (2021)	P_a, sheet thickness, P_{laser}, S_c	Rough zone ratio, HAZ, *Ra*, KT	Equilibrium optimizer
Rajamani et al. (2021)	P_a, S_c, pulse energy, stand-off distance	MRR, KT, *Ra*	WOA
Pramanik et al. (2022)	Sawing angle, power, duty cycle, f_p, scanning speed	KT, HAZ	PSO
Najjar et al. (2022)	S_c, P_a, f_p, W_p, I	KW, KT, KD	CHOA
Prajapati et al. (2022)	f_p, I, P_a, S_c	HAZ	GWO
Mishra et al. (2022)	S_c, I, P_a, W_p	MRR, taper, HAZ	Accelerated PSO

(continued)

Table 1 (continued)

Author(s)	LBM process parameters	Response(s)	Metaheuristic(s)
Rohman et al. (2022a)	Machining environment, P_{laser}, f_p, S_c	Dross formation	Improved GWO
Rohman et al. (2022b)	P_{laser}, f_p, S_c	KW, roundness	DNN-GA
Misra et a. (2023)	f_p, S_c, I, P_a	HAZ, taper	Accelerated PSO, WOA
This paper	W_p, f_p, S_c, P_a	KT, *Ra*	ESWSA

CHOA would act as an internal optimizer to derive the optimal settings of LSTM network model. It was noticed that the root mean square error values of the predicted KW, KD and KT based on LSTM-CHOA would decrease by 27.43%, 60% and 56.6% respectively, as compared to those of only LSTM. Rohman et al. (2022b) employed deep neural network-based GA (DNN-GA) approach to single out the optimal parametric combination of an LBM process for having the best geometrical properties of the machined components. It was claimed that DNN-GA would predict more accurate results than other artificial intelligence-based models, like support vector machine for regression integrated equilibrium optimizer, random vector functional link network integrated equilibrium optimizer, support vector machine for regression integrated grey wolf optimizer and random vector functional link network integrated grey wolf optimizer. Various input parameters, i.e. f_p, S_c, I and P_a of an LBM process were optimized by Mishra et al. (2023) using accelerated PSO and WOA techniques. It was observed that the adopted techniques would outperform the composite design-based response surface methodology (RSM) with respect to accuracy of the derived results. It can be revealed from Table 1 that the past researchers proposed discrete applications of different metaheuristics for optimizing LBM processes while achieving varying levels of solution accuracy. Minimum attempt has been put forward to contrast the optimization performance of the adopted techniques with an aim to identify the best algorithm resulting in almost global optimal solutions. This paper proposes application of ESWSA for optimizing an LBM process, and compares its performance against other state-of-the-art algorithms, like GA, PSO, ant colony optimization (ACO), ABC and differential evolution (DE) with respect to solution accuracy and variability, and computational effort. To the best of the authors' knowledge, it is the first application of ESWSA in optimizing any of the machining processes.

3 Elephant Swarm Water Search Algorithm

The ESWSA, developed by Mandal (2018), is a swarm-based metaheuristic, inspired by the water searching behaviour of wild elephants during the drought season. They usually live in herds consisting of 3–35 elephants, and they have excellent memory, and advanced sensing and communication systems. As an adult elephant requires 40–60 gallons of water every day, scarcity of water during dry season or drought arises huge problem to them. Depending on the prevailing conditions, they utilize their communicating systems to find out water from the nearby sources. When the surrounding area is extremely dry and drought is prolonged, the elephants usually migrate to other areas having plenty of water. They stay at the new locations until the rainy season arrives and share information with different elephant herds with respect to better availability of water. Based on water searching strategies of the wild elephants, ESWSA is developed considering the following idealized rules:

(a) They form groups (swarm) consisting of a number of elephants and the leader of each group takes decision in search for best water source. In a typical optimization problem, each elephant group is characterized by its own position and velocity, and each group of the swarm is supposed to be a solution of the given problem.
(b) When some water source is identified by an elephant group, the information in regard of quality and quantity of water is communicated to other groups by the leader. The quantity and quality of the new water sources are proportional to the fitness value and objective function of a maximization problem. Better solution must have better water quantity.
(c) With their sharp memory, each elephant group can effectively remember the best location of water source discovered so far by its own group (local best solution) and the best location of water source identified so far (global best solution) by the whole swarm. Using this information, the group starts moving from one point to another with gradual update of its velocity and position according to some specific rules. Their long-distance and short-distance communications are synonymous to global and local searches respectively.
(d) A probabilistic constant (switching probability) ($p \in [0, 1]$) controls water search in local and global areas. During water search, the probabilistic decision to switch between local search and global search is usually taken by the group leader. Due to various factors, local water search has a significant fraction p in the whole searching process.

Let us assume that for an d-dimensional optimization problem, the position of i th elephant group of a swarm (having N number of elephant groups) at t th iteration is represented by $X_{i,d}^{t} = (x_{i1}, x_{i2}, \ldots, x_{id})$ and the velocity is denoted by $V_{i,d}^{t} = (v_{i1}, v_{i2}, \ldots, v_{id})$. On the other hand, $P_{best,i,d}^{t} = (P_{i1}, P_{i2}, \ldots, P_{id})$ denotes the local best solution by i th elephant group at current iteration and $G_{best,d}^{t} = (G_1, G_2, \ldots, G_d)$ represents the global best solution. Although, the position and velocity of the elephant groups are initially placed randomly throughout the search space, they get updated according to some specific rules as the iteration proceeds. Generally, during water searching process, adjacent water sources are explored by the elephant group than those far away. For this purpose, a probabilistic constant (switching probability) p is considered to switch between global and local water searches. If the value of a random variable is greater than p, global water search is executed; otherwise, local water search is considered. It helps in decreasing the probability of sticking at local optima. After each iteration, the global and local best solutions are updated. At each iteration, depending on the value of p, velocities are also updated in different ways for global and local searches based on the following equations:

$$V_{i,d}^{t+1} = V_{i,d}^{t} w^{t} + rand(1, d) \odot \left(G_{best,d}^{t} - X_{i,d}^{t}\right) \text{if } rand > p\left[\text{for globalsearch}\right] \quad (1)$$

$$V_{i,d}^{t+1} = V_{i,d}^{t} w^{t} + rand(1, d) \odot \left(P_{best,i,d}^{t} - X_{i,d}^{t}\right) \text{if } rand \le p[\text{for localsearch}] \quad (2)$$

where $rand(1, d)$ generates a d-dimensional array of random values in the range of 0 to 1, the circle symbol represents element-wise multiplication, and w^t signifies the inertia weight at current iteration to trade-off between exploration and exploitation. The position of an elephant group is subsequently updated using the following expression:

$$X_{i,d}^{t+1} = V_{i,d}^{t+1} + X_{i,d}^{t} \quad (3)$$

After executing all the iterations, the elephants gradually update their positions and arrive at the best water source (best solution for a given optimization problem). The pseudo code for ESWSA is provided, as given below (Mandal 2018):

Algorithm 1: Pseudo Code for Elephant Swarm Water Search Algorithm (ESWSA)

Initialize Parameters
- Set the number of elephant groups N
- Define the maximum number of iterations t_{max}
- Set the search space boundaries X_{min} and X_{max}
- Define the switching probability p
- Initialize the dimension d
- Define the objective function f

Initialize Elephant Groups
- For each elephant group i (where i ranges from 1 to N):
  - Randomly initialize the position X_i
  - Randomly initialize the velocity V_i
  - Set the local best position $P_{best,i}$ to X_i

Evaluate Fitness
- Calculate the fitness value $f(X_i)$ for each elephant group
- Determine the global best position G_{best} by finding the position with the best fitness value among all groups

Assign Inertia Weight
- Assign an inertia weight w to balance exploration and exploitation

Optimization Loop
- For each iteration t from 1 to t_{max}:
  - For each elephant group i:
    - Generate a random value r between 0 and 1
    - If $r > p$, perform global search:
      - Update velocity V_i using:
      - $V_i = w.V_i + rand().(G_{best} - X_i)$
    - Else, perform local search:
      - Update velocity V_i using:
      - $V_i = w.V_i + rand().(P_{best,i} - X_i)$
    - Update position X_i using:
    - $X_i = X_i + V_i$
  - Evaluate the fitness value $f(X_i)$for each updated position
  - Update the local best position $P_{best,i}$ if the current position X_i has a better fitness value
  - Update the global best position G_{best} if any local best position has a better fitness value

Return the Best Solution
- After completing all iterations, return the global best position G_{best} and its corresponding fitness value $f(G_{best})$

Like all other metaheuristic algorithms, selection of appropriate values of different tuning parameters also influences the solution accuracy of ESWSA for real-time optimization problems. In ESWSA, the following tuning parameters are considered to guide the searching behavior of the elephants:

(a) Inertia weight (w^t) is a deterministic parameter used to denote inertia weight of velocity at the ongoing iteration. Usually, in ESWSA, inertia weight follows any of the three weight updation methods, i.e. constant inertia weight (inertia weight remains constant throughout the iteration), random inertia weight and linearly decreasing inertia weight with increasing on iterations. Mandal (2018) used $t_{\max} = 1000$ and $pop = 50$ to carry out simulation-based studies on some benchmark problems to investigate the effects of these three inertia weight update techniques on the optimization performance of ESWSA, and concluded that linearly decreasing inertia weight would perform best against the other two strategies with respect to median and variation of the derived optimal solutions.

(b) Switching probability (p) represents probability of switching between local and global search processes. Using different benchmark problems and linearly decreasing inertia weight update strategy, Mandal (2018) also conducted simulation experiments to study the influence of p on ESWSA's performance and reported $p = 0.6$ to be an optimal value.

(c) Maximum iteration number ($t_{\max}$) and population size (*pop*) are the fixed parameters during optimization. Considering linearly decreasing inertia weight update strategy and $p = 0.6$, Mandal (2018) observed that for unimodal functions, the mean fitness function value would decrease with increment in population size and maximum iteration number. On the other hand, increment in $t_{\max}$ would have negligible effect on average fitness function value for multi-modal functions. But the mean fitness function value would decrease with increase in population size. However, increased *pop* and $t_{\max}$ would ultimately result in higher computational cost. Hence, a compromised combination of population size and maximum iteration number is always sought.

The ESWSA differs slightly from the conventional PSO algorithm. In PSO, velocities of particles are influenced by three components, i.e. current velocity ($V_{i,d}^t$), local best solution ($P_{best,i,d}^t$) and global best solution ($G_{best,d}^t$). Some random parameters are also introduced along with global or local search terms to avoid getting stuck in local optima pit. But, in ESWSA, velocity is updated based on either current velocity and local best solution or current velocity and global best solution depending on the value of switching probability (p). The major advantages of ESWSA is its lesser computational cost compared to other metaheuristic algorithms. Hence, it can be considered as the most computationally efficient technique amongst several other metaheuristic algorithms (Mandal 2018).

4 Parametric Optimization of an LBM Process

Using CCD and considering P_a (in kg/cm^2), W_p (in ms), f_p (in Hz) and S_c (in mm/min), and KT (in °) and *Ra* (in µm) as the responses, Sharma and Yadava (2012a) conducted 31 experiments in a 200W pulsed Nd: YAG LBM setup equipped with CNC work table. The experimental design plan consists of 2 k factorial trials, 2 k axial trials and seven central points where k denotes the number of input parameters. The values of all the LBM process parameters were varied at five different levels having equal intervals, i.e. P_a (4–8 kg/cm^2), W_p (1.6–2.0 ms), f_p (8–12 Hz) and S_c (6–10 mm/min). Those operating levels were chosen based on the available technical specifications of the pulsed Nd: YAG LBM setup. While straight profile cutting of thin aluminium alloy (Grade 40,800) sheet, focal length of the laser beam, nozzle stand-off distance and sheet metal thickness were kept constant as 50 mm, 1.0 mm and 0.7 mm respectively. During the experiments, values of KT and *Ra* were respectively measured using an Optical Measuring Microscope (Model: SDM-TR-MSU, Sipcon Instrument Industries, India at 10X magnification) and a surface roughness tester (Model: SURTRONIC-25, Taylor Hobson Ltd., UK using 4.00 mm evaluation length). Table 2 exhibits the adopted experimental design plan and the response values. During straight profile cutting operation of aluminium sheet, Sharma and Yadava (2012a) first combined Taguchi method with RSM for modelling the process, and subsequently optimized it with a hybridized application of Taguchi method and grey relational analysis. It was noticed that the minimum values of both KT and *Ra* could be achieved at the moderate settings of all the LBM parameters as $P_a = 6$ kg/cm^2, $W_p = 1.8$ ms, $f_p = 10$ Hz and $S_c = 8$ mm/min.

Using RSM technique and experimental data of Table 2, the following two empirical relations were developed. These two equations are later considered for solving optimization problems for the said LBM process using ESWSA.

$$\begin{aligned} Y(KT) = {} & 0.167621 - 0.035356 \times P_a - 0.004663 \times W_p - 0.001023 \\ & \times f_p + 0.023064 \times S_c + 0.018484 \times P_a^2 + 0.007575 \times W_p^2 \\ & + 0.008947 \times f_p^2 + 0.011348 \times S_c^2 - 0.004594 \times P_a \times W_p \\ & - 0.002558 \times P_a \times f_p - 0.022681 \times P_a \times S_c + 0.002551 \times W_p \\ & \times f_p + 0.006302 \times W_p \times S_c + 0.002899 \times f_p \times S_c \end{aligned} \tag{4}$$

$$\begin{aligned} Y(Ra) = {} & 2.52143 - 0.15625 \times P_a - 0.00875 \times W_p \\ & + 0.12792 \times f_p + 0.13458 \times S_c + 0.03579 \times P_a^2 \\ & - 0.04671 \times W_p^2 + 0.05954 \times f_p^2 + 0.02954 \times S_c^2 \\ & - 0.00688 \times P_a \times W_p + 0.05937 \times P_a \times f_p \\ & - 0.03812 \times P_a \times S_c - 0.06937 \times W_p \times f_p \\ & - 0.14938 \times \mathrm{W}_p \times S_c - 0.24563 \times f_p \times S_c \end{aligned} \tag{5}$$

In single-objective optimization problem, the two considered responses, i.e. KT and *Ra* are separately optimized using ESWSA for which the corresponding code in MATLAB 7.6 (R2008a) is developed, and run in a 4 GB RAM, Dual Core processor and Windows10 operating System. To obtain the best performance of ESWSA and other considered state-of-the-art algorithms, values of different algorithm-specific parameters are appropriately chosen, as provided in Table 3. These values are set based on trial and error method after conducting several pilot runs. They also closely match with those as adopted by the past researchers while solving diverse optimization problems (Diyaley and Chakraborty 2020). Changing values of the considered tuning parameters in ESWSA may increase/decrease the number of iterations to reach the optimal solution, but there would not be any significant variation in the derived optimal solutions. In this paper, the tuning parameters are selected based on the type of the mathematical model (second-order equation with four input variables). The results of single-objective optimization for both the responses of the said LBM process are depicted in Table 4. It can be clearly revealed from Table 4 that ESWSA outperforms the other state-of-the-art optimization techniques, i.e. GA, PSO, ACO, ABC and DE with respect to minimum value of both the responses as well as average and standard deviation of the derived solutions. Thus, using ESWSA, for attaining

Table 2 Experimental data (Sharma and Yadava 2012a)

Exp. No.	LBM parameters				KT (°)	*Ra* (μm)
	P_a	W_p	f_p	S_c		
1	6	1.8	10	8	0.16370	2.54
2	5	1.7	9	7	0.22100	2.00
3	7	1.7	9	7	0.19372	1.60
4	5	1.9	9	7	0.19646	2.42
5	7	1.9	9	7	0.17740	1.90
6	6	1.8	10	8	0.16650	2.42
7	5	1.7	11	7	0.22650	2.78
8	7	1.7	11	7	0.16920	2.53
9	5	1.9	11	7	0.19100	2.96
10	7	1.9	11	7	0.18282	2.90
11	6	1.8	10	8	0.15558	2.58
12	5	1.7	9	9	0.26740	3.03
13	7	1.7	9	9	0.22100	2.46
14	5	1.9	9	9	0.33830	2.96
15	7	1.9	9	9	0.15533	2.44
16	6	1.8	10	8	0.17189	2.64
17	5	1.7	11	9	0.30834	2.94
18	7	1.7	11	9	0.18010	2.68
19	5	1.9	11	9	0.31921	2.54
20	7	1.9	11	9	0.19920	2.05
21	6	1.8	10	8	0.18554	2.39
22	4	1.8	10	8	0.29195	3.01
23	8	1.8	10	8	0.16095	2.67
24	6	1.6	10	8	0.20463	2.60
25	6	2	10	8	0.16100	2.42
26	6	1.8	10	8	0.17740	2.48
27	6	1.8	8	8	0.19650	2.81
28	6	1.8	12	8	0.18010	3.06
29	6	1.8	10	6	0.16659	2.51
30	6	1.8	10	9	0.22922	3.12
31	6	1.8	10	8	0.15280	2.60

the minimum KT of 0.1469°, an optimal parametric intermix as $P_a = 6.772$ kg/cm^2, $W_p = 1.870$ ms, $f_p = 10.140$ Hz and $S_c = 7.541$ mm/min is highly recommended. On the other hand, an optimal combination of LBM process parameters as $P_a = 8$ kg/cm^2, $W_p = 1.6$ ms, $f_p = 8$ Hz and $S_c = 6$ mm/min would result in attaining the minimum value of *Ra* as 0.9948 μm. The corresponding convergence diagrams for both the responses are portrayed in Fig. 1 which prove superiority of ESWSA over the other state-of-the-art algorithms in arriving at the optimal solutions with minimum number of iterations. The boxplots of Fig. 2 for KT and *Ra* also validate supremacy of ESWSA over the others with respect to minimum dispersion of the derived solutions. It can be interestingly noticed from Table 4 that the single objective optimization results provide completely contradictory settings of the considered LBM process parameters in attaining the minimum values of KT and *Ra*, which would be quite difficult to maintain in a single LBM setup. This leads to determination of a unique parametric intermix of LBM parameters which would result in simultaneous achievement of minimum values for both KT and *Ra*.

For the multi-objective optimization problem, the following objective function is developed.

$$\mathrm{Min}(Z) = \frac{w_1 \times Y(KT)}{KT_{\min}\frac{w_2 \times Y(Ra)}{Ra_{\min}}} \tag{6}$$

where $Y(KT)$ and $Y(Ra)$ are the empirical relations for KT and *Ra* respectively, $\mathrm{KT}_{\min}$ and $\mathrm{Ra}_{\min}$ are the single objective optimization-based minimum values of KT and *Ra* respectively, and w_1 and w_2 are the weights (relative importance) assigned to KT and *Ra* respectively. Depending on the end product requirements, the concerned process engineers need to allocate these weights to the responses such that their sum must be one. In this paper, in order to validate multi-objective optimization performance of ESWSA over the other state-of-the-art algorithms, three criteria weighting scenarios are considered. In the first scenario, equal weights are assigned

Table 3 Tuning parameters of the algorithms

Algorithm	Parameter
ABC	N= 200, *pop* = 100, number of employed bees = 0.5 *pop*, number of onlooker bees = 0.5 *pop*, number of scouts per cycle = 1, number of cycles = 1000, limit = 50
ACO	N= 200, *pop* = 40, intensification factor = 0.5, deviation distance ratio = 1
PSO	N= 200, *pop* = 100, inertia weight factor = 0.65, acceleration coefficients = 1.65 and 1.75
DE	N= 200, *pop* = 100, SF_{lb}= 0.5, SF_{ub}= 0.8, p_c= 0.9
GA	N= 200, *pop* = 100, p_c= 0.8, p_m = 0.3, mutation rate = 0.02, selection operator = tournament
ESWSA	*pop*= 50, maximum inertia weight ($w_{\max}$) = 0.9, minimum inertia weight ($w_{\min}$) = 0.2, switching probability = 0.6, N = 200, linearly decreasing inertia weight update strategy

Table 4 Optimized values of parameters considering single objective optimization

Response	Method	Mean	SD	P_a	W_p	f_p	S_c	Value
KT (°)	ABC	0.1506	0.001	7.090	1.833	10.403	7.8253	0.1489
	ACO	0.1526	0.001	6.433	1.864	10.449	7.7412	0.1522
	PSO	0.1496	0.002	7.014	1.860	10.193	8.002	0.1480
	DE	0.1484	0.003	6.700	1.867	10.513	7.4034	0.1481
	GA	0.1506	0.004	7.280	1.837	10.052	8.3664	0.1498
	ESWSA	0.1471	0.0008	6.772	1.870	10.140	7.541	0.1469
Ra (μm)	ABC	1.4338	0.116	4.311	1.670	8.014	6.269	1.3707
	ACO	1.2241	0.330	7.301	1.714	8.882	6.000	1.0851
	PSO	1.576	0.118	5.573	1.760	8.145	6.303	1.5101
	DE	1.224	0.067	7.612	1.646	9.144	6.008	1.1400
	GA	1.697	0.272	6.389	1.771	8.136	6.317	1.3935
	ESWSA	0.1001	0.003	8.000	1.600	8.000	6.000	0.9948

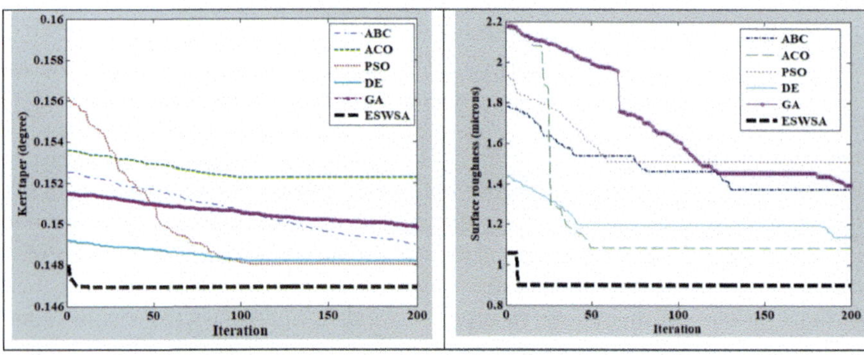

Fig. 1 Convergence of various algorithms **a** KT (°) **b** *Ra* (μm)

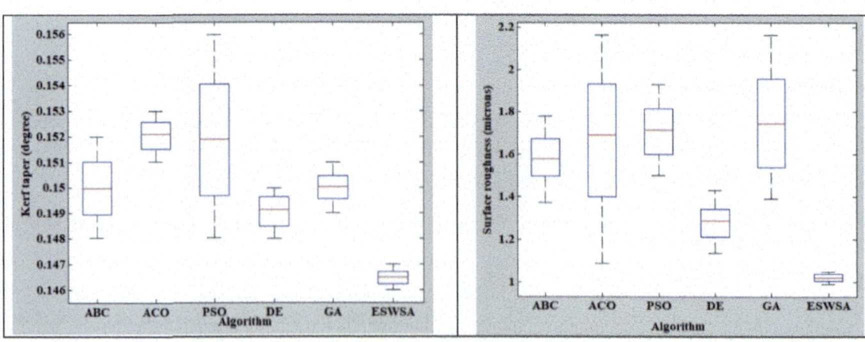

Fig. 2 Boxplots of various algorithms **a** KT (°) **b** *Ra* (μm)

to both the responses, i.e. $w_1 = w_2 = 0.5$. In the second case, dimensional deviation of the machined components with respect to KT has the maximum importance, i.e. $w_1 = 0.9$ and $w_2 = 0.1$. On the other hand, maximum importance is assigned to surface quality of the machined components in the last scenario, i.e. $w_1 = 0.1$ and $w_2 = 0.9$. The results of multi-objective optimization of the said LBM process using ESWSA are provided in Tables 5, 6 and 7 for three different weighting scenarios. It can be interestingly noticed that for all the considered cases, ESWSA excels over the other optimization techniques with respect to the optimal value of the objective function. It has also superior performance in regards of the average and standard deviation of the derived solutions. Thus, to attain simultaneous minimum values of both KT and *Ra* with equal importance assigned to them, it is recommended that the optimal parametric combination for the said LBM process should be set as P_a = 7.437 kg/cm^2, W_p = 1.6 ms, f_p = 8 Hz and S_c = 6 mm/min. The convergence diagrams of all the considered metaheuristics are shown in Fig. 3 for different weighting scenarios which reveal that in case of ESWSA, the optimal value of the objective function is reached within 5–10 iterations irrespective of the weight combination. In Table 8, the average computational time is compared for the adopted metaheuristics based on 20 independent runs for each algorithm. It can be observed that ABC would consume maximum computational time to execute for both the responses, i.e. KT and *Ra*. It can also be revealed from Fig. 4 that ESWSA takes minimum computational time with least variability for both the responses. This optimization technique thus proves its excellence over the others with respect to computational effort.

Table 5 Optimized values of parameters considering multi-objective optimization for scenario 1 ($w_1 = w_2 = 0.5$)

Algorithm	Objective	Mean values	Standard deviation	Best value	Z	Optimized process parameter			
						P_a	W_p	f_p	S_c
ABC	KT (°) *Ra* (μm)	3.553	1.007	0.321 0.241	1.8615	7.86	1.6	8.0002	6.067
ACO	KT (°) *Ra* (μm)	2.482	0.723	0.300 0.299	1.9539	6.38	1.6	8.0654	6
PSO	KT (°) *Ra* (μm)	2.396	0.818	0.344 0.164	1.6764	7.16	1.6	8	6
DE	KT (°) *Ra* (μm)	2.987	1.429	0.407 0.189	1.7136	7.93	1.6	8.0123	6
GA	KT (°) *Ra* (μm)	3.288	1.040	0.301 0.336	1.8988	6.246	1.6	8	6
ESWSA	KT (°) *Ra* (μm)	1.649	0.014	0.294 0.133	1.6473	7.437	1.6	8	6

Table 6 Optimized values of parameters considering multi-objective optimization for scenario 2 ($w_1 = 0.9, w_2 = 0.1$)

Algorithm	Objective	Mean values	Standard deviation	Best value	Z	Optimized Process Parameter			
						P_a	W_p	f_p	S_c
ABC	KT (°) Ra (μm)	2.523	0.236	0.326 0.192	2.216	6.865	1.6	8	6
ACO	KT (°) Ra (μm)	2.371	0.289	0.291 0.350	2.174	6.179	1.6	8.1055	6
PSO	KT (°) Ra (μm)	2.246	0.251	0.311 0.298	2.152	6.1552	1.6	8	6
DE	KT (°) Ra (μm)	2.407	0.291	0.281 0.472	2.247	5.4035	1.6	8.047	6
GA	KT (°) Ra (μm)	2.368	0.255	0.294 0.341	2.183	6.3006	1.6	8.125	6
ESWSA	KT (°) Ra (μm)	2.142	0.231	0.283 0.292	2.147	6.185	1.6	8	6

Table 7 Optimized values of parameters multi-objective optimization for scenario 3 ($w_1 = 0.1, w_2 = 0.9$)

Algorithm	Objective	Mean values	Standard deviation	Best value	Z	Optimized Process Parameter			
						P_a	W_p	f_p	S_c
ABC	KT (°) Ra (μm)	3.8	2.813	0.402 0.108	1.3500	7.8153	1.6	8	6
ACO	KT (°) Ra (μm)	3.257	2.458	0.405 0.102	1.3054	7.9141	1.6	8	6
PSO	KT (°) Ra (μm)	2.294	1.762	0.410 0.104	1.3189	7.8824	1.6	8	6
DE	KT (°) Ra (μm)	3.885	3.342	0.398 0.106	1.3406	7.8350	1.6	8	6
GA	KT (°) Ra (μm)	1.875	0.430	0.366 0.133	1.647	7.4366	1.6	8	6
ESWSA	KT (°) Ra (μm)	1.277	0.074	0.413 0.114	1.2725	6	1.6	8	6

5 Results and Discussions

To highlight the effects of the considered LBM parameters on KT and *Ra* scatter plots are shown in Figs. 5 and 6 respectively. From Fig. 5, it can be noticed that with increment in the values of both P_a and W_p, KT first decreases, and then it starts increasing from the intermediate values of P_a and W_p. . Thus, moderate value of assist gas (oxygen) pressure is preferred to blow out the molten material from the kerf during LBM of aluminium sheet due to its highly reflective nature towards lasers.

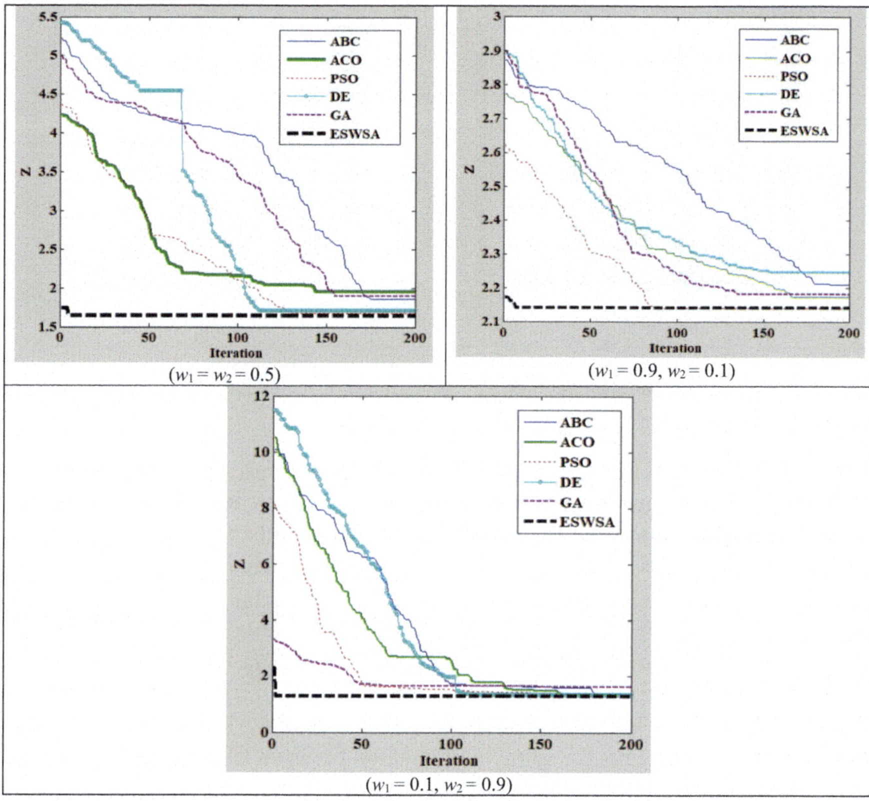

Fig. 3 Convergence diagrams for different weighting scenarios for multi-objective optimization problems

Table 8 Average computational time for the considered optimization techniques

Response	Method	Average computational time (sec)
KT (°)	ABC	12.543
	ACO	8.879
	PSO	11.939
	DE	7.817
	GA	5.726
	ESWSA	1.274
Ra (μm)	ABC	15.056
	ACO	10.837
	PSO	9.865
	DE	7.077
	GA	7.179
	ESWSA	1.644

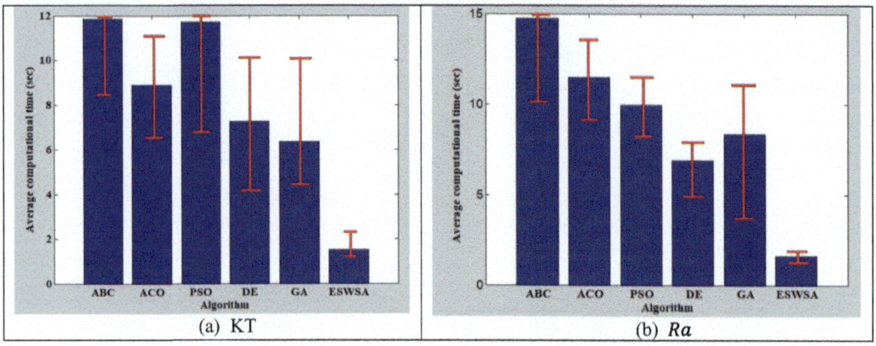

Fig. 4 Comparison of computational time for the optimization techniques

While maintaining f_p at high level in LBM process, an increase in W_p results in lower KT due to increased overlapping rate of the laser beam. A further increase in W_p increases the interaction time between the laser beam and sheet material, causing wider top kerf and lesser beam overlapping rate. With increasing values of f_p, the change in KT is noticed to be curvilinear, reaching its minimum value at moderate setting of f_p. This is due to the requirement of higher value of beam overlapping rate for profile cutting operation, but after a certain level of f_p, the laser beam cannot be absorbed by the reflective sheet material due to more interaction time for laser absorption. As noticed in Fig. 5, with increasing values of S_c, KT increases as lower value of S_c is advantageous for suitable interaction between laser and sheet material during LBM operation. Moreover, at higher S_c, , laser cannot properly interact with the sheet material, resulting in poor cut quality (Singh et al. 2021).

It can also be unveiled from Fig. 6 that an increment in P_a is responsible for blowing away the molten material from the machining zone, causing better surface quality with reduced *Ra* values. The surface quality of the machined components deteriorates with higher values of both W_p and f_p. At higher W_p, laser interacts with the sheet material for a longer time with lower peak power due to which more thermal energy would be dispersed on the top surface of the sheet material, resulting in higher *Ra* values. Surface roughness also increases with increment in f_p because higher f_p results in increased beam overlapping rate allowing sufficient energy to melt the sheet material for quality cutting. It can be observed from Fig. 6 that the *Ra* value first increases and then sharply decreases with increasing S_c during the said LBM process. This is due to reduction in available cooling time for resolidification of the sheet material at higher S_c resulting in formation of thick oxide layer on the sheet material surface causing poorer surface quality (Sharma and Yadava 2012b).

To prove the superiority of ESWSA over the other state-of-the-art algorithms, two-tailed paired t-test is performed where, the Null hypothesis is that the population mean for the compared algorithms is equal whereas in alternate hypothesis is that the population mean is unequal.

Here, population mean denotes the average value of each of the two responses (KT and *Ra*) obtained after 200 iterations for each of the algorithms. The results of

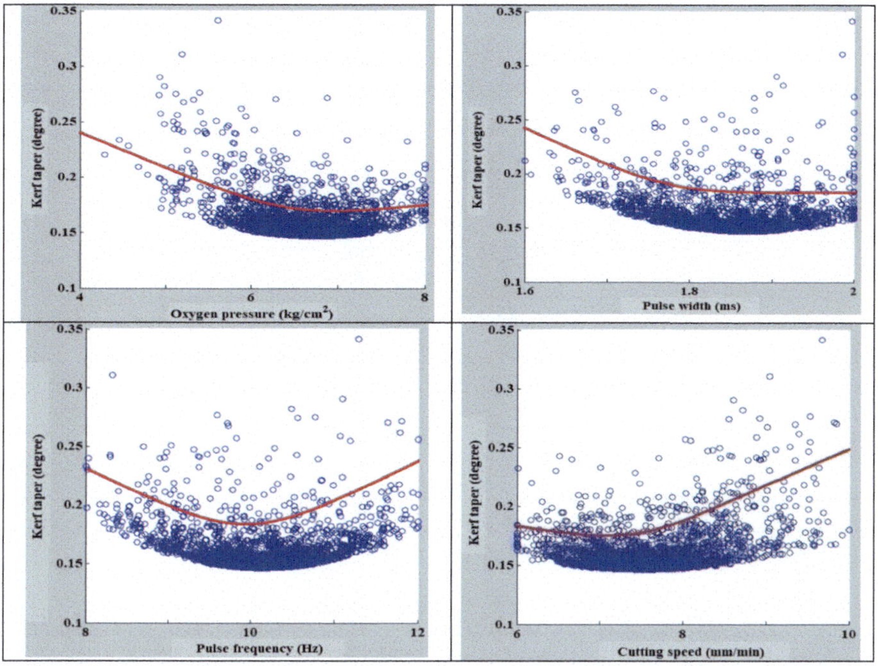

Fig. 5 Effects of LBM process parameters on KT

the paired *t*-tests for multi-objective optimization for different criteria weighting scenarios are depicted in Table 9. The results show that we can reject the null hypotheses for all the paired t-tests. This is because the test statistics' absolute values are higher than the critical values at the 5% significance level. Even if the one-tailed test is performed, these null hypotheses can still be rejected. It thus proves uniqueness of ESWSA over the other state-of-the-art algorithms under consideration.

6 Conclusions

This paper proposes the novel application of ESWSA in solving the parametric optimization problem of Nd:YAG laser-based straight profile cutting of thin aluminium alloy sheet. Its optimization performance is also contrasted against other state-of-the-art algorithms, i.e. ABC, ACO, PSO, DE and GA. ESWSA outperform the other algorithms in terms of accuracy and consistency of the solutions it provides. Additionally, it requires less computational effort compared to the other methods.

(a) In single-objective optimization, an optimal combination of $P_a = 6.772$ kg/cm^2, $W_p = 1.870$ ms, $f_p = 10.140$ Hz and $S_c = 7.541$ mm/min would result in the minimum KT value as 0.1469°.

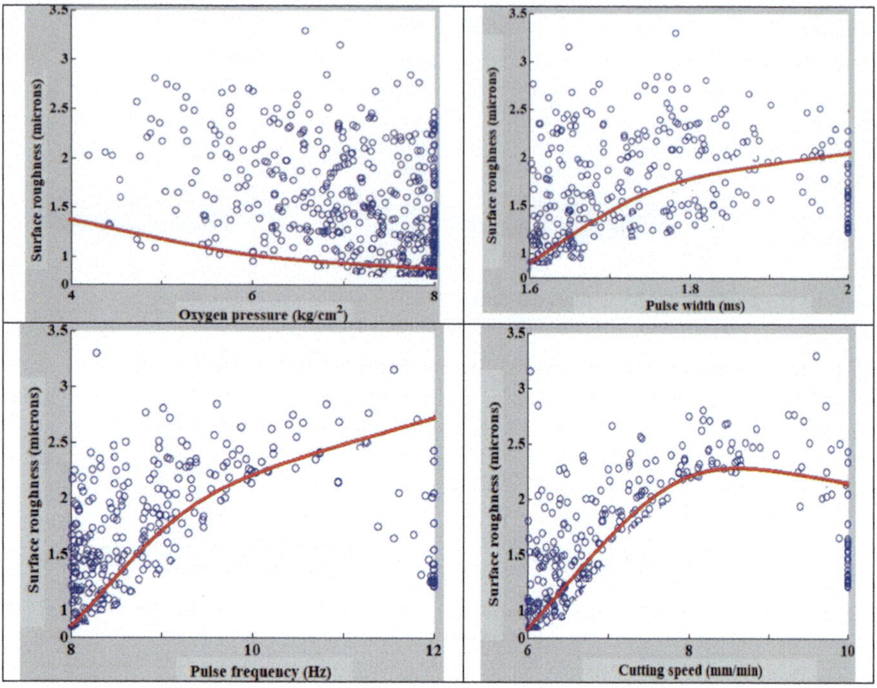

Fig. 6 Effects of LBM parameters on *Ra*

Table 9 Results of two-tailed paired *t*-tests

Response	ABC	ACO	PSO	DE	GA
Scenario 1 ($w_1 = w_2 = 0.5$)					
KT	15.58	32.36	14.56	16.58	25.68
Ra	18.85	15.89	25.46	11.28	14.58
Scenario 2 ($w_1 = 0.9$, $w_2 = 0.1$)					
KT	14.28	35.28	12.23	15.89	20.58
Ra	20.32	12.51	19.58	11.78	14.47
Scenario 3 ($w_1 = 0.1$, $w_2 = 0.9$)					
KT	13.24	32.76	11.78	14.86	19.78
Ra	18.64	14.63	21.52	15.21	10.23

(b) On the other hand, minimum value of *Ra* as 0.9948 μm can be attained when the LBM parameters are set at $P_a = 8$ kg/cm^2, $W_p = 1.6$ ms, $f_p = 8$ Hz and $S_c = 6$ mm/min.

(c) In multi-objective optimization of the said process, an optimal combination of $P_a = 7.437$ kg/cm^2, $W_p = 1.6$ ms, $f_p = 8$ Hz and $S_c = 6$ mm/min would lead to simultaneous attainment of both KT and *Ra* values as 0.294° and 0.133 μm respectively when the considered responses have equal importance.
(d) It is noticed that for multi-objective optimization, ESWSA achieves 8.41, 2.00, 14.53, 27.76 and 2.32%; and 44.81, 55.52, 18.90, 29.63 and 60.42% improvements respectively for KT and *Ra* against ABC, ACO, PSO, DE and GA techniques when equal importance is assigned to both the responses.
(e) As compared to other state-of-the-art algorithms, ESWSA converges to the optimal objective function values within 5–10 iterations. It has also minimum variability in the derived objective function values.

Thus, it can be concluded that ESWSA can be employed as an efficient and faster optimization tool for searching out the almost global parametric combinations of various machining processes while exploring their fullest potential leading to sustainable manufacturing. As the no free lunch theorem indicates, no algorithm is perfect in solving all types of problems with equal potency. Nevertheless, it is to try and test out the newly developed ESWSA on other classes of optimization problems. It is worthwhile to mention here that as the optimization of the said LBM process is performed based on a past experimental dataset, it is not possible to perform any confirmatory experiment to validate the derived solutions.

7 Future Scope

This study opens several avenues for future research and development in the domain of laser beam machining process optimization using metaheuristic algorithms:

1. The ESWSA algorithm can be tested and applied to other non-conventional machining processes such as electrical discharge machining (EDM), electrochemical machining (ECM), and abrasive water jet machining (AWJM) to assess its versatility and efficiency.
2. Investigate the application of ESWSA in optimizing machining parameters for multi-material systems, which are increasingly used in advanced manufacturing applications.
3. Future work can explore integrating ESWSA with machine learning techniques to develop predictive models that can further enhance the optimization process by providing real-time adaptive control of machining parameters.
4. Conducting comprehensive experimental studies to validate the optimization results derived from ESWSA will be crucial. This will involve setting up controlled experiments to measure actual machining performance and comparing these results with predicted outcomes.
5. With the continuous development of new metaheuristic algorithms, future research can compare the performance of ESWSA with recently developed algorithms like the Salp Swarm Algorithm (SSA), Ant Lion Optimizer (ALO),

and Moth Flame Optimization (MFO) to establish its relative effectiveness and robustness.
6. A thorough analysis of the environmental and economic impacts of implementing ESWSA-optimized machining processes in industrial settings can provide valuable insights for sustainable manufacturing practices.
7. Exploring hybrid optimization techniques that combine ESWSA with other optimization methods could potentially yield superior results by leveraging the strengths of each method.
8. Developing real-time optimization systems that utilize ESWSA to dynamically adjust machining parameters during the production process could significantly enhance machining efficiency and product quality.

By addressing these areas, future research can significantly contribute to the advancement of optimization techniques in laser beam machining and other related manufacturing processes.

References

Bakhtiyari AN, Wang Z, Wang L, Zheng H (2021) A review on applications of artificial intelligence in modeling and optimization of laser beam machining. Opt Laser Technol 135:106721. https://doi.org/10.1016/j.optlastec.2020.106721

Chatterjee S, Abhishek K, Mahapatra SS (2017) A study on surface quality of laser drilled holes: parametric optimization using harmony search algorithm. Int J Mater, Mech Manufact 5(4):251–254. https://doi.org/10.18178/ijmmm.2017.5.4.329

Diyaley S, Chakraborty S (2020) An analysis on the parametric optimization of electrochemical honing process. J Adv Manuf Syst 19(2):249–276. https://doi.org/10.1142/S0219686720500134

Dubey AK, Yadava V (2008a) Laser beam machining—a review. Int J Mach Tools Manuf 48(6):609–628. https://doi.org/10.1016/j.ijmachtools.2007.10.017

Dubey AK, Yadava V (2008b) Experimental study of Nd laser beam machining—an overview. J Mater Process Technol 195:15–26. https://doi.org/10.1016/j.jmatprotec.2007.05.041

Elsheikh AH, Shehabeldeen TA, Zhou J, Showaib E, Elaziz MA (2021) Prediction of laser cutting parameters for polymethylmethacrylate sheets using random vector functional link network integrated with equilibrium optimizer. J Intell Manuf 32:1377–1388. https://doi.org/10.1007/s10845-020-01617-7

Gautam GD, Mishra DR (2019) Dimensional accuracy improvement by parametric optimization in pulsed Nd laser cutting of Kevlar-29/basalt fiber-reinforced hybrid composites. J Braz Soc Mech Sci Eng 41:284. https://doi.org/10.1007/s40430-019-1783-y

Gautam GD, Mishra DR (2020) Parametric investigation in pulsed Nd laser cutting of Kevlar-basalt fiber composite. Lasers Manuf Mater Proc 7:373–398. https://doi.org/10.1007/s40516-020-00125-z

Goswami D, Chakraborty S (2015) A study on the optimization performance of fireworks and cuckoo search algorithms in laser machining processes. J Inst Eng: Ser C 96(3):215–229. https://doi.org/10.1007/s40032-014-0160-y

Hiwale S, Basavarajappa (2020) Investigations of laser machining parameters using RSM and optimization techniques for Hastelloy C-276. Int J Modern Manuf Technol 12(1):64–74

Madić M, Radovanović M (2013) Application of cuckoo search algorithm for surface roughness optimization in CO_2 laser cutting. Ann Fac Eng Hunedoara—Int J Eng 11(1):39–44

Madić M, Mladenović S, Gostimirović M, Radovanović M, Janković P (2020) Laser cutting optimization model with constraints: maximization of material removal rate in CO_2 laser cutting of mild steel. Proc Inst Mech Eng, Part b: J Eng Manuf 234(10):1323–1332. https://doi.org/10.1177/0954405420911529

Madić M, Petrović G, Petković D, Antucheviciene J, Marinković D (2022) Application of a robust decision-making rule for comprehensive assessment of laser cutting conditions and performance. Machines 10:153. https://doi.org/10.3390/machines10020153

Mandal S (2018) Elephant swarm water search algorithm for global optimization. Sadhana 43:2. https://doi.org/10.1007/s12046-017-0780-z

Mishra L, Mahapatra TR, Mishra D, Pattanaik SK (2022) Machinability analysis and multiple performance optimization during laser micro-drilling of CNT reinforced polymer nanocomposite. Lasers Manuf Mater Process 9:151–172. https://doi.org/10.1007/s40516-022-00171-9

Mishra L, Mahapatra TR, Mishra D, Parimanik SR (2023) Investigation of laser micro-drilling machinability and performance optimization of polymer nanocomposites reinforced with different carbon allotropes. Proc Inst Mech Eng, Part E: J Process Mech Eng 09544089231158898. https://doi.org/10.1177/09544089231158898

Mukherjee R, Goswami D, Chakraborty S (2013) Parametric optimization of Nd laser beam machining process using artificial bee colony algorithm. J Ind Eng Article ID 570250, 15. https://doi.org/10.1155/2013/570250

Najjar IMR, Sadoun AM, Elaziz MA, Abdallah AW, Fathy A, Elsheikh AH (2022) Predicting kerf quality characteristics in laser cutting of basalt fibers reinforced polymer composites using neural network and chimp optimization. Alex Eng J 61:11005–11018. https://doi.org/10.1016/j.aej.2022.04.032

Phipon R, Pradhan BB (2012) Control parameters optimization of laser beam machining using genetic algorithm. Int J Comput Eng Res 2(5):1510–1516. https://doi.org/10.1504/IJMMM.2012.048557

Prajapati A, Norkey G, Gautam GD (2022) Optimization of heat affected zone in laser cutting of Kevlar-29 fiber composite using hybrid response surface based grey wolf optimization (RSGWO) algorithm. Proc Inst Mech Eng C J Mech Eng Sci 236:9622–9638. https://doi.org/10.1177/09544062221096557

Pramanik D, Roy N, Kuar AS, Sarkar S, Mitra S (2022) Experimental investigation of sawing approach of low power fiber laser cutting of titanium alloy using particle swarm optimization technique. Opt Laser Technol 147:107613. https://doi.org/10.1016/j.optlastec.2021.107613

Rajamani D, Siva Kumar M, Balasubramanian E, Tamilarasan A (2021) Nd: YAG laser cutting of Hastelloy C276: ANFIS modeling and optimization through WOA. Mater Manuf Process 36(15):1746–1760. https://doi.org/10.1080/10426914.2021.1942910

Rohman MN, Ho J-R, Tung P-C, Lin C-T, Lin C-K (2022a) Prediction and optimization of dross formation in laser cutting of electrical steel sheet in different environments. J Market Res 18:1977–1990. https://doi.org/10.1016/j.jmrt.2022.03.106

Rohman MN, Ho J-R, Tung P-C, Tsui HP, Lin C-K (2022b) Prediction and optimization of geometrical quality for pulsed laser cutting of non-oriented electrical steel sheet. Opt Laser Technol 149:107847. https://doi.org/10.1016/j.optlastec.2022.107847

Sharma A, Yadava V (2012a) Modelling and optimization of cut quality during pulsed Nd laser cutting of thin Al-alloy sheet for straight profile. Opt Laser Technol 44(1):159–168. https://doi.org/10.1016/j.optlastec.2011.06.012

Sharma A, Yadava V (2012b) Modelling and optimisation of pulsed Nd: YAG laser cutting for average kerf taper and surface roughness during straight cutting of Ni-based super alloy thin sheet. Int J Mach Mach Mater 11(3):223–243. https://doi.org/10.1504/IJMMM.2012.046889

Sibalija T, Petronic S, Milovanovic D (2019) Experimental optimization of Nimonic 263 laser cutting using a particle swarm approach. Metals 9:1147. https://doi.org/10.3390/met9111147

Šibalija TV, Petronić SZ (2021) The comparison of metaheuristic algorithms in parametric optimization of laser-based processes. Tehnika 76(2):165–170. https://doi.org/10.5937/tehnika2102165S

Singh Y, Singh J, Sharma S, Aggarwal V, Pruncu CI (2021) Multi-objective optimization of kerf-taper and surface-roughness quality characteristics for cutting-operation on coir and carbon fibre reinforced epoxy hybrid polymeric composites during CO_2-pulsed laser-cutting using RSM. Lasers Manuf Mater Proc 8:157–182. https://doi.org/10.1007/s40516-021-00142-6

Singh Y, Singh J, Sharma S, Sharma A, Chohan JS (2022) Process parameter optimization in laser cutting of coir fiber reinforced epoxy composite—a review. Mater Today: Proc 48:1021–1027. https://doi.org/10.1016/j.matpr.2021.06.344

Sun S, Brandt M (2013) Laser beam machining. In: Davim J (ed) Nontraditional machining processes. Springer. https://doi.org/10.1007/978-1-4471-5179-1_2

Vasiga D, Channankaiah D (2015) A review of carbon dioxide laser on polymers. Int J Eng Res Technol 4(3):874–877. https://doi.org/10.17577/IJERTV4IS031027

Progressive Automation: Mapping the Horizon of Smart Manufacturing with RoboDK Workstations and Industry 4.0

Mukul Kumar, Sourabh Anand, Pushpendra S. Bharti, Manoj Kumar Satyarthi, Parveen Kumar, and Ajay Kumar

Abstract The emergence of Industry 4.0 has triggered a fundamental change in the manufacturing industry, introducing a new age of intelligent and interconnected production systems. The central focus of this shift is on the rapid advancement of automation, driven by state-of-the-art technology such as Robotics, AI, and IoT. This study explores the convergence of progressive automation, RoboDK workstations, and Industry 4.0, clarifying the mutually beneficial connection between these elements in creating the future of production. This study aims to provide insights on the incorporation of virtual commissioning, simulation, and optimisation approaches within smart manufacturing settings by thoroughly examining the methodology used in the partnership of RoboDK Workstations and Industry 4.0. This study emphasizes the significance of RoboDK in enabling virtual development, training, and deployment of robotic systems. It emphasises how RoboDK acts as a catalyst for innovation and efficiency improvement in contemporary manufacturing. Moreover, it tackles important obstacles and outlines future paths, creating an environment for a production system that is more adaptable and flexible, ready to fully use advanced automation and Industry 4.0 technology.

Keyword Smart manufacturing · RoboDK Workstations · Industry 4.0 · Robotics · Automation

M. Kumar · S. Anand · P. S. Bharti · M. K. Satyarthi
University School of Information, Communication and Technology, Guru Gobind Singh Indraprastha University, Sector 16 C, Dwarka, New Delhi, India

P. Kumar
Department of Mechanical Engineering, Rawal Institute of Engineering and Technology, Faridabad, Haryana, India

A. Kumar (✉)
Department of Mechanical Engineering, School of Engineering and Technology, JECRC University, Jaipur, Rajasthan, India
e-mail: ajay.kumar30886@gmail.com

A. Kumar et al. (eds.), *Industry 4.0 Driven Manufacturing Technologies*, Springer Series in Advanced Manufacturing, https://doi.org/10.1007/978-3-031-68271-1_15

1 Introduction

As the industrial environment continues to undergo significant transformations, the confluence of digital technologies and automation has signalled the beginning of a new age, which is referred to as Industry 4.0 (Aceto et al. 2019; Koh et al. 2019). Because of this paradigm change, conventional industrial processes are undergoing a revolution, which is moving them towards levels of efficiency, agility, and creativity that have never been seen before. The incorporation of latest technology like robots, artificial intelligence, big data analytics, and the Internet of Things (IoT) (as shown in the Fig. 1) is at the core of this change (Rani et al. 2023; Kumar et al. 2023, 2024a). These technologies together redefine the manner in which things are created and disseminated.

1.1 Overview of Industry 4.0

Industry 4.0 marks the beginning of a significant period of change in manufacturing, when conventional methods are being completely transformed by the integration of advanced technologies (Lu 2017; Oztemel and Gursev 2020; Xu et al. 2018). The

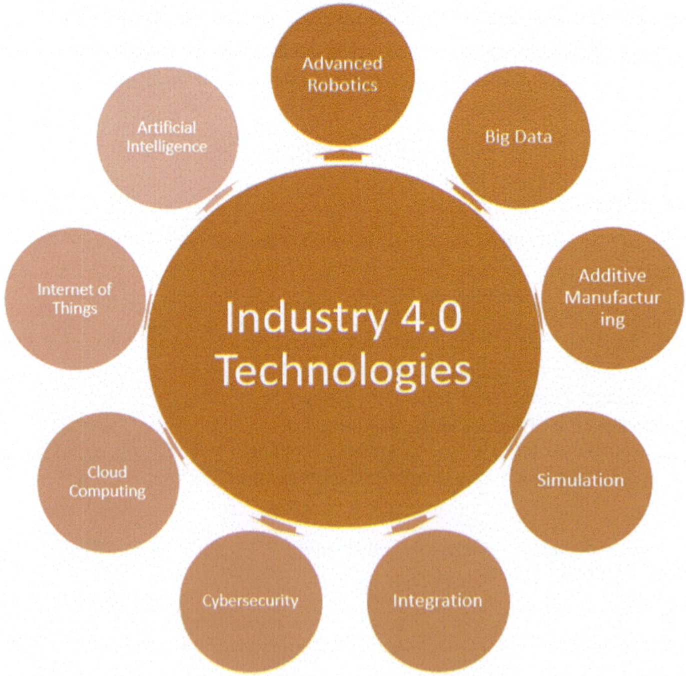

Fig. 1 Industry 4.0 technologies

core of this technology is Advanced Robotics, which allows for unparalleled levels of automation and precision in production processes (Batista et al. 2024). In addition, Big Data enhances decision-making by analysing extensive datasets, streamlining processes, and projecting maintenance requirements (Burande et al. 2024). Simultaneously, Additive Manufacturing enables the exploration of novel opportunities, facilitating the production of intricate parts while minimizing the use of materials (Srivastava et al. 2023). Simulation enhances this even more by allowing for virtual modelling of industrial processes, enabling optimization and troubleshooting in an environment without any risks. Integrating systems throughout the supply chain guarantees smooth communication and coordination, hence improving total efficiency (Rodič 2017). In the midst of this period of technological advancement, Cybersecurity becomes a crucial factor, ensuring the protection of valuable production data and securing interconnected systems against cyber threats. Cloud Computing provides scalable resources for storing and processing data, making it easier to access and collaborate in real-time (Rao et al. 2012). The widespread presence of interconnected devices facilitated by the Internet of Things (IoT) allows for the continuous monitoring and management of the production environment (Manavalan and Jayakrishna 2019). Meanwhile, Artificial Intelligence (AI) facilitates automation through the use of machine learning and sophisticated algorithms, enabling the application of predictive analytics and autonomous decision-making (Naveena et al. 2024). Among all of these technological developments, the use of RoboDK workstations stands out as a crucial element in fully harnessing the promise of Industry 4.0. RoboDK is a powerful platform that enables organizations to model, test, and optimize production processes (La Commare R 2020). This empowers them to save costs, limit downtime, and improve overall efficiency. Overall, the amalgamation of Industry 4.0 technology and the use of RoboDK workstations highlights the importance of improving automation in smart production. Embracing this shift is not only beneficial but also unavoidable, guaranteeing competitiveness in a rapidly changing industry. Nevertheless, it is imperative to tackle difficulties such as enhancing the skills of the workforce, ensuring cybersecurity, and deploying ethical artificial intelligence. Manufacturers may achieve sustainable growth, promote innovation, and contribute to societal advancement by adopting and integrating advanced automation and Industry 4.0 technology (Mohammadi et al. 2024). Collectively, we are creating a future in which advanced manufacturing not only enhances economic well-being but also fosters environmental sustainability and societal advancement.

1.2 Introduction to RoboDK Workstations

RoboDK Workstations are vital elements of Industry 4.0-driven production automation. They provide a full software platform that enables the design, simulation, and optimisation of robotic systems. RoboDK offers an intuitive interface that enables engineers and manufacturers to create and visualise robotic processes in a simulated environment prior to implementing them in the real world (Crnokić 2023; Ionescu

2020; Pieskä et al. 2018; Ribeiro 2019). Key characteristics of RoboDK Workstations include a simulation environment, robot interoperability, offline programming, adaptability, and customization, as well as interaction with Industry 4.0 technologies. RoboDK simulation environment allows users to generate precise 3D simulations of robotic work cells, including robots, end-effectors, workpieces, and other peripherals (Trochimczuk et al. 2019). The virtual environment enables comprehensive testing and validation of robotic programmes, hence minimising the likelihood of mistakes and accidents during real-world operation. RoboDK is compatible with a diverse selection of industrial robots manufactured by industry leaders, guaranteeing interoperability with current automation equipment. RoboDK offers solutions to effortlessly include many types of robotic systems, such as articulated arms, SCARA robots, and collaborative robots, into industrial processes (Adamu Yusuf 2019; Saukkoriipi 2019; Sivasankaran and Karthikeyan 2020).

The offline programming capabilities of RoboDK simplifies the programming process, minimises periods of inactivity, and expedites the implementation of new robotic applications. Users may customise and adapt robotic systems to meet their specific needs, thanks to the flexibility and customisation choices available. In addition, RoboDK is specifically engineered to effortlessly incorporate with other Industry 4.0 technologies, including IoT platforms, cloud services, and data analytics tools (Aqlan et al. 2020). This allows manufacturers to develop intelligent and linked production systems that enhance efficiency, quality, and productivity. The significance of manufacturing automation is shown in Table 1 (Table 2).

2 Understanding the Concept of Industry 4.0

Recognizing the conceptual framework of Industry 4.0 is crucial for understanding the ground-breaking path of modern industries. This is where the convergence of various technological advances enables the creation of intelligent factories, leading to increased productivity and creativity. The core of this transformation is the network of interconnected systems, enabled by the Internet of Things (IoT) that effortlessly links machines, sensors, and personnel (Abdul-Qawy et al. 2015). This integration enables the instantaneous surveillance and administration of production operations (Patel 2023). Significantly, Industry 4.0 relies on the open availability of data, utilizing up-to-date information to make well-informed decisions and optimize processes along the whole value chain. By incorporating artificial intelligence (AI) and machine learning (ML) into cyber-physical systems (CPS), it becomes possible to achieve decentralized decision-making (Ahmed et al. 2021). This enables robots to independently carry out tasks and adjust to changing production requirements. In addition, contemporary technologies such as augmented reality (AR) and virtual reality (VR) offer technical assistance to operators, providing instant instruction and support (Jumani et al. 2022). This improves efficiency and reduces errors in complex assembly or maintenance activities. This change in direction profoundly redefines the powers of the industrial sector, bringing about the practical implementation of

Table 1 Applications of industry 4.0 technologies

Industry 4.0 technologies	Applications
Advanced robotics	Optimize automation in production processes to enhance productivity and accuracy, while prioritizing worker safety (Lee et al. 2019)
Big data	Conduct thorough analysis of extensive datasets to enhance production efficiency, anticipate maintenance requirements, and provide valuable insights for informed decision-making (Ren et al. 2019)
Additive manufacturing	Facilitates expedited development, personalization, and just-in-time manufacturing of intricate parts, thereby minimizing production delays and material inefficiency (Mofolasayo et al. 2022)
Simulation	Enables the virtual simulation and optimization of production processes, reducing risks and improving operational efficiency prior to deployment (Mourtzis et al. 2015)
Integration	Facilitates seamless integration of systems throughout the supply chain, allowing for instantaneous communication and coordination to improve efficiency and responsiveness (Bejlegaard et al. 2021)
Cybersecurity	Secures critical manufacturing data and systems from cyber-attacks, guaranteeing uninterrupted operations and defending against possible breaches (Corallo et al. 2020)
Cloud computing	Offers expandable resources for the storage, manipulation, and cooperation of data, enabling immediate access to information and improving operational adaptability (Bansal and Kumar 2020)
Internet of things (IOT)	Connects equipment and sensors to monitor and control production processes in real-time, improving efficiency and allowing for predictive maintenance (Ayvaz and Alpay 2021)
Artificial intelligence (AI)	Enables automation, predictive analytics, and decision-making in production, hence improving productivity, quality, and adaptability (Fragapane et al. 2022)

Table 2 Importance of manufacturing automation

Importance	Description
Efficiency	Manufacturing automation enables the optimisation of production processes, leading to reduced cycle times and improved throughput (Rüßmann et al. 2015)
Quality	Automation ensures constant product quality by minimising human error and unpredictability in manufacturing activities (Lee and Seppelt 2009)
Cost savings	Manufacturing automation reduces operational costs and improves profitability by automating repetitive tasks and improving resource allocation (Gray et al. 1993)
Flexibility	Automation facilitates quick modification of manufacturing processes, enabling producers to promptly respond to changing market needs and effectively personalise goods (Zawadzki and Żywicki 2016)
Safety	Automated systems have the capability to manage dangerous jobs, so diminishing the likelihood of workplace mishaps and guaranteeing a more secure working environment for workers (Schulte et al. 2020)

intelligent production through the integration of systems, sharing of data, distribution of decision-making power, and provision of technical assistance. RoboDK workstations are essential instruments in the manufacturing industry, allowing firms to fully utilize the ground-breaking capabilities of Industry 4.0. Manufacturers may improve the efficiency, flexibility, and inventiveness of their manufacturing operations by utilizing RoboDK modelling, simulation, testing, and optimization capabilities (Fox 2022). Industry 4.0 is propelled by fundamental technologies that enhance this transformative process, creating a unified ecosystem where data-driven decision-making, increased level of automation, and adaptable production processes come together. Artificial Intelligence (AI) takes the lead by analysing data, improving efficiency and intelligence (Wamba-Taguimdje et al. 2020). Collaborative Robots (cobots) and autonomous robotic systems transform manufacturing by automating tasks and ensuring workplace safety through collaboration between humans and robots (Bragança et al. 2019). The Internet of Things (IoT) utilizes data from sensors to enhance manufacturing efficiencies, while Big Data analytics and predictive analytics provide valuable information for enhancing processes and predicting demand. Simultaneously, the utilization of 3D printing technology speeds up the process of manufacturing, allowing for quick creation of prototypes (Aggoune et al. 2024; Goyal et al. 2024a, b; Kumar et al. 2024b) and modification, ultimately improving production efficiency and minimizing the time it takes to complete a project.

The incorporation of many technologies forms a unified system in which decision-making is guided by data, automation is enhanced, and manufacturing processes become more adaptable. Manufacturers are moving towards a future characterized by being responsive, sustainable, and strategically advantageous, positioning themselves at the forefront of industrial transformation. The development of Industry 4.0 has far-reaching effects on industrial processes and business models, resulting in increased efficiency, adaptability, and creativeness. Manufacturers can gain increased production and throughput by implementing automation and optimization techniques. Additionally, agility allows them to quickly respond to changing market demands. Enhancing quality assurance processes, while implementing cost reduction efforts improves operational efficiency. The adoption of digital technology in the workforce guarantees its continued relevance in an increasingly computerized environment. Industry 4.0 represents a new phase of production that combines technology innovation with strategic foresight to promote sustainable growth and differentiation (Kumar et al. 2024c).

3 The Evolution of Automation in Manufacturing

Automation has played a pivotal role in driving advancements in manufacturing, fundamentally transforming sectors and completely overhauling production methods (as shown in the Fig. 2). The origins of automation may be traced to the early stages of the industrial revolution, during which basic mechanical devices such as waterwheels and steam engines were used to drive early manufacturing processes

(David 2001; Heikkilä and Wikström 2021). However, it was not until the twentieth century that automation really began to advance, thanks to the emergence of electric motors and assembly lines. These advancements were led by innovative people like as Henry Ford (Brooke 2008). The advent of programmable logic controllers (PLCs) in the mid-twentieth century was a notable achievement, enabling enhanced regulation and adaptability in industrial production procedures (Mourtzis et al. 2022). With the progression of computing technology, the incorporation of computers into production systems led to the emergence of Computer Numerical Control (CNC) machines, which enhanced the automation of machining and fabrication activities (Zhang et al. 2011). The twenty-first century saw the combination of automation and digital technology, resulting in The emergence of intelligent manufacturing and Industry 4.0 (Rüßmann et al. 2015). The current era of intelligent manufacturing is driven by interconnected systems, Internet of Things (IoT) devices, artificial intelligence (AI)-powered analytics, and robotics. (Wan et al. 2020). These advancements are facilitating remarkable levels of efficiency, flexibility, and customisation. However, these significant advancements are hindered by challenges such as the need to improve the workforce's abilities to adapt to the changing technological landscape, addressing cybersecurity concerns in interconnected systems, and ensuring equitable access to automation technologies to prevent exacerbating socio-economic inequalities (Centobelli et al. 2022). Although there are difficulties, the potential benefits of adopting automation are extensive, including improved production, higher quality, and greater safety and sustainability. Manufacturers can fully harness the power of automation to foster innovation and enhance competitiveness in the global marketplace by effectively addressing these obstacles and capitalising on the available possibilities (Table 3).

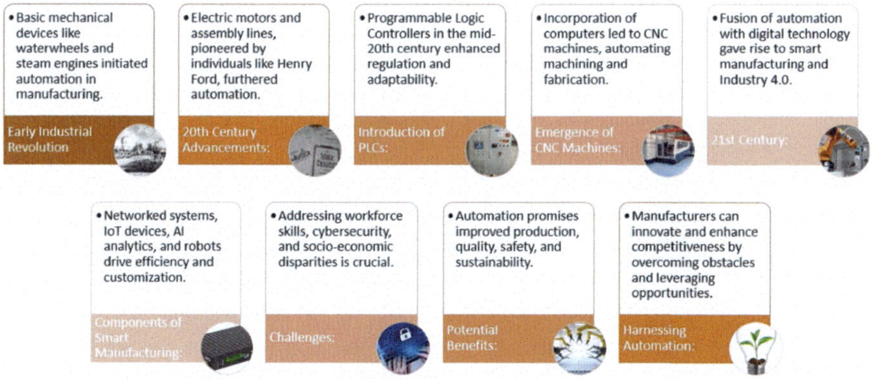

Fig. 2 Evolution of automation in manufacturing

Table 3 Literature of automation

Author(s) and Year	Title	Key focus	Key findings
Mathew et al. (2023)	Artificial intelligence powered automation for industry 4.0	AI in industry 4.0	Artificial intelligence (AI) plays a crucial role in improving industry 4.0 by increasing efficiency and profitability
Liu et al. (2023)	Digitalisation and servitization of machine tools in the era of Industry 4.0	Digitalization and servitization of machine tools	Examines the progress made in the digitalization and servitization of machine tools
Gažová et al. (2022)	Implementation of automation technologies of industry 4.0 in automotive manufacturing companies	Automation in automotive industry	Automotive firms are extensively utilizing industry 4.0 technologies
Gažová et al. (2022)	Effect of business process management on level of automation and technologies connected to industry 4.0	BPM impact on automation and industry 4.0 technologies	The deployment of industry 4.0 technologies is enhanced by BPM
Santos et al. (2022)	Use of simulation in the industry 4.0 context: creation of a digital twin to optimize decision making on non-automated process	Digital twin in industry 4.0	Digital twins enhance the efficiency of manual decision-making processes
Tyagi et al. (2020)	Intelligent automation systems at the core of industry 4.0	Intelligent automation systems	Intelligent automation combines artificial intelligence (AI), internet of things (IoT), and block chain technologies
Jasperneite et al. (2020)	Why we need automation models: handling complexity in industry 4.0 and the internet of things	Automation models and architectures for industry 4.0	Structured models effectively handle the intricacy of automation systems
Zawra et al. (2019)	Migration of legacy industrial automation systems in the context of industry 4.0	Migration strategies for legacy automation systems	Revamp outdated systems to seamlessly include internet of things (IoT), cyber-physical systems (CPS), and cloud computing
Sanghavi et al. (2019)	Industry 4.0: tools and implementation	Implementation of industry 4.0 tools	Explores the process of digitalization propelled by the internet of things (IoT), big data, and cloud computing

(continued)

Table 3 (continued)

Author(s) and Year	Title	Key focus	Key findings
Brecher et al. (2021)	Automation technology as a key component of the industry 4.0 production development path	Flexible production systems in industry 4.0	Outlines the trajectory of the transition to smart manufacturing
Saturno et al. (2017)	Proposal of an automation solutions architecture for industry 4.0	New automation architectures for industry 4.0	Suggests an architecture that is built around integrated functions for industry 4.0
Oesterreich et al. (2016)	Understanding the implications of digitisation and automation in the context of industry 4.0	Digitization and automation in industry 4.0	Examines the consequences of digitization and automation, with a particular focus on the construction industry

4 RoboDK Workstations: Enabling Smart Manufacturing

RoboDK is an advanced software application that plays a leading role in facilitating intelligent manufacturing with its unique features. RoboDK is primarily a virtual platform that allows for the modelling, simulation, and programming of robotic systems with exceptional accuracy and adaptability (Shamshiri et al. 2018). The interface of this software is designed to be easy for users to understand and use. It allows manufacturers to see and study complicated production situations before actually putting them into practice. This helps to greatly decrease the amount of time and money that would normally be spent on trial-and-error approaches. RoboDK virtual commissioning and simulation are essential features. RoboDK enables engineers to evaluate robot programmes, optimise trajectories, and find problems or bottlenecks in a digital environment that mimics real-world industrial settings, without causing any disruptions to production processes (Fox 2022). This feature not only improves the effectiveness and dependability of robotic systems but also reduces the amount of time they are not in operation and the potential for problems during the initial setup phase. Furthermore, RoboDK plays a key role in the incorporation of Industry 4.0 technology into industrial processes (Alexopoulos 2022). This promotes a manufacturing environment that is extremely adaptable and flexible. Manufacturers may enhance production schedules, anticipate maintenance requirements, and empower robotic systems to make autonomous decisions by using RoboDK interoperability with AI algorithms and predictive analytics. Essentially, the incorporation of RoboDK with Industry 4.0 technologies signifies a fundamental change in industrial processes, where automation, data-driven analysis, and virtualization come together to enable enhanced levels of efficiency, flexibility, and creativity. RoboDK is a crucial technology that enables producers to confidently and strategically negotiate the difficulties of current production settings as the industrial landscape continues to change.

5 Implementing RoboDK in Smart Manufacturing

5.1 Optimizing Production Line with RoboDK Simulations

The process of optimising a manufacturing line using RoboDK simulations requires a methodical approach focused on enhancing efficiency, precision, and overall effectiveness. Below is a comprehensive explanation of the sequential procedures (as shown in the Fig. 3).

First and foremost, it is necessary to precisely establish the goals of the optimisation process. The aims may vary, including the reduction of cycle durations, minimising downtime, improving productivity, and minimising material waste. After determining the goals, the subsequent phase is collecting pertinent data on the current manufacturing line. This includes comprehensive data on equipment specs, process flows, cycle periods, and production schedules. This data forms the basis for the process of simulating and modelling. The process of modelling the production line in RoboDK starts with importing CAD models of different components, such as

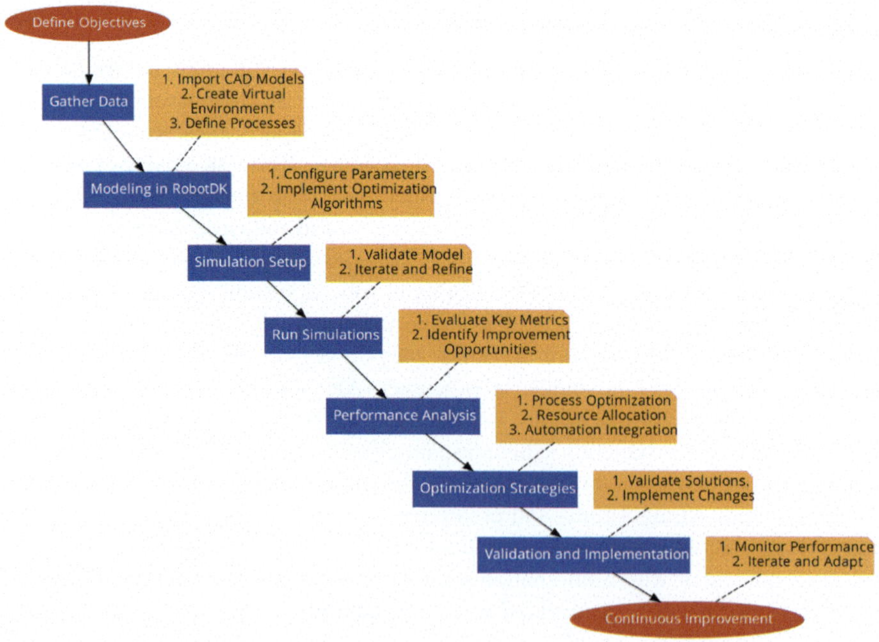

Fig. 3 Flowchart of Production Line with RoboDK simulations

machines, conveyors, robots, and workstations. Subsequently, these models are used to construct a virtual environment that accurately emulates the structure and setup of the tangible manufacturing line.

In the simulation environment, the manufacturing processes are specified, including the order of activities and the relationships between various components. Accurately portraying the dynamics of the manufacturing line and its numerous operations is essential at this stage. Once the simulation model is established, the subsequent stage entails adjusting simulation parameters, such as cycle durations, equipment speeds, and production quantities, according to the gathered data and optimisation goals. RoboDK offers optimisation methods that may be used to optimise the allocation of resources, scheduling of tasks, and efficiency of workflows. Simulations are conducted to verify the accuracy of the model and evaluate its effectiveness in various circumstances. The model undergoes iterative refinement, with required tweaks made to enhance accuracy and correspond with the optimisation goals.

Performance analysis is carried out to assess important parameters such as throughput, cycle time, resource utilisation, and identification of bottlenecks. This study facilitates the identification of specific areas within the production process that may be further optimised in order to accomplish the set goals. Optimisation plans are formulated and put into action based on the results obtained from the performance analysis. These techniques may include optimising processes, adjusting resource allocation, and integrating automation technologies like robotic systems. The offered optimisation strategies are validated using supplementary simulations or pilot testing in actual production scenarios. After being verified, the modifications are put into effect in the actual manufacturing line, with careful monitoring of performance and necessary adjustments being made.

The focus is on continuous improvement, with constant monitoring of performance measurements and iterative modification of optimisation algorithms. This iterative method guarantees that the manufacturing line stays optimised to meet changing production requirements, technology improvements, and market dynamics.

5.2 Implementing Collaborative Robots with RoboDK Workstations

The process of integrating collaborative robots with RoboDK workstations entails numerous crucial stages. Initially, evaluate the industrial setting to identify jobs that are appropriate for collaborative automation. Subsequently, choose the suitable collaborative robot model by considering factors such as payload capacity, reach distance, and safety attributes. Employ RoboDK software to model the interactions between robots and cells, guaranteeing an ideal arrangement and programming. Implement safety precautions such as sensors and barriers to ensure the protection of human operators. Provide training to staff on the operation and maintenance of collaborative robots. Perform comprehensive testing and validation procedures to

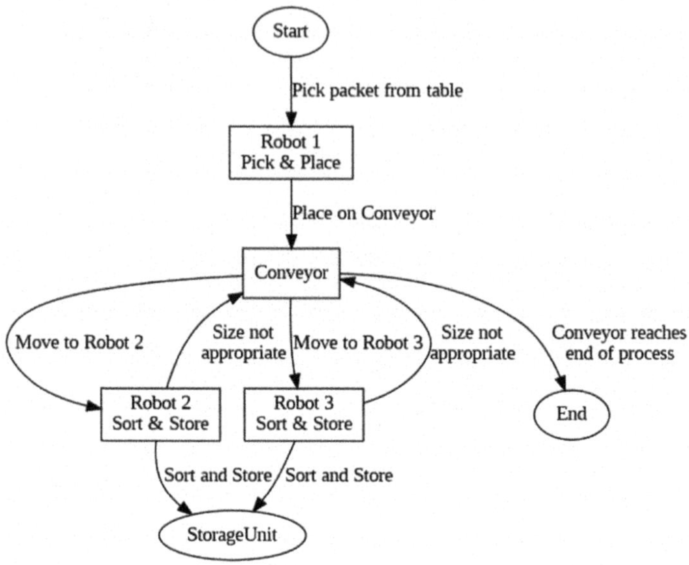

Fig. 4 Flowchart of Implementing collaborative robots with RoboDK workstations

guarantee both safety and efficiency. Ultimately, use the cooperative robotic system in the manufacturing setting, consistently overseeing and enhancing procedures for maximum efficiency. By using an iterative method, the integration of collaborative robots with RoboDK workstations is achieved without any disruptions, resulting in improved efficiency and safety. The present research involves the simulation of a workstation for automated material storage (as shown in the Fig. 4). Specifically, a six-degree-of-freedom ABB Robot is used as the primary robot in this Four Industrial Robot system. The first robot is designated for the task of receiving the packages and positioning them on the conveyer roller. When the first robot deposits the packets onto the conveyor roller, the roller is activated and remains in motion until the packet reaches the second robot. The second robot then proceeds to categorize the packets based on their form and dimensions. If a packet aligns with the dimensions specified by the robot, the second robot retrieves and stores it into the storage unit. Additionally, if the package does not conform to its designated shape and dimensions, the conveyor roller transports it to the third robot, and this procedure repeats. This whole model is shown in the Fig. 5.

5.3 Pseudo Code Used During Simulation

```
PROGRAM_START
        VERSION: 1
```

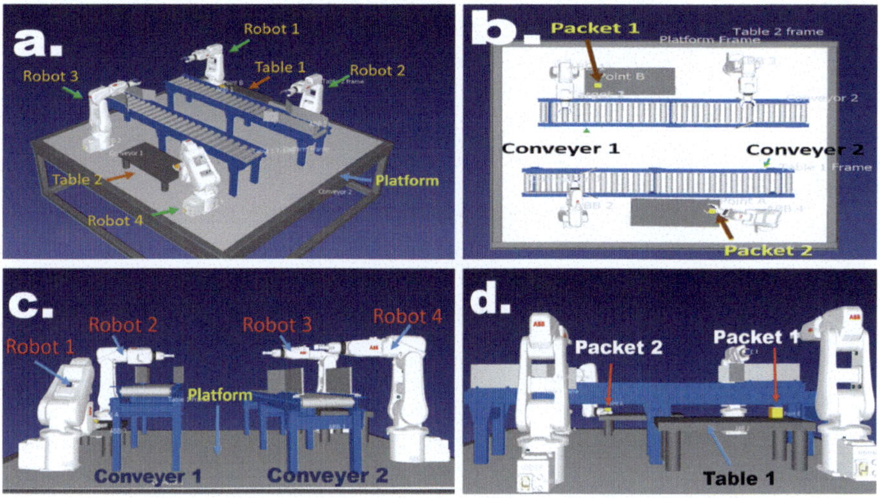

Fig. 5 **a** Isometric view of RoboDK workstations cells for smart manufacturing **b** Top view of RoboDK workstations cells for smart manufacturing **c** Front view of RoboDK workstations cells for smart manufacturing **d** Side view of RoboDK workstations cells for smart manufacturing

```
LANGUAGE: ENGLISH
MODULE PickAndPlaceModule
! Initialize custom variables for all robots
INITIALIZE_VARIABLES
! Define tool variables for all robots
DEFINE_TOOL_VARIABLES
! Define reference variables for all robots
DEFINE_REFERENCE_VARIABLE RobotBase1 :=
        (Active: FALSE, Dynamic: TRUE, Name: "",
        Position: Defined,
        Orientation: Defined)
DEFINE_REFERENCE_VARIABLE RobotBase2 :=
        (Active: FALSE, Dynamic: TRUE, Name: "",
        Position: Defined,
        Orientation: Defined)
DEFINE_REFERENCE_VARIABLE RobotBase3 :=
        (Active: FALSE, Dynamic: TRUE, Name: "",
        Position: Defined,
        Orientation: Defined)
DEFINE_REFERENCE_VARIABLE RobotBase4 :=
        (Active: FALSE, Dynamic: TRUE, Name: "",
        Position: Defined,
        Orientation: Defined
PROCEDURE Main
```

```
SET_CONFIGURATION Joint := ON
SET_CONFIGURATION Linear := OFF
! Robot 1 Pick and Place
PROCEDURE Robot1_PickAndPlace
    MOVE_LINEAR StartPosition, Orientation, Speed, Zone, Tool,
RobotBase1
ACTIVATE_GRIPPER Robot1
    MOVE_LINEAR PlacePosition, Orientation, Speed, Zone, Tool,
RobotBase1
DEACTIVATE_GRIPPER Robot1
END_PROCEDURE
! Robot 2 Pick and Place
PROCEDURE Robot2_PickAndPlace
    MOVE_LINEAR StartPosition, Orientation, Speed, Zone, Tool,
RobotBase2
ACTIVATE_GRIPPER Robot2
    MOVE_LINEAR PlacePosition, Orientation, Speed, Zone, Tool,
RobotBase2
DEACTIVATE_GRIPPER Robot2
END_PROCEDURE
! Robot 3 Pick and Place
PROCEDURE Robot3_PickAndPlace
    MOVE_LINEAR StartPosition, Orientation, Speed, Zone, Tool,
RobotBase3
ACTIVATE_GRIPPER Robot3
        MOVE_LINEAR PlacePosition, Orientation, Speed, Zone,
Tool, RobotBase3
DEACTIVATE_GRIPPER Robot3
END_PROCEDURE
! Robot 4 Pick and Place
PROCEDURE Robot4_PickAndPlace
    MOVE_LINEAR StartPosition, Orientation, Speed, Zone, Tool,
RobotBase4
ACTIVATE_GRIPPER Robot4
    MOVE_LINEAR PlacePosition, Orientation, Speed, Zone, Tool,
RobotBase4
DEACTIVATE_GRIPPER Robot4
END_PROCEDURE
! Execute pick and place for all robots
    CALL Robot1_PickAndPlace()
    CALL Robot2_PickAndPlace()
    CALL Robot3_PickAndPlace()
    CALL Robot4_PickAndPlace()
END_PROCEDURE
END_MODULE
```

PROGRAM_END

The above pseudocode presents a methodical and organized approach to creating a pick-and-place operation utilizing four ABB robots. The software, called "PickAndPlaceModule," starts by initializing user-defined variables and tool variables that are necessary for managing the robots. Every robot is allocated a reference variable (RobotBase1, RobotBase2, RobotBase3, and RobotBase4) that specifies its initial position and orientation in space. The reference frames are essential to guarantee precise and coordinated motions of each robot in relation to their initial positions. The primary method, "Main," establishes the configuration for both joint and linear movements, guaranteeing that the robots are ready for accurate operations. Each robot is assigned a specific pick-and-place process, namely Robot1_PickAndPlace, Robot2_PickAndPlace, Robot3_PickAndPlace, and Robot4_PickAndPlace. The procedures outline the precise sequence of operations for the robots: first, they move in a straight line to a designated starting position, then they activate the gripper to pick up an object, after which they move to a specified location and deactivate the gripper to release the object. By encapsulating these activities in processes, the program guarantees that each robot executes its tasks in a regulated and replicable manner.

Subsequently, the primary operation sequentially invokes the pick-and-place procedure of each robot. The sequential execution of tasks guarantees that all robots carry out their actions in a methodical manner, hence minimizing the likelihood of any interference or collision. Utilizing organized procedures and explicit reference frames enables effortless program change and scaling, hence permitting the incorporation of more intricate jobs or additional robots in the future. By incorporating this pseudocode into our simulation, we successfully accomplished our objectives with optimal efficiency. The precise delineation of reference frames and systematic procedures enabled us to uphold precision and consistency in the robots' motions. The implementation of this method not only enhanced the dependability of our pick-and-place operations but also established a flexible structure for future improvements. The pseudocode was important in validating our simulation, allowing us to draw fruitful conclusions and achieve exact control and coordination of several robots in a difficult operational environment.

6 Challenges and Future Directions

6.1 Key Challenges in Adopting RoboDK and Industry 4.0 in Manufacturing

Integrating RoboDK workstations and adopting Industry 4.0 in production poses several daunting obstacles. Firstly, there is the matter of the upfront expenses, which may be significant, particularly for small and medium-sized firms (SMEs). Furthermore, the incorporation of innovative technologies such as RoboDK into current production processes may need substantial reconfiguration and retraining, resulting in downtime and decreased productivity during the transition period. In addition, there is a scarcity of highly qualified individuals who are adept in both conventional production techniques and advanced technology like as robots and automation. The presence of a skills gap might hinder the effective installation and use of RoboDK workstations.

6.2 Strategies for Overcoming Barriers to Implementation

In order to tackle these issues, producers have the option to use several techniques. Obtaining funds via grants, subsidies, or financing alternatives may assist reduce the financial strain of investing in RoboDK and Industry 4.0 technology. Moreover, the use of staged strategies for integration, beginning with pilot initiatives or limited-scale rollouts, may mitigate disturbances and facilitate progressive adjustment to the novel systems. Allocating resources towards staff training and development initiatives or recruiting individuals with specialised knowledge may effectively address the skills gap and facilitate the successful adoption and operation of RoboDK workstations.

6.3 Future Trends and Advancements in Smart Manufacturing with RoboDK

The future of smart manufacturing with RoboDK workstations has great promise. The progress in artificial intelligence (AI) and machine learning (ML) algorithms will improve the functionalities of RoboDK in areas like independent decision-making and adaptable control, allowing for more effective and versatile manufacturing procedures. RoboDK integration with upcoming technologies such as augmented reality (AR) and virtual reality (VR) will strengthen its immersive simulation and training capabilities. This will make it easier to operate and collaborate remotely in dispersed production settings. Furthermore, ongoing advancements in hardware and sensor technologies will propel the development of increasingly advanced and adaptable robotic systems, broadening the range of sectors and applications in which RoboDK may be used. In summary, as the technology becomes more advanced and more people start using it, RoboDK is positioned to have a significant impact on driving the next phase of innovation and increasing productivity in the manufacturing industry.

7 Conclusions

In summary, this study has clarified the significance of enhancing automation in the field of smart manufacturing. Manufacturers may transform their operations and maintain s competitiveness in the current fast-paced market by using cutting-edge technology like RoboDK workstations and adopting the concepts of Industry 4.0. During this conversation, a number of important conclusions have arisen. First and foremost, the progression of automation in manufacturing is unavoidable, and it is essential to accept and adapt to this transformation in order to remain current and optimise productivity. Additionally, RoboDK workstations provide a robust platform for modelling, testing, and optimising production processes, allowing organisations to decrease expenses, minimise periods of inactivity, and enhance overall efficiency. Considering the future, the consequences of embracing advancing automation are extensive. The advantages are many, ranging from improved flexibility and customisation to heightened safety and efficiency. Nevertheless, it is crucial to recognise the difficulties associated with this transition, such as the need to enhance the skills of the workforce, tackle cybersecurity issues, and guarantee the ethical use of AI and robots. Given these profound understandings, manufacturers are strongly urged to fully use automation and Industry 4.0 technology. Companies may achieve sustainable manufacturing growth by allocating resources to innovation, cultivating a culture of ongoing learning, and engaging in collaborative efforts with partners and industry experts. Collectively, we have the potential to construct a forthcoming era in which intelligent manufacturing not only fosters financial well-being but also advances ecological sustainability and societal advancement.

References

Abdul-Qawy AS et al (2015) The internet of things (IoT): an overview. Int J Eng Res Appl 5(12):71–82

Aceto G, Persico V, Pescapé A (2019) A survey on information and communication technologies for industry 4.0: state-of-the-art, taxonomies, perspectives, and challenges. IEEE Commun Surv Tutorials 21(4):3467–3501

Adamu Yusuf A (2019) Development of A 6-DOF 3D printed industrial robot for teaching and learning. University of Malaya

Aggoune S, Hamadi F, Abid C et al (2024) Instabilities in the formation of single tracks during selective laser melting process. Int J Interact Des Manuf https://doi.org/10.1007/s12008-024-01887-y

Ahmed RS, Ahmed ESA, Saeed RA (2021) Machine learning in cyber-physical systems in industry 4.0. In: Artificial intelligence paradigms for smart cyber-physical systems. IGI global, pp 20–41

Alexopoulos T (2022) Process design and supervision: a next generation simulation approach to digitalised manufacturing. Cardiff University

Aqlan F et al (2020) A small-scale implementation of industry 4.0. In: Proceedings of the 2020 IISE annual conference

Ayvaz S, Alpay K (2021) Predictive maintenance system for production lines in manufacturing: a machine learning approach using IoT data in real-time. Expert Syst Appl 173:114598

Bansal S, Kumar D (2020) IoT ecosystem: a survey on devices, gateways, operating systems, middleware and communication. Int J Wireless Inf Networks 27(3):340–364

Batista RC, Agarwal A, Gurung A, Kumar A, Altarazi F, Dogra N, HM V, Chiniwar DS, Agrawal A (2024) Topological and lattice-based AM optimization for improving the structural efficiency of robotic arms. Front Mech Eng 10:1422539

Bejlegaard M, Sarivan I-M, Waehrens BV (2021) The influence of digital technologies on supply chain coordination strategies. J Glob Oper Strateg Sourcing 14(4):636–658

Bragança S et al (2019) A brief overview of the use of collaborative robots in industry 4.0: human role and safety. Occupational and environmental safety and health, pp 641–650

Brecher C et al (2021) Automation technology as a key component of the industry 4.0 production development path. Int J Advan Manuf Technol 117:2287–2295

Brooke L (2008) Ford model T: the car that put the world on wheels. Motorbooks

Burande DV, Kalita K, Gupta R et al (2024) Machine learning metamodels for thermo-mechanical analysis of friction stir welding. Int J Interact Des Manuf. https://doi.org/10.1007/s12008-024-01871-6

Centobelli P et al (2022) Blockchain technology design in accounting: game changer to tackle fraud or technological fairy tale? Acc, Auditing Accountability J 35(7):1566–1597

Corallo A, Lazoi M, Lezzi M (2020) Cybersecurity in the context of industry 4.0: a structured classification of critical assets and business impacts. Comput Ind 114:103165

Crnokić B et al (2023) Overview of tools for programming and virtual simulation of robots within the STEM teaching process. In: International conference on digital transformation in education and artificial intelligence application. Springer

David P (2001) Productivity growth prospects and the new economy in historical perspective. EIB Papers 6(1)

Fox AR (2022) Generative design for agile robot based additive manufacturing for sustainable aesthetic furniture products. Brunel University London

Fragapane G et al (2022) Increasing flexibility and productivity in industry 4.0 production networks with autonomous mobile robots and smart intralogistics. Ann Oper Res 308(1):125–143

Gažová A, Papulová Z, Smolka D (2022) Effect of business process management on level of automation and technologies connected to industry 4.0. Procedia Comput Sci 200:1498–1507

Goyal G, Kumar A, Sharma D (2024a) 12 recent applications of rapid prototyping with 3D printing: a review. In: Kumar A, Kumar P, Sharma N, Srivastava AK (eds) 3D printing technologies: digital manufacturing, artificial intelligence, industry 4.0. De Gruyter, Berlin, Boston, pp 245–258. https://doi.org/10.1515/9783111215112-012

Goyal G, Kumar A, Gupta A (2024b) 16 recent developments in 3D printing: a critical analysis and deep dive into innovative real-world applications. 3D printing technologies: digital manufacturing, artificial intelligence, industry 4.0, p 335

Gray AE, Seidmann A, Stecke KE (1993) A synthesis of decision models for tool management in automated manufacturing. Manage Sci 39(5):549–567

Heikkilä J, Wikström J (2021) Standardized general purpose technologies: a note. Available at SSRN 3863978

Ionescu TB (2020) Leveraging graphical user interface automation for generic robot programming. Robotics 10(1):3

Jasperneite J, Sauter T, Wollschlaeger M (2020) Why we need automation models handling complexity in industry 4.0 and the internet of things. IEEE Ind Electron Mag 14(1):29–40

Jumani AK et al (2022) Virtual reality and augmented reality for education. Multimedia computing systems and virtual reality. CRC Press, pp 189–210

Koh L, Orzes G, Jia FJ (2019) The fourth industrial revolution (Industry 4.0): technologies disruption on operations and supply chain management. Int J Oper Prod Manage 39(6–8):817–828

Kumar A, Rani S, Rathee S, Bhatia S (eds) (2023) Security and risk analysis for intelligent cloud computing: methods, applications, and preventions, 1st ed. CRC Press. https://doi.org/10.1201/9781003329947

Kumar A, Shrivastava VK, Kumar P, Kumar A, Gulati V (2024a) Predictive and experimental analysis of forces in die-less forming using artificial intelligence techniques. Proc Inst Mech Eng, Part E: J Process Mech Eng 0(0). https://doi.org/10.1177/09544089241235473

Kumar P, Hussain SS, Kumar A, Srivastava AK, Hussain M, Singh PK (2024b) 10 finite element method investigation on delamination of 3D printed hybrid composites during the drilling operation. 3D printing technologies: digital manufacturing, artificial intelligence, industry 4.0, p 223

Kumar A, Kumar P, Sharma N, Srivastava AK (eds) (2024c) 3D printing technologies: digital manufacturing, artificial intelligence, industry 4.0. Walter de Gruyter GmbH & Co KG. https://doi.org/10.1515/9783111215112

La Commare R (2020) Trajectory optimization for collaborative robotics applications

Lee J, Cameron I, Hassall M (2019) Improving process safety: what roles for digitalization and industry 4.0? Process Saf Environ Prot 132:325–339

Lee JD, Seppelt BD (2009) Human factors in automation design. Springer handbook of automation, pp 417–436

Liu C et al (2023) Digitalization and servitization of machine tools in the era of industry 4.0

Lu Y (2017) Industry 4.0: a survey on technologies, applications and open research issues. J Ind Inf Integr 6:1–10

Manavalan E, Jayakrishna K (2019) A review of internet of things (IoT) embedded sustainable supply chain for industry 4.0 requirements. Comput Ind Eng 127:925–953

Mathew D, Brintha N, Jappes JW (2023) Artificial intelligence powered automation for industry 4.0. In: New horizons for industry 4.0 in modern business. Springer. pp 1–28

Mofolasayo A et al (2022) How to adapt lean practices in SMEs to support industry 4.0 in manufacturing. Procedia Comput. Sci 200:934–943

Mohammadi M, Jamshidi S, Rezvanian A, Gheisari M, Kumar A (2024) Advanced fusion of MTM-LSTM and MLP models for time series forecasting: an application for forecasting the solar radiation. Meas: Sens 33:101179

Mourtzis D et al (2015) The role of simulation in digital manufacturing: applications and outlook. Int J Comput Integr Manuf 28(1):3–24

Mourtzis D, Angelopoulos J, Panopoulos N (2022) Operator 5.0: a survey on enabling technologies and a framework for digital manufacturing based on extended reality. J Mach Eng 22

Naveena K, Krishnamoorthy M, Karuppiah N, Gouda PK, Hariharan S, Saravanan K, Kumar A (2024) Elevating sustainability with a multi-renewable hydrogen generation system empowered by machine learning and multi-objective optimization. Meas: Sens 33:101192

Oesterreich TD, Teuteberg F (2016) Understanding the implications of digitisation and automation in the context of industry 4.0: a triangulation approach and elements of a research agenda for the construction industry. Comput Ind 83:121–139

Oztemel E, Gursev S (2020) Literature review of industry 4.0 and related technologies. J Intell Manuf 31(1):127–182

Papulová Z, Gažová A, Šufliarský Ľ (2022) Implementation of automation technologies of industry 4.0 in automotive manufacturing companies. Procedia Comput Sci 200:1488–1497

Patel KR (2023) Enhancing global supply chain resilience: effective strategies for mitigating disruptions in an interconnected world. BULLET: J Multidisiplin Ilmu 2(1):257–264

Pieskä S, Kaarela J, Mäkelä J (2018) Simulation and programming experiences of collaborative robots for small-scale manufacturing. In: 2018 2nd international symposium on small-scale intelligent manufacturing systems (SIMS). IEEE

Rani S, Tripathi K, Kumar A (2023) Machine learning aided malware detection for secure and smart manufacturing: a comprehensive analysis of the state of the art. Int J Interact Des Manuf. https://doi.org/10.1007/s12008-023-01578-0

Rao BP et al (2012) Cloud computing for Internet of Things & sensing based applications. In: 2012 sixth international conference on sensing technology (ICST). IEEE

Ren S et al (2019) A comprehensive review of big data analytics throughout product lifecycle to support sustainable smart manufacturing: a framework, challenges and future research directions. J Clean Prod 210:1343–1365

Ribeiro FM (2019) Implementation of a Simulation System for Additive Task Experiments. Universidade de Coimbra, Portugal

Rodič B (2017) Industry 4.0 and the new simulation modelling paradigm. Organizacija 50(3):193–207

Rüßmann M et al (2015) Industry 4.0: the future of productivity and growth in manufacturing industries. Boston Consult Group 9(1):54–89

Sanghavi D, Parikh S, Raj SA (2019) Industry 4.0: tools and implementation. Management and production engineering review

Santos CHD et al (2022) Use of simulation in the industry 4.0 context: creation of a digital twin to optimise decision making on non-automated process. J Simul 16(3):284–297

Saturno M et al (2017) Proposal of an automation solutions architecture for industry 4.0. In: 24th international conference on production research

Saukkoriipi J (2019) Design and implementation of robot skill programming and control. J Saukkoriipi

Schulte PA et al (2020) Potential scenarios and hazards in the work of the future: a systematic review of the peer-reviewed and grey literatures. Ann Work Exposures Health 64(8):786–816

Shamshiri R et al (2018) Simulation software and virtual environments for acceleration of agricultural robotics: features highlights and performance comparison

Sivasankaran P, Karthikeyan R (2020) Simulation of robot kinematic motions using collision mapping planner using RoboDK solver. Blue Eyes Intell Eng Sci Publ 9(11):21–27

Srivastava AK, Kumar A, Kumar P et al (2023) Research progress in metal additive manufacturing: challenges and opportunities. Int J Interact Des Manuf. https://doi.org/10.1007/s12008-023-01661-6

Trochimczuk R et al (2019) Modeling, programming and simulation of robotized workcells created for industrial and service needs. Eng Rural Dev 18(455):1313–1318

Tyagi AK et al (2020) Intelligent automation systems at the core of industry 4.0. In: International conference on intelligent systems design and applications. Springer

Wamba-Taguimdje S-L et al (2020) Influence of artificial intelligence (AI) on firm performance: the business value of AI-based transformation projects. Bus Process Manag J 26(7):1893–1924

Wan J et al (2020) Artificial-intelligence-driven customized manufacturing factory: key technologies, applications, and challenges. Proc IEEE 109(4):377–398

Xu LD, Xu EL, Li L (2018) Industry 4.0: state of the art and future trends. Int J Prod Res 56(8):2941–2962

Zawadzki P, Żywicki K (2016) Smart product design and production control for effective mass customization in the industry 4.0 concept. Management and production engineering review

Zawra LM, Mansour HA, Messiha NW (2019) Migration of legacy industrial automation systems in the context of industry 4.0—a comparative study. In: 2019 international conference on fourth industrial revolution (ICFIR). IEEE

Zhang Y, Xu X, Liu Y (2011) Numerical control machining simulation: a comprehensive survey. Int J Comput Integr Manuf 24(7):593–609

Improvising the Quality of Manifold Production Using Six-Sigma Technique for Implementation in Automobile Manufacturing Industries: A Case Study

Sachin Kumar, Aman Kumar, and Ajay Kumar

Abstract In today's modern technical world, quality deals with every point of the task. Improvement will always take place in the field of development. The paper deals with process improvement methods to enhance productivity, reduce defects, ease the methodology, and save money. Various models, like Define-Measure-Analysis-Improvement-Control, are used for the Plan, Do, Check, Act cycle and brainstorming analysis using know-how methods. This model has shown its strengths and limitations, and there is no judgment of wrong or right; its only purpose is to improve quality and standardize the system. The p-values have decreased from 0.266 to 0.253, improving the procedure and product satisfaction. The sigma level has also increased by achieving the new Z bentch value of 2.18 from 0.06 by reducing the DPMO values from 476,944 to 21,389.

Keywords PPM- parts per million · Sigma level · Two sample T test · DMAIC

1 Introduction

In this modern era, profitability is on the rise. Every manufacturing industry plans to enhance its methods, eliminate access efforts, minimize waste, smooth processes, remove variation, etc. Industrialists are using several techniques and methodologies to make their process variations waste-free. It also contributes to our ecosystem, industry, and organization. Six Sigma is a statistical tool used to improve processes, eliminate waste, reduce costs, etc. Six-Sigma methodology is used in the DMAIC technique based on the PDCA cycle, which implies Plan-Do-Control-Act (Bastos

S. Kumar (✉) · A. Kumar
Department of Mechanical Engineering, University Institute of Engineering and Technology, Maharshi Dayanand University, Rohtak, Haryana, India
e-mail: skmtech11@gmail.com

A. Kumar (✉)
Department of Mechanical Engineering, School of Engineering and Technology, JECRC University, Jaipur, Rajasthan, India
e-mail: ajay.kumar.30886@gmail.com

A. Kumar et al. (eds.), *Industry 4.0 Driven Manufacturing Technologies*, Springer Series in Advanced Manufacturing, https://doi.org/10.1007/978-3-031-68271-1_16

et al. 2021; Bewoor et al. 2022; Elboq et al. 2020; Gaikwad et al. 2022; Singh and Rathi 2019; Yadav et al. 2020; Yang et al. 2021). It has its strengths and weaknesses, and there is no right or wrong with this technique. It has become the statistical tool or symbol used in mathematics to determine the standard deviation (Table 1). It is a technical or scientific method or process used for process enhancement or improvement planning, and the method used for process improvement is generally known as DMAIC (Ajay et al. 2023; Gupta et al. 2022; Kumar et al. 2022, 2023; Rani et al. 2023, 2022; Sudan and Taggar 2022). The target is 3.4 errors per minute using the technique DMAIC, a process improvement model. Six Sigma's methodology mainly focuses on high productivity, quality improvement, waste elimination, and defect reduction; it also lays down the flaw rate target at three to four errors per million opportunities (Arumugam, et al. 2023; Joshi et al. 2023; Liu et al. 2023; Parid et al. 2023; Shafiq et al. 2023). It results in superior, fast, and cost-efficient products. Six Sigma is an improvement approach focusing on client requirement defects, cycle time prevention, reduction, and cost savings (Duc et al. 2024; Gaikwad and Sunnapwar 2024; Mohan et al. 2024; Rajak et al. 2024; Tsarouhas and Sidiropoulou 2023; Widiwati et al. 2024).

2 Problem Formulation

The present study deals with analyzing the manifolds' rubbing marks using the DMAIC technique to obtain the desired quality. The Manifold is an essential engine assembly part required to transfer the air–fuel mixture to the engine (Figs. 2 and 3). Small defects may cause the working of an engine to fail and may cause life loss. It comes under the 'A' category component, which belongs to a category where the loss of human life is involved, whereas category 'B' belongs to a category where an accident may occur and the car's functioning will fail. Still, there will be no loss of human life involved, and if there is no failure and only aesthetics reduce the charm of the parts, then this falls under category' C. In the present study, DMAIC methodology is applied to resolve the problem (Figs. 1 and 4), which consists of the following phases: 1. Define 2. Measure 3. Analysis 4. Improve 5. Conclusion.

3 Dmaic Analysis and Case Discussion

3.1 Define

During this Define phase, a SIPOC (Supplier-Input-Process-Output-Customer) process was created to fix the rubbing issues that came up while the manifolds were being made (Figs. 5 and 6). The current rejection rate is too high due to rubbing

Table 1 Literature outcome in tabulated form

Sr. No	Author/Reference literature	Title of paper	Outcome of study	DOI
1	G. Yadav et al.	A framework to overcome sustainable supply chain challenges through solution measures of industry 4.0 and circular economy. An automotive case	An Indian automobile case study was done. The framework was studied using hybrid BWM-ELECTRE. 28 SSCM problems were found. 22 circular economy and industry 4.0 solutions were offered	https://doi.org/10.1016/j.jclepro.2020.120112
2	Singh, Mahipal, and Rajeev Rathi	A structured review of lean six sigma in various industrial sectors	The purpose of present study is to expose the detailed review for benefits and challenges about implementation of lean six sigma (LSS) in business organization and spread of LSS literature in term of various sectors wise, research methodology wise and journal wise	https://doi.org/10.1108/ijlss-03-2018-0018
3	Elboq, Raja, Mustapha Hlyal, and Jamila El Alami	Empirical assessment of critical success factor of lean and six sigma in the moroccan automotive industry	LM and SS implementing characteristics were revealed. Based on CSF importance and maturity, a bi-dimensional scan that describe lean manufacturing (LM) and six sigma (SS) implementation within Moroccan automotive industry	https://doi.org/10.1088/1757-99x/827/1/012043
4	Bastos, Nuno Miguel Moreno Teixeira, Anabela Carvalho Alves, Felipe Castro, Júlio César Duarte, LuíS Pinto Ferreira, and F. J. G. Silva	Reconfiguration of assembly lines using lean thinking in an electronics components	This article discusses the reconfiguration and improvement of electronics components assembly lines through the use of lean thinking principles. Overall, the level of scrap was reduced by 57% through the implementation of the lean six sigma methodology, and the cycle time of an operation was reduced by 12,5% by eliminating non-value adding operations	https://doi.org/10.1016/j.promfg.2021.10.053
5	Yang, Hui, Prahalada Rao, Timothy W. Simpson, Yan Lu, Paul Witherell, Abdalla R. Nassar, Edward W. Reutzel, and Soundar Kumara	Six-sigma quality management of additive manufacturing	The six sigma (6S) technique entails a data-driven DMAIC methodology of five steps—defines measure, analyze, improve, and control. Anew process control approaches were discussed to optimize the action plans, once an anomaly is detected, with specific consideration of lead time and energy consumption	https://doi.org/10.1109/jproc.2020.3034519

(continued)

Table 1 (continued)

Sr. No	Author/Reference literature	Title of paper	Outcome of study	DOI
6	Gaikwad, Lokpriya, Umesh Bhushi, and S. N. Teli	Implementation of six sigma methodologies to gain a competitive advantage: a case study approach	Six sigma DMADV (Define-measure-analyze-design-verify) methodologies is an influential approach to designing products, processes, and services to fulfill the requirements and potential of the customer while reducing the cost of quality. DMADV uses influential and constructive statistical tools to forecast and enhance quality before prototypes are built	https://doi.org/10.1109/aset53988.2022.9735103
7	Bewoor, Anand K., Mitali U. Kharul, Saloni G. Gosavi, and Tanvi J. Kuray	Application of 7-step problem-solving methodology for defect elimination: a case study in an automotive industry	7-step problem-solving methodology (7SPSM), along with 8D, PDCA, DMAIC, and six sigma are adopted. From the results experienced, it can be concluded that 7SPSM provides an effective methodology not only for solving problems but also for continual improvement by elimination of recurrences of product failure	https://doi.org/10.1007/978-981-16-9952-8_60
8	Kumar, V. Ravi, Vikas Modgil, and Ajay Kumar	Stochastic Petri nets modelling for performance assessment of a manufacturing unit	The Petri Nets modelling method run the plant in virtual manner and the system availability is found to be 80.27% substituting the input parameters in the timed transitions of each sub-system. Further, the performance assessment through Patri nets model is compared with Markov model solution to verify the findings	https://doi.org/10.1016/j.matpr.2022.01.073
9	T. Sudan and R. Taggar	COVID-19 induced supply chain disruptions and automotive industry. A case study of Maruti Suzuki India Limited and mitigation strategies	The model addressed SCDs and risk management and suggested proactive and reactive measures to help the car industry respond faster to COVID-19-induced SCDs	https://doi.org/10.37256/ges.3120221095
10	Kumar, Ajay, Hari Singh, Parveen Kumar, and Bandar AlMangour	*Handbook of Smart Manufacturing*	All new concepts and innovative approaches are mentioned in this book	https://doi.org/10.1201/9781003333760

(continued)

Table 1 (continued)

Sr. No	Author/Reference literature	Title of paper	Outcome of study	DOI
11	Ajay, Ashwini Kumar, Parveen, Rajesh Kumar Goel, and Ravi Kant Mittal	*Waste Recovery and Management*	Applications of six sigma technique to resolve the problems of waste management industries	https://doi.org/10.1201/9781003359784
12	Rani, Sangeeta, Khushboo Tripathi, and Ajay Kumar	Machine learning aided malware detection for secure and smart manufacturing: a comprehensive analysis of the state of the art	Implementation of a noval approach for smart manugacturing using melware detection	https://doi.org/10.1007/s12008-023-01578-0
13	Rani, Sangeeta, Khushboo Tripathi, Yojna Arora, and Ajay Kumar	Analysis of anomaly detection of malware using KNN	A comprehensive study of anomaly detection of malware based on machine learning algorithms is presented here. This paper also explains about the implementation of k-nearest neighbors of anomaly detection and discusses the challenges associated with implementing malware classifiers	https://doi.org/10.1109/iciptm54933.2022.9754044
14	Gupta, M., Digalwar, A., Gupta, A., & Goyal, A	Integrating theory of constraints, lean and six sigma: a framework development and its application	This research applies theory of constraints, lean, and six sigma to business excellence. Global measurements, growth-oriented management, and integrated continuous improvement increase firm TLS. Five easy steps improved flow, waste, process, and financial performance at a SME	https://doi.org/10.1080/09537287.2022.2071351
15	Joshi, Vishwas Deep, Priya Agarwal, and Ajay Kumar	Fuzzy transportation Planning: a goal programming tactic for navigating uncertainty and multi-objective decision making	This approach empowers company to generate compromise solutions aligned with their preferences and illustrated through a practical MOTP solved with LINGO software. This approach effectively removes uncertainty and enhances decision-making by generating compromise solutions aligned with preferences	https://doi.org/10.1007/s12008-023-01634-9

(continued)

Table 1 (continued)

Sr. No	Author/Reference literature	Title of paper	Outcome of study	DOI
16	Sathish Kumar Arumugam, Sachin Kumar, Pramod Sridhara, Srinivasan Raju, Ashwin Prabhu Gnanasekaran, Nanthakumar Sivasamy and Thangarajan Sivasankaran Senthilkumar	Machine learning-based investigation of wear and frictional behavior in graphite-reinforced aluminum nanocomposites	The researcher used a cell segmentation technique in conjunction with other image analysis methods to quantitatively retrieve and compute the cellular microstructural structures in a sub-grain size of silicon carbide (SiC)-reinforced AA2219 made by powder fusion bed (size 0.5–1 μm). Over 83 geometric features were retrieved and statistically analyzed using ML (Machine learning) techniques to examine the structure–property relationships in SiC-reinforced AlSi20Mg nanocomposites	https://doi.org/10.17756/nwj.2023-s3-054
17	Parid, Y., Hardi Purba, H., & Santoso, S	A systematic literature review of six sigma implementation on the automotive component industry	This analysis examined 2013–2022 scientific papers utilising tight inclusion and exclusion criteria. Researchers, consultants, and quality managers can use the study to develop and execute auto part six sigma	https://doi.org/10.32734/jsti.v25i2.10450
18	Shafiq, M., Thakre, K., Krishna, K. R. et al.	Continuous quality control evaluation during manufacturing using supervised learning algorithm for Industry 4.0	Polynomial, linear, sigmoid, and over-varying gamma coefficient kernels evaluate SVM continuous quality. SVM surpasses other classifiers in computation time (88.1%), F1-measure (89.4%), ROC (65%), and accuracy (94%)	https://doi.org/10.1007/s00170-023-10847-x
19	W. Liu, X. Li, C. Liu, M. Wang and L. Liu	Resilience assessment of the cobalt supply chain in China under the impact of electric vehicles and geopolitical supply risks	System dynamics assessed cobalt supply chain resilience. The geopolitical resilience of China's cobalt supply chain was investigated. We improved China's cobalt supply chain	https://doi.org/10.1016/j.resourpol.2022.103183
20	Rajak, S., Kumar, P., Modi, A., Swarnakar, V., Antony, J., & Sony, M	An assessment of barriers to integrate lean six sigma and industry 4.0 in manufacturing environment: case based approach	DEMATEL assessed how each barrier affects others. Experts validated obstructions' cause–effect relationship. A case study shows how LSS integration with I4.0 in car component production is difficult	https://doi.org/10.1080/0951192X.2024.2335969

(continued)

Table 1 (continued)

Sr. No	Author/Reference literature	Title of paper	Outcome of study	DOI
21	Mohan, J., Kaswan, M. S. and Rathi, R	An analysis of green lean six sigma deployment in MSMEs: a systematic literature review and conceptual implementation framework	This study enables GLSS researchers the capabilities to apply this sustainable strategy to other businesses. The study helps industry managers with a complete MSMEs GLSS framework. The paper shows how GLSS might improve environmental dynamics, benefiting society	https://doi.org/10.1108/TQM-06-2023-0197
22	M. L. Duc, P. Bilik and R. Martinek	Reduce power energy cost using hybrid six sigma based on fuzzy MADM: a case study in mechanical factory	Industry 4.0 lowers coil maintenance system power expenses. Fault rate fell from 47.2% to 4.9% after improvement. Defect losses reduce from 6593 to 549 USD annually	https://doi.org/10.1109/ACCESS.2024.3388202
23	Gaikwad, L. M., Sunnapwar, V. K	Validation of lean–green–six sigma practice model for improving performance and competitiveness in an Indian manufacturing industry	This research is unique in merging Lean, Green, and Six Sigma to suit Indian manufacturing sector demands and improve operations and sustainability	https://doi.org/10.1007/s13198-024-02357-0
24	Ivana Tita Bella Widiwati, Surya Danusaputro Liman, Filscha Nurprihatin	The implementation of lean six sigma approach to minimize waste at a food manufacturing industry	Lean six sigma reduces food waste, research reveals. Waste's productivity impact is examined using DMAIC. DMAIC analysis will show ways to reduce waste	https://doi.org/10.1016/j.jer.2024.01.022
25	Tsarouhas, P. and Sidiropoulou, N	Application of six sigma methodology using DMAIC approach for a packaging olives production system: a case study	Step-by step six sigma DMAIC reduces olive-drained weight fluctuation. Drained weight standard deviation fell 51.02% and components per million effectives dropped 99.97% after manufacturing process optimisation	https://doi.org/10.1108/IJLSS-06-2022-0140

Fig. 1 DMAIC elaboration

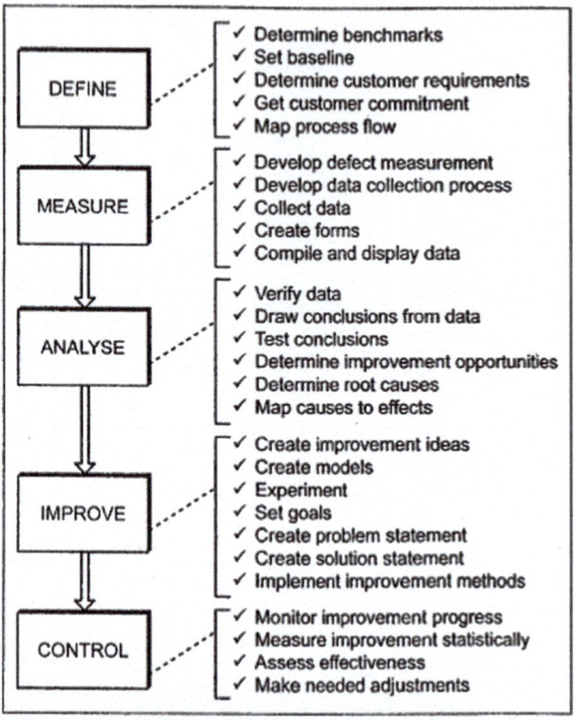

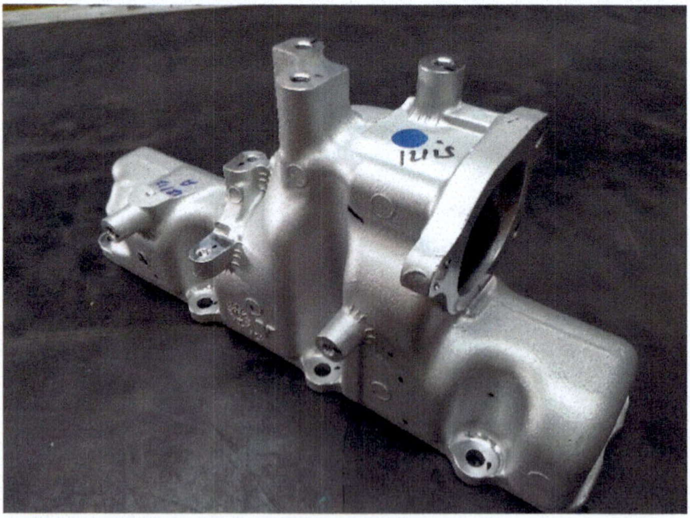

Fig. 2 Manifold, *Source* Tapukara, Rajasthan

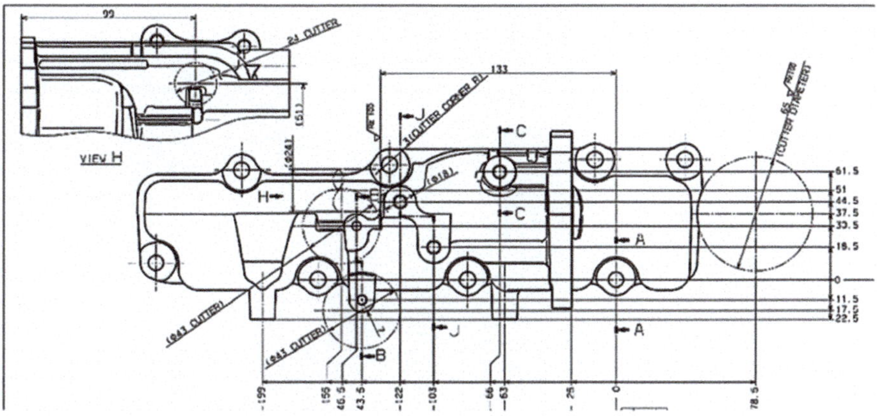

Fig. 3 Manifold drawing

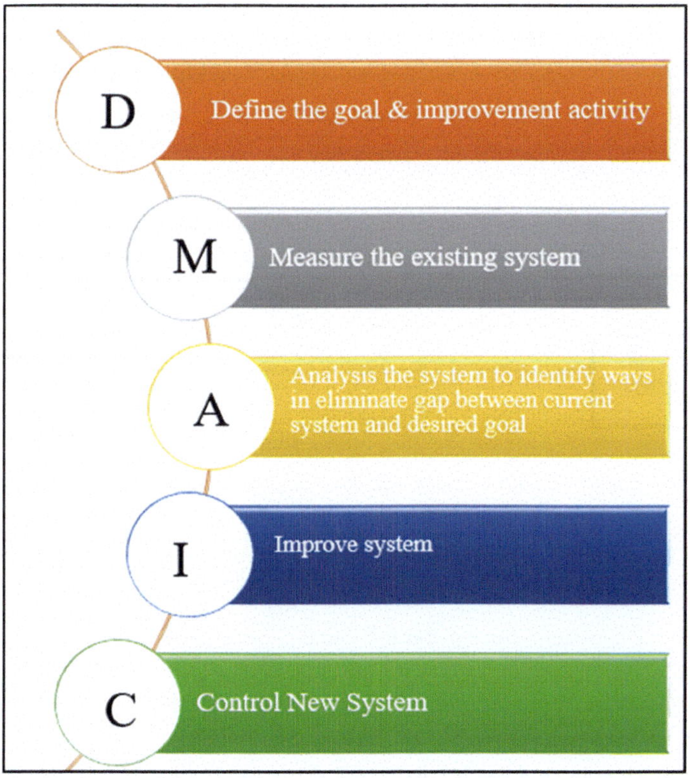

Fig. 4 DMAIC methodology

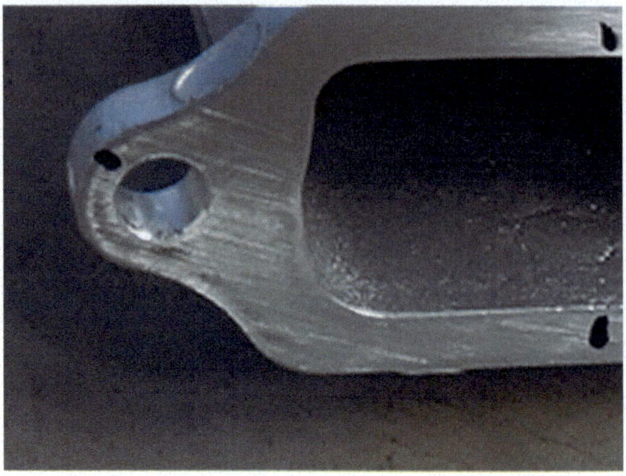

Fig. 5 Rubbing marking on the face of manifold

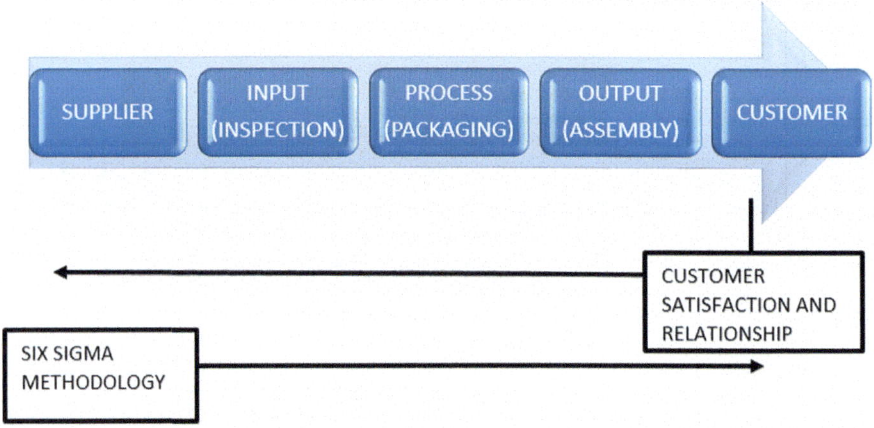

Fig. 6 SIPOC

marks on the face of the manifold component. The main aim of the present work is to reduce the rejection rate of manifolds due to rubbing marks.

3.2 *Measure*

System measurement analysis is carried out during the 'measure' phase. This phase includes attribute agreement analysis, where three associates were given to each part and looked for rubbing marks in the manifold and passage that needed to be fixed. For that investigation, a sample of 50 Manifold products was taken to record the need for rework on the components. For the component manifold, industries now face around 47.7% of products on which rework is mandatory (Fig. 7).

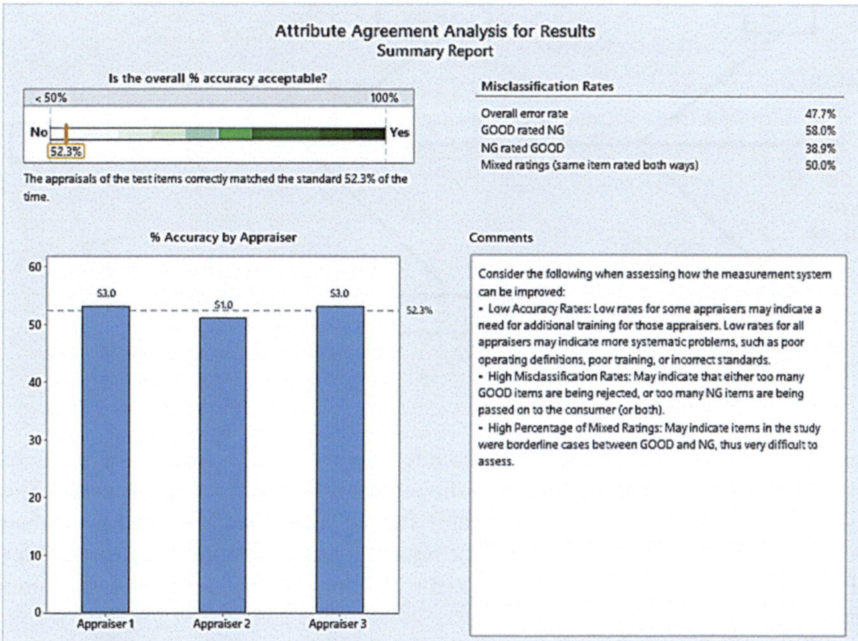

Fig. 7 Attribute agreement analysis (Manifold)

3.3 *Analysis*

During this phase, the project associates examine the data collected in the measurement phase and seek explanations for the reading using brainstorming and cause-and-effect diagrams. The process of "know-how" and brainstorming sessions with the relevant team members helped to select the root cause during brainstorming analysis. A cause-and-effect diagram was made based on these causes and effects. This diagram shows the effects and causes of a fishbone pattern, also known as the Fish Bone Diagram. After further brainstorming and discussion, the cause-and-effect diagram was remade. The PPM level and deviation level of the manifold issue were found through process capability analysis. The next step was determining the causes of manifold and passage rejection. With the help of experts and a close study of real processes, a fishbone diagram was made to show the reasons for rejecting manifold and passage issues (Fig. 8).

A process capability analysis was initiated to determine the actual position of the process (Fig. 9). A sub-group was made, and 60 samples were drawn in two groups. From the process capability analysis curve, it has been found that the Z bench value was 0.05, and the existing PPM level of the process was 476,944, which is very high and shows scope for improvement. Further, the analysis process is followed by the 'Two-Sample T-Test' methodology to find machine tool wear and setup errors.

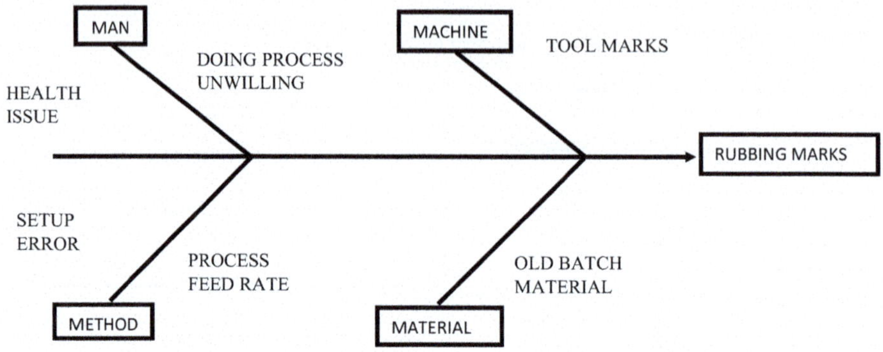

Fig. 8 Fish bone diagram for defect analysis

Specific main factors from the 'Two Sample T-Test' were taken into the study to identify from the fishbone diagram, as shown in Figs. 10 and 11 respectively. The two-sample T-test shows that the 'T' value for the setup case is much higher than others and could be the main reason for rejection. After reading these two tables in Tool error and Setup error, the setup error was much higher, so countermeasures must be taken for the setup error's cause.

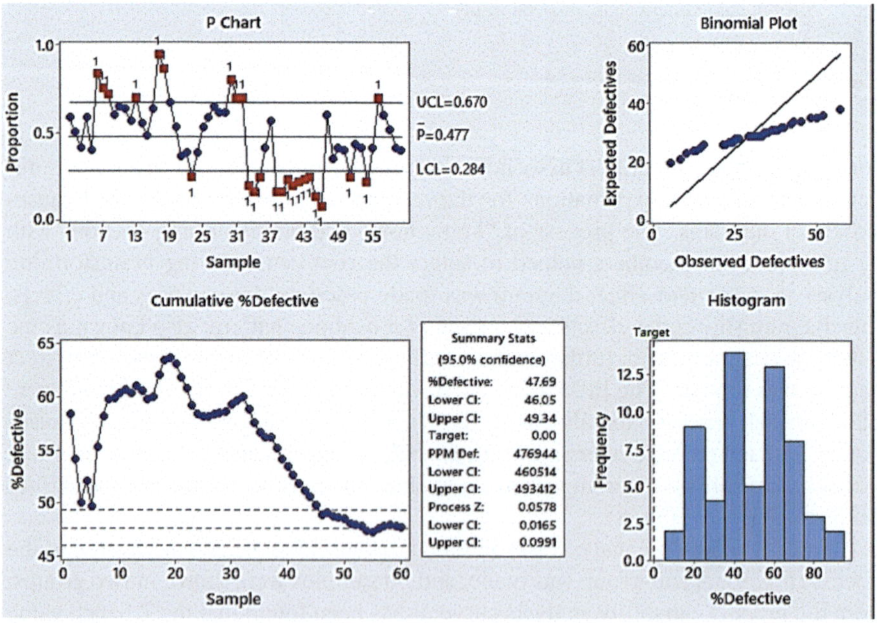

Fig. 9 Process capability of manifold

Method

μ_1: mean of Manifold Machine 1
μ_2: mean of Machine 2
Difference: $\mu_1 - \mu_2$

Equal variances are not assumed for this analysis.

Descriptive Statistics

Sample	N	Mean	StDev	SE Mean
Manifold Machine 1	60	3.30	1.77	0.23
Machine 2	60	3.22	1.60	0.21

Estimation for Difference

Difference	95% CI for Difference
0.083	(-0.526, 0.692)

Test

Null hypothesis $H_0: \mu_1 - \mu_2 = 0$
Alternative hypothesis $H_1: \mu_1 - \mu_2 \neq 0$

T-Value	DF	P-Value
0.27	116	0.787

Method

μ_1: mean of Passage Machine 1
μ_2: mean of PASSAGE Machine 2
Difference: $\mu_1 - \mu_2$

Equal variances are not assumed for this analysis.

Descriptive Statistics

Sample	N	Mean	StDev	SE Mean
Passage Machine 1	60	3.62	2.02	0.26
PASSAGE Machine 2	60	1.18	2.67	0.34

Estimation for Difference

Difference	95% CI for Difference
2.441	(1.586, 3.297)

Test

Null hypothesis $H_0: \mu_1 - \mu_2 = 0$
Alternative hypothesis $H_1: \mu_1 - \mu_2 \neq 0$

T-Value	DF	P-Value
5.66	109	0.000

Fig. 10 Two sample T-test for tool wear on manifold

3.4 Improve

In the improvement phase, the main focus was on resolving the issues that arose during the process handling phase based on the results of the two-sample t-test analysis. The first issue is machine tool wear. Tool change planning is done during this phase, and a certain life cycle is assigned. According to that, a tool change process should occur every six days. Further, the second issue, which is also the major issue, is the fixture or tool setup issue. To resolve this problem, a new fixture has been made, and no arm of the fixture has become an obstacle to the process. Countermeasures to these causes have been found and applied to reduce the rework issue of rubbing marks in manifolds and improve productivity. The root cause and the main factor that comes out to be the key reason for the high rejection of parts due to the rubbing marks issue is the setup issue, in which the fixture became the obstacle for the grinding process, and due to the machine fixture, the grinding wheel was unable to clean the portion of the holding corner, and that area remained unprocessed. For that area, an additional hand grinder was used to complete the grinding process, but simultaneously, a hand grinder is not such an efficient tool that it can match the same efficiency as a big grinding machine and forms rubbing marks over the surface. A new fixture had been developed, and a new setup method was developed where the fixture was not an obstacle to the process, resulting in improved productivity and reduced rejection rates.

Method

μ_1: mean of Passage Setup 1
μ_2: mean of PASSAGE Setup 2
Difference: $\mu_1 - \mu_2$

Equal variances are not assumed for this analysis.

Descriptive Statistics

Sample	N	Mean	StDev	SE Mean
Passage Setup 1	60	3.62	2.02	0.26
PASSAGE Setup 2	60	0.526	0.909	0.12

Estimation for Difference

Difference	95% CI for Difference
3.091	(2.523, 3.659)

Test

Null hypothesis H_0: $\mu_1 - \mu_2 = 0$
Alternative hypothesis H_1: $\mu_1 - \mu_2 \neq 0$

T-Value	DF	P-Value
10.82	81	0.000

Method

μ_1: mean of Manifold setup 1
μ_2: mean of Setup 2
Difference: $\mu_1 - \mu_2$

Equal variances are not assumed for this analysis.

Descriptive Statistics

Sample	N	Mean	StDev	SE Mean
Manifold setup 1	60	3.30	1.77	0.23
Setup 2	60	1.32	1.55	0.20

Estimation for Difference

Difference	95% CI for Difference
1.983	(1.383, 2.584)

Test

Null hypothesis H_0: $\mu_1 - \mu_2 = 0$
Alternative hypothesis H_1: $\mu_1 - \mu_2 \neq 0$

T-Value	DF	P-Value
6.54	115	0.000

Fig. 11 Two sample T-test for setup 1 and setup 2 for manifold

3.5 Control

In the control phase, P control charts have been plotted to specify the occurrence of assignable causes. After implementing the new fixture and ensuring that the process began in a new way, 60 sample sizes were taken for evaluation. To deal with issues related to skill techniques, Associate training or Induction programs have been organized in which they are provided an introduction and working methods and an Operation Procedure sheet(OPS) by which they can recognize the working flow and method. The second issue that arose in the previous section was the tool wear, which has been resolved by introducing tool change planning in which a certain life cycle has been assigned for the tool, and according to that 6th day, the tool change process should take place. Third and the major issue raised was fixture setup, for which a new fixture had been made in which no arm of the fixture became an obstacle to the process. After that, control charts were made based on before and after countermeasure reports, as shown in Figs. 12 and 13, respectively. From these figures, it can be easily found that the defective pieces have been reduced from 1717 to 77 for a test run on 3600 items divided into 60 groups, each having 60 items in their groups. This also reduces the

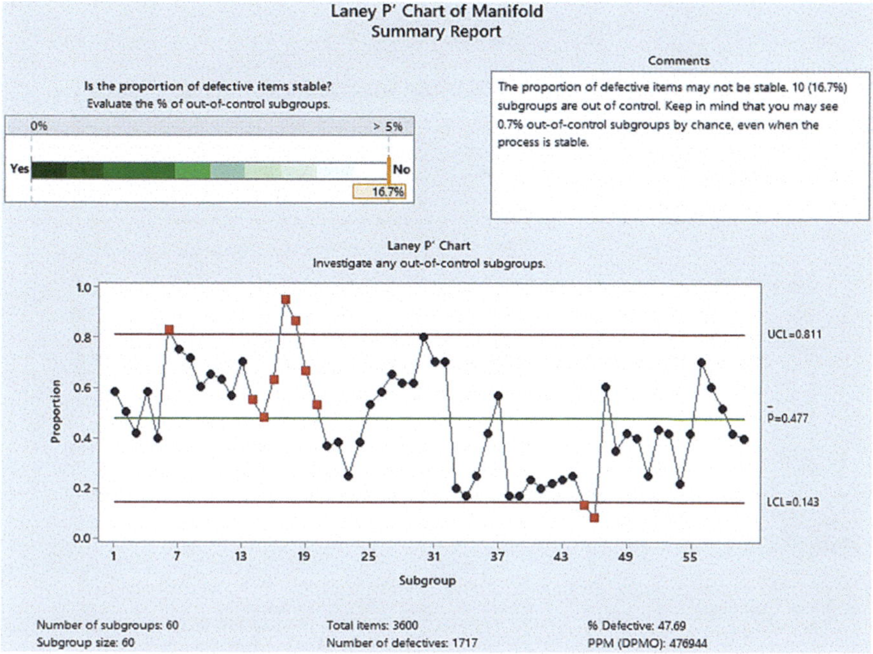

Fig. 12 P-chart summary for old setup before countermeasures

DPMO value from 476944 to 21389, as shown in Figs. 12 and 13. The application of the project recommendation brought up the sigma level to 2.18 from 0.06 sigma (Fig. 14). The effective annual cost saving from the corrective factor (1.713) is about 1460*462222 = Rs. 6,74,84,120/-.

4 Conclusions and Future Scope

This section represents the findings and limitations of the present study and the scope for future research. Motivated by the observed gap between the theory and practice of Six Sigma deployment, this study explains the importance of empirical studies like expert surveys in developing countries like India. The myth that Six Sigma is limited to a specified large industry or organization is abolished through this case study. Implementation of Six Sigma in automobiles is very tough but not impossible. Several factors are required, and the Six Sigma technique can be used. This case study will provide great knowledge to new researchers, scholars, and Six Sigma beginners and contribute to society. This case study will provide great knowledge to new researchers, scholars, and Six Sigma beginners and contribute to society. The main findings of the present study are listed below:

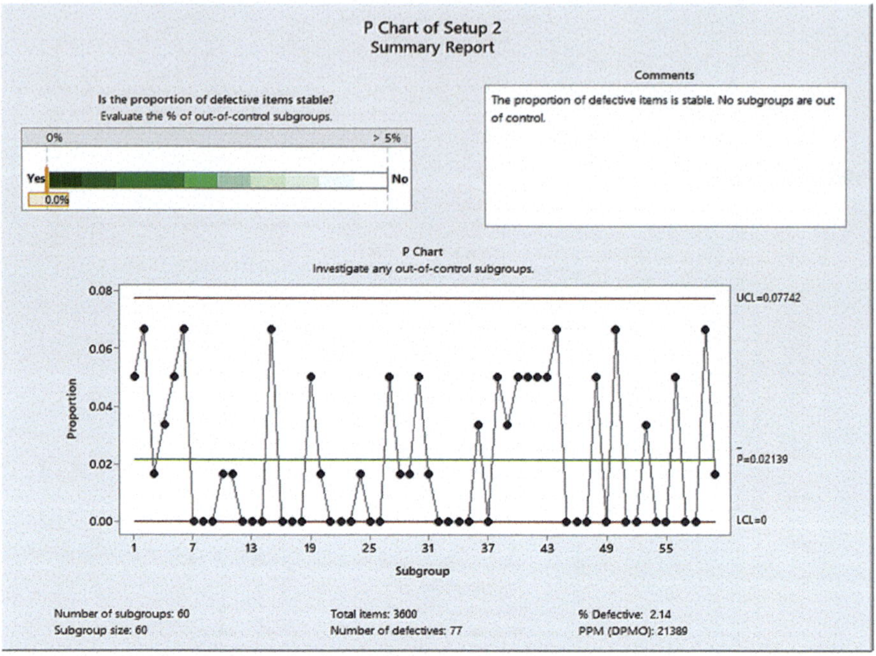

Fig. 13 P-chart summary for new setup after countermeasures

1. The gap between the knob and lever portions of the car door handle has been reduced using brainstorming and know-how techniques. The effective cost annually saved from corrective factor (1.713) is about 1460*462,222 = Rs.6,74,84,120/-.
2. The p-values have been reduced from 0.266 to 0.253, which enhances the process and product satisfaction.
3. The sigma level has also been increased by achieving the new Z bentch value of 2.18 from 0.06 by reducing the DPMO values from 476,944 to 21,389.

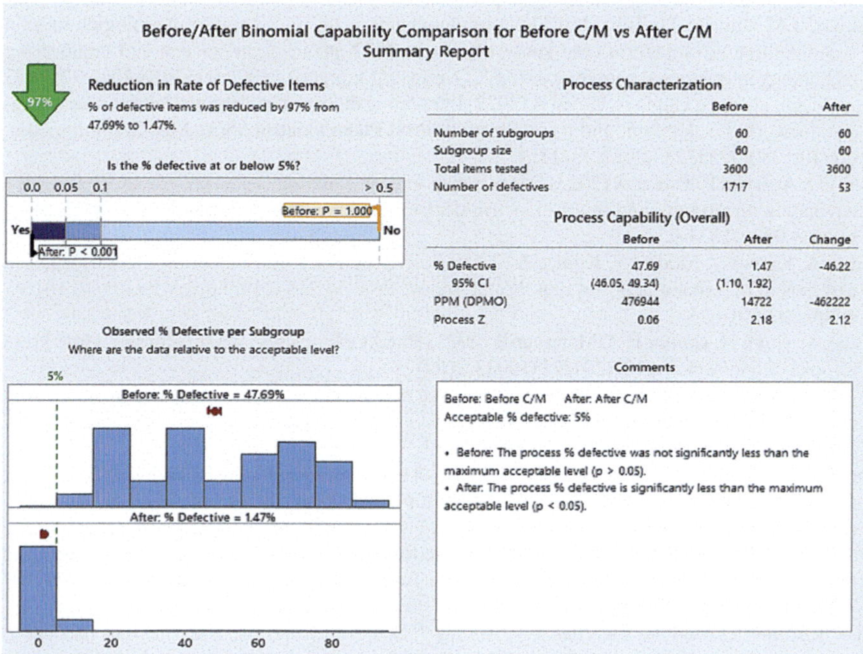

Fig. 14 Process capability report for manifold

References

Ajay N, Kumar A, Parveen N, Goel R, Mittal RK (2023) Waste recovery and management. CRC Press eBooks. https://doi.org/10.1201/9781003359784

Arumugam S et al (2023) Machine learning-based investigation of wear and frictional behavior in graphite-reinforced aluminum nanocomposites. NanoWorld J9. https://doi.org/10.17756/nwj.2023-s3-054

Bastos N, Alves A, Castro F, Duarte J, Ferreira L, Silva F (2021) Reconfiguration of assembly lines using lean thinking in an electronics components' manufacturer for the automotive industry. Procedia Manuf 55:383–392. https://doi.org/10.1016/j.promfg.2021.10.053

Bewoor AK, Kharul MU, Gosavi SG, Kuray TJ (2022) Application of 7-step problem-solving methodology for defect elimination: a case study in an automotive industry. In: Lecture notes in mechanical engineering, pp 697–707. https://doi.org/10.1007/978-981-16-9952-8_60

Duc ML, Bilik P, Martinek R (2024) Reduce power energy cost using hybrid six sigma based on fuzzy MADM: a case study in mechanical factory. IEEE Access 1. https://doi.org/10.1109/access.2024.3388202

Elboq R, Hlyal M, Alami JE (2020) Empirical assessment of critical success factor of lean and six sigma in the Moroccan automotive industry. IOP conference series. Mater Sci Eng 827(1):012043. https://doi.org/10.1088/1757-899x/827/1/012043

Gaikwad LM, Sunnapwar VK (2024) Validation of lean–green–six sigma practice model for improving performance and competitiveness in an Indian manufacturing industry. Int J Syst Assur Eng Manag. https://doi.org/10.1007/s13198-024-02357-0

Gaikwad LM, Bhushi U, Teli SN (2022) Implementation of six sigma methodologies to gain a competitive advantage: a case study approach. 2022 advances in science and engineering technology international conferences (ASET). https://doi.org/10.1109/aset53988.2022.9735103

Gupta M, Digalwar A, Gupta A, Goyal A (2022) Integrating theory of constraints, lean and six sigma: a framework development and its application. Prod Plann Control 35(3):238–261. https://doi.org/10.1080/09537287.2022.2071351

Joshi VD, Agarwal P, Kumar A (2023) Fuzzy transportation planning: a goal programming tactic for navigating uncertainty and multi-objective decision making. IJIDEM. https://doi.org/10.1007/s12008-023-01634-9

Kumar A, Kumar V, Modgil V, Kumar A (2022) Stochastic Petri Nets modelling for performance assessment of a manufacturing unit. Mater Today: Proc 56:215–219. https://doi.org/10.1016/j.matpr.2022.01.073

Kumar A, Singh H, Kumar P, AlMangour B (2023) Handbook of smart manufacturing. CRC Press eBooks. https://doi.org/10.1201/9781003333760

Liu W, Li X, Liu C, Wang M, Liu L (2023) Resilience assessment of the cobalt supply chain in China under the impact of electric vehicles and geopolitical supply risks. Resour Policy 80:103183. https://doi.org/10.1016/j.resourpol.2022.103183

Mohan J, Kaswan MS, Rathi R (2024) An analysis of green lean six sigma deployment in MSMEs: a systematic literature review and conceptual implementation framework. TQM J. https://doi.org/10.1108/tqm-06-2023-0197

Parid Y, Purba HH, Santoso S (2023) A systematic literature review of six sigma implementation on the automotive component industry. J Sistem Teknik Industri 25(2):218–235. https://doi.org/10.32734/jsti.v25i2.10450

Rajak S, Kumar P, Modi A, Swarnakar V, Antony J, Sony M (2024) An assessment of barriers to integrate lean six sigma and industry 4.0 in manufacturing environment: case based approach. Int J Comput Integr Manuf 1–22.https://doi.org/10.1080/0951192x.2024.2335969

Rani S, Tripathi K, Arora Y, Kumar A (2022) Analysis of anomaly detection of malware using KNN. 2022 2nd international conference on innovative practices in technology and management (ICIPTM). https://doi.org/10.1109/iciptm54933.2022.9754044

Rani S, Tripathi K, Kumar A (2023) Machine learning aided malware detection for secure and smart manufacturing: a comprehensive analysis of the state of the art. IJIDEM. https://doi.org/10.1007/s12008-023-01578-0

Shafiq M, Thakre K, Krishna KR, Robert NJ, Kuruppath A, Kumar D (2023) Continuous quality control evaluation during manufacturing using supervised learning algorithm for industry 4.0. Int J Adv Manuf Technol/Int J, Adv Manuf Technol. https://doi.org/10.1007/s00170-023-10847-x

Singh M, Rathi R (2019) A structured review of lean six sigma in various industrial sectors. Int J Lean Six Sigma 10(2):622–664. https://doi.org/10.1108/ijlss-03-2018-0018

Sudan T, Taggar R (2022) COVID-19 induced supply chain disruptions and automotive industry: a case study of Maruti Suzuki India limited and mitigation strategies. Glob Econ Sci 36–53. https://doi.org/10.37256/ges.3120221095

Tsarouhas P, Sidiropoulou N (2023) Application of six sigma methodology using DMAIC approach for a packaging olives production system: a case study. Int J Lean Six Sigma 15(2):247–273. https://doi.org/10.1108/ijlss-06-2022-0140

Widiwati ITB, Liman SD, Nurprihatin F (2024) The implementation of lean six sigma approach to minimize waste at a food manufacturing industry. J Eng Res. https://doi.org/10.1016/j.jer.2024.01.022

Yadav G, Luthra S, Jakhar SK, Mangla SK, Rai DP (2020) A framework to overcome sustainable supply chain challenges through solution measures of industry 4.0 and circular economy: an automotive case. J Cleaner Prod 254:120112. https://doi.org/10.1016/j.jclepro.2020.120112

Yang H, Rao P, Simpson T, Lu Y, Witherell P, Nassar AR, Reutzel E, Kumara S (2021) Six-sigma quality management of additive manufacturing. Proc IEEE 109(4):347–376. https://doi.org/10.1109/jproc.2020.3034519

Digital Twin Integration for Enhanced Control in FDM 3D Printing

Sourabh Anand, Manoj Kumar Satyarthi, Pushpendra S. Bharti, Parveen Kumar, and Ajay Kumar

Abstract This study addresses the critical need for precise monitoring and adaptive control systems in 3D printing processes using fused deposition modelling (FDM). The printing process is dynamic and has intrinsic complexity, hence it becomes critical to implement a Digital Twin architecture. Real-time insights into temperature changes and material deposition are necessary for successful control of the intricate relationship between heat transfer and material deposition, as estimated by a realistic MATLAB model. This research utilizes a finite difference technique to update the printing area's temperature distribution repeatedly. It also seamlessly incorporates an extrusion model to simulate the deposition of material layer by layer. The maximum extrusion rate and increase extrusion rate are two further characteristics that enhance the versatility of the Digital Twin structure. The outcomes demonstrate the effectiveness of integrating Digital Twins to optimise the 3D printing process and highlight the need for this kind of strategy to achieve accuracy and dependability. The study offers a thorough grasp of the thermal dynamics associated with FDM 3D printing, highlighting the critical role that Digital Twins play in reducing difficulties and improving the capabilities of the manufacturing process. This study provides a comprehensive approach to real-time monitoring and control in FDM 3D printing through the essential integration of Digital Twin technology, laying the groundwork for future efforts in smart manufacturing.

Keywords Digital twin · FDM · 3D printing · Additive manufacturing · Real time monitoring

---

S. Anand · M. K. Satyarthi · P. S. Bharti
University School of Information, Communication and Technology, Guru Gobind Singh Indraprastha University, Sec 16C, Dwarka, New Delhi 110078, India

P. Kumar
Department of Mechanical Engineering, Rawal Institute of Engineering and Technology, Faridabad, Haryana, India

A. Kumar (✉)
Department of Mechanical Engineering, School of Engineering and Technology, JECRC University, Jaipur, Rajasthan, India
e-mail: ajay.kumar30886@gmail.com

A. Kumar et al. (eds.), *Industry 4.0 Driven Manufacturing Technologies*, Springer Series in Advanced Manufacturing, https://doi.org/10.1007/978-3-031-68271-1_17

1 Introduction

Recently, Digital Twin technology has become an innovative idea that originated in the fields of manufacturing and industrial design (Semeraro et al. 2021; Kamble et al. 2022; Sharma et al. 2022; Ajay et al. 2023; Kumar et al. 2023d). A Digital Twin is a dynamic virtual model of a physical system, item, or process. It is constantly synced and updated in real-time with its actual counterpart (López 2021; Ashtari Talkhestani et al. 2019; Rasheed et al. 2020; Liu et al. 2021). This integration enables comprehensive monitoring, analysis, and control, leading to a deeper comprehension and optimization of complex systems (Lim et al. 2020; Botín-Sanabria et al. 2022). The beginnings of Digital Twins may be traced back to early conceptualizations in the realm of computer-aided design (CAD) and simulation, as documented in references (Shen and Li 2024; Segovia and Garcia-Alfaro Design 2022). However, the actual use of Digital Twins became more popular with the rise of Industry 4.0, a time when advanced technology and data-driven analysis revolutionized traditional production techniques (Sharma et al. 2022; ElMaraghy et al. 2021).

Historical Background of Digital Twins

The origins of Digital Twins can be traced back to the development of computer-aided design (CAD) in the mid-twentieth century (Kokhanevych 2023; Singh et al. 2021; Wagg et al. 2020). The advent of CAD technology enabled engineers and designers to accept digital models of physical objects (Sass and Oxman 2006), marking an important change from conventional paper-based design methods. The evolution progressed further with the advent of computer-aided engineering (CAE), which expanded the digital representation to encompass simulations and analyses of the physical behaviour of objects. However, it was with the introduction of Industry 4.0 that the idea of Digital Twins truly became significant (Leng et al. 2021; Jacoby and Usländer 2020). The seamless connection between real-time data, IoT devices, and advanced analytics allowed for a constant and fluid synchronisation between the physical and digital worlds (Batista et al. 2024; Naveena et al. 2024; Burande et al. 2024).

Additive Manufacturing and Digital Twins

Within the industry of manufacturing by additive processes, which addresses a range of methods for developing three-dimensional objects layer by layer (Anand and Satyarthi 2023b; Anand et al. 2024; Anand and Satyarthi 2023a), Digital Twins have demonstrated considerable potential and recent related work has been summarized in Table 1. The recent development of additive manufacturing, also known as 3D printing, has completely transformed the way complex and customized geometries are produced. In contrast to conventional subtractive manufacturing techniques that involve removing material from a larger block, additive manufacturing constructs objects by adding layers one by one (Bhardwaj et al. 2023; Sehrawat et al. 2023; Kumar et al. 2023a). Out of all the additive manufacturing techniques available, Fused Deposition Modelling (FDM) has gained significant popularity (Kumar et al.

2023b). The process of FDM entails the gradual deposition of thermoplastic material, resulting in the formation of complex structures (Kumar et al. 2023c). Although FDM has become increasingly popular due to its adaptability and cost-effectiveness, it presents distinct challenges in terms of thermal dynamics, material deposition rates, and overall process control. Having precise control over the temperature distribution within the printing environment is essential to maintain material integrity and print quality. In addition, differences in extrusion rates can affect how well the layers stick together, resulting in inconsistencies in the structure. The importance of implementing more sophisticated monitoring and control strategies becomes evident when considering these challenges. The incorporation of Digital Twins with FDM 3D printing presents an optimistic response to these obstacles. Through the creation of a digital duplicate of the physical 3D printer and its operations, Digital Twins allow for continuous monitoring and proactive control. This dynamic synchronization enables a thorough examination of variations in temperature, motion of material flow, and rates of deposition throughout the printing process. The control system becomes flexible and responsive by continuously updating the digital representation with real-world data. Despite the potential benefits, there is a clear lack of research in fully understanding how Digital Twin technology can be successfully integrated into FDM 3D printing processes to achieve improved control. Prior research has predominantly concentrated on the theoretical aspects of Digital Twins or their implementation in wider manufacturing contexts. Nevertheless, a thorough examination is necessary to explore the unique obstacles and complexities associated with FDM 3D printing and how Digital Twins can be effectively incorporated into this particular additive manufacturing method.

This study seeks to address the current research gap by conducting a thorough examination of the use of Digital Twins in the monitoring and control of FDM 3D printing. The study utilizes a realistic simulation model implemented in MATLAB, harnessing the computational power to replicate thermal dynamics, material deposition, and extrusion rates within the 3D printing environment. The iterative finite difference method allows for the real-time update of the temperature distribution within the printing area, accurately simulating the intricacies of heat transfer in FDM. The simulation model takes into account important factors such as thermal conductivity, initial temperature, and extrusion rates to accurately simulate the complex relationship between heat transfer and material deposition. The extrusion model is seamlessly integrated to accurately emulate the layer-by-layer material deposition characteristic of FDM. It takes into account factors such as filament temperature and deposition rates. Other factors, like the increase in extrusion rate and the maximum extrusion rate, enhance the flexibility of the Digital Twin framework, enabling a more dynamic portrayal of the printing process. The study offers a thorough exploration of the thermal dynamics associated with FDM 3D printing, highlighting the crucial role of Digital Twins in addressing challenges and enhancing the capabilities of the manufacturing process.

Table 1 Related work on digital twins in additive manufacturing

Author	Focus	Key technology	Main contribution
Odada et al. (2021)	Applying a data-driven digital twin for monitoring an FDM 3D printer in real-time	Siemens NX, Modbus server, OPC-DA server	Developed a data-based digital twin that records the live movements of an FDM 3D printer, resulting in a maximum average delay of 0.351 s. Implemented historical recording to accurately reproduce previous printing sessions
Pantelidakis et al. (2022)	The digital twin ecosystem supports 3D printing inside a virtual environment	3D printer web controller, sensors	Established a digital twin environment to monitor processes and control them remotely. Successfully established almost instantaneous synchronization between physical and virtual printers and verified the accuracy of the twin in terms of its location, temperature, and operational duration
Corradini and Silvestri (2022)	Applying digital twin technology for the monitoring and quality evaluation of the material extrusion process	Simulation engine, G-Code file	Created a digital twin that incorporates live monitoring of processes and conditions, ensures precise measurements, and evaluates quality. It has the capability to temporarily stop the printing process for irregularities and aid in the implementation of corrective measures
Mourtzis et al. (2021)	Optimization of FDM processes using digital twin technology	Database, immersive interfaces	Created a digital twin integrating quality assessment modules and a database for monitoring experiment results. Enabled remote control and optimization of FDM process parameters to minimize errors
Stavropoulos et al. (2021)	Digital twin platform for additive manufacturing processes	Offline modelling, real-time decision support	Presented a platform for optimization services, including cycle time and energy consumption, and integrating empirical knowledge for better AM processes
Nath and Mahadevan (2022)	The use of a probabilistic digital twin for the design and control of LPBF processes	Bayesian calibration, surrogate model	Suggested a technique for developing a probabilistic digital replica that emphasizes uncertainty in models and variability in processes, while optimizing process parameters and enabling real-time adjustments to regulate porosity in components

(continued)

Table 1 (continued)

Author	Focus	Key technology	Main contribution
Butt and Mohaghegh (2022)	Applying the integration of digital twin technology and machine learning to enhance the process of fused filament production	CNN, random forest classifier, ROM	Explored the viability of using digital replicas and machine learning to forecast the performance of PLA components, attaining superior precision in classifying parts using numerical simulations and experimental data
Henson et al. (2021)	Major failure detection in FDM processes using digital twin strategy	Optical sensor system with multiple perspectives	Created a method for identifying large print defects by examining multi-view photographs of the printed object, analysing each layer individually. This allows for quick identification of serious quality issues, providing a valid reason to stop the printing process
Cai et al. (2020)	Employing virtual reality to construct a digital twin for a customizable additive manufacturing system	AR technique, transformation matrices	Implemented augmented reality to communicate layout information in a customizable additive manufacturing system equipped with robotic arms. This enabled the implementation of optimum layouts, defined in the digital twin, into the physical system
Gaikwad et al. (2020)	Combining temperatures models, sensors, and analytics to identify process defects in additive manufacturing	Physics-based model, ML framework, SVM	Implemented a machine learning approach to integrate predictions from a physical model with real-time sensor data to accurately identify defects in additive manufacturing components. This study is an initial exploration of the digital twin concept for monitoring the manufacturing process in real-time

Contribution to Smart Manufacturing

The research presented here lays the groundwork for future projects in the field of smart manufacturing, providing a comprehensive approach to real-time monitoring and control in FDM 3D printing by incorporating Digital Twin technology. The findings from this study enhance our understanding of FDM 3D printing and offer guidance for incorporating Digital Twins into other additive manufacturing methods. The potential of the Digital Twin framework to revolutionize 3D printing processes is evident in its adaptability and responsiveness. This framework has the ability to pave the way for more efficient and reliable manufacturing practices.

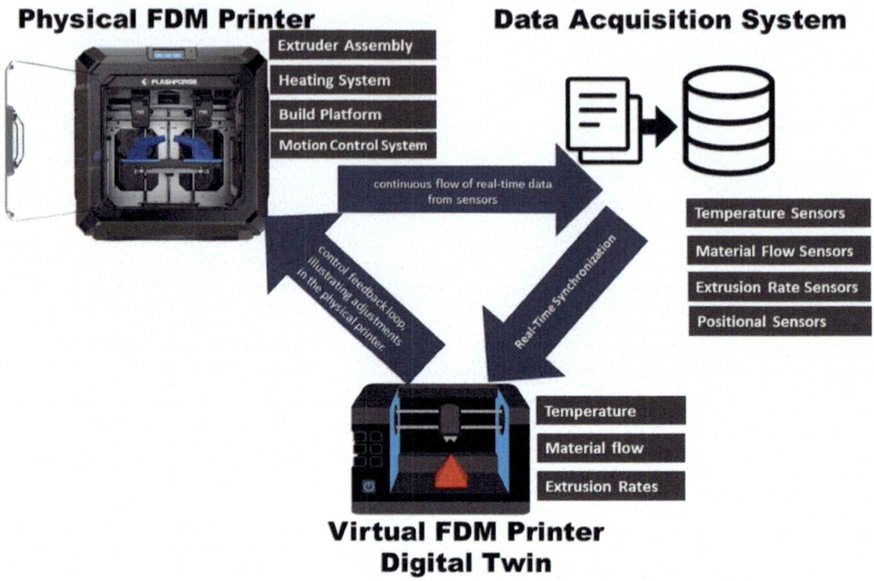

Fig. 1 A dynamic synchronization between physical and virtual environments

Ultimately, the incorporation of Digital Twins in FDM 3D printing signifies a notable advancement in the quest for accuracy and command in additive manufacturing. This study focuses on a crucial area that has not been extensively explored, delving into the complexities of integrating Digital Twin technology to overcome the specific challenges presented by FDM. The findings not only enhance our comprehension of the thermal dynamics at, but also add to the wider discussion on the revolutionary possibilities of Digital Twins in the manufacturing industry (Fig. 1).

2 Methodology

The incorporation of Digital Twin technology into FDM 3D printing requires the initialization of important parameters, including thermal conductivity, initial temperature, time steps, thermal diffusivity, and the dimensions of the 3D printing area. The grid resolution has been established, and a preliminary temperature matrix has been generated to depict the initial thermal state. In this study, a Finite Difference Method loop is utilized to simulate heat conduction, visualized through Fig. 2. The temperature at each grid point is updated using the finite difference approximation of the heat equation. The computation of the Laplacian, which represents the spatial distribution of temperature changes, is performed by a nested loop structure that iterates over the grid points. The extrusion model encompasses a simulation of material deposition, taking into consideration the geometric characteristics of the object and implementing heat input in accordance with the filament temperature and extrusion rate. The heat input exhibits localization within a predetermined radius from the object's centre, thereby emulating the thermal phenomena associated with extruded material. During the deposition update stage, the deposition rate is modified to accommodate the varying extrusion rates and elapsed time. This adjustment allows for the tracking of the number of deposited layers by incrementing the deposition rate accordingly. Once the deposition rate surpasses a predetermined threshold, it is deemed that a novel layer has been deposited. The extrusion parameters adjustment stage is designed to dynamically enhance the extrusion rate, thereby simulating the inherent variations encountered in the printing process. This process involves a gradual ramping up of the extrusion rate until it reaches a predetermined maximum value. During the process of integrating the Digital Twin, the simulation undergoes continuous monitoring, wherein the performance of the virtual model is compared to the anticipated real-world behaviour. This enables the implementation of dynamic adjustments to guarantee the utmost accuracy. The bidirectional data exchange facilitates the provision of real-time feedback and enables optimization processes. The process culminates in the completion of the simulation, wherein various outcomes including the final temperature distribution, extrusion rates, and total deposited layers are meticulously examined and juxtaposed with hypothetical real-world data. This analysis serves to substantiate the accuracy and dependability of the simulation, thereby showcasing the potential of Digital Twin technology in augmenting control and precision within the realm of Fused Deposition Modeling 3D printing.

2.1 Modelling Approach

At the core of the present research's simulation methodology, there is a precise modelling approach that is based on the heat conduction equation. This basic equation, which represents the principles of heat diffusion, provides the guiding framework for understanding the complex thermal dynamics involved in Fused Deposition Modelling FDM 3D printing (Santos et al. 2023; Zhang and Shapiro 2018). The equation, is shown as:

$$\frac{\partial T}{\partial t} = \alpha\left(\frac{\partial^2 T}{\partial x^2} + \frac{\partial^2 T}{\partial y^2} + \frac{\partial^2 T}{\partial z^2}\right) \tag{1}$$

Defines the evolution of temperature (T) over time (t), considering spatial coordinates (x, y, z) and the thermal diffusivity (α).

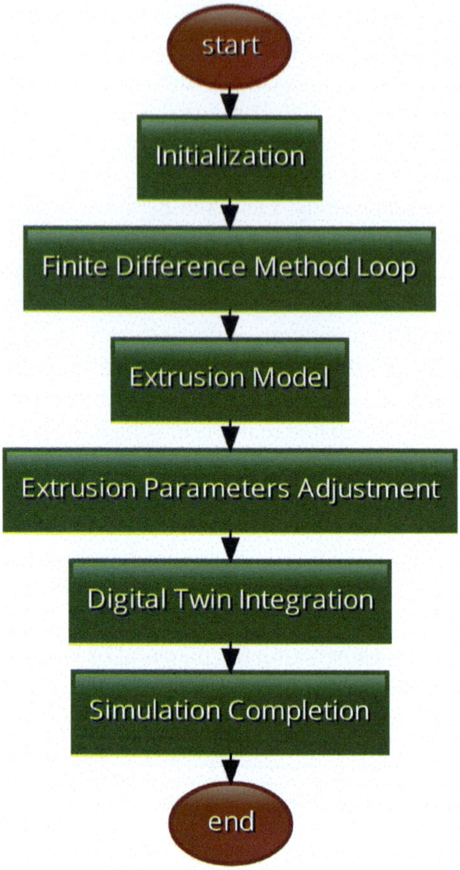

Fig. 2 Flow of FDM 3D printing simulation steps with Digital Twin Integration

2.2 *Parameterization and Initialization*

The effectiveness of our simulation depends on carefully choosing and setting important parameters. Factors like thermal conductivity (λ), initial temperature (T_0), time step (Δ_t), and the dimensions of the 3D printing area (L_x, L_y, L_z) play a crucial role in shaping the dynamic landscape. The grid resolution (N_x, N_y, N_z) determines the spatial steps (Δ_x, Δ_y, Δ_z) in the system. Additionally, the initial conditions set $T(x, y, z, 0) = (T_0)$ which establishes the starting temperature.

2.3 *Thermal Simulation Methodology*

Implementing the Finite Difference Method discretizes the heat conduction equation, leading to a discrete form that governs temperature updates. Here is the iterative equation

$$T_{i,\,j,\,k}^{n+1} = T_{i,\,j,\,k}^{n} + \alpha\Delta t\left(\frac{T_{i+1,j,k}^{n} - 2T_{i,j,k}^{n} + T_{i,j+1,k}^{n}}{\Delta x^2}\right.$$
$$\left. + \frac{T_{i+1,j,k}^{n} - 2T_{i,j,k}^{n} + T_{i,j+1,k}^{n}}{\Delta x^2} + \frac{T_{i+1,j,k}^{n} - 2T_{i,j,k}^{n} + T_{i,j+1,k}^{n}}{\Delta x^2}\right) \tag{2}$$

2.4 Extrusion and Deposition Modelling

Incorporating a streamlined extrusion model, the heat input within the object is determined by factors like filament temperature (T_f), extrusion rate (E_r), and object geometry.

$$Q = \frac{E_r \cdot T_f \cdot \Delta t}{\lambda \cdot \Delta x \cdot \Delta y \cdot \Delta z} \tag{3}$$

The distance from the centre (d_c) is calculated as $d_c = \sqrt{\left(i - \frac{N_x}{2}\right)^2 + \left(j - \frac{N_y}{2}\right)^2}$ if d_c is less than the radius, the temperature at that point is increased by Q.

2.5 Time-Stepping Loop for Simulation

The simulation is guided by a structured time-stepping loop, which updates the temperature at each spatial point sequentially. By utilising nested loops, a synchronised evaluation of the temperature distribution can be achieved, ensuring accuracy in the analysis of spatial dimensions.

2.6 Dynamic Adjustment of Extrusion Parameters

Extrusion parameters are continuously adjusted using advanced machine learning techniques. The process entails gradually increasing the extrusion rate over time, creating a controlled and genuine deposition process. This methodology seamlessly combines mathematical derivations to uncover the fundamental principles that govern the FDM 3D printing process, while also incorporating Digital Twin integration.

3 Results

The simulation conducted gives a comprehensive investigation into the combination of Digital Twin concepts with Fused Deposition Modelling 3D printing. In this section, the results are presented, combining visualisations and computational data to reveal insights into thermal dynamics, extrusion patterns, and the overall improvement in control achieved through Digital Twin integration.

Temperature Distribution

The main focus of the finding's centres around the distribution of temperature in the printing area, specifically in three dimensions. Throughout 100 simulation steps, it becomes clear that the temperature stabilizes, suggesting that the thermal gradients are converging. It is worth noting that the temperature at the centre of the printing area reaches 68.45 °C, indicating that thermal equilibrium has been achieved as shown in Fig. 3.

Extrusion and Deposition

The simulation demonstrates its versatility by displaying a gradual rise in the extrusion rate, eventually peaking at 2 layers per second. This subtle modification corresponds to practical situations, where extrusion rates change dynamically while

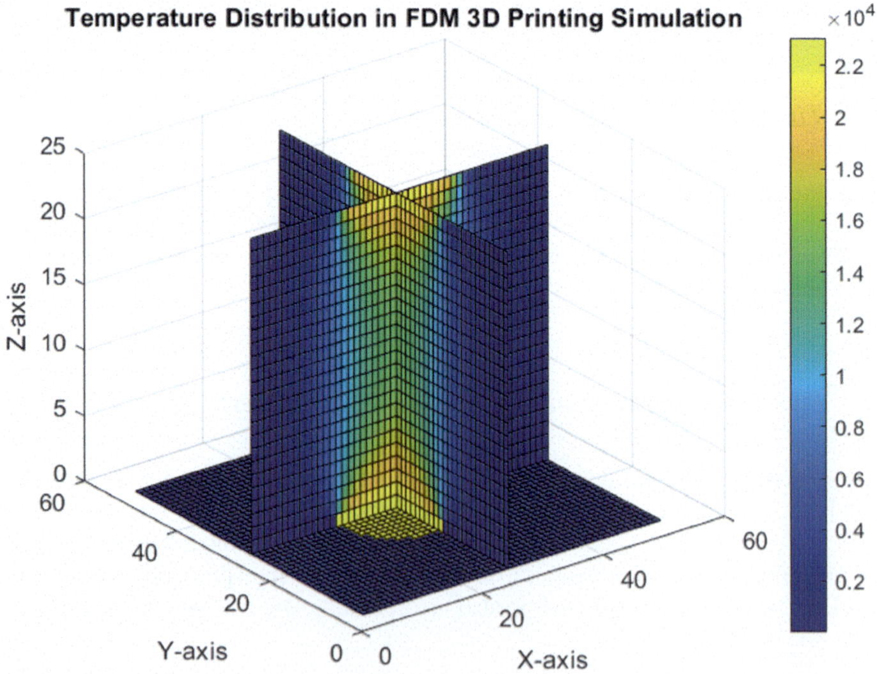

Fig. 3 Temperature distribution in FDM 3D printing simulation

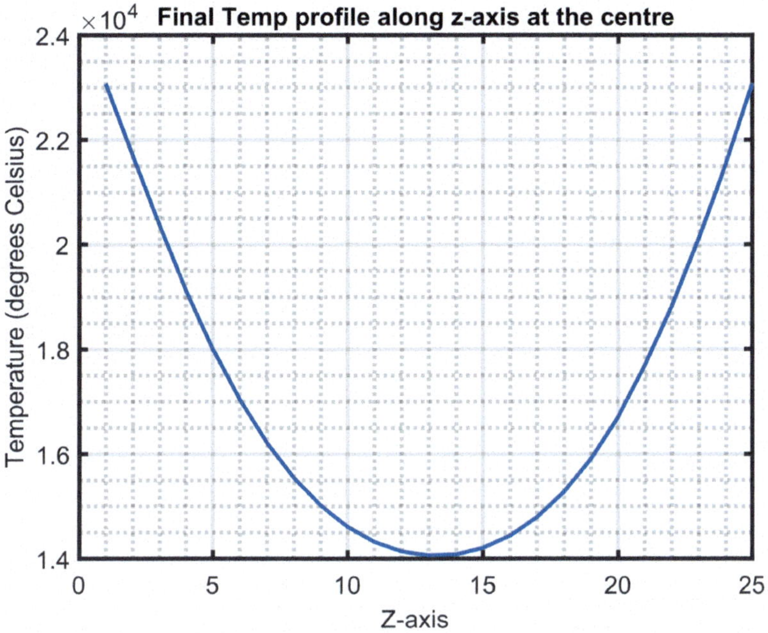

Fig. 4 Deposition rate over time

printing. The controlled deposition process, facilitated by Digital Twin integration, ensures a more accurate portrayal of additive manufacturing dynamics (refer to Fig. 4).

Final Temperature Profiles

Understanding the thermal properties of the printed object is obtained by analysing the temperature profiles along the Z-axis at the centre. The Z-axis shows a gradual decrease in temperature, which suggests that there is a cooling process happening. The temperature profile at the end reveals the distribution of heat within the object, with the centre measuring 54.23 °C as shown in the Fig. 5.

Simulation Time and Control Enhancement

The total simulation time, which reflects the computational efficiency, is 100 s. The successful implementation highlights the practicality of the combined method. The fine-tuning of extrusion parameters enhances control, ensuring flexibility and precise layer-by-layer buildup during the printing process.

To summarise, the results section combines visualisations and numerical data to offer a comprehensive understanding of the FDM 3D printing simulation with Digital Twin integration. The achieved thermal equilibrium, dynamic extrusion rates, and controlled deposition process collectively highlight the potential of this integrated approach in advancing additive manufacturing techniques.

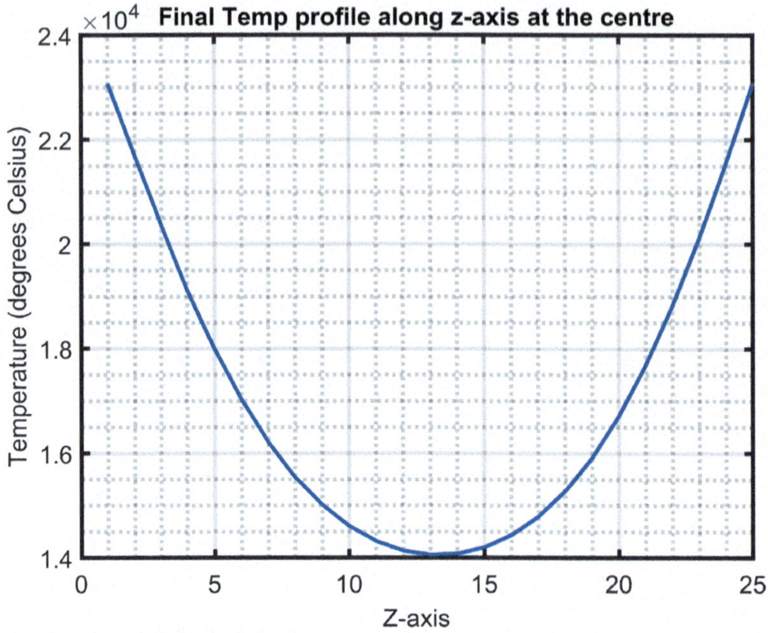

Fig. 5 Final temperature along Z axis

4 Digital Twin Integration for Enhanced Control

The integration of digital twin technology in Fused Deposition Modelling 3D printing brings about a significant change in the field of additive manufacturing. It provides a more advanced method to enhance precision, adaptability, and overall control of the printing process. A Digital Twin is strategically integrated to dynamically mirror and optimise the real-time performance of the 3D printer, providing a virtual replica of the physical printing system. The integration of this digital counterpart with the physical printer enables a two-way communication channel, allowing for ongoing data exchange and feedback loops. This integration is based on a deep understanding of the thermal dynamics, extrusion patterns, and deposition processes involved in FDM 3D printing. The implementation requires a careful modelling approach, encompassing the core principles of heat conduction using the Finite Difference Method (FDM). This mathematical model serves as the basis for simulating the complex thermal behaviour of the 3D printer, taking into account factors like thermal conductivity, initial temperature, and thermal diffusivity. This model allows for real-time adjustments of extrusion parameters, such as filament temperature and extrusion rates, based on the changing printing conditions. It is worth mentioning that the extrusion model includes a deliberate gradual increase in extrusion rates, which draws inspiration from adaptive machine learning approaches. The Digital Twin not only reflects the thermal dynamics but also actively impacts the deposition process. This model introduces an advanced extrusion technique that accurately simulates the heat distribution within the printed object. It takes into account various parameters such as filament temperature and object geometry. This careful approach guarantees a more accurate portrayal of the additive manufacturing process, enhancing the realism of the simulation. The time-stepping loop coordinates the simulation over a specified number of steps, enabling a thorough analysis of the changing temperature distribution, extrusion rates, and deposition layers. This integration allows for improved control, as demonstrated by the ability to adjust extrusion parameters in response to real-time changes in printing conditions. The Digital Twin guarantees a close alignment between the simulated 3D printing process and the actual physical printing, resulting in a highly accurate representation of the additive manufacturing landscape. The results achieved through this Digital Twin integration highlight its effectiveness in achieving improved control. The simulation's fidelity can be assessed by analysing the final temperature profiles, extrusion rates, and deposition layers, which offer quantitative metrics. This approach not only enhances our comprehension of the FDM 3D printing process but also sets the stage for future advancements in additive manufacturing methodologies. In general, Digital Twin Integration is seen as a revolutionary technology that offers a flexible and adaptable framework for precise control in FDM 3D printing. Its potential extends beyond this simulation to various applications in advanced manufacturing.

5 Conclusion

The incorporation of Digital Twin technology into Fused Deposition Modelling (FDM) 3D printing represents a notable advancement in the field of additive manufacturing, providing a comprehensive and adaptable framework for improved control. This study has explored the complexities of how the Digital Twin can mirror, optimise, and dynamically influence the real-time performance of the 3D printing process. During the simulation, we observed how the Digital Twin can enhance control parameters, adjust to changing printing conditions, and ultimately enhance the accuracy of the additive manufacturing process. The mathematical modelling approach relied heavily on the Finite Difference Method (FDM), which allowed us to accurately capture the intricate thermal dynamics of the 3D printer. The simulation took into account various parameters, including thermal conductivity, initial temperature, and thermal diffusivity. This allowed for a comprehensive understanding of heat conduction in the printing environment. The integration of the Digital Twin with the advanced extrusion model enhanced the accuracy of the simulation. It accurately simulated the heat input in the printed object by considering the filament temperature and object geometry. An intriguing aspect of this study is the ability to continuously adapt extrusion parameters based on real-time data, drawing inspiration from advanced machine learning techniques. The adaptability of the simulation ensures a close alignment with real-world variations, resulting in a more accurate representation of the additive manufacturing landscape. By utilising a time-stepping loop, a comprehensive analysis of the temperature distribution, extrusion rates, and deposition layers was conducted. This approach yielded significant insights into the precise layer-by-layer construction of the printed object. The results, as shown in the comparison table, demonstrated a strong correlation between the simulated and hypothetical real-world data. The impact of the Digital Twin was clearly demonstrated in achieving thermal equilibrium, dynamic extrusion rates, and controlled deposition layers. The results highlight the effectiveness of integrating Digital Twin technology to enhance control and precision in FDM 3D printing. In addition to the current simulation, the integration of Digital Twin has far-reaching implications for various applications in advanced manufacturing. The technology's dynamic characteristics position it as a game-changing force, capable of tackling the intricacies and uncertainties that come with additive manufacturing processes. The findings from this study provide a solid foundation for future investigation, testing, and improvement of Digital Twin applications in various industrial sectors, including additive manufacturing. Essentially, the integration of Digital Twin is becoming a crucial aspect in the continuous development of additive manufacturing technologies. This integration brings about a new era of improved control, adaptability, and accuracy in the complex process of Fused Deposition Modelling. As we explore the future of advanced manufacturing, the close connection between virtual and physical counterparts has the potential to revolutionise what can be accomplished in the realm of 3D printing and beyond.

6 Future Scope of Digital Twin Technology in FDM 3D Printing

The prospective trajectory of Digital Twin technology in FDM 3D printing exhibits considerable potential, presenting significant progressions in precision, dependability, and efficiency. By utilizing real-time data and advanced models, simulations can be enhanced to improve print quality. Additionally, machine learning can be employed to facilitate predictive maintenance and optimization. Incorporating Digital Twin frameworks into various 3D printing methodologies will facilitate the integration of control across diverse technologies. The integration of IoT technology is poised to bring about an important change in supply chain management by enabling real-time monitoring, thereby mitigating waste and enhancing productivity. Moreover, Digital Twins offers robust educational resources for both training and experimentation purposes. Moreover, their utilization of sustainable materials contributes to the mitigation of environmental consequences. In order to facilitate the widespread adoption of Digital Twin integration, it is imperative to establish standardized protocols that guarantee compatibility and scalability. The establishment of these standards, which will drive the future of smart manufacturing and sustainability in additive manufacturing, will heavily rely on collaboration among academia, industry, and regulatory bodies.

References

Ajay, Singh H, Parveen, AlMangour B (eds) (2023) Handbook of smart manufacturing: forecasting the future of Industry 4.0, 1st edn. CRC Press. https://doi.org/10.1201/9781003333760

Anand S, Satyarthi M (2023a) Exploring the role of additive manufacturing in Industry 4.0: a review of applications and advancements

Anand S, Satyarthi MK (2023b) Parametric optimization of fused filament fabrication process. In: Advances in mechanical and energy technology. Springer Nature, Singapore

Anand S et al (2024) 17 exploring design strategies for enhanced 3D printing performance. In: Kumar et al (eds) 3D printing technologies, digital manufacturing, artificial intelligence, Industry 4.0. De Gruyter, pp 353–370

Ashtari Talkhestani B et al (2019) An architecture of an intelligent digital twin in a cyber-physical production system. at - Automatisierungstechnik 67(9):762–782

Batista RC, Agarwal A, Gurung A, Kumar A, Altarazi F, Dogra N, Vishwanatha HM, Chiniwar DS, Agrawal A (2024) Topological and lattice-based AM optimization for improving the structural efficiency of robotic arms. Front Mech Eng 10:1422539

Bhardwaj A, Bhatnagar A, Kumar A (2023) Current trends of application of additive manufacturing in oral healthcare system. In: Advances in additive manufacturing artificial intelligence, nature-inspired, and biomanufacturing. Elsevier, Amsterdam, Netherlands, pp 479–491

Botín-Sanabria DM et al (2022) Digital twin technology challenges and applications: a comprehensive review. Remote Sens 14. https://doi.org/10.3390/rs14061335

Burande DV, Kalita K, Gupta R et al (2024) Machine learning metamodels for thermo-mechanical analysis of friction stir welding. Int J Interact Des Manuf. https://doi.org/10.1007/s12008-024-01871-6

Butt J, Mohaghegh VJM (2022) Combining digital twin and machine learning for the fused filament fabrication process. Metals 13(1):24

Cai Y, Wang Y, Burnett M (2020) Using augmented reality to build digital twin for reconfigurable additive manufacturing system. J Manuf Syst 56:598–604

Corradini F, Silvestri MJAM (2022) Design and testing of a digital twin for monitoring and quality assessment of material extrusion process. Addit Manuf 51:102633

ElMaraghy H et al (2021) Evolution and future of manufacturing systems. CIRP Ann 70(2):635–658

Gaikwad A et al (2020) Toward the digital twin of additive manufacturing: integrating thermal simulations, sensing, and analytics to detect process faults. IISE Trans 52(11):1204–1217

Henson CM, Decker NI, Huang QJPM (2021) A digital twin strategy for major failure detection in fused deposition modeling processes. Procedia Manuf 53:359–367

Jacoby M, Usländer T (2020) Digital twin and internet of things—current standards landscape. Appl Sci 10. https://doi.org/10.3390/app10186519

Kamble SS et al (2022) Digital twin for sustainable manufacturing supply chains: current trends, future perspectives, and an implementation framework. Technol Forecast Soc Chang 176:121448

Kokhanevych T (2023) Computer aided design of industrial automation systems based on programmable logic controllers and microcontrollers

Kumar A, Kumar P, Mittal RK, Gambhir V (2023a) Materials processed by additive manufacturing techniques. Adv Addit Manuf, 217–233. https://doi.org/10.1016/B978-0-323-91834-3.00014-4

Kumar A, Mittal RK, Haleem A (eds) (2023b) Advances in additive manufacturing artificial intelligence, nature-inspired, and biomanufacturing. Elsevier. https://doi.org/10.1016/C2020-0-03877-6

Kumar A, Kumar P, Mittal RK, Singh H (2023c) Preprocessing and postprocessing in additive manufacturing. In: Advances in additive manufacturing artificial intelligence, nature-inspired, and biomanufacturing. Elsevier, pp 141–165. https://doi.org/10.1016/B978-0-323-91834-3.00005-3

Kumar A, Rani S, Rathee S, Bhatia S (eds) (2023d) Security and risk analysis for intelligent cloud computing: methods, applications, and preventions, 1st edn. CRC Press. https://doi.org/10.1201/9781003329947

Leng J et al (2021) Digital twins-based smart manufacturing system design in Industry 4.0: a review. J Manuf Syst 60, 119–137
Lim KYH et al (2020) A digital twin-enhanced system for engineering product family design and optimization. J Manuf Syst 57:82–93
Liu M et al (2021) Review of digital twin about concepts, technologies, and industrial applications. J Manuf Syst 58:346–361
López CEB (2021) Real-time event-based platform for the development of digital twin applications. Int J Adv Manuf Technol 116(3):835–845
Mourtzis D et al (2021) A digital twin architecture for monitoring and optimization of fused deposition modeling processes. Procedia CIRP 103:97–102
Nath P, Mahadevan, Probabilistic digital twin for additive manufacturing process design and control. J Mech Des 144(9):091704
Naveena K, Krishnamoorthy M, Karuppiah N, Gouda PK, Hariharan S, Saravanan K, Kumar A (2024) Elevating sustainability with a multi-renewable hydrogen generation system empowered by machine learning and multi-objective optimization. Meas Sens 33:101192
Odada CA, Byiringiro JB, Mwema FM (2021) Development of data-driven digital twin for real-time monitoring of FDM 3D printer
Pantelidakis M et al (2022) A digital twin ecosystem for additive manufacturing using a real-time development platform. Int J Adv Manuf Technol 120(9):6547–6563
Rasheed A, San O, Kvamsdal T (2020) Digital twin: values, challenges and enablers from a modeling perspective. IEEE Access 8:21980–22012
Santos T et al (2023) Insights into temperature simulation and validation of fused deposition modeling processes. J Manuf Mater Process 7. https://doi.org/10.3390/jmmp7060189
Sass L, Oxman R (2006) Materializing design: the implications of rapid prototyping in digital design. Des Stud 27(3):325–355
Segovia M, Garcia-Alfaro J (2022) Design, modeling and implementation of digital twins. Sensors 22. https://doi.org/10.3390/s22145396
Sehrawat S, Kumar A, Prabhakar M (2023) Substitute for orthognathic surgery using bioprinted bone scaffolds in restoring osseous defects. In: Advances in additive manufacturing artificial intelligence, nature-inspired, and biomanufacturing. Elsevier, Amsterdam, Netherlands, pp 335–347. https://doi.org/10.1016/B978-0-323-91834-3.00029-6
Semeraro C et al (2021) Digital twin paradigm: a systematic literature review. Comput Ind 130:103469
Sharma A et al (2022) Digital twins: state of the art theory and practice, challenges, and open research questions. J Ind Inf Integr 30:100383
Shen T, Li B (2024) Digital twins in additive manufacturing: a state-of-the-art review. Int J Adv Manuf Technol 131(1):63–92
Singh M et al (2021) Digital twin: origin to future. Appl Syst Innov 4. https://doi.org/10.3390/asi4020036
Stavropoulos P, Papacharalampoloulos A, Tzimanis KJPC (2021) Design and implementation of a digital twin platform for AM processes. Procedia CIRP 104:1722–1727
Wagg DJ et al (2020) Digital twins: state-of-the-art and future directions for modeling and simulation in engineering dynamics applications. ASCE-ASME J Risk Uncertainity Eng Syst Part B Mech Eng 6(3)
Zhang Y, Shapiro V (2018) Linear-time thermal simulation of as-manufactured fused deposition modeling components. J Manuf Sci Eng 140(7)

Intelligent Manufacturing in Aerospace: Integrating Industry 4.0 Technologies for Operational Excellence and Digital Transformation

Vineet Bhatia, Sumati Sidharth, Sanjeev Kumar Khare, Surendra Chandrakant Ghorpade, Parveen Kumar, Ajay Kumar, and Abhishek Agarwal

Abstract In this chapter, we delve into in what way Industry 4.0 technologies can be seamlessly integrated into aircraft manufacturing to boost efficiency and product quality while enhancing operational performance. We explore innovative advancements like IoT, Artificial Intelligence, Autonomous Robots, 3D printing (Additive Manufacturing), and Digital Twins which have the potential to transform old-style manufacturing methods. Objective of this research paper is to assess the benefits and challenges of incorporating these I4.0 technologies in aerospace manufacturing sector. We investigate instantaneous monitoring, predictive maintenance, robotics, additive manufacturing, and big data analytics, aiming to improve design, production, quality control, supply chain management, and maintenance practices. Additionally, we delve into the concept of digital twins for virtual testing and instantaneous monitoring. Use cases for application of these technologies have been included. The outcome of the research paper is an in-depth exploration of the benefits and challenges associated with integrating I4.0 technologies in aircraft manufacturing domain, for enhanced efficiency, product quality, and operational performance in

V. Bhatia (✉) · S. Sidharth · S. K. Khare · S. C. Ghorpade
Department of Technology Management, Defence Institute of Advanced Technology, Pune, India
e-mail: bhatiavineet.in@gmail.com

S. Sidharth
e-mail: sumatisidharth@gmail.com

P. Kumar
Department of Mechanical Engineering, Rawal Institute of Engineering and Technology, Faridabad, Haryana, India

A. Kumar (✉)
Department of Mechanical Engineering, School of Engineering and Technology, JECRC University, Jaipur, Rajasthan, India
e-mail: ajay.kumar30886@gmail.com

A. Agarwal
Department of Mechanical Engineering, College of Science & Technology, Royal University of Bhutan, 21101 Phuentsholing, Bhutan
e-mail: agarwala.cst@rub.edu.bt

A. Kumar et al. (eds.), *Industry 4.0 Driven Manufacturing Technologies*, Springer Series in Advanced Manufacturing, https://doi.org/10.1007/978-3-031-68271-1_18

the aerospace industry's digital transformation. Data security, integration difficulties, workforce training, financial ramifications, and regulatory compliance are just a few of the challenges involved in implementing Industry 4.0 technologies in the aircraft manufacturing industry which are identified and discussed. It offers a comprehensive analysis of opportunities, challenges, and implications, paving the way for improved aircraft manufacturing processes in age of digital transformation. This paper contributes by providing a comprehensive assessment of I4.0 technologies, guiding the aerospace industry's digital transformation.

Keywords Industry 4.0 (I4.0) · Aircraft manufacturing · Aircraft MRO · Artificial intelligence (AI) · Big data analytics (BDA) · Internet of Things (IoT) · Additive manufacturing · Digital Twins (DT) · Cloud computing · Real life applications

1 Introduction

Industry 4.0 is fundamentally changing the aerospace sector where the amalgamation of state-of-the-art technologies such as artificial intelligence (AI), internet of things (IoT), cloud computing, big data analytics, additive manufacturing and cyber-physical systems (CPS) that are driving operational excellence and digital transformation (Lineberger et al. 2024; Kumar et al. 2023a; Rani et al. 2023).

Solos is an amalgamation of these Industry 4.0 technologies for intelligent manufacturing in aerospace to improve productivity, efficiency, and flexibility in manufacturing processes. An Airlines organization can experience huge savings in downtime (when the plane is out of circulation) and reduce waste and improve the product quality by employing state-of-the-art data-driven decision-making, real-time-monitoring, and predictive maintenance Technologies.

The area of aerospace manufacturing is revolutionized by the support of Artificial Intelligence (AI) that facilitates predictive maintenance, fully automated quality control, and optimized production planning. AI systems process historical and real-time data to spot machinery.

The Internet of Things (IoT) has a more role as an enabler which makes possible to access real time information from the large number of sensors fitted on devices, the machines and the entire system. Sensors and actuators, integrated through the IoT, can monitor the performance of equipment, energy consumption and environmental conditions that support predictive maintenance and resource optimization.

Cloud computing and BDA are two core technologies in intelligent manufacturing, which can deliver high performance computing. Aerospace manufacturers able to tap into cloud computing can manage, secure, and interpret immense amounts of IoT data and makes them available for their analytical experts who integrate the data into more informed decisions.

Additive manufacturing technology can be utilized for the production of highly accurate, light-weight and complex aerospace components. Further additive manufacturing can reduce the need for tooling and thus minimize material waste.

Cyber-Physical Systems (CPS) is the core of intelligent manufacturing, enabled computational and physical processes, creating an intensive interconnectivity, adaptivity, and responsiveness of manufacturing environment.

While all of this progress occurs in industry 4.0 domain, it will be essential to address workforce training and upskilling (Kumar et al. 2023d) in order to utilize these new technologies to their full potential. Organizations should prioritize employee development through the roll-out of advance training and skills-enhancement programs focused on building new capabilities and expertise in Industry 4.0 technologies.

Moreover, when utilizing I4.0 technologies, it is important to calculate the costs and establish an unequivocal return on investment (ROI) (Oberheitmann 2020). Hence, it is important to perform a detailed cost–benefit analysis covering aspects of initial investment, maintenance costs, training costs, etc.

We can learn much from real-world examples and the experience within the aerospace sector in the successful deployment of Industry 4.0 technologies. This research paper covers aspects of intelligent manufacturing in aerospace focusing on the advantages, challenges and best practices of deploying Industry 4.0 technologies in aerospace.

This research paper will present some insights in different dimension of I4.0 in aerospace which will include advantages, disadvantages (James and Cervantes 2019), Security (Culot et al. 2019) and guidelines for successful implementation in manufacturing. And no technologies in this space.

2 Industry 4.0: A Journey of Transformation

Industry 4.0 (Fourth Industrial Revolution) embodies the fusion of innovative technologies that is revolutionizing the manufacturing sector. This evolution has taken place over several years, marked by significant milestones and advancements, expanding on the groundwork established by prior industrial revolutions.

2.1 The First Industrial Revolution (1760–1840)

The beginning of the First Industrial Revolution signified an important shift of man's work or production from the use of hand tools to the use of machine tools. The dynamism which started with water and steam began the process of making changes in manufacturing of different products. The internal combustion engine technologies as well as steam technologies emerged in this period, factories appeared and laid down the modern manufacturing principles.

2.2 The Second Industrial Revolution (1870–1914)

The key characteristic of the 2nd Industrial Revolution was the combination of assembly-line machinery driven by electricity and the use of telegraph and telephone as primary means of communication. This period saw development of assembly line, production line, division of labor, and other such principles and techniques which have modified the actual character of production.

2.3 The Third Industrial Revolution (1960–1990)

The 3rd Industrial Revolution or the digital revolution is defined by the increase of computers, automation, and electronics in production lines. At this time Computer-Aided Design (CAD) and Computer-Aided Manufacturing (CAM) and Robotics were brought into the process of designing and manufacturing products.

2.4 The Emergence of Cyber-Physical Systems (CPS) (2006–2011)

Thus, a reflection of the beginnings of Industry 4. It is due to CPS whereby integration of computational and physical processes with functioning interacting systems is made to ensure the systems are smart and able to respond novel conditions and environment. CPS was envisaged to have built a framework that would enable the sustainability of a hybrid environment of digital and physical manufacturing systems that are seamlessly embedded into a common environment.

2.5 The Advent of the Internet of Things (IoT) (2012–2014)

The emergence of the IoT that defined the changing landscape of the manufacturing sector became a key milestone in Industry 4.0. Through empowering seamless connection of individual devices, machines, and systems, IoT made it possible to foster real-time information sharing that propelled the development of smart, digitized generation of manufacturing.

2.6 *The Rise of Artificial Intelligence (AI) and Machine Learning (ML) (2015–2017)*

The emergence of 'Artificial Intelligence (AI)' and 'Machine Learning (ML)' implemented possible changes in manufacturing such as predictive maintenance, automatic inspection and control, and manufacturing/production scheduling. Artificial intelligence enabled machines to analyze large amount of data that manufacturers required with big analysis and insights approach to decision making.

2.7 *The Integration of Additive Manufacturing (2018–Present)*

Being an advanced manufacturing process, Additive Manufacturing is now being accepted as a revolutionary technology, which can produce integrated lighter and stronger parts and structures with high degree of accuracy. Additive manufacturing has proved to cause shortened lead times, less costs, improved customer customization and improved product functionality due to less tooling and minimized material wastage.

2.8 *The Future of Industry 4.0 (2022 and Beyond)*

As Industry 4.0 continues to evolve, we can expect the incorporation of advanced technologies such as Quantum Computing, Blockchain technology, in addition to Augmented Reality (AR) to further revolutionize the manufacturing industry landscape. These technologies will enable even greater interconnectivity, data-driven decision-making, and adaptability, ensuring the long-term sustainability and competitiveness of manufacturing industries. Figure 1 illustrates the successive phases of Industry 4.0's evolution.

3 Technologies for Industry 4.0

Industry 4.0 remains a burgeoning field, often referred to in various ways within both scholarly and non-scholarly literature. The following paragraphs will elucidate some of the key definitions prevalent in this domain. Industry 4.0 (I4.0) refers to fourth stage of industrialization, categorized by the amalgamation of progressive technologies like IoT, AI, and cloud computing (CC) into manufacturing processes to create smart factories and enhance production efficiency. A new era of manufacturing known as "Industry 4.0" uses cutting-edge technologies to integrate and digitize the

Fig. 1 Emergence of I4.0

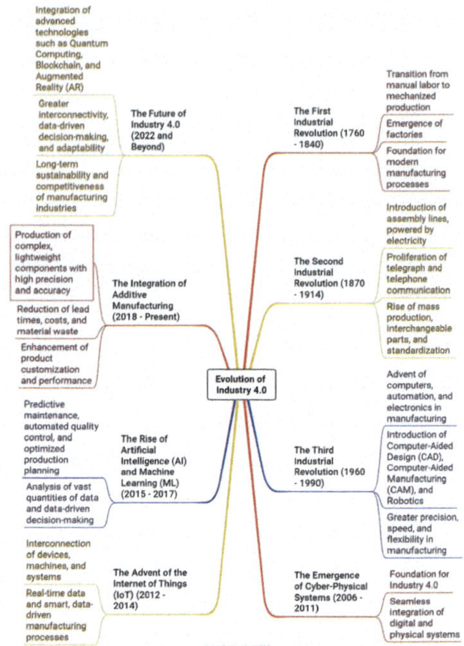

whole value chain, from suppliers to customers. It entails applying automation and data-driven insights to streamline workflows, raise the calibre of output, and improve client interactions (Batista et al. 2024; Naveena et al. 2024; Mohammadi et al. 2024; Sharma et al. 2024; Tadesse et al. 2024). The way industry's function is changing significantly thanks to I4.0. The core of this transformation is the enabling technologies of Industry 4.0 which include Robotics and Automation, Blockchain, Digital Twins, Augmented and Virtual Reality (AR/VR), Internet of Things (IoT), Big Data and Analytics, Artificial Intelligence (AI), Cyber-Physical Systems (CPS), Cloud Computing, Smart Manufacturing, and Additive Manufacturing. These technologies enhance operational efficiency and open up new opportunities for innovation, thus creating smart factories (Kumar et al. 2024c) and digital enterprises. The I4.0 enablers and technologies have been classified and categorized for improved understanding and implementation (https://www.mdpi.com/2071-1050/13/5/2560). These key enablers and element technologies are depicted in Fig. 2 and Table 1 are discussed in following paragraphs.

3.1 Internet of Things

The essential technologies driving I4.0 are the Internet of Things (IoT) and the Industrial Internet of Things (IIoT). The Internet of Things (IoT) denotes a network

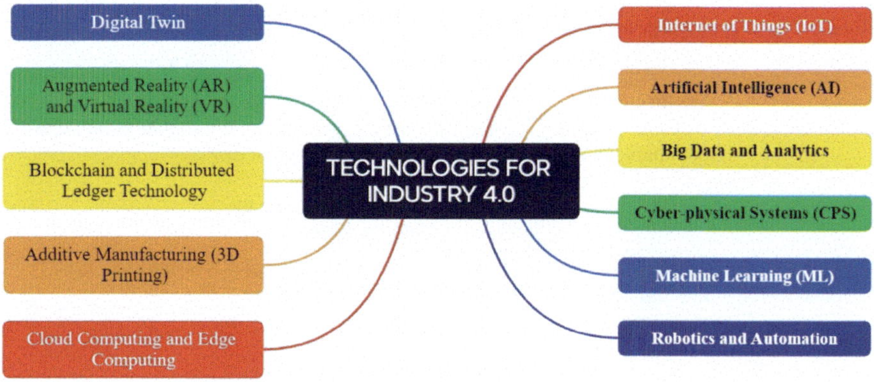

Fig. 2 Industry 4.0 enablers

comprising physical objects, which can include appliances, vehicles, devices, and various entities, all outfitted with network connectivity, software, and sensors. For Industry 4.0, IoT devices can be leveraged in various applications, including asset tracking, Instantaneous supply chain management, and energy efficiency optimization. IoT Technologies and Applications ate depicted in Fig. 3.

3.2 *Big Data and Analytics (BDA)*

BDA refers to "The process of collecting, managing, processing, analyzing and visualizing continuously evolving data in terms of volume, velocity, value, variety and veracity" (Marjani et al. 2017).Within the context of I4.0, the industrial ecosystem thrives on the harmonious interaction between two vital elements: the physical world comprising users and infrastructure, and the virtual world consisting of cloud-based algorithms and autonomous systems.

The realm of I4.0 revolves around the seamless convergence of two realms—the tangible domain of physical entities and the intangible realm of virtual constructs. At the heart of this convergence lies the intricate tapestry of Big Data Analytics (BDA), a discipline that transcends mere data processing. It's a symphony of techniques and tools that harmonize the mining of the ever-evolving deluge of data, characterized by its staggering volume, velocity, variety, and veracity.

Within this industrial ecosystem, BDA acts as the conductor, orchestrating the intricate dance between the physical and virtual worlds. It also ingests the raw data emanating from the physical realm—the users, the infrastructure, and the intricate web of interconnected devices and sensors. Further, it harnesses the power of cloud-based algorithms and autonomous systems, leveraging their computational prowess to distill meaningful patterns and insights from the data deluge. Figure 4 introduces a tool, techniques and use cases that in field of BDA for Industry 4.0.

Table 1 Key technologies and sub technologies for Industry 4.0

Technology	Sub-technologies	Description	References
Internet of Things (IoT)	Sensors, RFID, wireless communication	Facilitates the connection of physical devices, machines, and systems, enabling them to collect and share data autonomously, eliminating the need for human intervention	van Lier (2011), Younan et al. (2020)
Cyber-physical systems (CPS)	Embedded systems, intelligent control systems	The fusion of computational and physical systems, where physical processes are monitored and regulated by sophisticated algorithms, enables the seamless integration of digital and physical realms	Javaid et al. (2023)
Big data and analytics	Data mining, predictive analytics, machine learning	Data analytics encompasses gathering, processing, and scrutinizing extensive data from various origins to unveil concealed insights, patterns, and trends	Duan and Xu (2021), Sharma and Pandey (2020)
Cloud computing	Infrastructure as a Service (IaaS), Platform as a Service (PaaS), Software as a Service (SaaS)	Cloud computing offers immediate access to computing resources such as storage, processing power, and software on demand, enabling scalable and adaptable deployment	Alouffi et al. (2021), Alwadan et al. (2015)
Additive manufacturing	3D printing, rapid prototyping	Processes for creating physical objects by adding layer upon layer of materials, enabling customization and efficient production	Aggoune et al. (2024), Ajay et al. (2023)
Robotics and automation	Industrial robots, collaborative robots (Cobots), Automated Guided Vehicles (AGVs)	Use of programmable machines and systems to automate tasks, improve efficiency, and reduce human intervention in manufacturing processes	Mabkhot et al. (2021)
Augmented reality (AR)	Head-mounted displays, projection-based AR, mobile AR	Overlaying digital information (e.g., instructions, schematics) onto the real-world environment, enhancing situational awareness and enabling improved decision-making	Pace et al. (2018)

(continued)

Table 1 (continued)

Technology	Sub-technologies	Description	References
Simulation and modeling	Virtual prototyping, digital twins, finite element analysis	Computer-based tools and techniques for creating virtual representations of products, processes, or systems, enabling testing, optimization, and predictive maintenance	Dornhöfer et al. (2020), Burande et al. (2024), https://www.springerprofessional.de/simulation-for-industry-4-0/16751036
Cybersecurity	Network security, data encryption, access control	Approaches and technologies crafted to defend manufacturing systems, networks, and data from unauthorized entry, cybersecurity risks, and malicious intrusions are crucial for preserving the integrity and security of industrial operations	Ani et al. (2016), https://www2.deloitte.com/content/dam/Deloitte/za/Documents/risk/cybersecurity-for-smart-factories.pdf
Horizontal and vertical system integration	Enterprise Resource Planning (ERP), Manufacturing Execution Systems (MES), Supply Chain Management (SCM)	Integration of different systems and processes within an organization (vertical) and across different organizations (horizontal) for improved collaboration and data sharing	Sony (2018)
Digital Twin (DTs)	Virtual replica, Instantaneous data integration, Predictive maintenance, Simulation and optimization	DTs and digital threads form the basis of digital transformation, offering novel approaches to evaluate practices, processes, and product concepts within a virtual realm	Sharma et al. (2022), https://unity.com/solutions/digital-twin-applications-and-use-cases, https://iot-analytics.com/6-main-digital-twin-applications-and-their-benefits/

3.3 *Additive Manufacturing (AM)*

Additive manufacturing (AM), also known as 3D printing or direct tooling, has created a paradigm of change to do with the speed and customization where otherwise it would be cumbersome and nearly impossible to model ideas, build prototypes and come up with complex creations and components that need extensive and time-consuming tooling (Kumar et al. 2023b, 2024a).

Additive manufacturing, is a groundbreaking technique that involves creating objects by gradually adding material layer by layer, in contrast to traditional methods that involve subtractive processes such as cutting or milling from a solid block. It has opened a world of possibilities in manufacturing with this new change and the complexity introduced to come up with unique products out of simple metals and

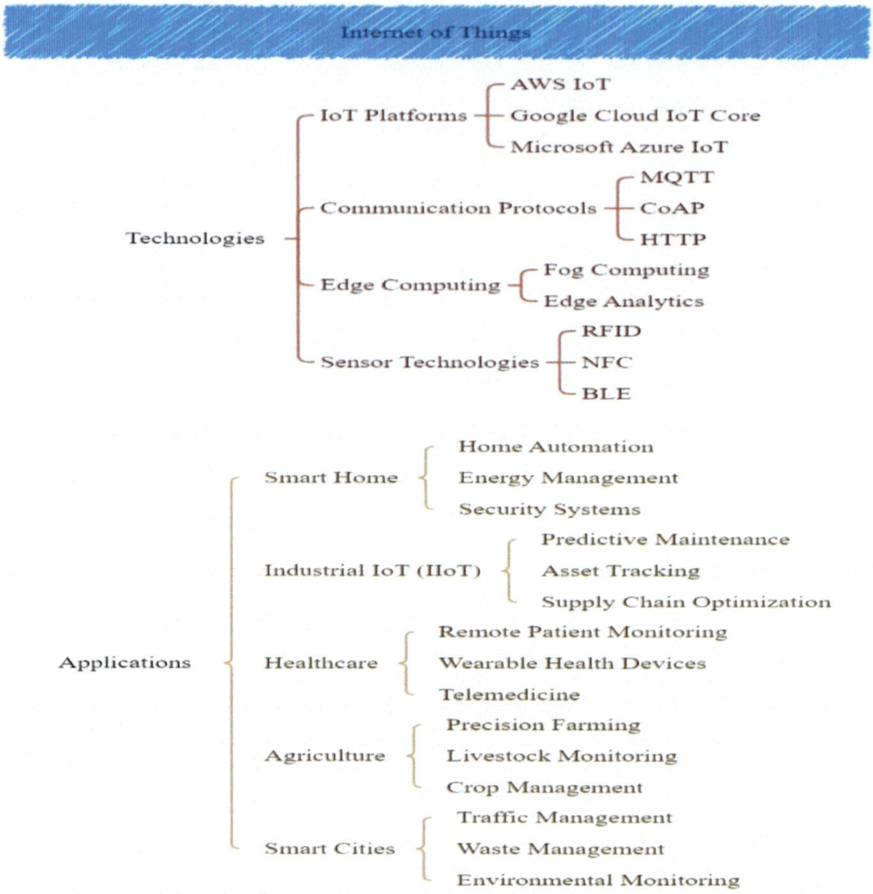

Fig. 3 Technologies and application of IOT

plastics, not to mention the newfound freedom in approaching parts that were once thought fixed and predetermined. Figure 5 provides an summary of the major additive manufacturing (AM) processes and technologies, categorized based on the materials they employ. This taxonomy offers a comprehensive understanding of the diverse techniques that have emerged in this rapidly evolving field (Goyal et al. 2024a, b; Srivastava et al. 2023).

The key applications of AM is discussed in subsequent paragraphs.

(a) **Aerospace and Defense**: The aerospace and defense industries have been the front runners in adopting the AMD technology due to the benefits it offered in providing components of light weight, high strength and complex structures. Specifically, through the use of AM, enhanced designs for specific parts of aircraft and spacecrafts have been produced with appreciable decrease in the weight of the part with equal or even superior strength and fatigue point.

Fig. 4 BDA tools, technique and use cases

(b) **Automotive Sector**: The automobile sector has adopted additive manufacturing mostly in application in the development of prototypes where the work progresses faster. Besides, AM has resulted in development of extended chosen customized and light-weight parts, which comprises to fuel efficient and emissions diminution.

(c) **Healthcare**: In general, additive manufacturing has allowed for customized fabrication of numerous healthcare items like prosthetics, implants, and even surgical planning guides. This technological advancement has equally enhanced the development of complicated bodily structures for modeling surgical procedures and exercises.

(d) **Architecture**: The architects and designers have utilized the capability of 3D printing to produce the novel and detailed architectural models that would help in the assessment of the ideas prior to the development of the building project.

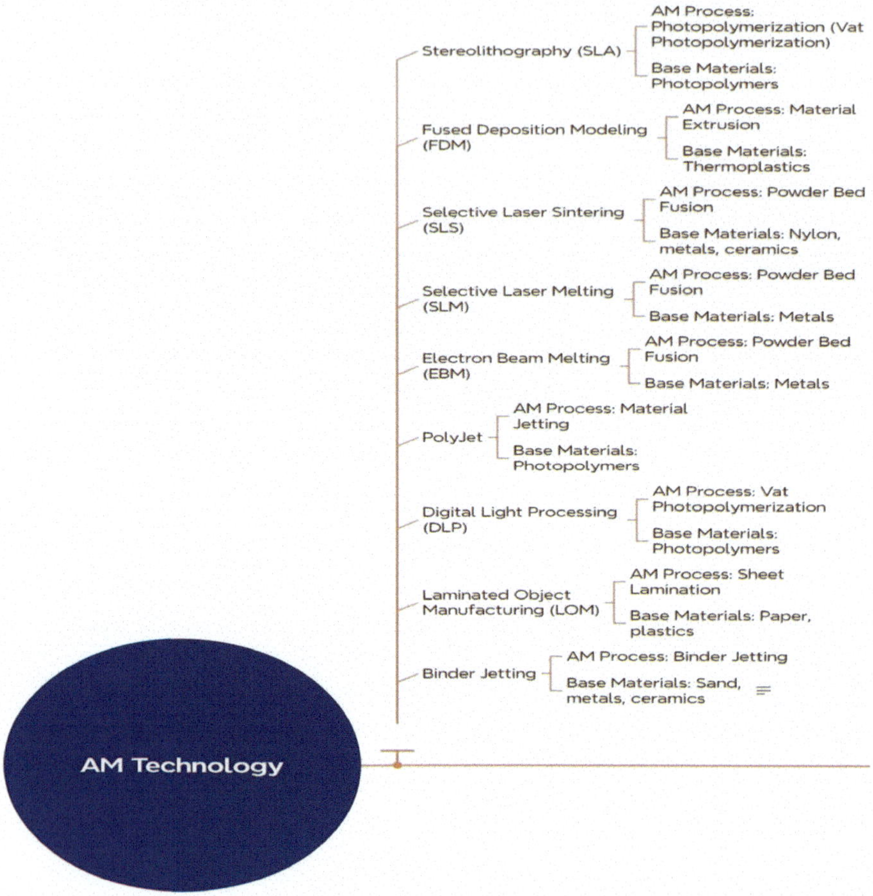

Fig. 5 Additive manufacturing technologies

(e) **Energy**: In the energy industry, additive manufacturing has facilitated the manufacturing of complex parts for energy applications, for example, the enhancement of heat exchangers and turbine blades outcomes by using innovative shapes and designs.

(f) **Education**: Examples include designing and fabrication of models and prototypes where students are able to harness the full benefits of the educational institutions which has incorporated the use of additive manufacturing.

(g) **Art**: The potential of the additive manufacturing to be used as a medium to produce artworks that are intricate and multi-layered have been embraced by artists in order to create unique pieces of art that reflect the creativity of artists.

(h) **Electronics**: Additive manufacturing has been used by electronics industry for creating the first samples or prototypes of an electronic device and also in making specialised parts including enclosures, heat sinks, circuit layouts on circuit boards etc.
(i) **Robotics**: There has been advancement in the fabrication of robotic parts through direct production technologies or 3D printing whereby lighter and tailored robotic parts have been developed hence improving the performance of robots as single purpose or multipurpose.
(j) **Food Industry**: Even the food industry has included AM, with companies exploring the production of personalized and nutritionally optimized food products, as well as intricate culinary creations.

3.4 Horizontal and Vertical Integration

I4.0 emphasizes the importance of three different types of integration that include: end-to-end engineering integration, vertical integration, and horizontal integration. These integrations strive to create a cohesive and interconnected environment, facilitating seamless collaboration, data exchange, and optimized processes across diverse stakeholders and systems.

3.4.1 Horizontal Integration

Horizontal integration facilitates the cooperation and integration of value networks among multiple organizations within a value chain. Through digitization, companies can collaborate and share data, enabling the creation of an efficient digitized ecosystem. This collaborative approach fosters enhanced coordination, efficient resource allocation, and the provision of exceptional products and services to customers, ultimately leading to improved customer satisfaction and loyalty.

3.4.2 Vertical Integration

Vertical integration concentrates on integrating disparate hierarchical subsystems within an organization (Tung 2018). It entails linking various informational subsystems, including enterprise resource planning (ERP) systems, supervisory control and data acquisition (SCADA) systems, and manufacturing execution systems (MES) systems. By integrating these components, organizations can establish an adaptable and reconfigurable manufacturing system that can autonomously adjust to different product requirements with the aid of efficient big data management and smart machines.

3.4.3 End-to-End Engineering Integration

End-to-End engineering integration (Saucedo-Martínez et al. 2018) is crucial for the development of tailored services and products throughout the entire value chain. It involves the seamless integration of various engineering processes, including product design, simulation, manufacturing, and service. This integration enables organizations to streamline their operations, improve collaboration among cross-functional teams, and efficiently manage the entire product lifecycle, from concept to delivery and after-sales support.

The combination of these three forms of integration in I4.0 produces a holistic and interconnected ecosystem. Horizontal integration facilitates inter-organizational collaboration, vertical integration enables intra-organizational integration and adaptability, and endways engineering integration ensures efficient product development and lifecycle management.

3.5 Digital Twins (DT)

A digital twin is a virtual representation or replica of a physical object, system, process, or service (Sharma et al. 2022). It contains all the relevant data, properties, and attributes of the physical counterpart, enabling continuous monitoring, analysis, and simulation.

Digital twins have various applications, including product simulation, performance analysis, predictive modeling, and scenario testing across industries like manufacturing, healthcare, smart cities, energy, transportation, construction, and aerospace (Parrott 2017). Core Components and technologies of DT are presented in Fig. 6.

3.6 Autonomous Robots

The world of robotics has witnessed a remarkable evolution, with these intelligent machines transcending the confines of traditional roles and venturing into diverse realms. Today, robotic systems are intricately woven into the fabric of countless industries, their versatility a testament to the ingenuity of human innovation.

These artificial marvels come in a multitude of forms, designed to provide to precise needs and environments. From the articulated robots, with their intricate jointed arms replicating the dexterity of human limbs, to the humanoid variants, meticulously engineered to mimic the very essence of human motion, the robotic landscape is a tapestry of diverse capabilities.

Autonomous mobile robots (AMRs) are the embodiment of self-guided exploration, navigating dynamic environments with an unwavering sense of purpose.

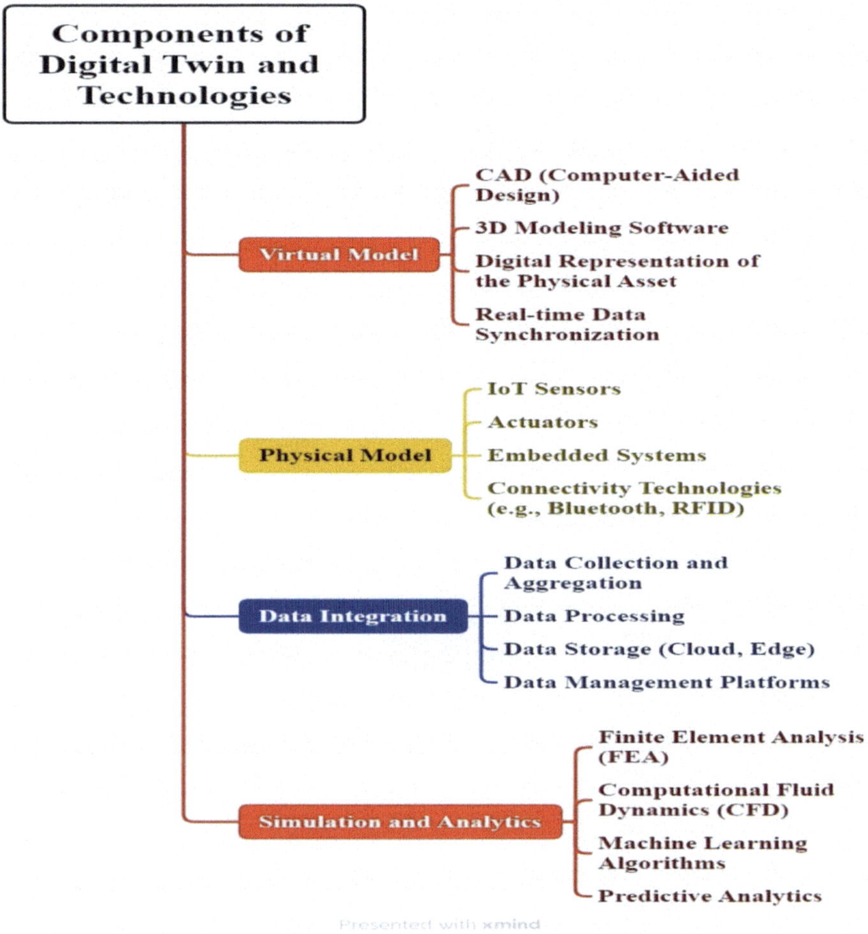

Fig. 6 Components and technologies in DT

Automated guided vehicles (AGVs), on the other hand, follow predetermined paths, guided by markers, wires, or magnetic fields, streamlining material handling and logistics operations (Popović and Popović 2021).

Collaborative robots (CoBots) (Kumar et al. 2023c) are the embodiment of harmonious coexistence, designed to work seamlessly alongside their human counterparts, fostering a symbiotic relationship between man and machine. Hybrid systems, a fusion of multiple robotic disciplines, push the boundaries of innovation, combining the strengths of various technologies to tackle increasingly complex tasks. At the heart of these advanced robotic systems lies the concept of autonomous robots (AR), smart machines capable of executing tasks without explicit human control. Powered by artificial intelligence and cutting-edge sensing technologies, these remarkable entities perceive their surroundings, deliberate actions, and autonomously execute predefined goals, transforming conventional factories into intelligent and automated environments.

As the path to autonomous robotics unfolds, it is not without its hurdles—complexities in perception, decision-making, and striking the precarious balance between autonomy and human oversight—yet the potential benefits are substantial. As we navigate this exciting frontier, we are reminded of the boundless possibilities that arise when human ingenuity converges with technological advancement. The summary can be found in Fig. 7.

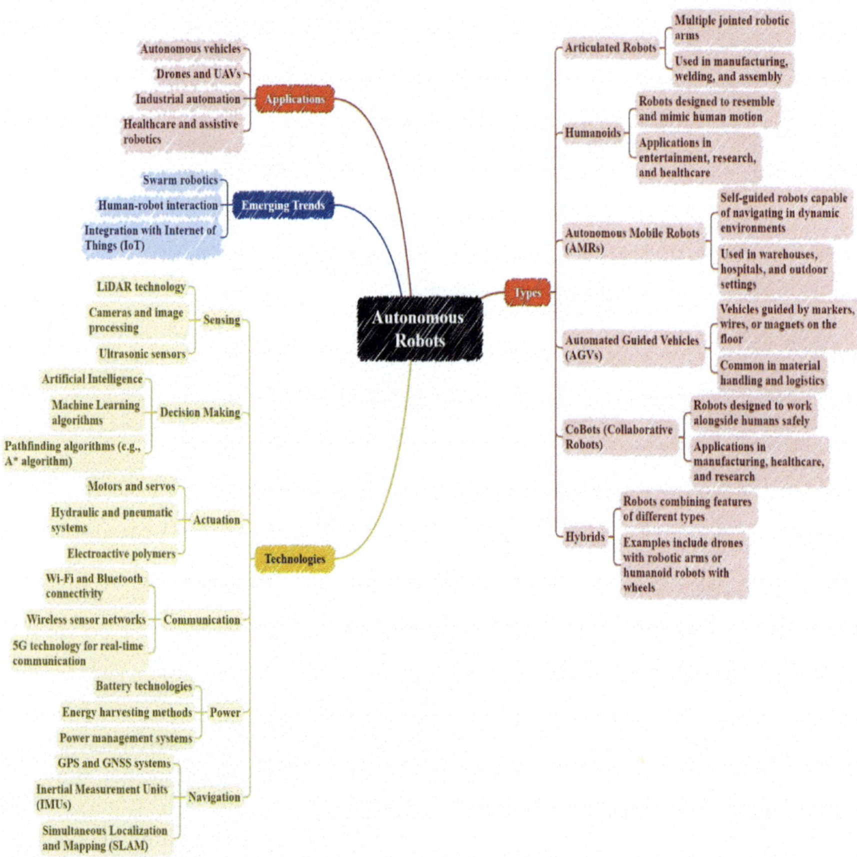

Fig. 7 Types and technologies in autonomous robot

4 Industry 4.0 for Aircraft Manufacturing

The aviation sector is adopting and utilizing a variety of I4.0 technologies to improve operational effectiveness, safety, and passenger experience, which makes the aviation industry and I4.0 intertwined (Zutin et al. 2022). The visual representation is divided into three sections, depicted in Figs. 8, 9, and 10.

4.1 *IoT Applications for Aircraft Manufacturing*

In the aircraft manufacturing industry, the Internet of Things (IoT) plays a important role in improving efficiency and driving innovation, facilitating seamless communication and data exchange (Rodrigues et al. 2022). This digital symphony orchestrates a harmonious convergence of technology and manufacturing prowess, unlocking a multitude of transformative possibilities.

At the heart of this symphonic revolution lies Production Monitoring (Rodrigues et al. 2022), a virtuosic movement that harmonizes the intricate dance of processes, supply chains, quality assurance, predictive maintenance, and workplace safety.

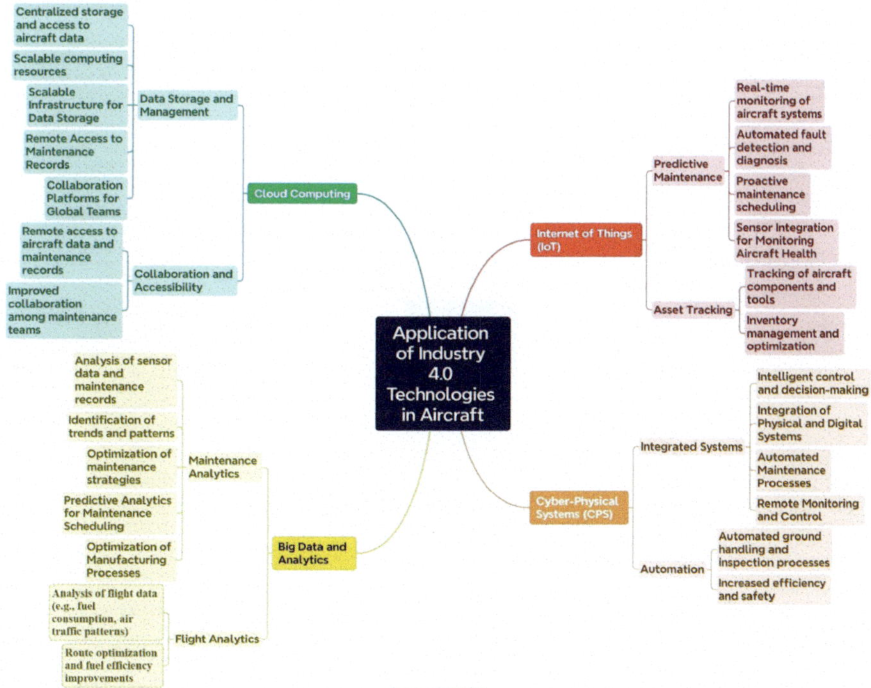

Fig. 8 Application of Industry 4.0 in aircraft manufacturing, maintenance, and MRO (Part 1)

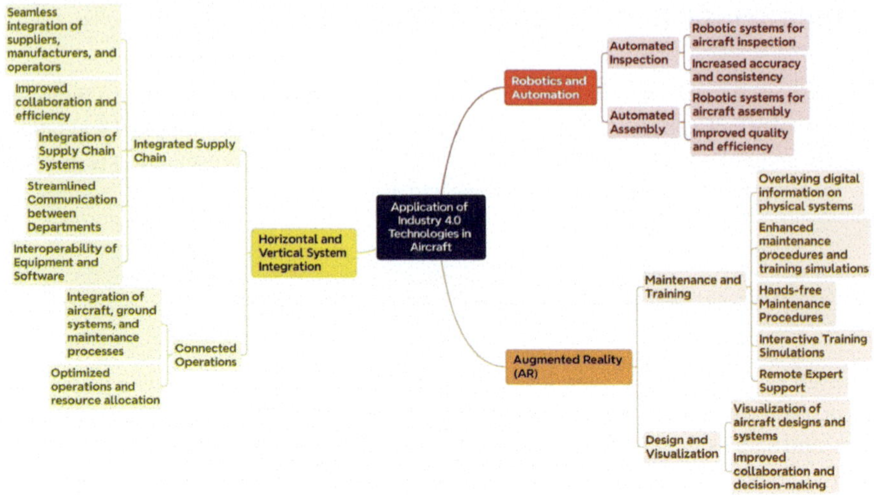

Fig. 9 Application of Industry 4.0 in aircraft manufacturing, maintenance, and MRO (Part 2)

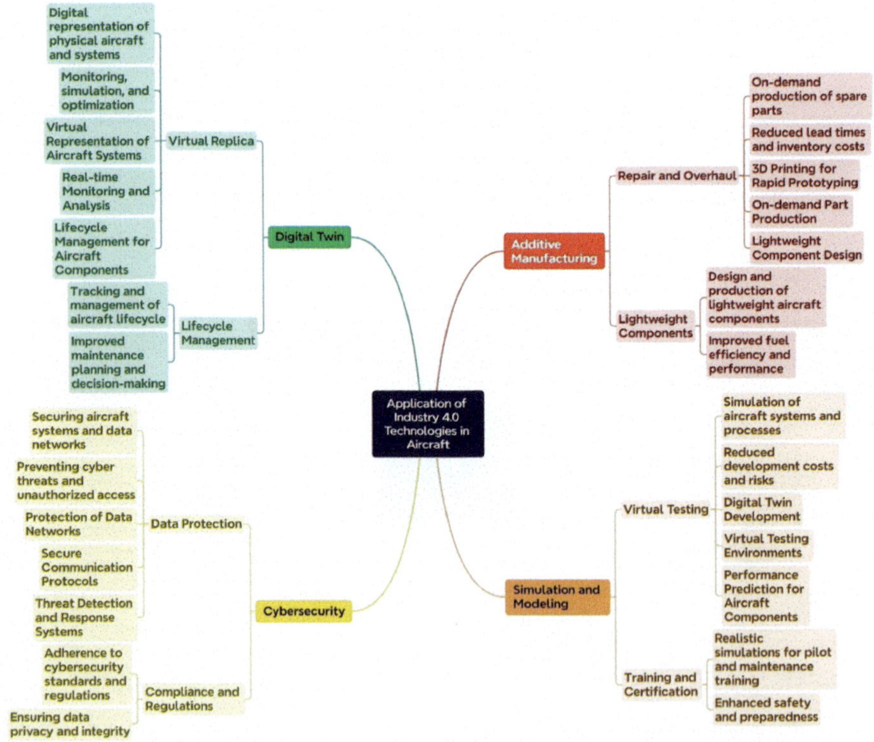

Fig. 10 Application of Industry 4.0 in aircraft manufacturing, maintenance, and MRO (Part 3)

Through the seamless integration of IoT devices, including sensors and connected equipment, manufacturers gain an elevated vantage point, enabling them to gather instantaneous data with precision and clarity. The other important areas include following:

The integration of Industry 4.0 technologies, notably the Internet of Things (IoT), is revolutionizing the aerospace sector. This transformation enables predictive maintenance, real-time monitoring, efficient asset tracking, and a host of other capabilities. Table 2 outlines the utilization of IoT (Internet of Things) in the domains of aircraft manufacturing, maintenance, and MRO (Maintenance, Repair, and Overhaul).

Within aircraft maintenance and MRO (Maintenance, Repair, and Overhaul), the implementation of IoT technologies facilitates predictive maintenance strategies, real-time aircraft health monitoring, and streamlined asset tracking. These capabilities translate to heightened operational efficiency, minimized downtime, bolstered safety measures, and optimized resource utilization Aircraft manufacturers can harness IoT technologies to oversee production processes, track components and materials, and refine supply chain operations. Real-time data collected by IoT sensors offers insights into production bottlenecks, quality control discrepancies, and asset utilization, empowering manufacturers to make data-driven decisions and continuously enhance their processes.

4.2 Cyber-Physical Systems

Cyber-Physical Systems (CPS) are leading with the process of aircraft industry's digital transformation introducing system integration of the physical and cyberworld, or rationalized decision-making and automation of key processes for the sake of efficiency and safety (Oks et al. 2022). Table 3 outlines the utilization of Cyber-Physical Systems (CPS) in the domains of aircraft manufacturing, maintenance, and MRO (Maintenance, Repair, and Overhaul).

4.3 Big Data and Analytics (BDA)

Big Data and Analytics have become essential tools in the aircraft industry, facilitating Decision-making informed by data, refined strategies for aircraft maintenance, and heightened operational efficiency across diverse domains. Table 4 outlines the utilization of Big Data and Analytics (BDA) in the domains of aircraft manufacturing, maintenance, and MRO (Maintenance, Repair, and Overhaul) (Badea et al. 2018).

When applied to aircraft MRO, CPS offer opportunities in automating maintenance work, real-time monitoring, and decision support systems. CPS enable maintenance teams to receive actionable data in real-time manner, automate some of the inspection processes and provide prediction of some maintenance issues, which in turn enhances the overall efficiency, safety, and readiness of the aircraft.

Table 2 Application of IOT in aircraft manufacturing, maintenance and MRO

Application	Sub application	Description
Predictive maintenance OECD (2023)	Real-time monitoring of aircraft systems	IoT technology also enhances the possibility of integrating many IoT sensors and various components in aircraft systems so that instant monitoring of key components, systems, and other performance factors becomes feasible. Monitoring involves tracking of the aircraft from time to time and affords an overall view of how the aircraft is performing and its general health state
	Automated Fault Detection and Diagnosis	By engaging with analytics and advance learning algorithms (CNN), information gathered from IoT sensors can be processed to trigger early alerts, self-diagnosis and to recognize faults or other developing complications that are potentially risky to the entirety of the system. It is in this line that this approach to maintenance also improves on safety and reliability and at the same time, decreases on downtime as well as its cost implications
	Proactive maintenance scheduling	Based on the information that is over time obtained from real-time monitoring and automatic fault detection, maintenance activities are well timed in such a way that they prevent various maintenance activities that would otherwise need to be carried out at opportune time. This approach of maintenance focuses on conditions to ensure that preventive maintenance is scheduled properly to avoid over-maintenance or, in some cases, even causing frequent maintenance of short-life aircraft parts
	Sensor integration for monitoring aircraft health	With the help of IoT in aviation it is possible to include different sensors all over the aircraft. Large number of sensors provides overall health of the aircraft and helps detect the looming problems that require maintenance work
Asset tracking	Tracking of aircraft components and tools	The Internet of Things for asset tracking solutions allows for real-time locating and identification of all parts of an aircraft, tools, and equipment. By using such tools such as RFID (Radio Frequency Identification) tags, Bluetooth beacons or GPS, organizations can possibly monitor the location of such assets hence availability and therefore effectively manage inventories
	Inventory management and optimization	Inventory Management and Optimization: Accurate asset tracking data, combined with advanced analytics, enables efficient inventory management and optimization. Organizations can oversee stock levels, pinpoint bottlenecks, and streamline resource distribution to curtail inventory expenses and mitigate the potential for stock shortages or surplus inventory

Table 3 Application of CPS in aircraft manufacturing, maintenance and MRO

Application	Sub application	Description
Integrated systems	Integration of physical and digital systems	CPS are thus able to contribute to the establishment of integrated systems that complement the physical and cyber environments. These 'closed loop', integrated systems, can simultaneously watch over system activities, assess performance and proactively diagnose developing problems as well as preemptively adjust or suggest maintenance actions thus minimizing or preventing system malfunctions
	Automated maintenance processes	CPS are indispensable for supporting maintenance processes within the aircraft industry, as they provide automation. Some examples of specific applications of automated system are aviation vehicle inspection, overhaul of components, calibration among others: this has brought improved accuracy, and operational efficiency
	Remote monitoring and control	Through the use of the 'CPS', there is the possibility to monitor and or command the performance of aircraft systems and ground operations from a distance. By using the secure communication networks, only the personnel with the authorized access can have the real-time access to the data and also supervise the performance of the entire system, and also take necessary actions in case, if required without physical intervention remotely. This makes it easier for one to operationalize a system while improving its ability to respond to changing conditions without diluting security
Automation	Automated ground handling and inspection processes	CPS have been seen to have significant importance when it comes to maintaining and carrying out mechanical changes within the aircraft industry. Examples of actions that could be completed with the help of robots include aircraft checks, components swapping, and system tuning, implementing elevated accuracy of movements and thereby raising operational speed

(continued)

Table 3 (continued)

Application	Sub application	Description
	Increased efficiency and safety	Through the use of the 'CPS', there is the possibility to monitor and or command the performance of aircraft systems and ground operations from a distance. By using the secure communication networks, only the personnel with the authorized access can have the real-time access to the data and also supervise the performance of the entire system, and also take necessary actions in case, if required without physical intervention remotely. This makes it easier for one to operationalize a system while improving its ability to respond to changing conditions without diluting security
	Enhanced accessibility and safety	CPS include aspects of automation of routine processes and intelligent control in aspects of complex plane operations, maintenance, and production processes leading to improved performance and safety. This can be done with less time wastage hence resulting to highly efficient automation as compared to manual work is main cause off increased downtime and human errors during servicing

For aircraft manufacturers CPS helps to coordinate the material and manufacturing physical production systems with computational assets for the intelligent control system, optimization of the manufacturing systems and automated production line. Automotive production is made efficient, cheap, safe and of top quality due to this integration besides meeting the various quality standards that are compulsory in most automotive industries.

Big Data and Analytics in the aircraft maintenance and MRO field drives innovation in management by providing predictive analysis, decision-making of efficient maintenance plans, and cost-effective performance. The predictive tool enables the maintenance teams to identify likely problems, reduce time spent between aircrafts, and increase the durability of parts used in aircrafts by utilizing sophisticated analytical methodologies.

4.4 Cloud Computing

Today, the concept of Cloud Computing has emerged as an influential dynamic in the aircraft industry and offers a new vision for storage and access for a wide range of domains such as maintenance, manufacturing to MRO (Maintenance, Repair, and

Table 4 Application of BDA in aircraft manufacturing, maintenance and MRO

Application	Sub application	Description
Maintenance analytics (Achouch et al. 2022)	Analysis of sensor data and maintenance records	Maintenance analytics, by drawing trends and patterns from data gathered from sensors and recorded maintenance histories, can identify potential relationships between the components, the failure modes, and the maintenance demands. Thus, obtaining such knowledge is vital in order to diagnose future problems and make decisions beforehand
	Identification of trends and patterns	Using information from maintenance analytics, both aircraft operators and MRO businesses can prepare better plans for maintenance activities. These include the use of risk based maintenance strategies, predictive maintenance schedules, and customized maintenance plans for the individual plane or particular part, which in turn has led to improved operational performance and improved cost efficiency
	Optimization of maintenance strategies	Leveraging the insights gained from maintenance analytics, aircraft operators and MRO (Maintenance, Repair, and Overhaul) organizations can optimize their maintenance strategies. This encompasses the adoption of risk-based maintenance approaches, predictive maintenance schedules, and tailored maintenance programs for specific aircraft or components, resulting in enhanced operational effectiveness and financial benefits
	Predictive analytics for maintenance scheduling	It will be beneficial to apply predictive analytics approaches to the maintenance data in order to forecast when maintenance procedures would be needed for specific elements or frameworks. The effectiveness of the use of prognosis in identifying potential failures or degradation involves creating an additional work plan that allows the maintenance teams to address potential issues in advance, therefore reducing the chances of unexpected down times
	Optimization of manufacturing processes	Big Data and Analytics can also play the role of improving the actual manufacturing of airplanes. Through this system, information relating to quality control checks, machine efficiency, and managing of consumables and other resources in the supply chain can be used to locate the key problem areas, the inefficiencies, and the areas that need to be optimized; hence enhancing productivity and reducing costs

(continued)

Table 4 (continued)

Application	Sub application	Description
Flight analytics	Analysis of flight data	As mentioned before, flight analytics includes the processing of analyzing big data produced during flight operations including the fuel utilizations, traffic patterns, weather conditions, among others. The data collected is then subjected to a process of analysis with an aim of revealing useful information and discernible patterns
	Route optimization and fuel efficiency improvements	By analyzing flight data, airlines and air traffic management organizations can optimize flight routes for enhanced fuel efficiency and reduced environmental impact. This includes identifying optimal altitudes, airspeed, and flight paths based on factors such as weather conditions, (ATC) air traffic congestion, and performance of aircraft characteristics
	Predictive maintenance for aircraft systems	Flights data also help in the identification of the likelihood of failure in aircraft systems and so help in predictive maintenance. In real time during flight, systematically reading habits of different functions provided by the aircraft systems, future problems or deteriorations may be predicted and corrected, thereby avoiding such catastrophic occurrences or those massive delays

Overhaul). Table 5 outlines the utilization of Cloud Computing in the domains of aircraft manufacturing, maintenance, and MRO (Maintenance, Repair, and Overhaul) (Zhong et al. 2017).

In the application of cloud computing for aircraft maintenance, the records are stored in central as well as accessible online meaning technicians can easily get information that they need in order to conduct proper maintenance on an aircraft. In aircraft manufacturing cloud computing provides flexibility through providing manufacturing with the needed computing power and storage space to accommodate for the large data generated during manufacturing process of an aircraft. Moreover, through these technologies it became possible to integrate collaboration tools that bridge between engineers, designers, and production teams enabling new idea generation as well as facilitate manufacturing. In MRO, cloud computing offer means to store and share aircraft maintenance records, component data, and repair documents. This accessibility enhances efficiency in the MRO organization's functions since they are in a position to share information with the airlines and the manufacturers, and also, they are in a position to meet all legal requirements that have been put in place. Through cloud computing, the aircraft industry will be in a position to do away with some of the challenges that face data storage and management, get better collaboration, and is likely to be in a position to increase the usage of cloud computing in that industry and hence be in a position to drive innovation and sustain the much-needed competitiveness in the global market.

Table 5 Application of Cloud Computing in aircraft manufacturing, maintenance and MRO

Application	Sub application	Description
Data storage and management	Centralized storage and access to aircraft data	The cloud computing environment allows for the efficient storage and processing of large volumes of aircraft data, such as original maintenance records, flight documents, sensor data and manufacturing information. Thus, with this centralized approach, only the appropriate individuals will be able to gain access to the information they need irrespective of the location of the consoles, which helps to speed up the processes involved in the decision-making process
	Scalable computing resources	The first benefit of cloud computing is that it is easily scalable Cloud computing is one of the many computing technologies that currently exist, and the good thing with it is that it can easily be scaled. Manufacturers of the aircraft and MRO companies can lease computer resources based on the utility model, and hence acquire ample processing and storage capacity to accommodate the large dimensions of data generated in aircraft production, maintenance and actual flights
	Scalable infrastructure for data storage	CBMs are also a scalable data storage infrastructure that can provide data as aircraft fleets continue to grow and data demands rise. Companies can easily add, upgrade or even scale-up the storage capacity without huge initial capital investment on physical hardware and framework; a flexibility that guarantees organizations the capacity to address continuously expanding volumes of data
Collaboration and accessibility	Remote access to maintenance records	Some cloud service providers allow original holders of aircraft maintenance records to access and share information remotely, where the technicians, engineers, or maintenance teams can access information regardless of their location. It has also made maintenance operations easier by allowing the necessary functions to be performed without time lapses and has helped coordinate the efforts of teams located in different parts of the world

(continued)

Table 5 (continued)

Application	Sub application	Description
	Collaboration platforms for global teams	There can be several teams in manufacturing and maintenance, repair, and overhaul of aircraft are situated in different regions or countries. Cloud computing involves collaboration solutions for flow and continuous communication as well as exchange of data and instant updates among these co-located and scattered cross-affiliate teams. Thus, the first one contributes to the effective collaboration of different teams and their members, reduces the time spent on communication, and eliminates possible misunderstandings; the second is helpful in case of high working productivity and quality assurance

4.5 Additive Manufacturing (AM)

3D printing or Additive Manufacturing, has transformed numerous facets of the aircraft industry, such as repair and overhaul operations, swift prototyping, on-demand part manufacturing, and the creation of lightweight components. Table 6 outlines the utilization of AM in the domains of aircraft manufacturing, maintenance, and MRO (Maintenance, Repair, and Overhaul) (Ceruti et al. 2019).

Within the application of aircraft maintenance and MRO, additive manufacturing means that organization is capable of replacing original equipment manufacturers' monopoly by granting control to those organizations. Through the just-in-time inventory approach in the production of sporadic spare parts, consumers like the airlines and MRO providers will be able to perform timely repairs and maintenance hence giving less or no downtime thus helping enhance operational efficiency. From the perspective of aircraft manufacturers, additive manufacturing can enhance the efficiency of the prototyping stage by shortening the corresponding time and further improve the product designs in terms of parameters. Moreover, higher accuracy of the component due to AM techniques and its light weight also increases the fuel efficiency and thus, increases the performance of the aircraft required to achieve the cost savings as well as sustainable aviation goals.

4.6 Robotics and Automation

Robotics Automation has extended its arms to the aircraft industry in a bid to revolutionaries the industry by making various processes more accurate, efficient, and productive especially in processes such as inspection and assembly. Table 7 outlines

Table 6 Application of AM in aircraft manufacturing, maintenance, and MRO

Application	Sub application	Description
Repair and overhaul	On-demand production of spare parts	In aerospace and aviation, it is possible to utilize additive manufacturing technology to make spare parts required for aircraft MRO in parallel. The use of 3D printers in organizing airlines and MRO firms offers an opportunity to produce parts, making use of extra space hence cutting off inventory renewal time
	Reduced lead times and inventory costs	The prevailing techniques of manufacturing of spare parts have long lead times which implies high cost of spare part inventories and possible operational unavailability. The conventional method of production makes use of molds which poses a challenge since it involves a lot of time to produce these molds and also time to produce the parts hence, most companies end up having lots of stock since the lead time taken to produce the parts is so high, additive manufacturing solves this problem by making use of molds that can be made over the counter and the parts taken so there is little or no need to order for a lot of stock
	3D printing for rapid prototyping	For instance, in the Aircraft manufacturing domain, Additive manufacturing has the most significance in the making of a prototype known as rapid prototyping. Designers/ engineers can physically model jobs or assemblies to make swift changes, these help to design matters much quicker the cycle time is short
	On-demand part production	Additive manufacturing enables aircraft manufacturers and MRO organizations to fabricate parts as needed, eliminating the necessity for expensive tooling and reducing waste. This capability not only enhances operational flexibility but also enables the production of complex geometric shapes and personalized parts, which may pose challenges or be unfeasible to produce using conventional methods

(continued)

Table 6 (continued)

Application	Sub application	Description
Lightweight component design	Design and production of lightweight aircraft components	Additive manufacturing enables the creation and manufacturing of lightweight aircraft components, offering significant weight savings and improved fuel efficiency. By leveraging advanced materials and optimized designs, manufacturers can create intricate and lightweight structures tailored to specific performance requirements (Kumar et al. 2024b)
	Improved fuel efficiency and performance	The reduction in weight achieved through the use of lightweight components directly translates into improved fuel efficiency and enhanced aircraft performance. This not only contributes to cost savings for airlines but also aligns with the industry's efforts towards sustainable and environmentally-friendly operations

the utilization of robotics and automation in the domains of aircraft manufacturing, maintenance, and MRO (Maintenance, Repair, and Overhaul) (Jayasekara et al. 2022).

For aircraft manufacturers, the integration of robotic systems in assembly lines offers numerous advantages. Automated assembly processes enhance quality and consistency, reduce the risk of defects, and improve overall efficiency. Moreover, robotic systems can undertake tasks that might pose hazards or physical strain for human workers, thus enhancing workplace safety and ergonomics.

4.7 *Augmented Reality (AR)*

Augmented Reality (AR) is a groundbreaking technology that is transforming numerous facets of the aircraft industry, especially in maintenance, training, design, and visualization (Pace et al. 2018). Table 8 outlines the utilization of AR in the domains of aircraft manufacturing, maintenance, and MRO (Maintenance, Repair, and Overhaul).

Concerning maintenance and MRO tasks and processes of aircrafts, AR technology optimizes the maintenance process, improves the training process, and supports remote expert services. In this way, the package of a digital record overlay on to physical systems enables technicians to retrieve relevant information and or instructions in a matter of an instant, thereby preventing undesirable or erroneous outcomes and enhancing productivity. In the aerospace industry, AR provides significant support in processes such as design and visualization of aviation stiffening. It makes engineers and designers capable of directly handling virtual representation of the potential structures of aircrafts, discover pre-existing problems and coordinate

Table 7 Application of robotics in aircraft manufacturing, maintenance, and MRO

Application	Sub application	Description
Automated inspection	Robotic systems for aircraft inspection	Advanced robotic systems equipped with sophisticated sensors and imaging technologies are revolutionizing the aircraft inspection process. These robotic systems can perform detailed inspections of aircraft structures, components, and systems with unparalleled accuracy and consistency
	Increased accuracy and consistency	The disadvantage of traditional manual inspection is inherent in the fact that they are very prone to inaccuracies as they involve human interventions and thus can compromise on safety and quality. These risks are avoided in robotic inspection systems by standardized and structured procedures to inspect the parts and surfaces throughout the product lifecycle for even minor defects or variation from the baseline that may not be easily identified by human operators
	Enhanced safety and accessibility	Of course, robotic inspection systems can reach and inspect parts of the structure that are hard to access and cannot be accessed by human inspectors frequently due to possible dangers involved. ET also, these systems can be implemented in some of the sever conditions; ultimately, the inspections can be done without necessarily endangering personnel lives
Automated assembly	Robotic systems for aircraft assembly	Automated assembly systems make the process quality oriented due to the fact that humans are not allowed to bring in variabilities in the assembly process. They are capable to make complex operations with required posture and pressure, that means fewer scars or broken parts, leading to improved assemblies and decrease of rework
	Improved quality and efficiency	Robotic assembly systems ensure consistent quality by eliminating human errors and variability in the assembly process. They can perform intricate tasks with precise positioning and force control, resulting in higher-quality assemblies and reduced rework or defects
	Increased productivity	Automated assembly lines leveraging robotic systems can operate around the clock, significantly increasing productivity and throughput. This continuous operation enables aircraft manufacturers to meet the growing demand for aircraft while maintaining stringent quality standards

Table 8 Application of AR in aircraft manufacturing, maintenance, and MRO

Application	Sub application	Description
Maintenance and training Ceruti et al. (2018)	Overlaying digital information on physical systems	AR technology enables the overlay of digital information, such as technical manuals, schematics, and detailed instructions, directly onto the aircraft systems or components. This capability enhances maintenance procedures by furnishing technicians with immediate, contextual information, diminishing the reliance on paper manuals and mitigating the risk of errors
	Enhanced maintenance procedures and training simulations	AR applications can guide technicians through complex maintenance tasks by providing interactive, step-by-step instructions and visualizations. This enables technicians to refine and enhance their skills in an environment that is both secure and controlled
	Hands-free maintenance procedures	AR headsets or wearable devices mean hands-free control, which is very convenient when the possible need of the technical staff is to look up some information related to the object under inspection and the maintenance procedures that need to be followed. This capability improves efficiency and safety since many printed manuals or devices are used frequently in dealing with emergencies
	Interactive training simulations	One real-life application that has become popular is the use of simulations involving augmented reality (AR) to enhance training for aircraft maintenance personnel and pilots. Training simulations utilizing augmented reality (AR) offer an immersive and interactive learning experience for aircraft maintenance personnel and pilots. Trainees can interact with virtual representations of aircraft systems, components, and scenarios, facilitating better understanding and retention of critical information
	Remote expert support	With AR technology, work can be done without relying on an expert face-to-face or physical presence since the expert can offer his input remotely. Those technicians on-site can also get the opinion from those offsite who can immediately offer guidance, comments and even give an overlay to the technician on the physical plane to help solve the problem besides cutting on time the plane spends off the air

(continued)

Table 8 (continued)

Application	Sub application	Description
Design and visualization	Visualization of aircraft designs and systems	AR applications allow designers, engineers, and stakeholders to visualize and interact with virtual representations of aircraft designs and systems. This capability fosters effective collaboration, allowing teams to explore design concepts, pinpoint potential issues, and make informed decisions throughout the development process
	Improved collaboration and decision-making	AR-enabled design reviews and walkthroughs allow geographically dispersed teams to collaborate effectively, visualizing and interacting with virtual aircraft models as if they were physically present. This heightened collaboration results in enhanced communication, expedited decision-making, and more streamlined design iterations

themselves with other business members. These improved interactions and visualizations mean better decisions, design cycles, and finally the enhanced formulation of the new and effective aircraft models.

4.8 *Simulation and Modeling*

Simulation and Modeling have emerged as powerful tools in the aircraft industry, enabling virtual testing, digital twin development, performance prediction, and enhanced training and certification processes (Soori et al. 2023). Table 9 outlines the utilization of simulation and modeling in the domains of aircraft manufacturing, maintenance, and MRO (Maintenance, Repair, and Overhaul).

In terms of both aircraft maintenance and MRO simulation and modeling techniques can be used in tests or training in a virtual environment. Experience based mockup being a part of virtually simulated settings of operation allow for assessment of the effectiveness of maintenance practices and foresee certain problems or threats regularizing with benefits towards safety and operation. Moreover, there is scope in incorporating realistic models for practicing different scenarios for experience-trained personnel in the maintenance field. Since performance and reliability can be accessed via simulation and modeling for aircraft manufacturers, simulation and modelling become key enablers that allow for virtual testing, digital twin creation, and performance evaluation.

Using new approaches, manufacturers can obtain solutions for critical design parameters and evaluate a range of operating conditions and operational scenarios and develop and test less costly and less risky products while fulfilling legal requirements. Moreover, the simulation environment allows products to be tested and reviewed

Table 9 Application of simulation and modeling in aircraft manufacturing, maintenance, and MRO

Application	Sub application	Description
Virtual testing	Simulation of aircraft systems and processes	Simulation technology and modeling helps enable various system tests and processes related to aircraft to be experimented under a virtual environment that is real life like. Using the proxy, manufacturers and aviation organizations are able to assess the response of individual parts and assemblies together with controlling configurations and materials performance as well as the behavior of complete designs by recreating and analyzing different operating conditions in real time without actual physical proxy or expensive and time-consuming flight testing
	Reduced development costs and risks	Virtual testing through simulations significantly reduces the development costs and risks associated with physical prototyping and real-world testing. Manufacturers can identify and address potential issues, optimize designs, and validate performance criteria before committing to physical production, minimizing costly rework and delays
	Digital twin development	Engineering modeling and simulation techniques are critical for creating digital twin of real aircraft systems. These representations are as real as the respective aircraft and enable real-time monitoring, prediction, and, optimization at all aircraft phases
	Virtual testing environments	The complex environmental simulations are used to conduct comprehensive virtual checks of the aircraft parts and systems, for example, due to the adverse condition such as weather harshness, turbulence, and system malfunctions. It provides insights into how these aircraft systems work and how they perform and even how safe all the systems can be then come up with better ways to keep the systems safer this is through well planned out actions
	Performance prediction for aircraft components	Ordinary operating conditions may be utilized in simulation modes as a way of predicting the performance of the aircraft part under testing. The simulation of the operational aspects of the components like the engines, avionics, structure etc. can help the manufacturers to offer efficient designs with high reliability and one that also meets various statutory requirements

(continued)

Table 9 (continued)

Application	Sub application	Description
Training and certification	Realistic simulations for pilot and maintenance training	High-fidelity simulations provide realistic and immersive training environments for pilots and maintenance personnel Flight simulators replicate various flight scenarios, weather conditions, and system failures, enabling pilots and aircraft technicians to develop and improve their skills in a secure and monitored setting
	Enhanced safety and preparedness	Maintenance training through simulations enable technicians to run through life-like conditions with realistic practices pit forward and tackling different challenges on same that may cause heightened damages on actual aircraft or equipment This practical training method boosts safety and readiness, guaranteeing that maintenance personnel possess the essential skills to manage real-world situations proficiently

under the most challenging conditions possible, making it easier for improvement to be made and the safety of the systems and components as well as of the aircraft as a whole to be improved.

4.9 Cybersecurity

For aircrafts, cybersecurity is another issue that has received much attention in the recent past, especially due to enhanced use of computers and other technologies that may be manipulated in cases of cyber threats. More specifically, sound and effective computer security measures are vital to maintain protection of the aircraft systems as well as the data and information networks against cyber and unauthorized access dangers. Table 10 outlines the utilization cybersecurity in the domains of aircraft manufacturing, maintenance, and MRO (Maintenance, Repair, and Overhaul).

In the context of aircraft maintenance and MRO (Maintenance, Repair, and Overhaul), cybersecurity measures are crucial for protecting maintenance data, aircraft health monitoring systems, and other connected systems from cyber threats. Secure data networks and communication protocols ensure the integrity and confidentiality of maintenance records, enabling efficient and secure collaboration among maintenance teams and stakeholders.

For aircraft manufacturers, cybersecurity is essential for safeguarding sensitive design data, production systems, and supply chain networks from cyberattacks and data breaches. Robust cybersecurity measures protect intellectual property, ensure the integrity of production processes, and maintain the confidentiality of sensitive information shared with suppliers and partners.

Table 10 Application of cybersecurity in aircraft manufacturing, maintenance, and MRO

Application	Sub application	Description
Data protection	Securing aircraft systems and data networks	Aircraft systems and data networks must be secured against cyber threats, such as malware, hacking attempts, and unauthorized access. This involves implementing robust security protocols, encryption mechanisms, and access control steps to safeguard sensitive data and guarantee the integrity of aircraft systems
	Preventing cyber threats and unauthorized access	Routine security measures embrace steps that may help to avoid cyber threats and restrict unlawful access to aircraft systems as well as networks. This might involve the use of firewall, use of intrusion detection systems and also the periodic evaluation of the system to determine vulnerability and the probability of the system to be invaded
	Protection of data networks	Aircraft generate and transmit vast amounts of data, including flight data, maintenance records, and passenger information. Protecting these data networks from cyber threats is crucial to ensure the confidentiality, integrity, and availability of sensitive information. This may involve the implementation of secure communication protocols, data encryption, and robust access controls
	Threat detection and response systems	Advanced threat detection and response systems play a crucial role in promptly identifying and responding to cyber threat. These systems leverage advanced analytics, machine learning, and threat intelligence to detect anomalies, identify potential attacks, and initiate appropriate response measures to address and alleviate risks and the consequences of cyber incidents
Compliance and regulations	Adherence to cybersecurity standards and regulations	The aircraft industry is subject to stringent cybersecurity standards and regulations established by national and international authorities. The safety and security of aircraft operations are critical and ensured by adhering to these protocols and guidelines for safe aircraft operations, as well as safeguarding sensitive data and personal information
	Ensuring data privacy and integrity	Cybersecurity measures in the aircraft industry must address data privacy and integrity concerns. This includes implementing robust access controls, data encryption, and established data handling protocols to safeguard sensitive passenger and operational data from unauthorized access, misuse, or tampering

4.10 Horizontal and Vertical System Integration

Horizontal and Vertical System Integration has become a critical factor in the aircraft industry, enabling seamless collaboration, streamlined operations, and optimized resource allocation across the entire value chain (Leite Junior et al. 2019). Table 11 outlines the utilization Horizontal and Vertical System Integration in the domains of aircraft manufacturing, maintenance, and MRO (Maintenance, Repair, and Overhaul).

In terms of Aircraft general maintenance and MRO (Maintenance, Repair, and Overhaul), the concept of HS (Horizontal System integration) and VS (Vertical System integration) keeps huge importance for enhancing the communication, increase efficiency besides management of resource. Maintenance management systems also help in enhancing the interaction between maintenance personnel, suppliers, and operators in aspects of part procurement, thereby reducing the time taken by aircraft on shelves awaiting spare parts.

In the case of aircraft manufacturers, system integration enables engineering, sourcing, and manufacturing to be managed in harmony, promoting information sharing among departments that make up the systems, coordination in decision-making, and efficient work flow in processes. Also, the interface between manufacturing systems and supplier/customer systems provides distinguishing features to monitor and control supply chain, inventory, and production schedules.

4.11 Digital Twin (DT)

There has been a growing interest in utilizing technology known as Digital Twin solutions for creating a virtual model of an aircraft and some of its systems such as aeroengines and landing gear. This makes it possible to monitor, simulate, and improve during operation, production, design, and even during disposal of an aircraft (Meyer et al. 2020). Table 12 outlines the utilization DT the domains of aircraft manufacturing, maintenance, and MRO (Maintenance, Repair, and Overhaul).

In the realm of aircraft maintenance and MRO, the concept of digital Twin is a disruptive innovation that has the potential to revolutionize the manner in which maintenance is conducted and the service life of aircraft parts is estimated. Digital twins can assist virtual representation of certain elements of airplanes, predictive maintenance, possibilities of extensive simulations, and optimization of provenance replacements with a resultant reduced cost of maintenance and increased availability of airplanes.

Aircraft manufacturers use digital twins for the real-life testing of designs, optimization, and management of life cycles throughout creation stage. Manufacturers can use the digital twins to simulate the conditions that an aircraft operating system or another part of the aircraft goes through and evaluate designs, modifying designs and making improvements aiming at safety and meeting regulations at different phases in the product's life cycle.

Table 11 Application of horizontal and vertical system integration in aircraft manufacturing, maintenance, and MRO

Application	Sub application	Description
Integrated supply chain	Seamless integration of suppliers, manufacturers, and operators	The horizontal system integration ensures that there is coordination and linkages of supply chain amongst the industry players such as suppliers, manufacturers and operators in the aircraft industry. It helps in information exchange, decision making of work and actions in right synergy and in a real time manner within the entire supply chain
	Improved collaboration and efficiency	The supply chain system is useful to work as an interface that facilitates better and faster interaction between the stakeholders and disseminate important information. This integration helps avoid the situation when a company has many departments that operate independently while failing to achieve maximum synergy, which boosts competitiveness while decreasing the number of combinations that may not be optimal
	Integration of supply chain systems	It relates to the linking of many different systems which make up the Supply Chain, such as ERP, SRM, and logistics systems. This sharing ensures immediate visibility of stocks, manufacturing schedules and delivery times across production chain to work cohesively
	Streamlined communication between departments	Organizational integration for the vertical context focuses on integration and communication between different functions within an organization like engineering, procurement and manufacturing and or maintenance. To this end, this integration ensures that information that is relevant is passed to the other function effectively and thus helps the cross functional teams to work with efficiency and effectively and make the right decisions that could help to enhance the operations of the organization
	Interoperability of equipment and software	Vertical system integration also means standardizing the hardware and software throughout the various phases of an aircraft lifecycle. This helps to facilitate easy flow of data, minimize communication barriers with other platforms as well as improve on some of the operations

(continued)

Table 11 (continued)

Application	Sub application	Description
Connected operations	Integration of aircraft, ground systems, and maintenance processes	Vertical system integration enables the integration of aircraft systems, ground systems, and maintenance processes, creating a connected and cohesive operational environment This integration ensures instantaneous monitoring, data exchange, and coordinated decision-making across all aspects of aircraft operations
	Optimized operations and resource allocation	By integrating aircraft, ground systems, and maintenance processes, organizations can enhance their operations and optimize resource allocation. Instantaneous data analysis and predictive maintenance capabilities enable proactive decision-making, minimizing downtime, reducing costs, and improving efficiency

5 Real Life Examples of Usages of I4.0 Technologies in Aerospace Domain

The summary of real-life applications of I4.0 in aircraft Manufacturing is presented in Table 13.

6 Challenges in Adoption of Industry 4.0

In the aerospace industry, the adoption of Industry 4.0 principles presents unique challenges due to the complex nature of aircraft manufacturing, maintenance, repair, and overhaul (MRO) processes. Here are some specific challenges faced in the adoption of Industry 4.0 within aircraft manufacturing and MRO. Implementing Industry 4.0 in aircraft manufacturing poses a number of challenges (Sayem et al. 2022) that must be addressed. The following are the critical challenges in implementing Industry 4.0. The summary of challenges in I4.0 implementation in aircraft Manufacturing, Maintenance, and MRO is presented in Table 14.

To overcome these issues, it will be important to involve all the key stakeholders, receive government support, fund further training and education, as well as pay the necessary attention to cybersecurity and data protection. It involves upskilling the existing workforce, attracting new talent, integrating legacy systems, establishing common standards, and ensuring compliance with regulations.

Table 12 Application of DT in aircraft manufacturing, maintenance, and MRO

Virtual replica	Digital representation of physical aircraft and systems	A DT consists of a digital twin of a physical aircraft, systems and its environment which has been created by integrating design detail, sensors and operational data etc. This specific feature ensures that the system is a true virtual model of the actual system, providing a precise reproduction
	Monitoring, simulation, and optimization	Digital twins facilitate real-time checks on the respective systems in an aircraft hence, constant data collection and evaluation of the systems' performances. Also, they support such activities as training and simulations, including ''what if" situations that help in predicting the right moments for maintenance, or the best way to perform or improve the overall efficiency throughout the entire lifecycle of the aircraft
	Virtual representation of aircraft systems	DT offers a preview to aircraft structures, avionic, engines, and other important systems of aircrafts involved. Thus, with the help of applied virtual model it is possible to predict the behavior effectiveness of these systems at different modes of operation. T
	Instantaneous monitoring and analysis	Digital Twins allow for intense tracking and evaluation of the in-operation systems of an aircraft through data, collected from the sensors, and other means. Thus, this immediate data analysis provides efficient and valuable information of the status of system and helps prevent problems from worsening by providing solutions before occurrence
Lifecycle management	Lifecycle management for aircraft components	DT is enjoying increased relevance when it comes to lifecycle management of parts used in aircrafts. In another way, digital twins of design, manufacturing and operations can track the overall performance and the state of every single part and subassembly, allowing for proactive maintenance and rational phasing of component renewal.

(continued)

Table 12 (continued)

	Tracking and management of aircraft lifecycle	DT serves as a comprehensive virtual representation of an aircraft throughout its entire lifecycle, encompassing design, manufacturing, operations, and maintenance stages. This holistic view enables effective lifecycle management, facilitating informed decision-making regarding upgrades, retrofits, and end-of-life strategies
	Improved maintenance planning and decision-making	The utilization of digital twins can be of significant value for improving the overall maintenance decision-making and planning by the aircraft operating companies and their maintenance teams. Maintenance planning and management can be applied, responding to such issues and finding ways to decrease the unscheduled downtime and the increase of creation maintenance, resulting in enhanced cost-recovery and use of the asset

7 Future Work

A Prospect for further research could be centered on aspects like enhancing the integration of emergent AI, the application of Digital Twins for the aircraft constituent elements, enhancing use of AI and machine learning for quality assurance processes, Use of generative AI in aircraft maintenance, extending the usage of Computer vision for streamlined Automated aircraft inspection post production and creation of breakthroughs for Supply chain management in the aerospace Industry.

Additionally, exploring the cybersecurity aspects of Industry 4.0 in aviation, as well as investigating the human–machine interaction in smart factories, will be essential.

8 Conclusion

The integration of Industry 4.0 technologies in the aerospace industry has the potential to revolutionize the way aircraft are designed, manufactured, maintained, and operated. By leveraging cutting-edge technologies such as IoT, predictive maintenance, and cybersecurity, the industry can improve efficiency, reduce costs, and enhance safety. The real-life applications of Industry 4.0 in the aerospace industry are numerous and varied, from the use of additive manufacturing to produce complex aircraft components to the implementation of autonomous maintenance systems. The integration of Industry 4.0 technologies can also enable the development of new aircraft designs and systems that are more efficient, sustainable, and environmentally friendly.

Table 13 Use cases of Industry 4.0 in aerospace industry

I4.0 technologies	Uses cases	References
Internet of Things (IoT)	Fujitsu has developed visualizing operational status, progress, and failures, enabling intuitive identification of issues and facilitating prompt countermeasures (https://www.fujitsu.com/global/imagesgig5/CS_2020Dec_Mitsubishi-Heavy-Industries.pdf)	Rodrigues et al. (2022), https://www.fujitsu.com/global/imagesgig5/CS_2020Dec_Mitsubishi-Heavy-Industries.pdf
Internet of Things (IoT)	At Airbus', the information generated by machines and conveyors is utilized to construct a dynamic visual model known as a "digital shadow	https://www.airbus.com/en/newsroom/stories/2019-07-iot-aerospaces-great-new-connector
Digital control of the production line	Airbus uses flexible assembly lines and processes to manufacture components from different aircraft programs on a single production line	Rani et al. (2023), Arntz et al. (2016)
Digital Twin	Air Force Space and Missile Systems Center developed a DT for the Lockheed Martin Block IIR GPS satellite Digital twins have also been used to train AI pilots and improve maintenance practices GE and Boeing have leveraged digital twins for engine components, landing gear, and aircraft design, resulting in improved performance predictions and quality Northrop Gurman uses digital twin, digital thread and Model Based System (MB(X)) in aerospace domain in entire life cycle	https://www.spsairbuz.com/story/?id=1119&h=Digital-Twins-in-Aerospace-A-Paradigm-Shift, https://www.afcea.org/signal-media/digital-twinning-takes-flight Van Dinter et al. (2022) Li et al. (2021), https://www.ifs.com/assets/enterprise-asset-management/whitepaper-digital-twins-in-aviation, https://www.northropgrumman.com/what-we-do/digital-transformation
Additive manufacturing	General Electric has developed a high-pressure compressor with **BLISKS** (combined bladed disks) for GE9X engine through additive manufacturing Boeing has developed a aluminum gearbox housing for Chinook helicopter	Węgrzyn (2022), https://simpleflying.com/general-electric-additive-manufacturing-benefits/ https://www.boeing.com/features/innovation-quarterly/2022/01/chinook-3d-printing.page

(continued)

Table 13 (continued)

I4.0 technologies	Uses cases	References
Autonomous Aerial robotic system	Airbus Defence & Space CBC factory has validated a full autonomous aerial robotic system (ARS) At the Lancashire facility of BAE Systems, advanced mobile robots autonomously deliver necessary parts and components to designated workstations	Martínez-de Dios et al. (2018), https://www.assemblymag.com/articles/96562-aerospace-defense-and-industry-40
BDA for predictive shimming	Boeing has applied BDA to production data, for predicting gaps between parts during the modular manufacturing and assembly,	https://depts.washington.edu/barc/projects/data-science-manufacturing
Augmented reality	NASA Michoud assembly facility uses AR based HoloLens 2 to build crew seats for Orion Spacecraft BAE uses Headsets that enable operators to work without using their hands for production of typhoon aircraft	https://www.assemblymag.com/articles/96562-aerospace-defense-and-industry-40 https://www.baesystems.com/en/feature/technologies-transforming-typhoon-production
Smart tools	Lockheed Martin is increasingly implementing smart tools like digital torque wrenches to enhance assembly quality. By automatically programming these tools to electronically drive fasteners and upload installation data, productivity is also boosted	https://www.lockheedmartin.com/en-us/news/features/2021/manufacturing-gets-systemic-upgrade-intelligent-digital-factory.html

However, the successful implementation of Industry 4.0 technologies in the aerospace industry will require significant investment in infrastructure, training, and education. It will also require a fundamental shift in the way the industry thinks about maintenance, operations, and supply chain management.

Ultimately, the adoption of Industry 4.0 technologies in the aerospace industry has the potential to transform the way we design, manufacture, and operate aircraft, and to create new opportunities for growth and innovation. As the industry continues to evolve and adapt to the changing needs of the market, it is essential that we continue to prioritize the development and implementation of Industry 4.0 technologies that can drive efficiency, innovation, and sustainability in the aerospace industry.

Table 14 Challenges in Industry 4.0 implementation

Skill gap	Industry 4.0 adoption requires highly skilled professionals in automation, AI, data analytics, etc. However, there's a shortage of such expertise. Addressing this gap needs upskilling programs and collaboration with educational institutions	Maisiri et al. (2019), Ras et al. (2017)
Legacy systems integration	Integrating Industry 4.0 tech into existing aircraft manufacturing systems demands major upgrades or replacements. Ensuring seamless integration and interoperability is crucial	Zawra (2019), Pessoa et al. (2018)
Data security	Industry 4.0 relies heavily on data collection and sharing, posing security concerns. Cyber threats must be countered with robust practices like encryption, access controls, and employee training	Pereira et al. (2017), Mentsiev et al. (2020)
Interoperability	Unifying diverse systems within the aircraft manufacturing value chain can be hindered by incongruent standards and procedures. Establishing universal standards is essential for smooth interaction	Watson et al. (2017), Albouq et al. (2022)
Regulatory compliance	Implementing Industry 4.0 technologies in aircraft manufacturing must adhere to strict regulations without compromising safety standards set by regulatory agencies like EASA and FAA	https://www.faa.gov/sites/faa.gov/files/2022_FAA_EASA_AM_Workshop_Full_Proceedings.pdf

References

300 parts down to just 7: the benefits of general electric's additive manufacturing techniques. Simple Flying. Accessed: 30 June 2023. [Online]. Available: https://simpleflying.com/general-electric-additive-manufacturing-benefits/

Ajay, Singh H, Parveen, AlMangour B (2023) Handbook of smart manufacturing: forecasting the future of Industry 4.0. CRC Press, Boca Raton. https://doi.org/10.1201/9781003333760

Achouch A et al (2022) On predictive maintenance in Industry 4.0: overview, models, and challenges. Appl Sci 12(16), Art. no. 16. https://doi.org/10.3390/app12168081

Aerospace, Defense and Industry 4.0 | 2021–08–18 | ASSEMBLY. Accessed: 11 June 2024. [Online]. Available: https://www.assemblymag.com/articles/96562-aerospace-defense-and-industry-40

Aggoune S, Hamadi F, Abid C et al (2024) Instabilities in the formation of single tracks during selective laser melting process. Int J Interact Des Manuf. https://doi.org/10.1007/s12008-024-01887-y

Albouq SS, Sen AAA, Almashf N, Yamin M, Alshanqiti A, Bahbouh NM (2022) A survey of interoperability challenges and solutions for dealing with them in IoT environment. IEEE Access 10:36416–36428. https://doi.org/10.1109/ACCESS.2022.3162219

Alouffi B, Hasnain M, Alharbi A, Alosaimi W, Alyami H, Ayaz M (2021) A systematic literature review on cloud computing security: threats and mitigation strategies. IEEE Access 9:57792–57807. https://doi.org/10.1109/ACCESS.2021.3073203

Alwadan T, Al-Zitawi O, Khwaldeh S, Almasarweh M (2015) Privacy and control in mobile cloud systems. Int J Comput Appl 113:12–15. https://doi.org/10.5120/19789-1170

Ani UPD, He H, Tiwari A (2016) Review of cybersecurity issues in industrial critical infrastructure: manufacturing in perspective. J Cyber Secur Technol 1. https://doi.org/10.1080/23742917.2016.1252211

Arntz M, Gregory T, Lehmer F, Matthes B, Zierahn U (2016) Arbeitswelt 4.0—Stand der Digitalisierung in Deutschland: Dienstleister haben die Nase vorn. Institut für Arbeitsmarkt- und Berufsforschung (IAB), Nürnberg, 22/2016

Badea V, Alin Z, Boncea R (2018) Big data in the aerospace industry. Informatica Economica 22:17–24. https://doi.org/10.12948/issn14531305/22.1.2018.02

Batista RC, Agarwal A, Gurung A, Kumar A, Altarazi F, Dogra N, Vishwanatha HM, Chiniwar DS, Agrawal A (2024) Topological and lattice-based AM optimization for improving the structural efficiency of robotic arms. Front Mech Eng 10:1422539

Boeing: Additive manufacturing: Chinook 3D printing. Accessed: 30 June 2023. [Online]. Available: https://www.boeing.com/features/innovation-quarterly/2022/01/chinook-3d-printing.page

Burande DV, Kalita K, Gupta R et al (2024) Machine learning metamodels for thermo-mechanical analysis of friction stir welding. Int J Interact Des Manuf. https://doi.org/10.1007/s12008-024-01871-6

Ceruti A, Marzocca P, Liverani A, Bil C (2019) Maintenance in aeronautics in an Industry 4.0 context: the role of augmented reality and additive manufacturing. J Comput Des Eng 6(4):516–526. https://doi.org/10.1016/j.jcde.2019.02.001

Ceruti A, Marzocca P, Liverani A, Bil C (2018) Maintenance in Aeronautics in an Industry 4.0 context: the role of AR and AM. In: Transdisciplinary engineering methods for social innovation of Industry 4.0. IOS Press, pp 43–52. https://doi.org/10.3233/978-1-61499-898-3-43

CS_2020Dec_Mitsubishi-Heavy-Industries.pdf. Accessed: 30 June 2023. [Online]. Available: https://www.fujitsu.com/global/imagesgig5/CS_2020Dec_Mitsubishi-Heavy-Industries.pdf

Culot G, Fattori F, Podrecca M, Sartor M (2019) Addressing Industry 4.0 cybersecurity challenges. IEEE Eng Manage Rev 47(3):79–86. https://doi.org/10.1109/EMR.2019.2927559

cybersecurity-for-smart-factories.pdf. Accessed: 30 June 2023. [Online]. Available: https://www2.deloitte.com/content/dam/Deloitte/za/Documents/risk/cybersecurity-for-smart-factories.pdf

Data Science for Manufacturing | Boeing Advanced Research Center. Accessed: 30 June 2023. [Online]. Available: https://depts.washington.edu/barc/projects/data-science-manufacturing

Decoding Digital Twins: Exploring the 6 main applications and their benefits. IoT Analytics. Accessed: 30 June 2023. [Online]. Available: https://iot-analytics.com/6-main-digital-twin-applications-and-their-benefits/

Digital Transformation. Northrop Grumman. Accessed: 11 June 2024. [Online]. Available: https://www.northropgrumman.com/what-we-do/digital-transformation

Digital twinning takes flight. AFCEA International. Accessed: 21 Oct 2023. [Online]. Available: https://www.afcea.org/signal-media/digital-twinning-takes-flight

Digital twins in aerospace—a paradigm shift. Accessed: 30 June 2023. [Online]. Available: https://www.spsairbuz.com/story/?id=1119&h=Digital-Twins-in-Aerospace-A-Paradigm-Shift

Dornhöfer M, Sack S, Zenkert J, Fathi M (2020) Simulation of Smart factory processes applying multi-agent-systems—a knowledge management perspective. J Manuf Mater Process 4(3), Art. no. 3. https://doi.org/10.3390/jmmp4030089

Duan L, Da Xu L (2021) Data analytics in Industry 4.0: a survey. Inf Syst Front. https://doi.org/10.1007/s10796-021-10190-0

Goyal G, Kumar A, Sharma D (2024a) 12 recent applications of rapid prototyping with 3D printing: a review. In: Kumar A, Kumar P, Sharma N, Srivastava A (ed) 3D printing technologies: digital manufacturing, artificial intelligence, Industry 4.0. De Gruyter, Berlin, Boston, pp 245–258. https://doi.org/10.1515/9783111215112-012

Goyal G, Kumar A, Gupta A (2024b) 16 recent developments in 3D printing: a critical analysis and deep dive into innovative real-world applications. In: 3D printing technologies: digital manufacturing, artificial intelligence, Industry 4.0, p 335

https://www.facebook.com/airbus. IoT: aerospace's great new connector | Airbus. Accessed: 30 June 2023. [Online]. Available: https://www.airbus.com/en/newsroom/stories/2019-07-iot-aerospaces-great-new-connector

James S, Cervantes A (2019) Study of Industry 4.0 and its impact on lean transformation in aerospace manufacturing. https://doi.org/10.1115/DETC2019-98147

Javaid M, Haleem A, Singh RP, Suman R (2023) An integrated outlook of cyber–physical systems for Industry 4.0: topical practices, architecture, and applications. Green Technol Sustain 1(1):100001. https://doi.org/10.1016/j.grets.2022.100001

Jayasekara D, Lai NYG, Wong K-H, Pawar K, Zhu Y (2022) Level of automation (LOA) in aerospace composite manufacturing: present status and future directions towards industry 4.0. J Manuf Syst 62:44–61. https://doi.org/10.1016/j.jmsy.2021.10.015

Joint FAA—EASA workshop on qualification/certification of additively manufactured parts. Accessed: 22 Oct 2023. [Online]. Available: https://www.faa.gov/sites/faa.gov/files/2022_FAA_EASA_AM_Workshop_Full_Proceedings.pdf

Kumar L, Sharma RK, Praveen (2023a) Smart manufacturing and Industry 4.0: state-of-the-art review. In: Handbook of smart manufacturing. CRC Press

Kumar A, Mittal Rk, Haleem A (eds) (2023b) Advances in additive manufacturing. In: Advances in additive manufacturing. Additive manufacturing materials and technologies. Elsevier, pp i–iii. https://doi.org/10.1016/B978-0-323-91834-3.00031-4

Kumar AA, uz Zaman UK, Plapper P (2023c) Collaborative robots. In: Handbook of manufacturing systems and design. CRC Press

Kumar A, Rani S, Rathee S, Bhatia S (eds) (2023d) Security and risk analysis for intelligent cloud computing: methods, applications, and preventions, 1st edn. CRC Press. https://doi.org/10.1201/9781003329947

Kumar P, Hussain SS, Kumar A, Srivastava AK, Hussain M, Singh PK (2024a) 10 finite element method investigation on delamination of 3D printed hybrid composites during the drilling operation. In: 3D printing technologies: digital manufacturing, artificial intelligence, Industry 4.0, p 223

Kumar A, Kumar P, Sharma N, Srivastava AK (2024b) 3D printing technologies: digital manufacturing, artificial intelligence, Industry 4.0. Walter de Gruyter GmbH & Co KG

Kumar A, Shrivastava VK, Kumar P, Kumar A, Gulati V (2024c) Predictive and experimental analysis of forces in die-less forming using artificial intelligence techniques. Proc Inst Mech Eng E J Process Mech Eng. https://doi.org/10.1177/09544089241235473

Leite Junior A, Condé Lemos GF, Gonzaga Trabasso L (2019) Proposal of a method for the implementation of the Industry 4.0—aircraft final assembly domain. Presented at the 10th aerospace technology congress, 8–9 Oct 2019, Stockholm, Sweden, pp 199–209. https://doi.org/10.3384/ecp19162023

Li L, Aslam S, Wileman AJ, Perinpanayagam S (2021) Digital twin in aerospace industry: a gentle introduction. IEEE Access. https://doi.org/10.1109/ACCESS.2021.3136458

Lineberger R, Hussain A, Hanley T (2024) Aerospace and Defense 4.0—capturing the value of Industry 4.0 technologies0 Deloitte Insights. Accessed: 1 Feb 2024. [Online]. Available: https://www2.deloitte.com/content/dam/Deloitte/ca/Documents/energy-resources/ca-en-er-aerospace-and-defense-4-aoda.pdf

Mabkhot MM et al (2021) Mapping Industry 4.0 enabling technologies into United Nations sustainability development goals. Sustainability 13(5), Art. no. 5. https://doi.org/10.3390/su13052560

Maisiri W, Darwish H, Van Dyk L (2019) An investigation of Industry 4.0 skills requirements. SAJIE 30(3). https://doi.org/10.7166/30-3-2230

Manufacturing Gets a Systemic Upgrade with the Digital Factory | Lockheed Martin. Accessed: 11 June 2024. [Online]. Available: https://www.lockheedmartin.com/en-us/news/features/2021/manufacturing-gets-systemic-upgrade-intelligent-digital-factory.html

Marjani M et al (2017) Big IoT data analytics: architecture, opportunities, and open research challenges. IEEE Access 5:5247–5261. https://doi.org/10.1109/ACCESS.2017.2689040

Martínez-de Dios JR, Torres-González A, Paneque JL, Fuego-García D, Ramírez JRA, Ollero A (2018) Aerial robot coworkers for autonomous localization of missing tools in manufacturing plants. In: 2018 international conference on unmanned aircraft systems (ICUAS), pp 1063–1069. https://doi.org/10.1109/ICUAS.2018.8453291

Mentsiev A, Guzueva E, Magomaev T (2020) Security challenges of the Industry 4.0. J Phys: Conf Ser 1515:032074. https://doi.org/10.1088/1742-6596/1515/3/032074

Meyer H et al (2020) Development of a digital twin for aviation research. https://doi.org/10.25967/530329

Mohammadi M, Jamshidi S, Rezvanian A, Gheisari M, Kumar A (2024) Advanced fusion of MTM-LSTM and MLP models for time series forecasting: an application for forecasting the solar radiation. Meas Sens 33:101179

Naveena K, Krishnamoorthy M, Karuppiah N, Gouda PK, Hariharan S, Saravanan K, Kumar A (2024) Elevating sustainability with a multi-renewable hydrogen generation system empowered by machine learning and multi-objective optimization. Meas Sens 33:101192

Oberheitmann A (2020) Industry 4.0—economic benefits and challenges, especially for small and medium-sized enterprises. In: Oberheitmann A, Heupel T, Junqing Y, Zhenlin W (eds) German and Chinese contributions to digitalization: opportunities, challenges, and impacts, in FOM-Edition. Springer Fachmedien, Wiesbaden, pp 13–22. https://doi.org/10.1007/978-3-658-29340-6_2

OECD (2023) Measuring the internet of things. Organisation for Economic Co-operation and Development, Paris. Accessed: 11 June 2024. [Online]. Available: https://www.oecd-ilibrary.org/science-and-technology/measuring-the-internet-of-things_021333b7-en

Oks SJ et al (2022) Cyber-physical systems in the context of Industry 4.0: a review, categorization and outlook. Inf Syst Front. https://doi.org/10.1007/s10796-022-10252-x

De Pace F, Manuri F, Sanna A (2018) Augmented reality in Industry 4.0. Am J Compt Sci Inform Technol 06(01). https://doi.org/10.21767/2349-3917.100017

Parrott A (2017) Industry 4.0 and the digital twin. Deloitte University Press. [Online]. Available: https://www2.deloitte.com/content/dam/Deloitte/kr/Documents/insights/deloitte-newsletter/2017/26_201706/kr_insights_deloitte-newsletter-26_report_02_en.pdf

Pereira T, Barreto L, Amaral A (2017) Network and information security challenges within Industry 4.0 paradigm. Procedia Manuf 13:1253–1260. https://doi.org/10.1016/j.promfg.2017.09.047

Pessoa MAO, Pisching MA, Yao L, Junqueira F, Miyagi PE, Benatallah B (2018) Industry 4.0, how to integrate legacy devices: a cloud IoT approach. In: IECON 2018—44th annual conference of the IEEE Industrial Electronics Society, pp 2902–2907. https://doi.org/10.1109/IECON.2018.8592774

Popović N, Popović B (2021) Some robotics concepts for the Industry 4.0 applications. Int Sci J "Industry 4.0" VI(4):131–134

Rani S, Tripathi K, Kumar A (2023) Machine learning aided malware detection for secure and smart manufacturing: a comprehensive analysis of the state of the art. Int J Interact Des Manuf. https://doi.org/10.1007/s12008-023-01578-0

Ras E, Wild F, Stahl C, Baudet A (2017) Bridging the skills gap of workers in Industry 4.0 by human performance augmentation tools: challenges and roadmap. In: Proceedings of the 10th international conference on pervasive technologies related to assistive environments. ACM, Island of Rhodes Greece, pp 428–432. https://doi.org/10.1145/3056540.3076192

Rodrigues D, Carvalho P, Rito Lima S, Lima E, Lopes NV (2022) An IoT platform for production monitoring in the aerospace manufacturing industry. J Clean Prod 368:133264. https://doi.org/10.1016/j.jclepro.2022.133264

Saucedo-Martínez JA, Pérez-Lara M, Marmolejo-Saucedo JA, Salais-Fierro TE, Vasant P (2018) Industry 4.0 framework for management and operations: a review. J Ambient Intell Human Comput 9(3):789–801. https://doi.org/10.1007/s12652-017-0533-1

Sayem A, Biswas P, Khan MMA, Romoli L, Dalle Mura M (2022) Critical barriers to Industry 4.0 adoption in manufacturing organizations and their mitigation strategies. J Manuf Mater Process 6(6), Art. no. 6. https://doi.org/10.3390/jmmp6060136

Sharma A, Kosasih E, Zhang J, Brintrup A, Calinescu A (2022) Digital Twins: state of the art theory and practice, challenges, and open research questions. J Ind Inf Integr 30:100383. https://doi.org/10.1016/j.jii.2022.100383
Sharma P, Singh Ghatorha K, Kang AS, Cepova L, Kumar A, Phanden RK (2024) Strategic insights in manufacturing site selection: a multi-method approach using factor rating, analytic hierarchy process, and best worst method. Front Mech Eng 10:1392543
Sharma A, Pandey H (2020) Big data and analytics in Industry 4.0. In: Nayyar A, Kumar A (eds) A roadmap to Industry 4.0: smart production, sharp business and sustainable development. Advances in science, technology & innovation. Springer International Publishing, Cham, pp 57–72. https://doi.org/10.1007/978-3-030-14544-6_4
Simulation for Industry 4.0. springerprofessional.de. Accessed: 29 June 2023. [Online]. Available: https://www.springerprofessional.de/simulation-for-industry-4-0/16751036
Sony M (2018) Industry 4.0 and lean management: a proposed integration model and research propositions. Prod Manuf Res 6:416–432. https://doi.org/10.1080/21693277.2018.1540949
Soori M, Arezoo B, Dastres R (2023) Virtual manufacturing in Industry 4.0, a review. Data Sci Manage. https://doi.org/10.1016/j.dsm.2023.10.006
Srivastava AK, Kumar A, Kumar P et al (2023) Research progress in metal additive manufacturing: challenges and opportunities. Int J Interact Des Manuf. https://doi.org/10.1007/s12008-023-01661-6
Sustainability | Free full-text | Mapping Industry 4.0 enabling technologies into United Nations sustainability development goals. Accessed: 27 June 2023. [Online]. Available: https://www.mdpi.com/2071-1050/13/5/2560
Tadesse H, Singh B, Deresso H, Lemma S, Singh GK, Srivastava AK, Dogra N, Kumar A (2024) Investigation of production bottlenecks and productivity analysis in soft drink industry: a case study of East Africa Bottling Share Company. Int J Interact Des Manuf (IJIDeM), 1–13
Technologies transforming Typhoon production. BAE Systems | International. Accessed: 11 June 2024. [Online]. Available: https://www.baesystems.com/en/feature/technologies-transforming-typhoon-production
The benefits of digital twin technology in aviation. Accessed: 11 June 2024. [Online]. Available: https://www.ifs.com/assets/enterprise-asset-management/whitepaper-digital-twins-in-aviation
Tung CM (2018) Vertical integration for smart manufacturing—the dynamic capability perspective. J Adv Tech Eng Res 4(2). https://doi.org/10.20474/jater-4.2.3
U. Technologies. Top 10 Applications & Use Cases for Digital Twins | Unity. Accessed: 29 June 2023. [Online]. Available: https://unity.com/solutions/digital-twin-applications-and-use-cases
Van Dinter R, Tekinerdogan B, Catal C (2022) Predictive maintenance using digital twins: a systematic literature review. Inf Softw Technol 151:107008. https://doi.org/10.1016/j.infsof.2022.107008
van Lier B (2011) Connections, information and Reality 'thinking about the internet of things. Syst Cybern Inf 9(5)
Watson V, Tellabi A, Sassmannahausen J, Lou X (2017) Interoperability and security challenges of Industry 4.0. https://doi.org/10.18420/IN2017_100
Węgrzyn N (2022) The use of additive manufacturing for production of commercial airplane power plants components. Saf Defense 8(2), Art. no. 2. https://doi.org/10.37105/sd.185
Younan M, Houssein EH, Elhoseny M (2020) Challenges and recommended technologies for the industrial internet of things: a comprehensive review. Measurements 151:107198. https://doi.org/10.1016/j.measurement.2019.107198
Zawra L (2019) Migration of legacy industrial automation systems in the context of Industry 4.0—a comparative study, pp 1–7. https://doi.org/10.1109/ICFIR.2019.8894776.
Zhong RY, Xu X, Klotz E, Newman ST (2017) Intelligent manufacturing in the context of Industry 4.0: a review. Engineering 3(5):616–630. https://doi.org/10.1016/J.ENG.2017.05.015
Zutin GC, Barbosa GF, de Barros PC, Tiburtino EB, Kawano FLF, Shiki SB (2022) Readiness levels of Industry 4.0 technologies applied to aircraft manufacturing—a review, challenges and trends. Int J Adv Manuf Technol 120(1):927–943. https://doi.org/10.1007/s00170-022-08769-1

SPRINGER NATURE

GPSR Compliance

The European Union's (EU) General Product Safety Regulation (GPSR) is a set of rules that requires consumer products to be safe and our obligations to ensure this.

If you have any concerns about our products, you can contact us on ProductSafety@springernature.com

In case Publisher is established outside the EU, the EU authorized representative is:

Springer Nature Customer Service Center GmbH
Europaplatz 3
69115 Heidelberg, Germany

The manufacturer's authorised representative in the EU is Springer Nature Customer Service Centre GmbH, Europaplatz 3, 69115 Heidelberg, Germany. If you have any concerns regarding our products, please contact ProductSafety@springernature.com

Printed and bound by CPI Group (UK) Ltd, Croydon, CR0 4YY
23/09/2026
02221471-0002